# TRENDS IN OBJECTIVE GEOLOGY

# TRENDS IN OBJECTIVE GEOLOGY

FOR CIVIL SERVICES AND
OTHER COMPETITIVE EXAMINATIONS

THIRD EDITION

**Ausaf Sayeed**
M.Sc., Ph.D.
Indian Foreign Service

**CBS Publishers & Distributors** Pvt Ltd

New Delhi • Bengaluru • Chennai • Kochi • Kolkata • Lucknow • Mumbai
Hyderabad • Jharkhand • Nagpur • Patna • Pune • Uttarakhand

*Disclaimer*

Science and technology are constantly changing fields. New research and experience broaden the scope of information and knowledge. The author has tried his best in giving information available to him while preparing the material for this book. Although, all efforts have been made to ensure optimum accuracy of the material, yet it is quite possible some errors might have been left uncorrected. The publisher, printer and the author will not be held responsible for any inadvertent errors or inaccuracies.

**Trends in Objective Geology**

**ISBN:** 978-81-239-0802-1



**Third Edition: 2002**

**Reprint : 2003, 2004, 2008, 2009, 2010, 2011, 2012, 2014, 2016, 2017, 2018, 2019, 2020, 2024**

**First Edition: 1991**

**Second Edition 1995**



Published by Satish Kumar Jain and Produced by Varun Jain for

**CBS Publishers & Distributors** Pvt Ltd

4819/XI Prahlad Street, 24 Ansari Road, Daryaganj, New Delhi 110 002, India.4819/XI Prahlad Street, 24 Ansari Road, Daryaganj, New Delhi 110 002, India.

Ph: 23289259, 23266861 Website: www.cbspd.com e-mail: delhi@cbspd.com

*Corporate Office:* 204 FIE, Industrial Area, Patparganj, Delhi 110 092

Ph: 011-4934 4934 Fax: 011-4934 4935 e-mail: publishing@cbspd.com; publicity@cbspd.com

*Branches*

- **Bengaluru:** Seema House 2975, 17th Cross, K.R. Road, Banasankari 2nd Stage, Bengaluru 560 070, Karnataka
  Ph: +91-80-26771678/79 Fax: +91-80-26771680 e-mail: bangalore@cbspd.com
- **Chennai:** 7, Subbaraya Street, Shenoy Nagar, Chennai 600 030, Tamil Nadu, India
  Ph: +91-44-26680620/26681266 Fax: +91-44-42032115 e-mail: chennai@cbspd.com
- **Kochi:** 42/1325, 1326, Power House Road, Opp KSEB, Power House, Ernakulam 682 018, Kochi, Kerala, India
  Ph: +91-484-4059061-65, 67 Fax: +91-484-4059065 e-mail: kochi@cbspd.com
- **Kolkata:** 147, Hind Ceramics Compound, 1st Floor, Nilgunj Road, Belghoria, Kolkata-700056, West Bengal, India
  Ph: +033-25633055, 033-25633056 e-mail: kolkata@cbspd.com
- **Lucknow:** Basement, Khushnuma Complex, 7 Meerabai Marg (Behind Jawahar Bhawan), Lucknow-226001, UP, India
  Ph: +91-522-4000032 e-mail: tiwari.lucknow@cbspd.com
- **Mumbai:** PWD Shed, Gala no 25/26, Ramchandra Bhatt Marg, Next to JJ Hospital Gate no. 2, Opp. Union Bank of India Noorbaug, Mumbai-400009, Maharashtra, India
  Ph: 022-66661880/89 e-mail: mumbai@cbspd.com

*Representatives*

- **Hyderabad** 0-9885175004
- **Patna** 0-9334159340
- **Jharkhand** 0-9811541605
- **Pune** 0-9664372571
- **Nagpur** 0-8692091830
- **Uttarakhand** 0-9716462459

Printed at: SRK Graphics, Shahdara, Delhi, India

*To my parents*

# FOREWORD

It is an acknowledged fact that the spirit of competitive examinations has caught the imagination of the student community in general so much that these examinations have become a popular vehicle to achieve the goal of securing employment in several reputed organizations. But this widespread enthusiasm has been dampened by the dearth of good quality books, especially in the subject like geology, that can cater to the students' needs. The laudable work of Mr. Ausaf Sayeed, IFS, has come at an appropriate time to alleviate the hardships and to remove the handicap of those students who are desirous of taking the Competitive Examinations in the field of Earth Science. I am quite confident that this book will meet their high expectations and receive their critical acclaim and wholehearted appreciation. The dedication and dynamism with which Mr. Ausaf Sayeed has attempted to frame and compile the questions, can be felt when one endeavours to make a journey through the pages of this engrossing book.

**V. Madhavan**, M.Sc., Ph.D.
Reader, Department of Geology,
Osmania University, Hyderabad, A.P.

# PREFACE TO THE THIRD EDITION

Substantial modifications have been made in the Third Edition. The short notes on Geochemistry & Mineralogy, Paleontology and Petrology have been rewritten and updated.

In Section II, the objective questions on Paleontology, Igneous Petrology, Sedimentary Petrology, Metamorphic Petrology, Economic Geology and Applied Geology have been revised and updated. Solved previous papers of Civil Services Preliminary Examinations of 1996 and 1997 have also been added.

I hope that this edition fulfils the requirements of the students and help them in better preparation for their examinations.

**Dr. Ausaf Sayeed**

# PREFACE TO THE FIRST EDITION

The recruitment to various reputed services is largely done on the basis of competitive examinations, most of which are objective in nature. These examinations have undergone rapid structural changes over the last few years and the questions are being framed in such a manner as to assess the candidates' overall abilities including his mental alertness, conceptual clarity and depth of knowledge. This demands total dedication and a great deal of potential from the candidates in order to come out successfully in these examinations.

Geology, as a branch of Earth Sciences, is a subject of considerable importance from the point of view of examinations. It has emerged as one of the safest and rewarding optional for Civil Services and States Services examinations. Besides these, examinations in Geology are also conducted by UPSC for recruitment of Geologists, and by various PSEs like ONGC, OIL, etc., for recruitment of executive trainees. However, the students of geology are hampered greatly by unfamiliarity with examination patterns and inadequacy of effective question banks on this subject. Books in this field, especially in the objective pattern, are relatively few and the existing ones do not fully conform to the standards of the present examinations.

The main aim of this book is to bridge this gap and eliminate the hardships of the students by presenting them a comprehensive and stimulating question bank on Geology. While covering all the branches of the subject, it lays greater emphasis on certain important questions. Explanatory answers are provided for difficult and confusing questions. Questions based on Reasoning and Assertion, which have now become an integral part of every competitive exam, also finds a place in this book.

A section consisting of short notes on selected topics of importance is given at the beginning of the book. It provides valuable information to the candidates in a concise and crisp form. The book also includes a whole section based on the latest competitive examinations which would give the candidates an insight into the differing styles and standards of these examinations.

The Author expresses his sincere gratitude to Dr. V. Madhavan, Reader, Department of Geology, Osmania University, Hyderabad, for giving his valuable suggestions and also for writing a foreword for this book.

Sincere thanks are also accorded to Messrs. D.B. Kumar and M. Srinivas, Research Scholars, Osmania University, for contributing the Civil Services (1990) and ONGC (1990) papers respectively.

Lastly, the Author thanks all others who have lent a helping hand in the successful completion of this book.

New Delhi

AUSAF SAYEED

# CONTENTS



## Section I : Short Notes



## Section II : Objective Questions

## Section : III (Previous Papers)

## Section I : Short Notes

# 1

# *General Geology and Geotectonics*

**The Earth.** Surface area is $510 \times 10^6$ km², water occupies about 70·78% and land occupies about 29.22%. Its mean radius is 6371 km and the difference between the polar radius and the equatorial radius is 21 km (14 miles).

(*a*) **The Crust.** It constitutes less than 1% of E's volume and 0.4% of its mass. Average density is 2.85. Comprises of continental, transitional and oceanic crust. The continental crust is composed of sediment layer (0–5 km), upper layer (10–20 km) and the lower basaltic layer (15–25) km ; together with the transitional crust it comprises 79% of the total crust by volume. The oceanic crust is again composed of 3 layers : Layer-1—composed of unconsolidated sediments ; Layer-2—variable in terms of thickness and velocities ; basaltic ; cause of oceanic magnetic anomalies ; Layer-3—most uniform in thickness and velocities ; partly serpentinized peridotite and amphibolite.

(*b*) **The Moho.** It is defined by a sudden increases in P-wave velocity from 6·5 to 8 km/s ; occurs from 10–12 km beneath oceans to 30–50 km beneath continents ; Moho temperatures beneath continents range from 500–700°C and those beneath oceans from 150–200°C.

(*c*) **The Mantle.** Occupies 83% of E's volume and 69% of its mass ; thickness of 2900 km or 1800 miles. (1) Upper mantle—Ranges from 21–370 km has two layers :

(*i*) The Low Velocity Zone (Anamalous Upper Mantle or Gutenberg layer)—Extends from depths of 60–70 km till 100–300 km ; top marks the base of the lithosphere ; has lower S and P wave velocities than the material above and below it ; most prominent under the ocean basins ; generally not detected beneath precambrian shields ; zones of well-developed LVZ has+anomalies. It is also characterized by low Q and high electrical conductivity :

Asthenosphere—Located beneath depths of 70–700 km ; characterized by plastic flow ; convection is the dominant mode of heat transfer ; causes the plate movement.

(*ii*) Golitsyn layer—Very steep increase in velocity with depth.

In the upper mantle as a whole the P-wave velocity increases from about 8 to 11.5 km/s (2) Lower mantle—Extends from 1000—2900 km ; steady increase in density and velocities, probably due to pressures ; consists of 2 layers . 'D[1] (950—2700 km) and D[2] (2700—2900 km). The seismic velocities reach the maximum value for earth at the D' layer (*i.e.*, 13.7 km/s for P and 7.3 km/s for S waves).

(*d*) **Core-Mantle boundary**. (Weichert-Gutenberg discontinuity). Lies at a depth of 2900 km (1800 mi) and has a radius of 3485 km. It is the most important and drastic compositional boundary within the earth ; abrupt fall in P and S velocities ; the density increases downwards across the boundary from 5500 to 10,000 kgm$^{-3}$.

(*e*) **The Core.** Occupies 17% of E's volume and 34% of its mass ; density is 10.7. The outer core extends from 2885 to 5144 km, forming 95·6% of the core by volume and 95% by mass sulphur and oxygen are the abundant light elements of the outer core. The Ingleman discontinuity (5200 km) separates the outer core from the inner core ; the inner core extends from 5144—6371 km ; radial extent is 1227 km.

(*f*) Other important limits of the Earth are : Reptii discontinuity at 970 km ; Dahm discontinuity at 2780 km and the inversion surface at 2730 km.

**Hypothesis for the Origin of the Earth**

1. **Nebular Hypothesis.** Proposed by Kant and Laplace ; proposes a 'hot' origin for the earth ; earth formed from a single gaseous incandescent nebula ; it explained the single direction of rotation of sun and the planets and the near circular shapes of the planetary orbits, etc. It could not explain the retrograde movement of certain planets and their satellites nor could it explain the observed differences in the densities, size and mass of planets.

2. **Biparental Hypothesis**. Dynamic encounter theory of Buffen, the Planetesimal hypothesis of Moulton and Chamberlin ; and the Tidal hypothesis of Jeans and Jeffreys are included here.

(*a*) Planetesimal Hypothesis—Moulton (1901) and Chamberlin (1905). Planets formed from the little fragments given off from the solar surface when a large star approached a wandering sun. It explained the rigidity of the mass of the earth and the phenomenon of meteorites.

(*b*) Tidal Hypothesis—'Hit and Run' or the 'Catastrophic' theory—Jeans (1919) and Jeffreys (1929)—Planets formed from the filament thrown out of the sun when a huge star passed closed to it. It explained the coplanar placement of planets, but could not explain the distribution of the angular momentum.

3. The Meteorite Hypothesis of O.J. Schmidt (1944), C.F. Weizacker (1944) and G.P. Kuiper's hypothesis belong to the category of the 'cold' origin of earth.

**Tectonic Elements of the Crust**

(*a*) **Continental Shelf.** Part of the ocean ranging in depth from 0–200 m with a very gentle slope (1°) ; occupies 7.6% of the ocean's area.

(*b*) **Continental Slope.** Part of the ocean ranging between 200—2000 m, Slope varies from 4—15°. 15% of ocean area ; very complex relief. It is the main structural boundary between the continental and oceanic crust.

(*c*) **Ocean floor.** Ranges from 2000—6000 m ; most extensive part of the ocean by area (76.2%). The slope is very slight, being measured in minutes. In this part of the ocean there are mid-oceanic ridges, volcanic massifs and broad abyssal plains.

1. **The Mid Oceanic Ridges (MORs).** Linear belts occurring near the centres of the oceans ; topographic highs, tectonically unstable, being the areas of shallow focus seismicity. They occupy ⅓ of the area of the oceans and rise between 2—4 km from the floor. Their width varies typically in the range of 500—1000 km and the ratio of their vertical to horizontal dimensions is around 1 : 500. They are isostatically compensated, the excess relief being balanced by hotter less denser material beneath. All MORs constitute a single system crossing the ocean floors. These are composed of 2 parts : the crest provinces comprising the rift valley, rift mountains and the fractured plateaus ; and the Flank provinces comprising of several steps bounded by scarps.

Characteristics of MORs–Crustal thickness ranges from 3–6 km ; sites of shallow focus earthquakes of low magnitude ; associated with the central graben produces +ive Bouger anomalies. Heat flows varies between $5.1\times10^{-6}$ and $7\times10^{-6}$ cal/cm$^2$ s. Low-K tholeiites are most voluminous ig. rks. produced beneath the ridges.

The Mid-Atlantic Ridge (MAR)—It is asymmetrical, both topographically, thermally and magnetically ; it is S-shaped and in line with the general configuration of the ocean coasts ; its height ranges between 2000–3000 m ; it is a young rock structure that in all probability is continuing to develop at the present time and altering essentially ; it is associated with great seismicity, extensive volanism and high heat flow ($7\times10^{-6}$ cal/cm sq s). It give rise to Iceland, St. Helena, Ascension, the Azores and the St. Paul islands. Iceland is the largest 'outcrop' of any oceanic ridge.

The East-Pacific Rise—arcuate in outline ; compared with MAR, it is broader and less elevated above the abyssal plains and has smooth topography. The most astonishing feature is the E—W fracture zones (transcurrent faults) which have sliced the rise into long crustal slabs. Heat flow measurements show abnormally high values in a narrow zone over the crest and abnormally low values beneath the crest. The Easter Island tops this ridge.

The Indian Oceanic Ridges–90°E Ridge–It is 4800 km long, inert/aseismic ; marks the boundary of the oceanic plate on which the Indian plate rests.

Laccadive-Maldives Ridge—Runs in the middle of the Indian ocean and is partly continental. The Carlsberg ridge trends NW—SE, is an active spreading centre and separates the two basins of Laccadives and Seychelles.

Ridges of the Arctic Ocean—The Lomonosov, the alpha and the arctic are the MORs of the arctic ocean. They have a very thick E's crust (15—18 km) containing regions of granitic layer ; in particular, the Lomonosov ridge rises on the continental crust.

2. **The Oceanic Trenches.** These are arcuate depressions on the ocean bottoms associated either with island-arc system or young mountain belts. They mark the beginning of the subduction zones. They range in depth from 5—8 km ; are generally around 100—150 km wide and 30—50000 km long ; crustal thickness ranges from 3—12 km. The volcanic and orogenic activity occur landward of the trench axis ; paired metamorphic belts occur on their landward side ; they are associated with low heat flow and negative gravity anomalies. The interactions in the trenches give rise to calc-alkaline volcanism and plutonism in the adjacent island arcs.

3. **Island Arcs.** These are arcuate chain of islands formed by volcanic processes and which are invariably paralled by oceanic trenches to the landward. They occur above active or recently extinct subduction zones. Each island arc is generally made up of two parallel arcs of islands 50—150 km apart, the outer one comprising folded and thrust sediments lacking volcanic activity and the inner one composed of volcanic islands. Also called "Festoon islands". They mark the convergence of the two plates.

Structure—The convex side of the arc faces the oceans and concave towards the land. Deep oceanic trenches (foredeeps) occur on the convex side and deep basins on the concave side. The zone between the trench and the volcanic island is known as the 'Arc-Trench Gap'. The small ocean basins occuring continentward of the arcs are known as the Marginal sea basins. These structures have been attributed to secondary ocean floor spreading behind the island arcs (Back-Arc Spreading). Examples are : Japan sea, the China sea, the Phillipine sea, etc.

Characteristics—(*a*) Crustal thickness—Average is 22 km ; a grantitic-metamorphic layer is often present ; crust is of the transitional type.

(*b*) Seismicity—Both shallow and deep-focus earthquakes. In most cases the focii of the earthquakes define a steeply inclined seismic zone which emerges near the trench the slopes under the volcanic arc. These zones are described as the 'Benioff zones' or 'seismic shear zones. In terms of plate tectonics, this zone is represented as a brittle region in the upper 10—20 km of a descending lithospheric plate.

(*c*) They show intense positive gravity anomalies.

(*d*) Lithology—Plutonic rocks are dominantly composed of

granodiorites ; Andesites are most important volcanic rocks ; Graywackes showing graded bedding are common sedimentary rocks ; two types of metomorphisms occur : HP/LT developing at the outer margins of the arc-trench gap and the HT, variable pressure metamorphism on the inner margins. The composition of the Archaen Greenstone belts closely parallel the composition of many of the present day island-arc systems. Thus, Island arcs are geological analogues and they give evidence of the palaeotectonic history of the earth.

4. **Continental Margins.** They mark the abrupt transition between the continents with their shallow shelves and the deep oceans. The deep structure associated with margins extends down into the upper mantle and locally reaches the mantle transition zone.

(*a*) **Passive Margins.** Seismically inactive margins where the adjacent continental and oceanic lithosphere are welded together as parts of the same plate. They originated by continental splitting but once formed they ceased to be plate boundaries. These are of two types : Rifted Margins which result from continential splitting at newly formed divergent plate boundaries. A typical rifted margin is morphologically subdivisible into continental shelf, continental slope and continental rise *e.g.*, eastern margin of N. America.

Offset (sheared) Margins : these are formed by continental separation along transform faults.

(*b*) **Active Margins.** They mark the juxtaposition of continental and oceanic lithosphere at convergent boundaries. A deep trench occurs on the oceanward side of the margin and an Andean type mountain range with predominantly andesite volcanism on the continent side. Island arcs are similar to active margins but the convergent plate margin at them separates two portions of the oceanic lithosphere *i.e.*, it is separated from the mainland by a marginal sea.

5. **Geosynclines.** Linear areas of subsidence in which sedimentation may occur. These are found at or near the junctions of the plates or within the plate interiors where large-scale subsidence occurs due to the sedimentary loading on a thinning crust. Kay (1944, 1947) classified the tectonic elements as follows ; (*a*) Orthogeosynclines—linear, deeply subsiding, shallow to deep-water and located between cratons. Subdivided into : (*i*) Euogeosynclines—Actively subsiding with associated volcanics and (*ii*) Miogeosynclines—less active with the absence of volcanics. (*b*) Parageosynclines (Intracratonic Basins)—ovate, less actively subsiding basins located within the cratons or adjacent to it. Types : (1) Exogeosynclines (Transverse Basin)—These are tongue—like extensions from orthogeosynclines ; gaining the detritus from them. *e.g.*, Deltas, Belt Geosyncline (Precambrian) ; Siwaliks (2) Autogeosynclines (Isolated basins)—ovate basins in which the rate of subsidence decreases from the centre downward. These are isolated areas within the cratons : consists of fine clastics, carbonate rocks and

evaporites. *e.g.*, Michigan Basin, Papaghani sub-basin, Kaladgi basin, etc. (3) Zeugogeosynclines (Yoked basins). These are ovate or linear areas of subsidence in the craton, adjacent to complimentary uplifts in the cratons. Composed of subgreywackes, silty shale and limestone ; also Arkose. *e.g* Nallamalai and Srisailam sub-basins (4) Taphrogeosynclines—Riftogenic basins of continental origin and bounded by faults on both the sides deriving their detritus from the craton. *e.g.*, Godavari graben, Paland basin, etc.

Geosynclines and Plate tectonics : Many authors such as Mitchell and Reading (1969), Dickinson (1971) and Hsu (1972) modified the geosynclinal concept to reconcile it with plate tectonics theory. Six different types of geosynclines are recognized. There are : (*a*) Red-Sea type Geosynclines—these are areas of sedimentation in divergent zones formed during the initial stages of continental break-up. Characterized by alkaline volcanism. Aulacogens are examples. (*b*) Andean-type Geosynclines—these are found on the continental margins associated with ocean—continent lithospheric collision. It shows calc—alkaline volcanism of plutonic—epigenetic affiliation. The type area is western South American margin. (*c*) Japan Sea-type Geosynclines—These include the marginal sea basins occuring continentward of the island arc. (*d*) Atlantic-type Geosynclines—Type area is the eastern part of North America. They form on the continental margins in the interior of the plates and forms a miogeosynclinal—eugeosynclinal couple. Evaporites and metalliferous muds are the chief deposits. (*e*) Mediterranean—type Geosynclines—Originate in plate margins showing dominantly lateral plate motions. (*f*) Island—Arc type Geosynclines—It consists of two types of mineral depositions : one which is formed outside the arc—trench environment and transported to it by plate motions (allochthonous) *e.g.*, podiform chromites and the other originating within the geosynclinal complex itself (autochthonous).

6. **Miogeoclines and Eugeoclines.** Study of sediments along the eastern continental margin of North America by seismic refraction profiling has revealed the presence of thick sequences of sediments in two parallel, elongated zones. One of these is located under the continental slope and the other at the base of the shelf. The rate of deposition is more than 13 m/ma but the basement does not downwarp into a syncline, but appear to dip uniformly seaward at a low angle like a monocline. Robert Dietz proposed that the zone in which thick sediments underlie the shelf be called monogeocline and the zone in which thick sediments underlie the base of slope be called a eugeocline.

**Some Hypotheses**

1. **Continental Drift Theory.** Proposed by Alfred Wegener (1912) and later supported by du Toit (1937). About 250 m.a. there was only one supercontinent called Pangea and only one universal ocean Panthalasa. Pangea had an area of 150 mi sq. km. and it

spread more or less equally between the two hemispheres. It consisted of N. America (with Greenland) and Eurasia (without India and Arabia) in the extreme North ; below it, South America and Africa (with Arabia) ; and further down, Antartica, Australia and India. It broke in the late Triassic by two rifts. The northern rift cut the Pangea from E-W and created Laurasia in the north and Gondwana in the south. It also opened up the Atlantic ocean. The southern rift divided Gondwana into two parts : (*i*) South America and Africa (with Arabia) and (*ii*) Antartica, Australia and India. This rift opened up the Indian ocean. India was liberated from the Antartica complex by a Y-shaped rift about 135 M.Y. ago and started drifting towards north Some 65 MY ago, N. America separated from Eurasia and South America from Africa. Later, the two Americas were united by the Isthmus of Panama, while Australia separated from Antartica and moved northwards. About 20 MY ago, Arabia split from Africa to merge into Asia. This brought into existence the Red Sea and the Gulf of Aden. India moved northwards and finally joined South Asia about 45 MY ago. The northern margin of the Indian plate dipped into the Tethys sea and slid under the Asiatic plate This subduction produced vast geological transformations in South Asia. (*a*) It filled up the Tethys Sea and a land mass was formed in its place, which later on downwarped to form Indo-Gangetic Plains. (*b*) The western part of the Tethys remained unaffected and gave rise to the Mediterranean sea. (*c*) The Tibetan plateau and the Himalayan mountains were formed.

Mechanism of Drift. Tidal forces of the Sun and Moon and precision effects (pohlfluct forces) caused the general drift of the continents towards West

Supporting Evidences. (1) Matching of the coastlines of South America and Africa (at 2000 m contour) ; (2) Presence of Permo—Carboniferous tillites in widely separated regions of the present day. *e.g.* Talchir tillites (India), Pagoda tillites (Antartica), Dywka tillites (S. Africa), Tubaro Tillites (Brazil) etc., (3) Paleontological evidences : (*a*) Presence of Mesosaurus (Permain) in S. America and Africa, (*b*) Occurrence of marsupial opossums in Australia and S. America, (*c*) Distribution of lumbricid and megascolecid earthworms in different continents, (*d*) Distribution of Gondwana plant fossils in all the southern continents. (4) Similiarity of stratigraphic sequence of rock units in Africa, India, S. America, Australia and even Antartica, (5) The continuity of geosynclinal mountain belts which encircle the continents ; (6) Paleomagnetic reconstruction based on Polar wandering curves.

2. **Sea-Floor Spreading**. Formulated by R.S. Dietz (1961) and Prof. Harry Hess (1962). According to this hypothesis, the oceanic lithosphere splits at oceanic ridges and moves away from ridge axes with the motion like that of a conveyor belt as new lithosphere is created and fills the resulting cracks. Oceanic lithosphere is consumed in the asthenosphere at subduction zones to accommodate the newly created lithosphere such that the surface

area of the earth remains constant. The important junction in the hypothesis is not the crust-mantle interface, but the lithosphere-asthenosphere boundary about 100 km down. The mechanism of sea-floor spreading is thought to be the slow 'displacement' or convection of the material of the mantle under the action of temperature difference between its floor and root.

Supporting evidences. (*a*) MORS ; The lavas are symmetrically placed along the axis of the MORs, and their age increases from the ridge crest. No material greater in age than 100 MY is found in the oceanic basin.

(*b*) Marine magnetics ; The puzzling bands of magnetic anomalies occurring parallel to the ridges in the Indian and Pacific oceans are also accounted for by the sea-floor spreading. These anomalies had resulted from periodic reversals in E's magnetic field and are a result of alternating strips of normally and reversely magnetized lavas that had been extruded at the crests of MORs. Rocks at the ridge crest younger than 0.69 M.Y. are normally magnetized whereas those further from the axis formed between 0.69 and 0.89 M.Y. are reversely magnetized. The sea-floor spreading acts as a gigantic 'tape recorder' and the history of the E's magnetic field is permanently recorded in the ocean floor.

(*c*) Direct sampling by drilling the ocean floor by the JOIDES (Joint Oceanographic Institution for Deep Earth Sampling) project using oil-well drilling equipment in a ship called Glomar Challenger had revealed that the spreading has been continuous throughout the Cenozoic at an essentially constant rate of 2 cm/yr/ridge flank.

3. **Plate Tectonic Theory**. Proposed by W.J. Morgan in 1967. It is a unifying working hypothesis which provides a kinematic model of the upper layer of the earth *i.e.* it deals with motions and their observable effects. The basic assumptions of this theory are : (*a*) Large areas of the E's outer portion act as rigid caps on a sphere and that these rigid plates undergo no significant internal deformation. (*b*) Each plate is in relative motion with respect to the other plates and that the plates undergo significant deformation only at their boundaries with each other.

Plates—These are the inert aseismic regions of the earth bounded by narrow mobile belts. They are composed of crust and parts of the U. Mantle. Their average thickness is 100 km. They may be entirely continental (Eurasian plate), entirely oceanic (Pacific) or a combination of both (African plate). They vary in size from $10^4-10^8$ km$^2$. The marginal part of a particular plate is called a plate margin and the zone of motion between two plates is called a plate boundary. The plate boundaries are situated wherever convection in the mantle produces upcurrents or down-currents. There are 7 major plates on the E's surface.

Types of Plate Boundaries ; (*a*) Divergent (Constructive) Boundaries or the Obduction zones lie along the oceanic ridges where the adjoining parts are moving away from each other. (*b*) Convergent (Destructive) Boundaries or the Subduction zones are the zones where two plates meet and one is subducted below the other. They follow either the deep oceanic trenches or the young mountain chains. The following collisions are possible at these boundaries · (1) Ocean-Continent collisions—Oceanic lithosphere is consumed in the subduction zone *e.g.* Pacific and Asian plates ; (2) Continent-Continent collision—it results in the onset of orogeny and rise of great mountain chains, nappes and thrusts *e.g.* collision of Indian and Eurasian plates ; (3) Ocean-Ocean Collision—One of the plate returns to the mantle and is destroyed. A pattern of deep trench and island arc develops which results in the generation of predominantly basic magmas. (*c*) Neutral (Conservative) Boundaries or Parallel or Transform fault boundaries consists of steep faults with a lateral (strike-slip) sense of displacement. Here plate material is neither created nor destroyed but is conserved. These boundaries can join either divergent or convergent plate boundaries in different combinations.

Driving Mechanism for plates : It is still a matter of considerable debate. Probably a combination of several factors are responsible for the plate motion like : convection currents in the solid mantle ; ridge-push mechanism ; slab-pull (gravitational effect of cooler denser material in and around the sinking slab) or the Mantle drag (where the plate is pulled by lateral convective flows within the asthenosphere).

4. **Contraction Hypothesis**. Proposed by Elie de Beaumont (1829). It assumed that the earth is cooling as a result of losing heat and this cooling is accompanied with a contraction of its volume. It postulates that mountains developed by crumpling due to cooling and shrinking of a hot earth, which was confined to the region between depths of 100—700 km.

5. **Expansion Hypothesis**. Proposed by O. Hilgenberg (1933) and I.V. Kinilov (1957). It assumes that : (*a*) originally the earth had a higher density (of 9·13 g/cm$^3$) and shorter radius (of 5430 km) than the present, (*b*) A drop in the density caused an expansion of the earth, (*c*) This expansion was accompanied by the broadening of ocean floors and the formation of new ocean basins. As a result, the blocks of continental crust were separated by the oceanic crust and gradually moved away.

6. **Tetrahedral Theory**. Proposed by Lowthian Green (1875). It assumes that the earth started as a hot sphere and as it cooled, it contracted attaining a tetrahedral form, since of all regular bodies the tetrahedron has the minimum volume for a given surface. It assumed that the corners of the tetrahedron correspond to the continents, while the faces of the tetrahedron correspond to the oceans. It also assumed that the continents have always been stable.

7. **Isostasy**. Dutton (1889)—"There exists in the earth a certain level of compensation above which all columns of materials

having unit cross-sectional area must have the same mass". Elevated rocks are characterized by rocks of low density and depressed basins by rocks of higher density.

(*a*) **Pratt's Hypothesis.** "All blocks of the same weight and cross-section sink to the same depth in a suitable liquid-medium". Salient features are : compensation takes place by lateral density variations because the crust varies laterally in density ; thickness of the crust is constant only at the bottom ; crust synonymous with lithosphere.

(*b*) **Airy's Hypothesis.** "All blocks of the same material but different heights will sink to different depths in a liquid medium and irregularities will be on both the surfaces *i e.* top and bottom". Salient features are : compensation takes place by adjustment of crustal base ; density of the crust and the subtratum is constant throughout ; thickness of the crust varies by the top and bottom while that of the substratum varies only by the top ; lithosphere comprises both crust and part of the upper mantle.

**Transform Faults**

These are named by J. Tuzo Wilson in 1965. These form a type of plate boundary called Conservative or Parallel plate boundary in which only transform motion (large horizontal displacement) occurs *i.e.* the plates move past each other with no destruction or construction. At these faults the displacement suddenly stops or changes the direction. These are horizontal shear faults which terminate abruptly at both ends but may show great displacement of the crustal blocks. These are parallel to the direction of relative motion of the plates on either side of the plates and are therefore controlled by the relative velocity of the two plates. They commonly trend at right angles to the MORs and appear to offset them. Their motion is opposite to that of the transcurrent faults (offset increases with transcurrent motion but it remains constant with transform faults). In between the offset ridges, transform faults show shallow seismicity i.e. earthquakes are confined to the area between the offset ridges. Another important feature is that transform motion occurs only in the area between the offset ridges, and beyond the ridges the motion again changes to that of a transcurrent fault. In many areas they connect belts of of plate generation with zones of plate destruction, *e.g.* Mendocino fault, San Andreas fault, etc. Transform faults are of 3 types : Ridge-Ridge which are most common and retain a constant length as a function of time and Ridge-Trench and Trench-Trench which increases or decreases in length as they evolve.

**Triple Junctions.** These are Y-shaped junctions where three plates share a corner. They are classified as stable or unstable, depending on whether or not they preserve their geometries as they evolve. The geometric conditions for stability are described with vector velocity triangles. Only RRR triple boundaries are stable for all orientations of plate boundaries. *e.g.* Pacific, Coccos and

Nazca plate combination. There are 16 possible combinations of ridge (R), trench (T), and transform fault (F) triple junctions.

**Aulacogens**. These are sediment-filled troughs formed by normal faulting in the early stages of spreading of a MOR. They have been described as the failed third arm of a spreading ridge (or cratonic-rift triple junction). Aulacogens are mantle-activated rifts *i.e* they are produced by doming and cracking of the lithosphere, where the doming is produced by upwelling asthenosphere or rising mantle plumes. In their early stages aulacogens typically intersect spreading ridges at a 60° angle where the spreading ridge itself make a 60° bend. These are characterized by large volumes of volcanic rocks and very thick and dominantly clastic sedimentary deposits. **E.g. the East African rift system and the Benue Trough.**

**Magma Associations**

Most magmas appear to be generated at plate margins, the exception being transform faults. The chief environments of magma production are : Platemargin settings (subduction zones and oceanic ridges) and intraplate environments. Three principal magma series are recognized in the world.

1. Tholeiite series—It is characteristic of extensional stress regimes such as ridges, marginal seas and continenial rifts. The dominant rock type is tholeiite (an olivine-free basalt).

2. Calc-alkaline series—Characterized by a predominance of Andesite in volcanic terranes and of granodiorite in plutonic terranes. Calc-alkaline volcanics is typical of small back-arc and fore-arc basins and subduction zones which are associated dominantly with compressive stress.

3. Alkali series—Least abundant of the three, is characterized by alkali olivine basalts, and other alkali rocks such as nephlinite, trachyte, latite, phonolite, etc. It is frequently found in the regions of minor stress (compressional or extensional) such as ocean basins and cratonic areas.

**Ores of Volcanic Affiliation**

Show a progression from Cyprus type ⟶ Besshi type ⟶ Kuroko type depending on the stages of development of island—arcs.

(*a*) **Cyprus Types**. Associated with basic volcanics (ophiolites) ; formed at oceanic ridges, back-arc basins and arc-trench gap ; essentially consists of cupriferous pyrite bodies.

(*b*) **Besshi Type**. Associated with the main calc-alkaline stage of island arc formation ; found in complex structural settings with thick grey-wacke successions. Commonly carry Cu and Zn.

(*c*) **Kuroko Type**. These are metal associations associated with the felsic volcanics developed during the last stages of island-arc formation. They are Cu—Zn—Pb ores often with Au and Ag. Large amounts of barytes, quartz and gypsum may also be found.

**Miscellaneous**

1. **Andesite Line.** It is a boundary between the basalts of the oceanic crust and island arcs and the andesite-dacite-rhyolite volcanic rocks of the circum-pacific belt. It is generally located along the trenches, and folding, faulting, earthquakes and andesite volcanism are concentrated on one side of this line. As such it constitutes one of the fundamental geological boundaries of the earth and is petrographically and tectonically of the highest significance. It is a line where oceanic lithosphere is subducted at Benioff zones.

2. **Suture Line.** It is a line marking the final closure of continents after the collision of two plates. It is marked by the presence of ophiolites in orogenic belts. It also marks the site of former oceans. Thus, it is important in recording older plate movements in the geological record of the continents.

3. **Hot Spots.** The volcanically active areas at the end of linear chains of volcanic islands are termed hot spots.

4. **Mantle Plumes.** These are upwellings of hot, plastic (but mostly solid) mantle material underlying the hot spots. Plumes have more or less circular cross sections and are thought to originate deep in the mantle. As a plate moves over a mantle plume, a linear chain of volcanoes could be produced.

5. **Sea Mounts** These are isolated submarine volcanic features 2—100 km wide and more than 1000 m high, whose surfaces lie less than 2000 m below sea level, many of which form volcanic islands. *e g.* Hawaii and Azores.

6. **Guyots.** Seamounts which are flat-topped and whose summits lie more than 2000 m below sea-level are called guyots. They have steep-sides (12°—35°) and give the impression of wave-bevelled platforms submerged largely by sea-floor subsidence, and partly by the post—glacial rise of sea level. It is generally believed that the flat tops are due to marine planation Guyots indicate one of the following : (*a*) a rise of sea level, (*b*) a sinking of a volcano or (*c*) a sinking of a large part of ocean floor.

7. **Orocline.** It is an orogenic belt with a change in trend (*e.g.* from the Alps to the Apennines) which is interpretable as an impreesed strain.

8. **Sphenochasm.** A sphenochasm is a wedge-shaped sector of oceanic crust separated by continental blocks which originated by rotation or pulling apart of the later.

9. **Stienmann's Trinity.** It is the association of Ophiolites, greywackes (turbidites) and radiolarian cherts in an orogenic belt.

10. **Orogenic Cycle.** Krynine (1941). It is a tectonic cycle with three stages of intensity, representing the cyclical repetition of

geosynclinal accumulation, orogenic disturbances, and widespread peneplanation. The type of sandstone formed during each stage is characteristic of the degree of tectonism involved. Thus, quartzites (with highly stable minerals) mark the peneplanation stage, Greywacke marks the geosynclinal stage and Arkose (with abundant felspar and other unstable minerals) marks the orogenic stage.

11. **Some Orogenies.** Caledonian orogeny—U. Silurian—L. Devonian (British Isles, Scandinavia) ; Tactonic Orogeny—M. Ordovician ; Acadian Orogeny—M. Dev ; Alleghenian orogeny—L. Carboniferous—Permian (Appalachian Mts) ; Antler orogeny—M. Devonian—Carboniferous (Western US) ; Sonoma orogeny—L. Permian—E. Triassic (UK).

12. **Ophiolite Complexes.** The sequence of rock types, from deep-sea sediments downwards through pillow lavas, sheeted dykes, gabbros (layered and cummulate) and Peridotite (serpentinite) are called Ophiolite complexes. Most geological and geophysical evidences suggests that ophiolite complexes represent fragments of older oceanic crust that have been tectonically emplaced in continental orogenic belts. If so, ophiolites represent samples of oceanic crust and upper mantle produced at ancient oceanic ridges. The gabbro-ultramafics contact in ophiolite complexes is usually interpreted as a "fossil moho".

13. **Paired Metamorphic Belts.** The subduction complex is flanked towards the inner side by the volcanic arc with the basalt-andesite-rhyolite association, but if the arc is developed in the continental block then hypabyssal plutonic association of granite-granodiorite is also present. This is a zone of high heat flow and steep geothermal gradient exceeding 25°C/km, which would give rise to Amphibolite facies of regional metamorphism. Thus, this HT amphibolite facies evolves in the inner side of the subduction complex and is parallel to the outer belt of HP/LT blueschist facies and both are broadly contemporaneous in age. These have been described as the Paired Metamorphic Belts by Miyashiro (1973).

14. **Flysch.** It is used to describe synorogenic sediments produced by the erosion of uprising and developing fold structures ; and are composed of rhythms of sandstone, shale and occasionally limestone beds. Flysch sequences are marked by their great thickness and predominance of argillaceous rocks. These are composed of rocks of marine origin. They are most frequently of Cretaceous and Palaeocene age (130—40 M.Y.), less frequently of Jurassic age. Most flysch sequences are characterized by turbidite facies. *e.g.* Guimal and Chikkim sandstone of Himalayas.

15. **Molasse.** The term 'molasse' is used to describe sediments produced by the erosion of mountain ranges after the final phase of an orogeny. They are most typical of orogenic troughs and consists of thick conglomerates and sandstones intercalated with members of shales, red beds and clayey rocks. These can be divided into lower and upper molasses. The lower molasse is generally composed of bedded sandstones alternating with clayey sediments and limestones.

It also contains coal measures. The upper molasse consists of enormous quantities of coarse debris, including red beds. Molasse sequences are of continental, lagoonal—continental lacustrine, or near-shore marine origin. They range in age from upper Oligocene to Miocene *i.e.* 40—20 M.Y. *e.g.* the Siwaliks.

**16. Radioactive Decays and Half life Periods**

$U^{238}$ ————→ $Pb^{206}$ (Half life=$4.5\times10^{9}$ years)

$U^{235}$ ————→ $Pb^{207}$ (Half life=$0.71\times10^{9}$ years)

$Th^{232}$ ————→ $Pb^{208}$ (Half life=$13.9\times10^{9}$ years)

$Rb^{87}$ ————→ $Sr^{87}$ (H.L.=$50\times10^{9}$ years)

$K^{40}$ ————→ $Ar^{40}$ (H.L.=$11.8\times10^{9}$ years)

$Sm^{147}$ ————→ $Nd^{143}$ (H.L.=$10^{16}$ years)

$C^{14}$=half life is 5570 years

Radium has a half life of 1622 years.

# 2

# *Physical Geology and Geomorphology*

**Arid Geomorphic Cycle.** The arid cycle is completed in three stages; Erosion, transportation and deposition. (1) Erosion—It takes place by three processes: deflation, abrasion and attrition. (*a*) Deflation—is the lifting and transport of solid particles of primary rock by wind. Its typical landform features include: blow-outs; desert pavements and hohlwegs (which result from furrow deflation); (*b*) Abrasion—is the wear and tear of the country rocks by wind-borne particles; dry regions with scant vegetation are favourable for abrasion. Ventifacts or brazil nuts are angular rocks with faceted and polished surfaces. Dreinkanter is a ventifact having 3 faces. Pedestal rocks or mushroom rocks have slender columns and wide tops. (*c*) Attrition—involves the wear and tear of wind-borne particles themselves. It forms rounded sands called millet sands. (2) Transportation—is effected by saltation and suspension.

**Wind Deposits**

1. **Sand Dunes.** Mounds or hills of sand formed when the velocity of the wind is checked by some obstacles. The areas where these are deposited are called wind-shadow areas. Slip face is an area where the sand particles slip and rest at an angle of repose varying between 20°—40° depending upon the coarseness of grains. Types: (*a*) Transverse dunes—are developed with their long axis at right angles to the wind direction. (*b*) Longitudinal dunes—are long ridge like dunes developed parallel to the wind. (*c*) Seifs—special type of longitudinal dunes looking like gigantic railway embankments; they originate from barchans. (*d*) Barchans—crescent shaped sand dunes with the arms of the crescent trailing downwards *i.e.*, facing the leeward direction. They are transverse dunes; isolated, stationary barchans are called dome dunes. (*e*) Parabolic dunes—U-shaped sand dunes whose tips point towards the wind. (*f*) Star dunes—are the dunes with multiple faces. (*g*) Draas—very large whale back dunes with extreme heights and wavelengths; these are longitudinal.

2. **Deserts** (*a*) Sandy deserts—consists of vast horizontal sheets of sand; "sand seas"; deflation is dominant; also called ERG or KOUM. (*b*) Stony deserts—consists of extensive sheets of angular fragments; also known as REG, SERIR or HARRAS. (*c*) Rocky deserts—predominance of bare rocks; also known as HAMADA. (*d*) Mountain deserts—consists of dissected high plateaus of complex forms.

3. **Loess**—is fine-grained, porous, buff-coloured, silt-grade, unstratified wind deposit, forms fertile soil, LIMON and ADOBE are similar to loess. Loess deserts are known as adyry.

4. **Inselbergs**—are isolated pillar like masses in arid regions; they have rounded tops and steep slopes; also called "island of mountains".

5 **Zeugens**—ridges left after the erosion of softer strata in deserts.

6. **Yardangs**—are ridges left between successive U-shaped troughs formed within softer rocks. Deflation is a dominant process in their formation; similar stream-lined features have been noticed in Mars.

**Fluvial Geomorphic Cycle.** In the cycle of erosion by running water three stages are distinguishable: (*a*) Youth stage corresponding to the upper course of a river, (*b*) Mature stage corresponding to the valley course and (*c*) Old stage corresponding to the plain course.

**Drainage Patterns.** Refers to the particular plan or design which the individual stream courses collectively form. (*a*) Dendritic—characteristic of homogenous rock surfaces of uniform resistance implying a notable lack of structural control; pinnate pattern is a variation. (*b*) Trellis—subparallel streams aligned along the strike of rock formations; characteristic of uniformly dipping beds of varying rock hardness; reflects marked structural control of most stream courses. (*c*) Radial—typical of domal uplifts, slopes of volcanoes, large basins and gulf depressions. (*d*) Rectangular—both the mainstream and its tributaries display right-angled bends. They reflect control exerted by joint or fault systems. Angulate pattern is a variant.

**River Valleys.** Consequent valley: its course is determined by the initial slope of the land; subsequent valley: their courses have been shifted from the original consequent ones to belts of more readily erosible rocks. They represent structurally adjusted stream courses. Insequent valleys: are those which show no apparent adjustment to structural or lithological control. Obsequent valleys drain in a direction opposite to the original consequent valleys. Resequent valleys flow in the same direction as the original consequent but at lower topographic levels.

**Deltas.** Low-lying land-masses formed at the river mouths due to deposition. The delta structure is composed of topset beds (coarse and fine material; gently inclined), foreset beds (coarse material; steeply-inclined and cross-bedded) and bottomset beds (fine sediments; roughly horizontal).

The delta surface is often marked by distributaries and gravel/ sand is often deposited in between them. It is then known as Arcuate (Lobate or Fan-shaped) delta *e.g.*, Nile, Ganges, Indus, etc. But when the distributaries are only few and the material deposited is composed of fine sediments a Bird's foot digitate delta is formed *e.g.*, Mississippi.

Gorges, canyons, V-shaped valleys, waterfalls, cascades, cataracts and river piracy are associated with the youth stages Meanders and ox-bow lakes are associated with the middle stage. Extensive flood plains and alluvial fans and cones are associated with the old stage.

**Mesas and Buttes.** Flat-topped hills or small mountains formed by fluvial erosion are called mesas. Isolated masses without flat tops are called buttes.

**Hogbacks and Cuestas.** A hogback is an erosional ridge with a long, gentle slope on one side and a short, steep slope on the other. A cuesta has a long, gentle slope on one side and an abrupt slope or even a cliff on the other.

**Waterfalls.** Waterfalls are formed when a stream plunges over a vertical or nearly vertical rock face. When water descends over a step like structure it is called a Cascade : and if the volume of water in a waterfall is huge, it is called a Cataract. Where a stream rushes over a steep slope in its bed it forms a Rapid, *e.g.* Jog falls or the Gersoppa falls are the highest waterfalls in India consisting of Raja. Rocket, Roarer and Dame Blanche falls.

### Glacial Geomorphic Cycle

**Glaciers.** Glaciers are the river like masses of ice that flow downward under the influence of gravity. Valley glaciers occupies the valleys and flow downwards and are fed by the snow-fields ; also known as Alpine or mountain glaciers. Piedmont glaciers are formed by the union of the spreading ends of several mountain glaciers on a large plain surface.

### Features Associated with Glaciers

1. Roches mountonees—Low, rounded and abraded mounds on the valley floor ornamented with fine striations. The stross (uphill) side is gently sloping and smooth and is effected by abrasion, whereas the Lee (downhill) side is steep and roughened and is effected by plucking action.

2. Drumlins—Unstratified mound like deposits with long axes parallel to the direction of glacial movement. Uphill sides blunt ; downhill sides are smooth and gently sloping. They, thus, show features which are opposite to those of roches mountonees.

3. **Cirques**—Basin like structures which serve as the starting points for glaciers. These are formed due to the process of nivation. They are called Corries (Scotland), Cirques (France) and CWMS (Wales).

4. Moraines—These are the unstratified glacial drift. It may be deposited on the basement of a valley (ground moraine); at the end of a glacier (terminal moraine); or along the sides of a glacier (lateral moraine) or in the centre of the two valleys (median moraine).

5. Kettles, moulins and glacial mills are cylindrical hollows formed by melting of ice.

6. Kame terrace—Stratified glacial drift deposited in the form of terrace along the valley walls.

7. Eskers—Are glacio-fluvial stratified deposits occurring in the form of a network of meandering ridges. They are mainly composed of sand and gravel. Also referred as Osser or Oss.

8. Aretes or combs—Also knife-edge—Where two corries exist on a mountain, the glacier gradually cuts them back and a sharp edge is formed which is known as arete or comb.

9. Pyramidal peaks or horns—Very sharp aretes formed on a steep-sided mountain due to the enlargement of three or more cirques.

10. Outwash plains (frontal apron)—These are smooth sloping plains lying in front of the glacial margins.

11. Nunataks—Very high peaks projecting above and ice.

12. Crag and Tail—When ice meets an obstruction, it is diverted over and around the rock projection and leave a gently sloping "tail" of moraine in its lee, which is described as crag and tail.

13. Knob and Kettle topography—a rough irregular surface of knobby hills and deep hollows.

14. **Glacial Lakes.** (*a*) Tarns—Lakes occupying the cirques. (*b*) Ribbon lakes—Long, deep, narrow lakes occupying the excavated floors of glaciated valleys.

14. Mollisol/Pergelisol—Thickness of the surface layer liable to alternate freezing and thawing is called mollisol. Permanently frozen thickness beneath the mollisol is called pergelisol.

**Indicators of Glacial Climate**

1. Scratched, grooved and facetted rock fragments.

2. Smooth and scratched bed rock.

3. Unsorted deposits and tillites.

4. Undecomposed felspars in the drift (it may also suggest an arid climate).

5. Absence of red ferruginous and black carbonaceous matter in the drift.

**The Karst Topography.** It is characterized by the development of underground solution networks, disorganized surface drainage and a surface terrain characterized by a hierarchy of holes and closed depressions resulting from solution and collapse. The term is derived from the Karst Plateau on the Adraitic Coast of Yugoslavia. Ideal conditions for karst development are : a considerable thickness of pure limestones with high crushing strength and low permeability ; considerable uplift of the land above the sea level ; and high annual precipitation.

**Karst Features.** 1. Terrarosa—Reddish-clayey soil characteristic of moderate to gentle slopes in limestone terranes, specially in tropical and temperate regions ; often contain fossils of Pliestocene age.

2. Lappies or Karren complex—An etched, pitted, grooved and rugged surface formed in areas of high relief is called karren, lapies or Griekes. Best developed in non-porous limestones with steep surface slopes.

3. **Sink Holes**—Assymetrical cup-shaped hollows occurring above the water table in limestone terranes.

(*a*) Dolines—are the sink holes that exhibit steep-sided, rocky and abruptly descending forms. They are a result of collapse of near surface caves.

(*b*) Uvalas—are larger depressions formed by the collapse of extensive roof sections.

(*c*) Poljes—are large, more or less closed depressions with steep sides and flat alluvial floors, frequently dotted with conical residual hills.

(*d*) Ponors (Swallow holes)—are vertical channels leading to UG karstic caverns within the limestone.

(*e*) Dripstones—Calcareous deposits formed by percolating water containing $CaCO_3$ in solution and occurring in the form of solid columns are called dripstones. It consists of downward extending stalactites and upward—growing stalagmites.

(*f*) Hums—These are the erosicnal remnants of a limestone upland occurring as residual hills analogous to monadnocks in the fluvial cycle. These are also described as Pepino hills (Haystack hills), Mogotes or Temoines.

(*g*) Helictites—are crystalline formations of most varied shapes arranged in spirals or loops in limestone caves.

(*h*) Anthodites—deposition of calcite in the roof of caverns in the form of delicate flower-like formations.

**Coastlines.** It is the boundary between the coast and the shore. It is the seaward limit of the coast.

Coastlines of Submergence. 1. These are produced either by subsidence of land or rise in sea level ; (2) It is very irregular with bays and headlands ; (3) Typical features include : sea caves, blow holes, arches, stacks, inlets, etc.

Coastlines of Emergence. (1) These are produced either by the uplift of the sea bottom or gradual fall in the sea level : (2) It is very regular and does not offer shelter from wind and waves ;

(3) Typical features include : benches, wave-built terraces, offshore bars, spits and lagoons.

**Types of Submergence Coastlines**

1. Fiord Coastline—It is a result of partial submergence of a glaciated area of high relief.

2. Ria Coastline—It is generally caused by the drowning of the river mouths. It is formed by the partial submergence of highlands unaffected by glacial erosion.

3. Dalmatian Coastline—These are developed when the mountain ridges running parallel to the coast are submerged.

4. Haff Coastline—It consists of large sandy strips running parallel to the coast and enclosing between them shallow lagoons known as "haffs".

**Volcanoes.** Mere openings of any size and shape on the earth's surface acting as an avenue for the magma to rise and spread in the form of lava flows. The central pipe through which magma makes its way up is called conduit; depressions on the top of volcanic cones are called craters ; and craters of very large diameter are described as Calderas.

**Types. 1. Shield Cones—These are gently sloping mountains formed when a mobile lava spreads in the form of shields. These are formed by basic lavas lacking pyroclastic materials *e.g.*, Mauna Lao, Mt. Etna.**

**2. Plug Cones or Spines or Domes—These are steeply sloping cones formed by a viscous acidic lava. *e.g.*, Mt Pelee.**

**3. Cinder (Pyroclastic) Cones—Perfect cones with steep slopes in which pyroclastic materials are arranged at an angle of repose round the vent.**

**4. Strato Volcano (Composite Cones)—Cones formed by alternate layers of pyroclastic materials and lava flows. Most of the world's great volcanoes are composite cones.**

Pyroclastic—These are particles of various sizes that are ejected from the volcanoes. These are described as bombs or blocks if the particle size is 32 mm ; lapilli when the size range is 4—32 mm ; ash when the size range is 0.25 mm, and fine ash if the size is 0.25 mm the last two are often described as volcanic tuffs.

Fumaroles—These are fissures or vents through which volcanic gasses are ejected. Steam forms 92% of the total volume of erupted gases. Other gases that are emitted are : $CO_2$, HCl, $Cl_2$, $SO_2$, $H_2S$, $H_2$, $CH_4$, etc. Fumaroles emitting appreciable amounts of sulphurous vapours are called Solfataros ; those emitting carbondioxide are called Mofettes and those emitting boric acid vapours are called Saffioni.

**On the degree of volcanic activity, the volcanoes are classified into Hawaiian, Strombolian, Vulcanian, Vesuvian, Plinian (Krakotoan) and Pelean types with the volcanic activity increasing from left to right.**

# 3

# Crystallography

**Face.** A natural flat surface on a crystal (XI), that develops as a result of internal structure.

**Edge.** Intersection of two faces belonging to one or more forms. Given by

$$E=\frac{1}{2}(f_1.s_1+f_2s_2+......) \qquad \text{where, } f=\text{Faces, } s=\text{Sides.}$$

**Solid Angles.** ('Corners') : Intersection of 3 or more faces. Given by

$$C=E+2-F \qquad \text{Where, E=Edge,} \qquad \text{F=total No. of faces}$$

**Form.** A set of faces related to each other by the symmetry of the crystal.

(*a*) Simple Forms—Xls. bounded by all like faces *e.g.* Cube.

(*b*) Combination Form—Xls. consists of more than one simple form.

(*c*) Congruent Forms—Any two faces of a form are called congruent, if one face can be brought into the position of the other by **ROTATION.**

(*d*) Enantiomorphic Forms—If one face can be brought into the position of the other by **REFLECTION** or **INVERSION.**

(*e*) Total No. of naturally occurring Forms is 48. 33 are in non-isometric systems and 15 are in isometric system.

(*f*) **Closed and Open Forms.** (*i*) **Closed.** All 15 isometric forms and 15 non-isometric forms *i.e.* Rhombohedron (1), Scalenohedrons (2), Bisphenoids (2). Bipyramids (7), Trapezohedrons (3) are closed forms. Totally 30 closed forms. They may have 4, 6, 8, 12, 24 or 48 faces.

(*ii*) **Open forms**—18 non-isometric forms—these forms may exhibit 1, 2, 3, 4, 6, 8 or 12 faces.

(*g*) **General Forms.** Faces of any 'general form' in any crystal class has no special relationship to any symmetry element, *i.e.* the faces do not lie either perpendicular to or parallel to one or more symmetry axes or planes present in the holohedral class of the system.

Triclinic System—No symm. axes or planes, so All forms in this system are general forms.

(*h*) **Unit Form or Fundamental Form or Parametral Form :** It is the form whose face is taken as intersecting the axes at unit lengths.

(*i*) **Limiting Forms**. They have fixed Miller Symbols. In any class there are only 3 limiting forms.

Cubic system—Cube (100), Dodecahedron (110) and Octahedron (111).

Hexagonal—Basal pinacoid (0001), Prisms ($10\bar{1}0$) and ($1\bar{1}20$).

Tetragonal—Basal pinacoid (001), Prisms (110) and (100).

Orthorhombic, Monoclinic and Triclinic—(100), (010) and (001).

(*j*) **Bevel and Bevelling.** If the edge between two faces of a form is replaced by two small faces, each of which makes an equal angle with its adjacent larger face, then the pair of small faces is said to bevel the edge

(*k*) **Truncate (Truncation)**. A small face equally inclined to two adjacent larger faces of a form is said to truncate the edge that would otherwise occur between the larger faces. Similarly a small face equally inclined to 3 or more faces of a form is said to truncate the vertex at which they would meet.

**Nomenclature of forms**

(*a*) One-faced forms : Open, all non-isometric systems. *e.g.* PEDION.

(*b*) Two-faced forms : Open. 3 two-faced forms are there :

(A) Pinacoid—All systems except Cubic.

(B) Dome—Monoclinic and Orthorhombic system.

(C) Sphenoid—Monoclinic.

(*c*) Prisms ; All Open—have 3, 4, 6, 8 or 12 faces—Totally 7 prisms occur.

(A) Trigonal—3F—Cross-section (C.S.) is Equilateral triangle.

(B) Tetragonal—3F—Cross-section (C.S.) Square

(C) Rhombic—4F—C.S. Rhomb (Orthorhombic and Monoclinic systems).

(D) Hexagonal—6F—C.S. Regular Hexagon.

(E) Ditrigonal—6F—C.S.—equal sides and alternate equal angles.

(F) Ditetragonal—8F.

(G) Dihexagonal—12F.

(*d*) Pyramids : All Open—totally 7 as above—3, 4, 6, 8 or 12 (*e*) Bipyramids : Do not occur in Cubic, Monoclinic and Triclinic systems. All Closed—totally 7—have 6, 8, 12, 16 or 24 faces .

(*f*) Rhombohedron : 6F—Trigonal System—Closed form C.S. Eq. triangle—Only vertical symm. planes, No Hor. symm. planes. Only 1 type.

(*g*) Scalenohedron : Closed, 8F in Tetragonal and 12F in Ditrigonal—2 types.

(*h*) Trapezohedron : 3 types, all Closed—Hexagonal, Trigonal and Tetragonal—Characterized by the absence of inversion axes, alt. axes, symm. planes and centre of symmetry.

(*i*) Bisphenoids : 2 types, each of 4F, occurs in Tetragonal and Orthorhombic.

**Symmetry Element.** A point, line or plane which is not moved by a symmetry operation and with which the particular operation is associated.

(*i*) Plane of Symmetry—A plane along which a Xl. is divided into 2 halves such that one half is the mirror image of the other. Maximum no. of planes possible in a Xl. is 9 (3 AP and 6 DP) *e.g.* Cube.

**Axial plane of symmetry (A.P.S.)** -includes two *c*-axes in the plane of symmetry.

**Diagonal plane of symmetry (D.P.S.)**-bisects the angle between two *c*-axes.

(*ii*) Axis of Symmetry—It is a line about which a Xl. may be rotated so as to bring it into identical orientation 2, 3, 4 or 6 times in a single rotation of 360°. A degree of symmetry greater than 6 is impossible so also a pentad axis cannot exist.

(*iii*) Centre of Symmetry—The point symmetry element about which the operation is applied that inverts every point $x, y, z$ to $-x, -y, -z$.

**(*iv*) Axis of Rotary Inversion (Composite Symmetry)—The symmetry element corresponding to the operation that both rotates through 360°/*n* and then inverts through the origin. Designated as $\bar{1}\ \bar{2}\ \bar{3}\ \bar{4}\ \bar{6}$ etc.**

**Hermann-Mauguin Symbols.** (HMS) Shorthand symbols used to designate the characteristic symm. of each of the 32 Xl. Classes. Each symbol is made of 3 or less parts. Each part refers to a definite axial direction within the Xl. The parts are always arranged in a definite order. Meaning of each part in the symbol varies from system to system. Combines both axes of rotary inversion as well as axes, planes and centre of symmetry. Thus,

X=rotation axis alone $\bar{X}$=inversion axis alone.

X/m=rotation axis normal to a plane of symmetry

Xm=rotation axis with a vertical plane of symmetry

$X_2$=rotation axis with a diad axis normal to it

X/mm=rotation with both kinds of planes of symmetry.

In the Cubic Sys. the first position of the H-M-S refers to an a-axis, the 2nd position to an axis coincident with the body diagonal of the cube, or the normal to the (111) face ; and the third to the axis normal to (110). All Xl. is this Sys. have 4 axes of 3-*f* symm. hence, in H-M-S the no. 3 is always in the second position. This arrangement is not found in any other Xl. class. In the Tetragonal sys. the first position refers to the *c* axis, the second to $a_1$ or $a_2$, the third to the axis which bisects the 90° angle between $a_1$ and $a_2$. In this sys. the first unit in the H-M-S. is always either 4 or $\bar{4}$. In the Hex. Sys. the 3rd position refers to an axis which bisects the 60° angle between adjacent '*a*' axes.

$\bar{1}$ corresponds to ... a Centre of Symmery (C)

$\bar{2}$ corresponds to . a Reflection plane (m)

$\bar{3}$ corresponds to ... a 3-*f* axis combined with a centre ($\bar{3}=3+\bar{1}$)

$\bar{6}$ corresponds to ...a triad axis combined with a perpendicular mirrior plane ($\bar{6}=3+m$).

**The Seven Crystal Systems**

| *Crystal system* | *Symmetry elements* | *Characteristic symmetry* | *No. of classes in system* |
|---|---|---|---|
| Triclinic | $a \neq b \neq c$<br>$\alpha \neq \beta \neq \gamma \neq 90°$ | monad | 2 |
| Monoclinic | $a \neq b \neq c$<br>$\alpha = \gamma = 90° \neq \beta$ | 1 diad (//*y*) | 3 |
| Orthorhombic | $a \neq b \neq c$<br>$\alpha = \beta = \gamma = 90°$ | 3 diads (//*x, y, z*) | 3 |
| Tetragonal | $a = b \neq c$<br>$\alpha = \beta = \gamma = 90°$ | 1 tetrad (//*z*) | 7 |
| Hexagonal | $a = b \neq c$<br>$\alpha = \beta = 90°\ \gamma = 120°$ | 1 hexad (//*z*) | 7 |
| Trigonal | $a = b = c$<br>$\alpha = \beta = \gamma \neq 90°$ | 1 triad (//[111]) | 5 |
| Cubic | $a = b = c$<br>$\alpha = \beta = \gamma = 90°$ | 4 triads (//<111>) | 5 |

1. **Cubic System**—5 classes—15 different and closed forms.

(*a*) Hexoctahedral Class (Normal Class)—Galena Types 9 planes—3 AP, 6 DP ; 13 Axes—3 $A_4$, 4 $A_3$, 6 $A_2$, C, A.R.I.—Nil.

**Forms.** 1. Cube—Closed—6 F—each square—(110).

2. Rhombdodecahedron—Closed—12 F—Rhomb—(110).

3. Octahedron—8 F—Eq. Triangle—(111)—Unit or Fundamental form. (1), (2) and (3) are the Limiting Forms of Cubic System.

4. Tetrahexahedron—24 F—Isos. Triangle—(hko) or (210)—Variable—Limiting forms are Cube (100) and Rhombdodecahedron (110).

5. Trisoctahedron—24 F—Isos. Triangle—(hhl) or (221)—Limiting forms are Octahedron (111) and Rhombodecahedron (110).

6. Trapezohedron—24 F—Trapezium—(hll) or 211—Limiting forms are Octahedron (111) and Cube (100).

7. Hexoctahedron—48 F—Scalene triangle—(hkl) or 321-General form.

**Examples.** (*a*) Galena : Cube/Octahedron.

(*b*) Spinel and Magnetic—(111).

(*c*) Fluorspar, Sphene and Rock Salt : Cubes.

(*d*) Leucite and Analcite : Trapezohedron.

(*e*) Garnet : (110) or Trapezohedron or both.

(*b*) **Diploidal Class.** Pyrite Type—H-M-S 2/$m^3$ 2 $m^3$

3 Planes—3$A_2$, 4$A_3$—C—Axis of Rotary Inversion (ARI)=Nil.

**Forms** 1. Pyritohedron—12F—Pentagonal—(hko) or (210)—Positive, if the long axis is vertical ; Negative, if long axis horizontal—It is developed from the alternate faces of Tetrahexahedron (210). Hemihedral from of Tetrahexahedron.

2. Diploid—24 F—Trapezium—(321)—General from—Hemihedral form of Hexoctahedron and is developed from its alternate faces.

All other forms have lower molecular symmetry.

**Examples** Pyrite, Cobaltite, Smaltite, Chloanthite, etc.

(*c*) **Hextetrahedral Class.** Tetrahedrite type—H-M-S is 6DP, 4 $A_3$, A.R.I. 3 $A_4$ No Centre.

1. Tetrahedron—4 F—Eq. Triangle—(111)—The c-axes join the centres of the opp. edges and one axis is vert. Developed from alt. octants of octahedron. Hemihedral form of octahedron.

2. **Deltoid—Dodecahedron—12 F—Trapezium—(221) or (hhl)—Hemihedral form of Trisoctahedron—Developed by alternate octants.**

3. **Tris—Tetrahedron—12 F—(hll) or (211)—developed from Trapezohedron.**

4. **Hex—Tetrahedron—24 F—Scalene triangle—(321) or (hkl).** It is developed from Hexoctahedron by the development of alt. octants.

**Examples** Tetrahedrite, Chabazite, Boracite, Zinc Blende, etc. (*d*) Gyroidal Class—3 $A_4$, 4 $A_3$, 6 $A_2$—No C—A.R.I.—Nil—H.M.S.—432 (*e*) Tetartoidal Class—Only 3 $A_2$, 4 $A_3$—H-M-S—23.

2. **Tetragonal System.** 7 classes.

(*a*) **Ditetragonal dipyramidal class. ZIRCON TYPE—Normal** Class 4/m 2/m 2/m 5 Planes—3 AP (2 V, 1 H), 2 DP ; 5 Axes—1 $A_4$ (c-axis), 4 $A_2$—Centre.

**Forms.** 1. **Basal Pinacoid—2 F—(001)—open.**

2. **Tet. Prism of II order—4 F—Open—(100)—Horizontal axis arise from the centres of the faces.**

3. **Tet. Prism of I order—4 F—Open—(110)—Horizontal axis arise from the centres of vertical edges. First order prism is smaller than II order.**

4. **Ditetragonal Prism.** 8 F—(210)—Open.

5. **Tetragonalbipyramid of II order—8 F—(101)—Closed-Isos-triangle—Axes arise from the centres of opp. hor. faces.**

6. **Tetragonal bipyramid of I order-8 F—(111)—Closed—Axes originate from opp. solid angles**

7. **Ditetragonal bipyramid—16 F—Closed—Scalene Triangle (hkl) h and k should be different, like (212).**

**Examples** Zircon, Cassiterite, Rutile, Apophyllite, etc.

(*b*) Tetragonal Scalenohedral class—2 planes, 2 $A_2$—No C—H.M.S. $\bar{4}$2 m A.R.I.—1 $A_4$ e.g. Chalcopyrite, melilite.

(*c*) Ditetragonal Pyramidal class—4 planes ; 1 $A_4$, No C ; e.g. Diaboleite H.M.S. 4 mm.

(*d*) Tetragonal Trapezohedral class—A.R.I.—nil ; 4 $A_2$ ; 1 $A_4$ ; No C—H.M.S.—422 *e.g.* Phosgenite.

(*e*) Tetragonal Dipyramidal class—C ; 1 plane ; 1 $A_4$ ; H.M.S. 4/m—Scheelite.

(*f*) Tetragonal Pyramidal class—only 1 $A_4$; H.M.S. 4 ; No example.

(*g*) Tetragonal Disphenoidal class—only A.R.I. 1 $A_4$ ; *e.g.* Cahnite ; H.M.S.—$\bar{4}$.

3. **Orthorhombic System.** 3 classes—Barytes type—3 AP ; 3 $A_2$ ; Centre.

**Forms.** (1) Basal pinacoid—(001)—2 F ; (2) Macropinacoid—2 F—(100) ; (3) Brachypinacoid—2 F ; 010 ; (4) Prism ("Prism of the third order" 4 F—(110 (5) Macrodome—4 F, (101) ("Prism of the second order or hol"). (6) Brachydome—4 F—Open—011—("Prism of the first order or okl). (7) Pyramid—8 F—(111)—closed.

E.g. Barytes, Sulphur, Olivine, etc.

(*a*) Rhombic dipyramidal class—includes the barytes type mentioned above—Hermann—Mauguin symbol is mm (2/mm).

(*b*) Rhombic pyramidal class—No C—2 planes—1 $A_2$—H.M.S. mm (2 m)—Hemimorphite.

(*c*) Rhombic disphenoidal class—only 3 $A_2$ H.M.S. is 222—e.g. Epsomite.

4. **Monoclinic System.** 3 classes—H.M.S. refers to b-axis

(*a*) Prismatic class—Normal class—Gypsum type—H.M.S. is 2/m 1 plane ; containing '*a*' and '*c*' axis ; 1 $A_2$—(b-axis) and centre.

**Forms.** (1) Basal Pinacoid—open, 2 F—001 (2) Orthopinacoid—2 F—100 (3) Clino pinacoid—(side pinacoid)—2 F—010 (4) Prism—4 F—110 (5) Hemiorthodome—2 F (because orthopinacoid is not a plane of symm).

101—"Horizontal Prism"-Two forms : Positive and Negative ;

When faces lie in Acute β (*i.e.* 80° 42′)—negative

When faces lie in Obtuse β (*i.e.* 113° 50′)—positive

It is also called 'Pinacoid of the second order'.

(6) Clinodome—4 F open—011 (7) Hemipyramids—4 F—111.

**Examples** Gypsum, Orthoclase, Epidote, HBl, Mica, Sphene, etc.

(*b*) Domatic class—H.M.S. m or $\bar{2}$ ; 1 Plane ; A.R.I.—1 $A_2$ ; Kaolinite.

(*c*) Sphenoidal class—H.M.S. 2 ; only 1 $A_2$ ; e.g. Clinohedrite.

5. **Triclinic System.** H.M.S. $\bar{1}$ ; C ; A.R.I. 1 ; AXINITE TYPE-PINACOIDAL CLASS.

**Forms.** (1) Basal pinacoid 001 (2) Macro Pinacoid—2 F-100 (3) Brachy pinacoid—2 F-010 (4) Hemi Prism—2 F—110 (5) Hemi-macrodome-101 (6) Hemi-Brachydome-011 (7) Quarter Pyramids 2 F—111.

(*b*) Pedial class. H.M.S. is 1 ; Asymmetric class—absence of any symm. elements. *e.g.* Calcium thiosulphate, parahilgardite.

6. Hexagonal System. 7 classes

(*a*) Dihexagonal dipyramidal class—Normal—Beryl type—6/mm 7 planes-4 AP (3 V, 1H), 3 DP; 7 Axes-1 $A_6$ (c-axis), 6$A_2$ ; C.

**Froms.** (1) Basal pinacoid—2 F—(0001) ; Open

(2) Hex. Prism or II order—6 F—open—11$\bar{2}$0 ; Axes emerge from the centres of the opp. faces.

(3) Hex. Prism of I Order—Open 6 F—10$\bar{1}$0

(4) Dihex. Prism—open—12 F—21$\bar{3}$0

(5) And (6) Hex. bipyramids-closed-12 F-I Order is 10$\bar{1}$1 ; II order 11$\bar{2}$1. (7) Dihex. bipyramid—24 F—closed—21$\bar{3}$1

**Example** Beryl, etc.

**Other Classes.** Ditrigonal dipyramidal ($\bar{6}$ m 2) *e.g.* Benetoite ; 4 P, 3 $A_2$ ; $\bar{4}$ Dihex. Pyramidal—(6 mm) ; 6 P ; 1$A_6$ ; *e.g.* Zincite ; Hex. Trapezohedral (622) 6 $A_2$, 1 $A_6$ ; *e.g.* Beta—Quartz. ; Hex. dipyramidal (6/m) ; C ; 1 P ; 1 $A_6$ ; *e.g.* Apatite Hex. Pyramidal (6) only 1 $A_6$ ; *e.g.* Nepheline ; Trig.dipyr.—($\bar{6}$) ; 1 P ; $\bar{6}$.

7. **Trigonal System.** 7 classes (*a*) Hexagonal Scalenohedral—Calcite type—H.M.S. ($\bar{3}$2/m ; C 3 VDP ; 3 $A_2$, 1$A_3$ ($\bar{3}$) ; *e.g.* Calcite, Siderite, Corundum.

**Forms.** (*a*) Rhombohedron—6 F ; Rhomb—Positive is (10$\bar{1}$1) ; Negative is (01$\bar{1}$1) ; Development by alt. sectants of hex. pyr. of I order (Hemihedral).

(*b*) Scalenohedron : 12 F, (2131) ; closed ; Hemihedral of Dihex. Pyramid

(*b*) Ditrigonal Pyramidal class—Tourmaline type—(3 m) No C ; 3 VDP ; 1 $A_3$.

(1) Pedion or basal pinacoid (0001) ; (2) Hex. Prism of II order 11$\bar{2}$0) (3) Trig. Prism of I order (10$\bar{1}$0) (4) Ditrig. Prism—6 F—(11$\bar{2}$0) (5) Hemimorphic hex. Pyr. (11$\bar{2}$1) ; (6) Trig. Pyr (10$\bar{1}$1)—corresponds to Rhombohedron—4 Trigonal Pyr.

(7) Ditrig. Pry. (21$\bar{3}$1)—Corresponds to Scalenohedron.

(*c*) Trigonal Trapezohedral class—Quartz type—H.M.S. is 32.

No planes, no Centre, 3 $A_2$, 1 $A_3$ ; *e.g.* Alpha-Qtz, Cinnabar, etc.

(1) Trig. trapezohedron—6 F—(21$\bar{3}$1)—Tetartohedral form of Dihex. Pyr.

(2) Trig. Pyramid—6 F, (11$\bar{2}$1) ; corresponds to Hex. Pyr. of II Order.

(*d*) Rhomdohedral class—H.M.S. is $\bar{3}$ ; C ; A.R.I. $\bar{3}$ ; *e.g.* Dolomite, Dioptase.

(*e*) Trigonal Pyramidal—H.M.S. is 3 ; only 1$A_3$ e.g. doubtful

**Axial Elements.** To know fully about the axial elements, the following has to be determined :

(*a*) Tetragonal—c/a ratio (*b*) Orthorhombic—a/b and c/b ratios (*c*) Monoclinic—a/b and c/b ratios angle β (*d*) Triclnic—a/b and c/b ratios and α, β and γ angles.

**Zone.** A set of faces which are all parallel to a line and intersect in parallel edges.

**Zone Axis.** The direction which is parallel to all the faces in a zone (Singular Axis).

(*a*) Cubic—Absence of a singular axis.

(*b*) Tetragonal, Hexagonal and Trigonal—one singular axis.

(*c*) Orthorhombic—3 singular axis.

(*d*) Monoclinic—One singular axis with innumerable singular axes perpendicular to it.

(*e*) Triclinic—All axes singular.

**Zone Symbol.** Set of 3 whole Nos. [uvw] enclosed in square brackets, such that the coordinates of any point on the z.a. are in the ratio ua : vb : wc.

**Holohedral (Holosymmetric) classes.** These are the normal classes baving the highest symmetry.

**Merohedral (Merosymmetric) Classes.** Have lower symmetry than Holohedral class.

1. **Hemihedral Classes.** The forms in these classes have 1/2 the no. of faces of the corresponding holohedral form, and to each holohedral from there correspond 2 similar and complimentary hemihedral forms called positive and negative. They have same indices as holohedral, belong to same crystal system but the classes are different *e.g.* Pyritohedron (210)-12F of Tetrahedron (210)-24 F.

2. **Tetartohedral Classes.** They have 1/4 the no. of faces as the holohedral forms—4 forms to each holohedral. Same indices—same XI Sys. E.g. Trigonol trapezohedron ($21\bar{3}1$)—6 F of Dihex. Pyr. ($21\bar{3}1$)—24 F.

3. **Ogdohedral Classes.** It is formed by omitting 7/8 of the faces and extension of the remaining faces—8 forms to the holohedral one.

*E.g.* only example is trigonal polar class (3) of trigonal system. Trigonal pyramid of III order is the Ogdohedral form of dihexagonal bipyramid.

| *Crystal System* | *Holohedral Class* | *Hemihedral Classes* | *Tetartohedral Classes* |
|---|---|---|---|
| Cubic | Hexoctahedral (4/m $\bar{3}$2/m) | Gyroidal (432)<br>Hextetrahedral ($\bar{4}$3 m)<br>Diploidal (2/m $\bar{3}$) | Tetartoidal (23) |
| Tetragonal | Ditetragonal dipyramidal | Tet. trapezohedral (422) | Tetragonal—Disphenoidal ($\bar{4}$) |

| | | | |
|---|---|---|---|
| | ($\bar{4}$/m $\bar{2}$/m 2m) | Tet. Scalenohedral ($\bar{4}$2 m)<br>Ditet. Pyramidal (4 mm)<br>Tet. Dipyramidal (4/m) | Tetragonal—Pyramidal (4) |
| Hexagonal | Dihexagonal Dipyramidal (6/m 2/m 2/m) | Hex. trapezohedral (622)<br>Ditrig. Dipyr. (6 m²)<br>Dihex. Pyr. (6 mm)<br>Hex. Dipyr. (6/m) | Trig. Pyr. (6)<br>Hex. Pyr. (6)<br>Rhombohedral ($\bar{3}$)<br>Ditrig. Pyr. (3 m)<br>Trig. Trap. (32) |
| Orthorhombic | Rhombic. Dipyr. (2/mmm) | Rhomb. Pyr. (mm²)<br>Rhombic Disphenoidal (222) | — |
| Monoclinc | Prismatic (2/m) | Sphenoidal (2)<br>Domatic (m) | — |
| Triclinic | Pinacoidal ($\bar{1}$) | Pedial (1) | — |

**Hemimorphic Classes.** Crystal having only one singular polar axis are called Hemimorphic. All systems except cubic and triclinic. All have open forms—8 hemimorphic classes—Forms include—Pedion, Pinacoids, Prism, Pyramid and Sphenoids.

(*i*) Hexagonal—Dihexagonal Pyramidal (6 mm) and Hex. Pyramidal (6)

(*ii*) Trigonal—Ditrigonal Pyramidal (3 m) and Trigonal Pyramidal (3)

(*iii*) Tetragonal—Ditetragonal Pyramidal (4 mm) and Tet. Pyramidal (4)

(*iv*) Orthorhombic—Rhombic Pyramidal (mm 2)

(*v*) Monoclinic—Sphenoidal (2)

**Enantiomorphic Classes.** Crystals having no symmetry elements other than rotary axes of symmetry are enantiomorphic. Devoid of inversion and alternating axes of symmetry, planes of symmetry and centre of symmetry. Incongruent forms, *e.g.* gyroid tetartoid, rhombic bisphenoid, sphenoid, pedion, trapezohedron and pyramids—11 enantiomorphic classes.

1. Cubic—Gyroidal (432) and Tetartoidal (23)

2. Tetragonal—Tetragonal trapezohedral (422) and Tet. pyramidal (4)

3. Hexagonal—Hexagonal trapezohedral (622) and Hex. Pyramidal (6)

4. **Trigonal—Trigonal trapezohedral (32) and Trigonal pyramidal (3)**

5. **Orthorhombic—Rhombic Disphenoidal (222)**

6. **Monoclinic—Sphenoidal (2).**

7. **Triclinic—Pedial (1).**

**Centrosymmetrical classes.** Xls. having a centre of symmetry Total=11. *i.e.* 4/m $\bar{3}$2/m ; 2/m $\bar{3}$ ; 6/mm ; 6/m ; $\bar{3}$2/m : $\bar{3}$ ; 4/mm ; 4/m ; 2/m ; $\bar{1}$.

**Bravais Lattices.** 14—(*a*) Cubic—P, I, F (*b*) Tetragonal—P, I (*c*) Orthorhombic—P, C, I, F (*d*) Monoclinic—P, C (*e*) Triclinic—P (*f*) Trigonal—R (*g*) Hexagonal (and trigonal)—C.

**Unit Cell.** The geometrical block-like outline enclosing a unit of pattern in a crystal's structure.

| *Xl.* | *Cell Shape* | *Symmetry of shape of cell* |
|---|---|---|
| Triclinic | General parallelepid | $\bar{1}$ |
| Monoclinic | Right prism on parallelogram as base | 2/m |
| Orthorhombic | Rectangular parallelepiped | 2/m 2/m 2/m |
| Tetragonal | Square prism | 4/m 2/m 2/m |
| Cubic | Cube | 4/m $\bar{3}$ 2/m |
| Trigonal | Cube deformed along one diagonal | $\bar{3}$ 2/m |
| Hexagonal | Prism on 60°—120° base | This cell does not itself have hex. symm. but the complete lattice does. |

**Unit Net.** The unit net is obtained by joining the nearest points in a net in such a manner as to show the highest symmetry. In a primitive unit cell the points are only at the corners.

**Symmetry exhibited by a Unit Net.** Oblique—2 Hexagonal—6 mm Square—4 mm Centered rectangular—2 mm.

**Screw Axis.** A symmetry element in a repeating pattern whose operation is a combination of a rotation about the axis by 360°/n and a shift parallel to the axis by m/n of the lattice translation, where $1 \leqslant m < n$.

**Glide Plane.** A symmetry element in repeating pattern whose operation is a combination of reflection in the plane and shift of the reflected image parallel to the plane by half of a lattice translation.

**Hopper Crystals.** Growth of xls. from a supersaturated solution. In such xls. the edges are more developed than the centres of faces.

**Important Twins.** 1. Quartz—Japan twins (C.P.—$11\bar{2}2$) ; Brazil twins—R+L $1\bar{1}20$ ; Dauphine twins (Electrical twins)—R+R or L+L.

2. Calcite—Glide twins (Butterfly or heart-shaped twins)—Rhomb (01$\bar{1}$2) T.P.

3. Staurolite—Maltese cross (T.P. 032), Skew (232) ; Saw-Horse twins.

4. Gypsum—Swallow tail (Arrow-Head-twins)—T.P. Orthopinacoid (100).

5. Aragonite—Cyclic (Penetration twins)—T.P. prism (110)

6. Rutile—Geneculated (Knee-shaped) twins—T.P. and C.P. (101).

7. Pyrite—Iron cross (Penetrated)—T.P. (110).

# 4

# *Geochemistry, Mineralogy and Optical Mineralogy*

Cosmic Abundance of Elements : Hydrogen is the most abundant element in the Universe and is followed by He, Li, Be, B, C and N. The elements show a rapid exponential decrease for elements of lower atomic weights, followed by an almost constant value for the heavier elements. Elements of even atomic number are more abundant than those of odd atomic numbers. (Oddo- Harkins Rule).

Isotopes have the same number of protons (Z) in the nucleus with differing neutrons (N) and differing atomic numbers (A). Isotones have the same N but different values of A and Z. Isobars are different elements with the same atomic numbers (A) but differing values of neutrons (N) and protons (Z).

**Composition of the Crust** : The mineralogy of the earth's Crust is ultimately controlled by the abundance and distribution of the chemical elements. Although 87 elements occur naturally in minerals, eight of these i.e., O, Si, Al, Fe, Ca, N, K and Mg are present in amounts greater than 1% and account for 98.5% by weight of' the crust. In terms of the atomic percentage, nine out of ten ions are either $O^{2-}$, $Si^{4+}$ or $Al^{3+}$. $O^{2-}$ alone takes up nearly 94% of the volume of the Crust. Thus, the crust as a whole, is essentially a packing of oxygen anion, bonded by silicon and the ions of common metals. Oxygen, silicon and aluminum together make up 90% of the atoms in the earth's crust.

Another important fact is that many exotic or unfamiliar elements are far more abundant than many technologically common elements. Thus, Rb is more common than Ag or Zn, Ce is more common than Ne, Zr is more abundant than Cu, Pb is comparable in abundance to Ga, Hg is rarer than most of the so called Rare Earths.

This discrepancy in the abundance of various elements is due to the fact that some ions are similar enough to other more common ions in size and charge that they occur mainly in the minerals of the common ions, and are rarely concentrated in their own phases. Such elements are called *dispersed elements.* Prime examples of dispersed elements are Ga which follows Al; Ne which follows Si; and Hf, which follows Zr.

**Most Common Elements of the Continental Crust**

| Element | Wt% | Atom % | Volume |
|---|---|---|---|
| O | 46.60 | 62.55 | 93.77 |
| Si | 27.72 | 21.22 | 0.86 |
| Al | 8.13 | 6.47 | 0.47 |
| Fe | 5.00 | 1.92 | 0.43 |
| Mg | 2.09 | 1.84 | 0.29 |
| Ca | 3.63 | 1.94 | 1.O3 |

**Oxide composition:** $SiO_2$ = 59.26%; $Al_2O_3$ = 15.35%; $Fe_2O_3$ = 3.14%; FeO = 3.74%; CaO = 5.08%; $Na_2O$ = 3.81%

**Relative Abundance of Elements:**

| | | |
|---|---|---|
| The Universe | : | H, He, O, Ne, N, C, Si |
| The Earth (whole) | : | Fe, O, Mg, Si, Ni, S, Ca |

Crust : O, Si, Al, Fe, Ca, Na,
Moon : O, Si, Mg, Fe, Ca
Sea water : O, Cl, H, Na, Mg, S, Ca
Atmosphere (gases) : Nitrogen, Oxygen, Argon, Carbondioxide

**Geochemical Classification of Elements.** Geochemically and mineralogically, the elements are divided into the following four groups:

1. Lithophiles : They occur mainly in oxygen compounds and have 8 electrons in the outermost shell and form electrostatic bonds. They include alkali metals, alkaline earths and rare earths. e.g., Li, Na, Rb, K, P, V, Nb, O, Mg, Ca, Sr, Ba, B, Al, Cr, H, F, Cl, Br, I, Sn, etc.

2. Chalcophiles : These elements belong to the sub-group B in the Periodic Table and have 18 electrons in the outermost shell. They ionize to from covalent compounds with sulphur. They occur mainly as sulphides and show a tendency to migrate towards siderophiles. e.g., Cu, Ag, Zn, Hg, Pb, As, Sb, Bi, S, Se, Te, Ni, Co, Mo, etc.

3. Siderophiles : They belong to the Group VIII in the Periodic Table and have incompletely filled outermost shells. They are characterised by metallic bonds, that is why they are able to exist in their native forms. In the presence of sulphur (i.e. in the absence of oxygen), they have the capacity to combine with sulphides in covalent bonds, but ionic bonds are totally absent. e.g., Pt, Ir, Os, Au, Ru, etc.

4. Atmophiles : These are gaseous elements which do not form compounds readily and are therefore present mainly in the atmosphere. *e.g.,* He, Ne, Ar, N, etc.

**Mineralogical Phase Rule** : A Phase is a homogenous, physically distinct, and mechanically separate part of a system. It may be solid, liquid or gaseous. A mineral is an example of a solid phase, and each mineral species in an assemblage or system is a separate phase. Components are those constituents whose concentration can undergo independent variation in the different phases. The components of a system are chosen from the constituents as the smallest number necessary to represent all phases in a system.

The Phase Rule is an absolutely rigorous and completely general statement about the numbers of different phases that may coexist under conditions of heterogenous equilibrium. In any system which is at equilibrium, the number of phases (P) plus the number of degrees of freedom (F) is equal to the number of components (C) plus two, or $P+F=C+2$ or $P=C+2-F$ or $P=C+2$ (if $F=0$). Most mineral systems are formed over a range of temperatures and pressures, so that P and T are not fixed for any real mineral assemblage. Thus, most mineral assemblages are formed under conditions where $F=2$. By substituting the value of F in the equation : $P=F+C-2$, we get $P=C+2-2$ or $P=C$. This expression is known as Goldschimdt's Mineralogical Phase Rule.

**Types of Bonds**

1. *Ionic or Heteropolar Bonds*: In this type of bonds, the atoms are in the ionized state i.e., they acquire either a positive or a negative charge. These are held together by electrical attractions between oppositely charged bodies. These are non-directional bonds. These are common in inorganic compounds and are therefore important in the structure of minerals. They never occur between similar elements. No molecules are formed. These minerals are mechanically strong and have high melting points and good electrical conductivity.

2. *Covalent or Homopolar Bonds* : In this type of bonding, stability is attained by the sharing of electrons between the two atoms. These are directional bonds in which molecules are formed. These are common in organic compounds but are rare in minerals (*e.g.* diamond). Minerals of this type are good electrical insulators.

3. *Metallic Bonds* : In this type of bonding, the metallic atoms attain a positive charge by losing electrons. These electrons are free to move about the atomic nuclei and the aggregate cloud of electrons binds the nuclei together. No directional forces are involved and no molecules are formed.

4. *Vanderwaal's or Residual Bonds* : The molecules of some substances, particularly organic compounds and elements that normally are gaseous at ordinary temperatures may be tightly bonded by covalent bonds, but the individual molecules are loosely held together by cohesive forces called Vanderwaal's bonds.

## Ionic-Atomic Substitutions

(a) *Camouflage* : It is a type of substitution that takes place in two atoms having same charge and similar radius.

e.g. $Al^{+3} \rightleftharpoons Ga^{+3}$
(0.51) (0.62)

(b) Capture : It takes place in atoms having: (i) higher charge and similar radius; or (ii) same charge and lower radius

e.g. $K^{+} \rightleftharpoons Ba^{+2}$
(1.33) (1.34)

(c) *Admittance*: It takes place in atoms having: (i) lower charge and similar radius ; and (ii) same charge and higher radius.

**e.g.** $Mg^{+2} \rightleftharpoons Li^{+}$
(0.66) (0.68)
$Ca^{+2} \rightleftharpoons Sr^{+2}$
(0.99) (1.12)

**Rules for Ionic Substitution :** 1. If two ions are to replace each other in a crystal, the difference between their ionic radius must not exceed 15%. e.g. Olivines, $Fe_2SiO_4$ — $Mg_2SiO_4$
($Fe^{+2}$=0.74 A; $Mg^{+3}$= 0.66 A).

2. When two ions have the same charge but their ionic radii are different, the one with the smaller radius tends to be incorporated.

3. When two ions have the same radius but their electrical charges are different, the one with the greater charge tends to be incorporated. *e.g.* when feldspars are formed, anorthites ($Ca^{2+}$ = 0.99 Å) appear before albites ($Na^{+}$ = 0.98 Å) because they carry stronger charge.

**Isomorpbism** : It is the phenomenon wherein two minerals have the same general internal structure, closely similar external morphology and similar X-ray diffraction patterns. The cause of the phenomenon is that anions and cations of the same relative size and in the same numbers tend to crystallize in the same structure type. It is common in plagioclase feldspars and pyroxenes. Good examples are Carbonate minerals: Calcite-Aragonite, and Magnesite-Siderite. Minerals in which the structure is exactly analogous, and in which each and every ion in one mineral has a counterpart in an identical environment in the other, are called *Isotypes*. Minerals in which the structure is basically the same, but not all ions in one mineral have counterparts in the other, are termed *Isostructural*. In isomorphism, the similarity in the size of the two ions is an important factor rather than a chemical similarity. Thus, many calcium ($Ca^{2+}$ =0.99Å) and magnesium ($Mg^{2+}$ =0.66 Å) compounds are not isomorphous although their chemical behaviour is similar.

**Solid Solution** : In compounds in which an atom, an ion, or radical of one kind substitutes for an atom, an ion or radical of another kind, mixed crystals are formed. Thus, solid solution can be simply defined as a homogenous crystalline solid of variable composition. In *Normal or Extensive* solid solution there is a continuous replacement of ions (e.g. olivine and plagioclase), whereas, in a *Limited* solid solution there is a partial replacement of one ion by another in a coupled substitution (e.g. Albite-Orthoclase). In the *Interstitial* solid solution, foreign ions or atoms fit into the interstices of the crystal lattice (e.g zeolites). In the *Defect or Omission* solid solution there is a systematic omission of one ion coupled with a valence change in another *(e.g.* Pyrrhotite). A solid solution in which two different elements are completely inter-replaceable at high temperatures but not at lower temperatures will tend to break down on cooling into two separate phases, one rich in the first mineral and the other rich in the second mineral. This breakdown of a homogenous solid solution is known as *Exsolution* or "Unmixing".

**Diadochy** : It is the ability of different elements to occupy the same lattice position in a crystal. Thus, Mg, Fe and Mn are diadochic in the structure of dolomite.

**Polymorphism** : The phenomenon of the existence of a compound in more than one crystal form possessing different physical properties but identical chemical properties is known as polymorphism. Two polymorphs contain the same number of atoms, but with different alignments in the crystal structure. When polymorphism occurs in elements it is described as *Allotrophy*. $CaCO_3$ is dimorphic (Calcite-Aragonite); $TiO_2$ is trimorphic (Anatase, Rutile, Brookite). As different polymorphs of the same substance are formed under different P-T conditions and chemical environment, they serve as very good geologic thermometers. A reversible polymorphic transformation at a transition temperature is described as *Enantiotropy.*

*e.g.* Quartz $\overset{867°C}{==}$ Tridymite $\overset{>>867°C}{==}$ Cristobalite

An irreversible polymorphic transformation is referred as *Monotropy* e.g. Diamond → Graphite. *Polytypism* (One-dimensional polymorphism) occurs when two polymorphs differ only in the stacking of identical two-dimensional sheets *e.g.* mica. When two different ions occupy the same structural site, it gives rise to *Order-Disorder* type of polymorphism. This is most important in the mineralogy of silicates, where $Al^{3+}$ substitutes for $Si^{4+}$ in the tetrahedral coordination.

**Pseudomorphism** : Compounds that exhibit external shapes which are inconsistent with the internal arrangement of ions are called Pseudomorphs. Pseudomorphism may occur due to substitution, incrustation, infiltration or alteration. *Paramorphism* is the change in the molecular constitution without change of chemical substance. *e.g.* Aragonite ($CaCO_3$) → Calcite ($CaCO_3$).

**Silicate Structures**

The fundamental unit in silicate structures is a $(SiO_4)_{-4}$ tetrahedron. Each oxygen atom is five times as large as the silicon atom. The distance between any two oxygen atoms is 2.8 A or 2.8x $10^{-8}$ cm.

(a) Nesosilicates (Orthosilicates) : They consist of independent tetrahedral group with no sharing of oxygen ions. The Si: O ratio is 1: 4 and the resultant composition is $(SiO_4)_{-4}$. The excess charge of –4 is balanced by two bivalent cations like

$2Mg^{+2}$ or $2Fe^{+2}$. These minerals form equidimensional crystals and are generally hard and have high specific gravities and high indices of refraction (R.L). These silicates are chemically diverse. ***e.g.*** epidote, piedmonite, zircon, phenakite group, andalusite, sillimanite, kyanite,staurolite, sphene, etc.

(b) Sorosilicates : They are characterised by double tetrahedral structures in which one oxygen is shared between the two silicon ions. The resultant composition is ($Si_2O_7$). They are characterized by high specific gravities, high R.I. and tabular habit. Chemically, Na, K and Al are generally lacking. *e.g.* hemimorphite, melilite, lawsonite, danburite, etc. Epidote, Zoisite and Vesuvianite (Idiocrase} have a combination of both $SiO_4$ and $Si_2O_7$ groups.

(c) Cyclosilicates (Metasilicates) : These are characterised by ring silicate structures formed by sharing of two oxygen atoms of each tetrahedron with the neighbouring tetrahedra. Rings of 3, 4 or 6 tetrahedra are known and these have the formulae : $(Si_3O_9)_{-6}$; $(Si_4O_{12})_{-8}$ and $(Si_6O_{18})_{-12}$ respectively. A two-member ring is unlikely because of the necessary close approach of two highly-charged silicon ions. They exhibit either trigonal or hexagonal symmetry. *e.g.* Benitoite and Tourmaline (3-membered rings), *Axinite (4-membered ring) and Beryl (6-membered ring)*.

(d) *Ionosilicates* : In these silicates, the tetrahedra are joined together to produce chains of indefinite extent. In the Single-Chain structures, each tetrahedron is joined to its neighbours through 2 oxygen atoms to form a single open chain. The Si : O is 1 : 3. e.g pyroxenes. Double Chain silicates are formed when the two single chains are placed side by side. Their composition is $Si_4O_{11}$. e.g. Amphiboles. Ionosilicates are characterized by prismatic habits of all kinds.

(e) *Phyllosilicates (Sheet silicates)* : The $SiO_4$ tetrahedra are linked by three of their corners and extend indefinitely in a two-dimensional network or sheet. The Si:O ratio is 4 : 10. Adjacent sheets are united through ions of other elements. Most of these silicates are characterised by the presence of (OH) F or Cl. They exhibit perfect basal cleavage. The sheets forma hexagonal planar network and thus, exhibit pseudo-hexagonal symmetry. e.g. micas, chlorites.

(f) *Tectosilicates* : These are Three-dimensional networks in which all the four oxygen atoms are shared. The (Si : Al) : O ratio is 1: 2. Minerals are massive and hard. e.g. silica, feldspars, feldspathoids, zeolites, etc.

In the silicate structures, the charges are balanced in one of the following ways :- (1) Cations may be used to neutralize the unit, (2) Sharing of one or more oxygen atoms with adjacent units, (3) Oxygen sharing and charge balance by cations, (4) Substitution of ions of different valencies for oxygen or silicon, (5) Presence of additional groups such as (OH), F or Cl.

## Certain Physical Properties of Minerals

1. *Colour* : Colouration results from the absorption and reflection of light by the mineral. It is due to the presence of pigments called Chromophores (Va, Ti. Cu, Fe, Cr, Ni, etc). Idiochromatic minerals have a characteristic colour. They neve' occur in a colourless state e.g. gold, magnetite. Allochromatic minerals exhibit more than one shade of colour e.g. Quartz.. Pseudochromatic minerals exhibit false colouration.

(a) *Tarnish* : It refers to the formation of iridescent colours on the exposure to the atmosphere. e.g. Erubescite (bornite), 'the Peacock ore".

(b) *Play of Colours* : It refers to a changing series of colours in different directions due to high dispersion. e.g. diamond.

(c) *Change of colours* : It is caused due to interference of light reflected from thin plates of other minerals enclosed in parallel planes within the feldspars. e.g. feldspars.

(d) *Opalescence* : It is formed due to reflection of light from within the mineral. e.g. opal.

(e) *Iridescence* : It refers to the display or prismatic colours due to interference of light in minute fissures containing thin film of. air or liquid. e.g. quartz, calcite, mica, labradorite, limonite, etc.

(f) *Aesterism* : It is the concentration of light along lines radiating from the centre. e g. sapphires and rubies.

2. *Luster (Reflectivity)* : It is defined as the appearance of a material in reflected light. The higher the refractive index, the higher is the luster.

(a) Metallic : The Refractive Index (R.I.) is 3 or >3 e.g. negative metals and most sulphides.

(b) Submetallic : The Refractive Index (R.I.) is 3.0 to 2.6 e.g. cuprite, cinnabar, hematite.

(c) Adamentine : The Refractive Index (R.I.) varies from 2.6 to 1.9 e.g. Diamond, zircon, cassiterite, sulphur, sphalerite, rutile, etc.

(d) Vitreous : The Refractive Index (R.I.) varies from 1.9 to 1.3. Vitreous lustre is exhibited by nearly 70% of all minerals including all silicates, carbonates, phosphates and sulphates.

3. *Diaphaneity* : It is the property relating to the transmission of light. On the basis of diaphaneity, minerals are classified as transparent, translucent and opaque.

4. *Hardness* : It is the resistance offered by a mineral to a point tending to cause abrasion on its surface. It increases with (a) density of packing in the structure, (b) increase in the valency, (c) decrease in ionic size. Determined by a *Sclerometer*. It is measured in terms of an arbitrary scale called the *Moh's Scale*.

5. *Specific Gravity* : It is the ratio of the weight of the body to that of the equal volume of water. It depends on the atomic weight of the constituents and the density of packing. It is determined by heavy liquids, chemical balance, Walker's steelbar, Jolly's spring balance, etc. A *Pycnometer* or a Specific gravity bottle is used in the case of friable minerals and small fragments of liquids.

Heavy liquids : Bromoform (Sp. gr.=2.90), Methylene iodide (Sp.gr=3.3), Thoulet's solution (Sp.gr.=3.17), Kiems' Solution (Sp.gr.=3.24), Clerici's solution (Sp.gr.=4.0).

6. Fusibility : Minerals can be arranged according to their relative fusibilities (melting points) by Blowpipe techniques. Fusibility is measured in terms of Von Kobell's scale: Stibnite (5250C), Natrolite (9650C), Almandine (10500C), Actinolite (l2000C), Orthoclase (13000C) and Bronzite (14000C).

7. Luminence : It is the emission of light other than incandescence.

(a) Fluorescence : It refers to glow in he presence of the source e.g. Fluorite exhibits blue colour due to the substitution of rare-earth elements for a small amount of calcium. Calcite shows red, pink or yellow fluorescence due to the presence of manganese or organic impurities (porphyrins); Scheelite shows white or bluish white colours due to the replacement by $MO^{-6}$ of W+6.

(b) Phosphorescence : It refers to glow in the absence of the source.

(c) Triboluminence : It is induced by pressure, crushing, scratching or rubbing. e g. aragonite, calcite, barite, dolomite, fluorite, gypsum, etc.

8. *Chatoyancy* : It refers to the changeable silky sheen which is a result of fibrous structure.

**1. The Olivine Group** : Minerals of the Olivine Group belong to the Nesosilicates and has Mg, Fe and Mn as cations. They exhibit complete diadochy between atomic pairs Mg and Fe+2 and $Fe+^2$ and Mn. They have high Refractive Index and strong birefringence. They represent an Isomorphous series: Forsterite ($Mg_2SiO_4$ ) -Hyalosiderite - Hortonite - Ferrohortonite -Fayalite ($Fe_2SiO_4$) - Knebelite-Tephroite ($Mn_2SiO_4$). These are characterised by the absence of aluminum. In the series, Forsterite - Olivine - Fayalite, from left to right, the refractive index and the specific gravity increases and the melting point decreases. The order of crystallization is Forsterite - Olivine - Fayalite. Olivine alters to serpentine, chlorite, amphibole, carbonates and talc. It also alters to Iddingsite and bowlingite. Fayalite alters to grunerite and fosterite alters to antigorite. Olivine forms under magmatic and metamorphic conditions. Forsterite is formed by the metamorphism of dolomitic limestone. *Varieties* : Peridot (gem); Chrysolite (20% Fe); Hyalosiderite (50% Fe). Monticellite is isomorphous with Olivine. The Humite group resembles olivine but contain (OH) and F in the structures.

**2. The Pyroxene Group** : The minerals belonging to this group are Ionosilicates having single chain structures. These are the silicates of Ca, Na Mg, Fe and Al. The fundamental form is the prism (110) in monoclinic and 210 in the orthorhombic system. The crystals are short prismatic with interfacial angles of about 87° and 93°. Enstatite and hypersthene belong to the orthorhombic system; diopside, hedenbergite, pigeonite, spodumene and jadiete belong to the Monoclinic system while rhodonite crystallizes in the Triclinic system.

Enstatite-Orthoferrosilite : These are ortho-pyroxenes and represent silicates of Mg and Fe. Enstatite is $En_{100}$ - $En_{90}$ ; Bronzite is $En_{90}$ - $En_{70}$; Hypersthene is $En_{70}$ - $En_{50}$. They alter to bastite or schillarspar and to uralite. These are the common constituents of the ultra-basic rocks.

Jadiete : Jadiete (Na (Al Fe) $Si_2O_6$) is of much petrographic significance, being intermediate in composition between Albite and Nepheline. It has high density (3.3) as compared to other Na-Al silicates {Albite (2.6), Nepheline (2.6) and Analcime (2.3)}. It is a HP/LT mineral and occurs in metamorphic rocks, meta-graywackes and arkoses.

Omphacite : It the characteristics clino-pyroxene of eclogites. It can be considered as a solid solution of jadiete, augite and acmite (Jd 25-75;. Aug 25~75; Ac 0~25).

Pigeonite : It is a Ca-poor pyroxene which is intermediate in composition between diopside- hedenbergite and clinoenstatite. It has low optic axial angles. It is restricted to quickly chilled lavas and is unknown in metamorphic rocks.

Aegirine (Acmite) : It is optically negative and has low extinction angles (2-60) and exhibit strong pleochroism.

*Varieties* : (i) Diopside : Mansjoite (F-bearing); Coccolite (white); Malacholite (translucent, colourless). Diallage is the foliaceous or fibrous green variety of diopside and augite.

(ii) Spodumene: Hiddenite (emerald green); Kunzite (pink); Triphane (colourless/yellow).

(ii) Chlromelanite is dark green, Fe-bearing variety of Jadeite.

**3. The Amphibole Group** : These are Ionosilicates exhibiting double chain structures. These are the silicates of Mg, Fe, Ca and Na with or without Al. They differ from pyroxenes in the presence of (-OH ) i.e. One (-OH) for every 11 oxygen atoms. They exhibit long prismatic crystals with prism angles of 120°. They are characterised by acicular or fibrous crystals, which are pseudomorphic and show perfect prismatic cleavages at 56° and 124°. They are characteristic of plutonic rocks. Anthopylute and ferrosiderite crystallize in the orthorhombic system, Cummingtonite, grunerite, tremolite, Hornblende, Edenite, pargasite, Tschermakite, glaucophane and riebeckite crystallize in the monoclinic system while Cossyrite (anigmatite) crystallizes in the triclinic system.

Orthorhombic amphiboles are unknown in igneous rocks and are associated with talc/cordierite. Tremolite is the low-Fe end of the Tremolite-Actinolite series. It breaks down into enstatite, diopside and quartz. It occurs in impure crystalline limestones and calc-silicate hornfelses. Actinoite is a typical mineral of greenschist facies.

Hornblende is a characteristic mineral of intermediate plutonic rocks.

Basaltic hornblende (Lamprobolite/Oxyhornblende) : It is Ca-Na-Ti hornblende. It has high $Fe^{+3}$: $Fe^{+2}$ ratio and low (-OH) and differs from the ordinary hornblende in higher refractive index, higher birefringence, smaller extinction angles and dark brown colour. It commonly occurs in volcanic rocks.

*Varieties* : Anthopyllite : Amosite (fibrous) and Gedrite (positive).

Tremolite : Mountain leather (fibrous), Nephrite (acicular).

Hornblende : Edenite (Fe-bearing), Dashkessanite (Cl-bearing)and Pargasite (Fe-bearing, positive).

Crossite is a soda amphibole which is intermediate between glaucophane and riebeckite. Kataphorite and Barkevikite are varieties of Arfvedsonite. Asbestos is the fibrous variety of Actinolite. Uralite is a pseudomorph formed on replacing pyroxene. It has pyroxene form and amphibole cleavages. Gedrite, pargasite and cummingtonite are optically positive.

**4. The Feldspar Group :** Feldspars are the most abundant of all the minerals and constitute 60% of the earth's crust. These are Tectosilicates. Orthoclase (Monoclinic) and Microcline (Ticlinic) are the Potash feldspars. Albite is a Soda feldspars. Anorthite is a Lime feldspar. Hyalophane and celsian are barium feldspars. Properties of Feldspars : Their hardness is around 6 and specific gravity ranges from 2.5 to 3.0. They exhibit two kinds of cleavages : parallel to 001 and 010 at right angles. Feldspars commonly alter to sericite or kaolin. Lime feldspars alter to Saussurite, a mixture of albite and epidote. Symmetry in feldspars (Order-disorder) : (a) If the ions are distributed throughout the lattice as a whole, but without any order and occupy any place appropriate to them, the lattice obtained will be of high symmetry (disorder). (b) When ions occupy strictly defined positions in the lattice, the symmetry is lowered. The greater the ordering, the lower is the symmetry. Sanidine is the least ordered feldspar and microcline is the most ordered.

| Sanidine | → | Orthoclase | → | Microcline |
|---|---|---|---|---|
| (Monoclinic) | | (Monoclinic) | | (Triclinic) |

Symmetry decreases →

*Varieties* : Alkali feldspars : Sanidine is a high temperature (HT) variety of Orthoclase and is common in K-rich volcanic rocks such as trachytes and rhyolites. Adularia is a low temperature (LT) variety of Orthoclase and occurs as a vein mineral. Moonstone is an opalescent adularia showing play of colours. Murchisonite is a red variety of Orthoclase.

Plagioclase feldspars : Albite : It crystallizes in the Triclinic system. It is optically positive. It occurs in granites and pegmatites. Anorthoclase is a soda-microcline which crystallizes in the triclinic system and is optically negative. Peristerite (play of colours), Clevelandite (lamellar) and Pericline (white) are the varieties of albite.

**5. The Feldspathoid Group** : These minerals exhibit tectosilicate structures. These are silica deficient feldspars. These are silicates of K, Na, Al and Ca and may contain molecules of $CaCO_3$, NaCl, ; $Ca(SO_4)_2$; $Na_2SO_4$ and $Na_2S$. They occur in igneous rocks which are low in silica such as syenites and phonolites.

Leucite : ($KAlSi_2O_6$) It is Pseudo-isometric and dimorphous and exhibit polysynthetic twinning, which disappears at 600°C. It commonly occurs as phenocrysts in lavas but is rare in coarse-grained igneous rocks.

Nepheline : It crystallizes in the Hexagonal system and is optically negative. It occurs in soda-rich igneous rocks. It alters to zeolites, sodalite and muscovite. Kalsilite is naturally occurring nepheline. Ijolite (dark) and Cancrinite (with $CaCO_3$) are the other varieties.

Sodalite (with NaCl); Hauyne (with $CaSO_4$); Nosean (with $Na_2SO_4$) and Lazurite (with $Na_2S$) are all isometric.

**6. The Silica Group :** Silica ($SiO_2$) occurs in nature as quartz, tridymite, cristobalite, coesite, stishovite, lechatelerite and natural hydrated silica, i.e. opal. Forms of silica: The minerals of the silica group exhibit different forms such as crystalline (e.g. quartz, tridymite, cristobalite), cryptocrystalline (e.g. chalcedony, jaspar, flint), cryptocrystalline hydrous (e.g. opal) and amorphous (e.g. Lechateleirite).

Quartz : It is the most common of all the vein minerals. It has many varieties: (a) Amethyst (violet), (b) Rose quartz (red/pink with titanium), (c) Morion (black) , (d) Citrine (yellow); (e) Tiger's eye (yellow fibrous), (f) Cat's eye (fibrous with play of colours), (g) Averturine (with mica and hematite), (h) Cairngorm stone or 'smoky quartz is a smoky black or yellow variety (due to an exposure to radioactive minerals), (i) Rock Crystal is a coarse, colourless, crystalline variety. Milky quartz is Rock Crystal with liquid inclusions.

Chalcedony : It is the principal constituent of cherts and jaspers. It is a fine crystalline form of quartz. Carnelian (red/brown), Chrysoprase (yellow), Sard, Plasma (green), Heliotrope (with jasper) are the varieties of chalcedony. Jasper is opaque chalcedony. Chert is a massive form. Agate is banded. Onyx is a straight banded agate. Opal is a solidified colloidal silica. Hyalite, hydrophane, geyserite and menilite are the varieties of opal.

Each mineral of the silica group exhibit polymorphism. Each form has its high and low temperature forms. For example,

(a) Low Quartz $\underset{573°C}{\rightleftharpoons}$ high Quartz

(b) High Tridymite $\underset{117°C}{\rightleftharpoons}$ low Tridymite

(c) High Cristobalite $\underset{200\text{-}268°C}{\rightleftharpoons}$ low Cristobalite

High-low transformations are reversible. This change is due to displacement and rotation which alters the symmetry of the structure without breaking the $SiO_4$ tetrahedral links.

Coesite (monoclinic) and Stishovite (tetragonal) are very rare forms of silica formed by the instantaneous high pressure of the meteoric impact. Stishovite is characteristically found in the mantle transition zone.

**7. The Mica Group** : These are Phyllosilicates in which the Si: O ratio is

4 : 10. These minerals are isomorphous and exhibit perfect basal cleavage (001) with prism angles 60° or 120°. They exhibit pearly or metallic lustre. These minerals produce percussion figures on applying pressure. The individual species in this group include a number of polymorphs. The general formula is $X_2Y_4\text{-}_6Z_8$ (OH, F)$_4$ where X—K, Na or Ca (also Ba, Rb, Cs); Y=Al, Mg or Fe (also Mn, Cr, Ti, Li); Z=Si or Al (also Fe+3 and Ti). Micas are said to be 'Dioctahedral' when the number of Y ions is 4 (e.g., Muscovite, Pargasite and Glauconite) and 'Trioctahedral' when it is 6 (e.g. Phlogopite, Zinwaldite, Lepidolite). In 'Brittle' Micas' e.g., Margarite, Clintonite and Xantbophyllite, 'X' is largely Ca. Optically micas are strongly birefringent, exhibit parallel extinction and their interference colours show speckled appearance ('bird's eye effect').

Muscovite : It occurs as a detrital mineral in elastic sedimentary rocks and also in granites and granitic pegmatites. It is a characteristic product of fluorine metasomatism (greisenization). Some Na is always replaced by K. This replacement increases with an increase in temperature. Thus, the Na content of muscovite may be a useful geological thermometer. Varieties : Fuchsite or Mariposite, (green with $Cr^{+3}$), Roscoelite (Va-bearing) ; Sericite or hydromica is a minutely crystalline variety with less K and more $H_2O$. In Phengites, the Si: Al ratio is 3:1.

Phlogopites : They occur in ultrabasic rocks, metamorphosed dolomites and dolomitic sediments. Biotite is distinguished from phlogopite by their Mg: Fe ratio of 2:1. Lepidomelene, Manganophyll and Haugtonite are the varieties of Biotite.

## Optical Mineralogy

Properties observed under ordinary light : Crystalline form, cleavage, colour and refractive index.

Properties observed under polarized light : Pleochroism, pleochroic holes and twinkling.

Properties observed under crossed nicols : Isotropic/anisotropic character, extinction, polarization colours, twinning, alteration and elongation.

Properties observed under convergent light : Interference figures and optic sign.

**Extinction** : It occurs when the vibration planes of the minerals are parallel to the nicol planes. A section of anisotropic minerals extinguish or show darkness between crossed nicols four times during a complete revolution of the stage.

Parallel or straight extinction : Extinction occurs with the cleavage parallel to the vibration planes of the polars.

Inclined or oblique Extinction : It occurs when the cleavages or crystal boundaries lie at oblique angles to the planes of vibration of two polars.

Symmetrical extinction : Extinction occurs when the planes of vibration of the polars are parallel to the diagonals of the pattern.

**Cubic system** : All sections are isotropic.

**Tetragonal and Hexagonal systems** : Basal sections, being perpendicular to the optic axis are isotropic. Vertical sections give straight extinction.

**Orthorhombic system**. Pinacoidal sections give straight extinction.

**Monoclinic system** : Ortthopinacoidal and basal pinacoidal sections give straight extinction. Clinopinacoidal sections give oblique extinction.

**Triclinic system** : All sections give oblique extinction.

## Isotropic and Anisotropic Substances

*Isotropic Substances* : Substances in which light is transmitted with the same velocity in all directions are called isotropic substances. The properties such as electrical conductivity, thermal conductivity, mechanical strength and the refractive index are the same in all directions and only single refraction is found. The wave normal is circular, wave-surface is a sphere while the wave front is circular. They exhibit constant refractive index. Isotropic substances appear dark when examined under crossed nicols. These are characteristic of the Cubic system. Examples include glass, rubber, liquids, gases, garnet, spinel and fluorite.

*Anisotropic Substances* : Substances in which light is transmitted with different velocities in different directions are called anisotropic. Their physical properties are different in different directions. They show strong double refraction and have two refractive indices. The wave normal and the wave surface are ellipsoidal. The wave front is either an ellipse inside a circle or a circle inside an ellipse. Anisotropic substances appear coloured under crossed nicols. They belong to all the crystal systems except the Cubic System. Examples include tourmaline, apatite, hornblende and biotite.

(i) *Optically Uniaxial Minerals* : Minerals crystallizing in the Tetragonal and Hexagonal systems show double refraction in all directions, except one i.e. parallel to C-crystallographic axis. In this direction the light ray shows single refraction and the velocity of the ordinary and the extraordinary rays are the same. This direction is called the Optic axis and these minerals have only one optic axis. The extra-ordinary ray vibrates in the plane containing the ray and the optic axis. The ordinary ray always vibrates at right angles to the plane containing the ray and the optic axis. Examples include calcite, nepheline, quartz, beryl and apatite.

(a) Positive Uniaxial Minerals : The velocity of the extraordinary ray is less than that of the ordinary ray. The refractive index of the extraordinary ray is greater than that of the ordinary ray. Wave front is an ellipse inside the circle. Example : Quartz (*E* -1.553 Å, *W*- 1.544 Å).

(b) Negative Uniaxial Minerals : The velocity of the extraordinary ray is greater than that of the ordinary ray. The refractive index of the extraordinary ray is less that of the ordinary ray. Wave front is a circle inside an ellipse. Example : Calcite (*E* -1.486 Å, *W*- 1.658 Å).

(ii) *Optically Biaxial Minerals* : In the crystals belonging to the orthorhombic, monoclinic and triclinic systems, there are two optic axes or two directions along which there is no double refraction and along which light travels with a single velocity. There are three principal vibration directions : X - fastest, Y - intermediate and Z - slowest. Examples include pyroxenes, axinite, olivine, diaspore, stuarolite, kyanite and borax.

## Optical Properties of Some Minerals

**Quartz :** (1)Quartz is colourless and transparent, (2) Absence of cleavage and alteration, (3) No shagreen surface or relief, (4) Lack of twinning, (5) Quartz has higher refractive index than orthoclase or microcline or the Canada balsam, (6) Quartz is distinguished from nepheline or albite by its uniaxiality and optic sign even in most obscure cases, (7) Birefringence of quartz is 0.009 with *E* - 1.553 Å, *W*- 1.544 Å. It is optically positive.

**Orthoclase** : (1) Monoclinic orientation, (2) Biaxial negative, (3) Large optic axial angle (69-72°), (3) Crystals are sometimes flattened parallel to 010, (4) Refractive index is lower than that of quartz or balsam, (5) birefringence is low, (6) polarization colours are grays of the first order, (7) It exhibits simple twins of the Carlsbad, Baveno and Menabach types, (8) maximum extinction on 010 is 5-12°. Extinction on 001 is parallel.

# 5

# Palaeontology

## I. Fossils

Fossils are traces or remains of plants and animals.

**Types of Fossilization :** (1) Preservation of entire organism : This mode of fossilization is very rare e.g. mammoths and rhinoceroses of Siberia. (2) Preservation of hard parts : It is the commonest type of fossilization in which the composition of hard parts control the degree of fossilization. Siliceous parts are altered easily; e.g. Sponges, Radiolarians. Hard parts of calcium phosphate are virtually unaltered. e.g. Brachiopods, Arthropods. (3) Altered hard parts : (a) Carbonization - In carbonization, the percentage of oxygen, hydrogen and nitrogen decreases while that of carbon increases e.g. Jelly fishes, worms, leaves; *(b)* Petrifaction - It involves the preservation of original form and structure. It is also called Per-mineralization. *(c)* Replacement (Mineralization) - In this, the original structure and form are retained but the original material is replaced. e.g. Replacement of SiO2 by CaCO3 (Sponges); Pyritization (crustaceans, brachiopods); replacement by Glauconite (Foraminifers), etc. *Beekite rings* - In this mode of fossilization, silica is deposited in the form of concentric rings around a nucleus. (4) Preservation of original form only : This include *Casts* and *Moulds*. (5) Traces of organisms : This includes tracks and trails, burrows and borings, Coproliths (fossil fecal pellets), Gastroliths (found in body cavities) and pseudo-fossils.

**Rocks in which fossils occur :** (1) Land deposits : Old soils, peat-bogs, cave deposits, loess, inter-glacial deposits, volcanic ash, etc. (2) Lake deposits : Lake deposits are very favourable for fossilization (3) Marine deposits : These are the largest repositories of fossils.

**Important Fossils of Different Periods of Time**

| Group | System | Chief Fossils |
|---|---|---|
| Archaen | Archaen (3600 m.a.) | Lifeless |
| Proterozoic or Precambrian | Precambrian (2500 m.a.) | Soft-bodied animals and plants |
| Paleozoic or Primary | Cambrian (605 m.a.) | Abundance of *Trilobites*. |
| | Ordovician (505 m.a.) | Abundance of *Trilobites* and *Graptolites*. |
| | Silurian (445 m.a.) | First Fishes and Land Plants; *Graptolites* disappear at the end. |
| | Devonian (410 m.a.) | Abundance of *Corals*, *Brachiopods*; First Amphibians and Lung fishes appear |
| | Carboniferous (350 m.a.) | Many non-flowering plants; First Reptiles appear. |
| | Permian (290 m.a.) | Trilobites disappear |

| | | |
|---|---|---|
| Mesozoic or Secondary | Triassic (240 m.a.) | Ammonites, reptiles and amphibians abundant; Arid climate. |
| | Jurassic (200 m.a.) | Ammonites abundant; Birds and flowering plants appear; sea-urchins common. |
| | Cretaceous (140 m.a.) | Giant reptiles and ammonites disappear at the end; Flowering plants numerous. |
| Tertiary or Cenozoic | Eocene (65 m.a.)<br>Oligocene (38 m.a.)<br>Miocene ( 25 m.a.)<br>Pliocene (8 m.a.) | Mammals, Molluscs and flowering plants dominant. |
| Quaternary | Pliestoceoe (1 m.a.)<br>Recent | Living animals; many mammals die off due to glaciation; Man appears |

**Characteristic Fossils of Different Systems**

| *System* | *Characteristic Fossils* |
|---|---|
| Middle Cambrian | *Paradoxides* |
| Ordovician | *Trinucleus* |
| Silurian | *Pentamerus Oblongus* |
| Carboniferous | *Goniatites* |
| Permian | *Glossopteris and Gangamopteris* |
| Triassic | *Ceratites* |
| Eocene | *Nummulites* |

**Precambrian Life** : Most of the Precambrian sedimentary terrain of the world contain rocks like limestones, dolomites, sandstones, shales, silts, clays and quartzites. It is from these rocks that the first evidence of life was discovered. However, the evidence of life in them is relatively scanty and poorly preserved which has created numerous conflicting hypotheses on Precambrian life.

The most abundant signs of life in the Precambrian times are the stromatolitic structures that occur in calcareous dolomitic rocks. There have been several reports of stromatolitic structures in Archaen Greenstone belts, the oldest among them being from the 2900-3200 m.a. old Bulawayan Group of Zimbabwe. The first positive organic evidence of Precambrian life, however, comes from plant-related forms which embrace the microscopic organic remains of bacteria, algae, fungi and associated spore-like (Sporomorph or Acritarch) remains. The earliest record is that from the Fig Trees series of Swaziland (3400-300 m.a.) in South Africa.

In the *Prokaryotes*, the cell is devoid of nuclei and the DNA is not arranged in chromosomes. These are believed to be the first reef-forming organisms e.g. Pilbara stromatolites, western Australia; Bulawayan stromatolites, Zimbabwe and those found in the Vaimpalli dolomites and Kheinjua formation of the Semri group. In *Eukaryotes*, the chromosomes are bound by nuclei and organelles. E.g. Bitler Spring Formation, Central Australia, which shows evidence of the earliest known sexually reproducing organisms.

## I. Phylum Protozoa

The Protozoans are microscopic animalcules, representing the simplest and the most primitive of all life forms. They are unicellular or acellular, containing one or more nuclei and organelles, with little or no histologic differentiation into tissues.

**(a) Foraminifers** : Foraminifers ("hole-bearers") are characterised by their thread-like pseudopodia and the presence of a test or shell, which may be calcareous, arenaceous, chitinous, siliceous or gelatinous. The shells are of different shapes and range in size from 0.01- 190mm. *Allogromia* is the most primitive of all Foraminifers and has an egg-shaped chitinous shell. In *Saccamina* (Carboniferous), the spherical shell is lined with chitin and consists externally of closely cemented foreign particles. In *Spirillina*, the shell is devoid of foreign substances and is wholly calcareous. In its vitreous or perforated forms, the shell is glass-clear or hyaline and is pierced by numerous pores for the passage of pseudopodia. E.g. *Nodosaria, Cristellaria, Rotalia.* In the porcellanous or imperforate forms, the shell is opaque and white and is not perforated by canals, instead it is provided with one or two large apertures through which the pseudopodia pass out. E.g. *Miliola, Alveolina.* Foraminifers range in age from Ordovician to Recent.

*Formation of Shells* : Some Foraminifers like Elphidium are characterised by multi-chambered shells. The chambers communicate with one another through septa or large openings present between them. The shells may be coiled in straight line (*Nodosaria*), curved line (*Dentalina*), Plano-spiral (*Cristellaria, Elphidium*), Helicoid Spiral (*Rotalia*), Irregular Spiral (*Globigerina*) or Conical Spiral (*Turrispirilina*). Many higher forms of Foraminifers such as *Elphidium, Globigerina, Cornuspira*, etc exhibit the phenomenon of *Dimorphism* i.e. a single species exists in two forms. In the megascleric form, the proloculus of the first chamber is large and more or less spherical while in the microscleric form the proloculus is small. This phenomenon was first noticed in the *Nummulites* of the Eocene deposits.

**(b) Radiolarians** : The Radiolarians are characterised by minute, complex and beautiful siliceous shells. Radiolarians are marine, with stiff, radiating pseudopodia. They have numerous shapes and different varieties of ornamentation. The protoplasm is differentiated into three layers: the central capsule; intra-capsular protoplasm and the extra-capsular protoplasm, through which the pseudopodia radiate out.

## II. Phylum Coelentrata

The Ceolentrates (or Cnidarians) include the polyps, hydroids, jelly fishes, sea-anemones and corals. These are the first multi-cellular animals to possess true tissues, a digestive cavity and two distinct germ layers.

**(a) Graptolites** : Graptolites constitute an extinct group of marine animals which used to live in the form of colonies during the Lower Paleozoic era and are presently known only as fossils. These are found mostly in argillaceous rocks, especially in black carbonaceous shales and are comparatively rare in sandstones and limestones. The entire skeleton of a simple colony of graptolites is known as *Rhabdosome* (or *Stock* or *Polypary*), while that of the compound colony is described as *synrhabdosome*. The rhabdosome is made up of one or more branches or *stipes*, with a row of cups or *hydrothecae* on the ventral side. *Nema* or *Virgula* is a hollow chitinous thread or rod within the peridem of the common canal, on the dorsal side of the rhabdosome. *Sicula* is the small, hollow, chitinous cone attached to the nema. A rhabdosome with only one row of theca on the stipes is called *uniserial* (e.g. *Monograptus*); those with two rows are called *biserial* (e.g. *Diplograptus*). The most ancient graptolites have been found to occur within the rocks of Upper Cambrian age (e.g. *Tetragraptus, Clonograptus, Bryograptus,* etc). The floating forms of the graptolites were distributed all over the world during the Ordovician (e.g. *Didymograptus, Phyllograptus, Dichograptus, Diplograptus, Leptograptus,* etc) and Silurian (e.g. *Monograptus, Cyrtograptus, Rastrites,* etc) periods and serve as very useful index fossils.

**(b) Corals** : The corals belong to the class Anthozoa and orders Zoantharia and Alcyonaria. Corals are basically deposits of CaCO3 occurring in the form of colonies. Each colony consists of a number of individual animals or polyps. The coral polyps secrete an epidermal calcareous skeleton. The epidermal cells secreting the skeleton are termed *calicoblasts*. The entire skeleton is known as *corallum* while the skeleton of a single polyp is referred as *corallite*. The peripheral walls of the corallite are called the theca. *Septa* are the vertical plates dividing the visceral chamber while *Costae* are the projections of septa outside the theca. *Tabulae* are the plates which cross the septa in the visceral chamber. *Palli* are the vertical plates which originate from the columella and radiate outwards. *Synapticulae* are calcareous bars or rods that connect one septum with another. Thin calcareous plates joining adjacent septae are known as *dissepiments*. When individual corrallites in a compound coral are free and diverge from one another, the corallum is termed *dendroid*; when they are nearly parallel to one another, it is called *fasciculate*. If the corallites are polygonal in outline, the corallum is termed *massive*. When in a massive coral, the corallites are nearly parallel to one another, the corallum becomes *basaltiform*. In some massive corals, the boundaries of the corallites tend to be indefinite; such corals are termed *astroeiform*.

Corals are known to have existed since the dawn of the Paelozoic era. During the post-Cambrian period, corals were extremely abundant and built up extensive reefs and banks known as Bioherms. The corals with hard and stony skeletons appeared during lower Ordovician period. These belonged to the sub-classes Tubulata, Schizocorallia and Rugosa. *Favosites, Halysites, Cyathophyllum*, etc are some of the well-known corals of the Paleozoic age. The Alcyonarian corals developed during the Silurian. The Zoantharian corals originated during the Triassic and flourished during the Mesozoic and Tertiary eras.

## III. Phylum Arthopoda

These are characterised by 'jointed legs'. The body is bilaterally symmetrical and metamerically segmented i.e. formed of a series of segments. Some of the anterior segments undergo cephalization, forming a distinct head. The body is triploblastic and is externally covered with a thick, tough and non-living . chitinous cuticle, forming the exoskeleton. They exhibit the phenomenon of *Moulting* or *Ecdysis* i.e. casting of the old skeleton and growth of new one from the underlying epidermis.

**Trilobites** : *Morphology* : Trilobites are exclusively marine and range in size from 5 to 7.5cm, though some like *Terataspis* attained a length of 67.5 cm while *Agnostus* was only 6mm in length. The body is oval, dorso-ventrally flattened and divided into the head, thorax and *pygidium*. The most distinctive feature is the presence of two longitudinal shallow furrows or grooves, all along the dorsal surface, dividing the body into three distinct longitudinal lobes. There is an elevated median or *axial lobe*, containing the main organs of the body, and two depressed lateral or pleural lobes, which are respiratory and locomotory in function. The dorsal surface the body is protected by a strong calcareous exo-skeleton called the *dorsal shield*. The part that covers the head is usually semi-circular or triangular in shape and is called the *Cephalon*. It is made up of a convex axial or median lobe known as *Glabella* and two lateral lobes called the *Cheeks*. The glabella is separated from the Cheeks by means of an *axial furrow*. The last segment of the glabella known as the *Occipital (or Neck) Lobe* is separated from the rest of the glabella by an *Occipital or Neck Furrow*. The posterior angles of the cheek, known as the *genal angles* may be rounded as in *Calymene* of produced into spines, the *genal spines*, as in *Paradoxides*. *Hypostome* or *labrum* is an oval plate situated below the mouth. *Metastome* is a small lower lip plate lying behind the mouth e.g. *Triarthus*. The eyes may contain only a single lens (e.g. *Trinucleus*) or two or three lenses (e.g. *Harpes*) while in some species like *Agnostus, Microdiscus, Ampyx, Conoryphr*, etc, the eyes are absent. In *Cyclopyge* the eyes are unusually large while in *Olenellus*, the eyes are crescentric in form. The thorax consists of a series of segments which vary in number from two (e.g. *Agnostus*) to forty two and are movable upon one another. The pygidium is formed of a variable number of segments which are fused together and immovable.

*Geologic Distribution* : Trilobites occur as fossils only within rocks of Paleozoic age. They first appeared in the Lower Cambrian beds and attained their culmination during the Ordovician period. They became less important in Devonian and only one family survived into the Carboniferous. They were totally extinct by the close of the Paleozoic era. *Examples* : **Lower Cambrian** : *Olenellus, Holmia, Agnostus, Microdiscus*; **Middle Cambrian**: *Paradoxides, Solenopleura, Agnostus*; *Upper Cambrian*: *Olenus, Parabolina, Peltura, Niobe*; **Ordovician**: *Agnostus, Ampyx, Trinucleus, Ogygia*; **Silurian**: *Calymene, Phacops, Homalonotus*; **Devonian**: *Homalonotus, Bronteus, Protus*; **Carboniferous**: *Phillipsia*; *Permian*: *Neophillipsia, Neoproteus.*

## IV. Phylum Echinodermata

This includes all marine animals with well-developed water-vascular system. The body is radially symmetrical and pentamerous.

**Classification :** This Phylum includes two Subphyla: Eleutherozoa and Pelmatozoa. The Eleutherozoa has three classes: Asterozoa, Echinoidea and Holothuroida. The Pelmatozoa has four classes: Crinoidea, Cystidea, Blastoidea and Edrioasteriodea.

**Echinoids** : There are two orders in Echinoids : Regularia and Irregularia.

**Regularia** (Endocyclic) : These have spherical tests with mouth and anus exactly at the centre. They have a radially symmetrical test. It includes two suborders: Endobranchiata and Ectobranchiata.

**Irregularia** (Exocyclic) : These have bilaterally symmetrical tests. The anus lies outside the apical disc. There are three sub orders : Holectypina, Clypeastrina and Spatangia.

***Morphology* :** The Echinoids or *Sea-urchins* usually have a globular, heart-shaped or discoidal body, covered with spines. The test or shell consists of numerous calcareous plates or spines, which are usually hemispherical, but rarely ellipsoidal or discoidal in shape. The lower of the ventral side is commonly flattened while the upper side or dorsal side is generally convex in shape. The mouth lies on the ventral side and is covered by a number of plates known as the *Peristome*. The anus lies on the dorsal side of the test and is surrounded by a membrane bearing small plates known as *Peripoct*. The peripoct is encircled by a ring formed of ten plates, known as the *apical system* or *disc*, five are called *genitals* and the other five are called *oculars*. Commonly the genital plates are larger and lie in contact with peripoct, forming the inner part of the ring. The ocular plates lie in alternation with the genital plates and constitute the outer part of the ring. One of the genital plates is provided with a number of pore-like openings described as the *Medreporic plate*. The other genital plates and all the ocular plates are provided with one perforation. When the oculars separate the genitals and touch the periproct they are said to be *insert*; when they do not touch the periproct they are described as *exsert*. In *Salenia* and *Peltestes*, there is an extra plate in the apical disc, in front of periproct, which is described as *sur-anal plate*. *Corona* is the main part of the test and consists of 20 columns of plates, five pairs arise from Ocular plates and are described as *Ambulacral plates* while the other five which arise from the genitals are called *Inter-ambulacral plates*. Each pair of columns form an *Ambulacral area*. Ambulacral plates are smaller in size, larger in number than the inter-ambulacral plates and are also perforated. *Aristotle's lantern* is a masticatory apparatus composed of 40 calcareous pieces and five powerful teeth arranged in a circlet around the mouth. It is found in all Regular echinoids and in some Irregular forms like *Holectypina* and *Clypeastrina*. It, however, disappeared in the *Spatangids* due to the burrowing mode of life. The *Perignathic girdle* is comprised of vertical plates attached at the margin of peristome.

## Evolution of Echinoids

1. Perischoechinoidea (U. Ordovician): All are regular echinoids. They became extinct by the end of the Paleozoic.

2. Bothriocidaroidea (U. Ordovician ): Miocidaris (Carbonaceous-Jurassic)- Cidaroids (Triassic-Present) - Ectobranchiates (Triassic-Present)- Irregularia (L. Jurrasic -Present). Thus it can be seen that the Irregular Echinoids were derived from Regular echinoids.

**Evolutionary Changes :** (a) Symmetry changed from radial to bilateral. Bilateral symmetry is very conspicuous in Spatangia. (b)Test shape changed from round to heart-shaped, (c) Anus and mouth were central in regularia but in Irregularia, the anus moved backwards and the mouth forwards; (d) Aristotle's lantern and Perignathic girdle disappeared in the Spatangids; (e) Ornamentation : Spines of variable sizes were found in members belonging to the Regularia but in Irregularia the spines were of the uniform size, (f) Ambulacral areas changed from simple (in Regularia and *Holectypina*) to petalloid (in Clypeastrina and Spatangina); (g) Ambulacral plates changed from simple to compound (except in *Clypeastrina* and *Spatangina*).

**Burrowing Echinoids :** These are characterized by distinctly elongated test, marked bilateral symmetry, movement of the mouth towards the anterior margin, absence of Aristotle's lantern and perignathic girdle and simple ambulacral plates. e.g. *Spatangina*

***Geologic Distribution***: The earliest known fossil echinoid, Bopthriocidaris, has been found from the Middle Ordovician rocks of Russia. Bothriocidaris, Echinocystis, Lepidocentrus, Archaeocidaris and Miocidaris are some of the common echinoids of the Paleozoic era. Most of the Paelozoic echinoids belonged to the order Regularia. These echinoids became more important during the Mesozoic era and became very abundant during the Jurassic. The echinoids belonging to the order Irregularia appeared for the first time during the Lower Jurassic period. The echinoids attained the acme of their development during the Upper Cretaceous.

## V. Phylum Brachiopoda

Brachiopods are bivalved organisms having two unequal, equilateral valves, with the ventral valve being bigger in size than the dorsal valve. The larvae of brachiopods are described as Lophophore. Brachiopods are described as Pygocaulia when the pedicle is formed from the tail segment of the larva. They are described as Gastrocaulia when it develops from the ventral mantle lobe. Protegulum is the embryonic shell. Umbo is the tapering beak found on the posterior end of the valves. It is more pronounced in the ventral valve. The ventral valve is provided with an opening called Pedicle or Foramen. In *Lingula*, the pedicle is shared by both the valves. In *Orthis*, it occurs as a triangular notch called Delthyrium (Mixoperipheral shell margin). If the delthyrium is partly closed by two plates, it is called Deltidium. If it is closed by only a single plate, as in Protremates, it is called Pseudo-deltidium.

Structural changes began in the brachiopods during the Lower Cambrian when teeth and sockets appeared. Hinge margin is long and straight in Orthis, Strophomena and Productus while it is short and curved in *Syntrophia, Conchidium* and *Pentamerus*. Spondylium evolved later. During Ordovician and Silurian, the brachidium was evolved. It is absent in all Inarticulates and some Articulate forms of Protremata such as *Productus* and *Chonetes*. In Rhynconella, the brachidium consists of two short curved processes known as the Crura. The brachidium is loop-like in *Centronella*. The two valves of the brachiopods are opened by divaricator muscles and closed by adductors.

## V. Lamellibranchia

These are marine, brackish water and fresh water molluscs. These are also referred as *Pelecypoda* (hatchet-footed), *Bivalvia* (*Didisoconch*) and *Acephala* (absence of head). The embryonic shell is called *Prodisissoconch. Veliger larva* is typical. These are characterised by two equal and equilateral valves. Burrowing (Cavicole) lamellibranchs such as *Lutraria, Ensis, Pholas*, etc. have tongue-shaped foot. The shells are compressed or cylindrical with anterio-posterior elongation. Some lamellibranchs such as *Mytilus, Pinna, Arca, Avicula*, etc. attach by means of a gland called *Byssus*. The shells in these cases show certain modifications: (a) the byssus, the foot and the byssal opening tends to shift forwards, (b) the organs in front of the byssus undergo reduction whereas those at the back freely develop, (c) the foot and the anterior adductor muscle may diminish or altogether disappear, (d) the shell may become equilaterally symmetrical.

**Ligament :** It is an elastic band or rod of conchiolin which facilitates in the opening of the valves. The most primitive ligament was described as *Amphidetic* (equal extension in front of as well as behind the umbo); a *Prosodetic* ligament lies wholly in front of the umbo and an *Opisthodetic* ligament lies wholly behind the umbo. A *Resilium* is similar to ligament and lies in a central pit below the hinge line. It is an organ of compression in contrast to ligament, which is an organ of tension.

**Dentition** : The earliest dentition in Lamellibranchs appeared in the Ordovician. Later on two patterns evolved: ***Taxodont (Ctenodont)*** with teeth converging from the hinge *e.g. Ctenodonta and Nucula* and ***Actinodont*** with teeth radiating downward from the umbo e..g. *Actinodonta*. The taxodont dentition did not diversify whereas the actinodont dentition showed a long series of modifications. ***Heterodont*** dentition is characterised by cardinal and lateral teeth. The Paleozoic fore-runners of these Heterodontia possess ***Preheterodont*** teeth e.g. *Venus*. Among the Trigonacea, the dentition is least modified and is described as ***Schizodont*** e.g. *Trigonia.* The ***Isodont*** dentition is characterised by strong teeth of equal size which fit into the corresponding sockets in the other valve e.g. *Spondylus.* The ***Dysodont*** teeth are simple teeth developed from internal ribs at the margin of the valves e.g. *Mytilus* and other byssated members. The ***Pantodont*** dentition is characterised by grooved cardinals and more than two laterals e.g. *Allodesma* and *Orthodonticus*. The ***Cyclodont*** dentition is characterised by a flat hinge pate and the bending of teeth e.g. *Cardium* and *Tridacna.* The ***Asthenodont*** dentition is typically found in burrowing animals such as *Mya* and *Pholas* in which obsolete teeth are present. When the teeth are very small or absent the dentition is described as ***Anamalodont*** e.g. *Allorisma.* Shells without dentition are called ***Edentulous*** *e.g. Achrax, Anodonta* and *Solenopsis*.

**Classification** : Lamellibranchs are classified into two groups: *Monomyaria* (single muscle-scar) and *Dimyaria* (two muscle-scars); the later is subdivided into *Isomyaria* (scars of the same size) and *Heteromyaria* (anterior scar smaller). When the sinus is absent the condition is called *Integopalliate*, when sinus is present, it is called *Sinopalliate.* Wood's classification of lamellibranchs identifies three orders :

(a) **Taxodonta** : Hinge taxodont, two equal adductors, siphons absent. *e.g. Nucula, Arca, Ctenodonta.*

(b) **Anisomyaria** : Anterior adductor smaller; hinge isodont or dysodont; byssus present *e.g. Mytilus, Pinna, Spondylus, Pecten, Ostrea* and *Exogyra.*

(c) **Eulammelibranchia** : Two equal adductors; siphons present, hinge schizodont or heterodont. *e.g. Trigonia, Unio, Cardita, Mya, Venus*, etc.

## VI. Gasteropoda

The name Gasteropoda means "stomach foot" referring to the fact that the animal appears to walk on its stomach. These are marine, freshwater and terrestrial molluscs with a size ranging from 1mm-1cm. They have calcareous uni-valved shells. The embryonic shell of Gasteropods is called *Protoconch.* The shells are of various forms. The body is spirally coiled and is made up of many whorls. *Columella* forms the axis of the shell. In some Gasteropods, the whorls are coiled around a hollow axial cone called *umbilicus*. Gasteropod shells with umbilicus are called *perforate* and those with Columella are called *imperforate.* The aperture of the Gasteropod may be variously shaped and its margin is described as the *peristome*. *Operculum* is a plate-like cover for the aperture. It may be multi-spiral (e.g *Trochus*), paucispiral (e.g. *Littorina*), concentric (e.g. *Viviparus*) or *unguiculate* (e.g. *Fusinus*). Certain Cambrian Gasteropods are characterised by non-coiled shells e.g. *Scenella* and *Stenotheca.* This condition is also exhibited by modern Limpets e.g. *Patella.* The *Helicoid* type of coiling was established as early as the Cambrian e.g. *Raphistoma.* Gasteropods are specially abundant in warm regions and in comparatively shallow waters. They were the index fossils for the Tertiary. They are divided into 3 subclasses: Prosobranchia (Streptoneura) ; Opisthobranchia (Euthyneura) and Pulmonata.

## VII. Ammonoids

These are entirely marine Molluscs having uni-valved shells made up of aragonite which range in size from 1 cm-10 m. The type Ammonoid for Paleozoic is *Goniatites*. It showed a simple type of suture called the *Goniatitic suture* with rounded saddles and angular lobes. It attained its peak development in the Carboniferous. During the Middle Triassic, the *Ceratite suture* appeared which had rounded saddles and denticulated lobes. In the Mesozoic, the *Ammonite sutures* appeared which are the most complicated of all the suture patterns. It has minutely folded and crinkled lobes and saddles *e.g. Dactioceras*. The Goniatitic suture (Paleozoic) shows great resemblance to the early type of suture development of the Mesozoic Ammonoids like *Dactlioceras*. This reflects the principle of *Palingenesis* i.e. the Mesozoic Ammonoids repeats many features that are manifested during the major part of the development of Paleozoic Ammonoids. It also illustrates the principle of Tachygenesis or acceleration in development i.e all those changes that took a lifetime to accomplish in the Paleozoic type were passed through the Mesozoic individual whilst it was in its infancy. The tendency to uncoil became very marked in the Cretaceous. These retrogressive changes are specially evident in *Rhabdoceras* of Triassic.

## IX. Evolution of Elephants

The evolution of elephant is characterized by : increases in size, lengthening of limb-bones and broad feet ; extraordinary growth of skull ; shortening of neck ; elongation of lower jaw and growth of trunk, hypertropy of second incisors, etc. The evolution continued through the Mastodons which appeared in the Lower Oligocene. *Paleomastodon* and *Phiomia* were primitive mastodons. The earliest representative of elephants was the *Stegodon* (L. Pliocene-Pliestocene). *Mammoths* (Elephas) arose from the Stegodons by the end of the Pliocene. They existed in all continents except South America. They became extinct by the end of Pliestocene. Now, only two species of elephants exist: the African Elephant (*Laxodoma*) and the Asian Elephant (*Elephus*).

## X. Evolution of Horses

The first known horses belonged to the early Eocene (*Eohippus* or *Hyracotherium*). The evolutionary changes in the horses include: increase in size, lengthening of legs and feet, reduction of lateral toes with the emphasis on middle toe, straightening and stiffening of the back ; increase in the height of the crowns of the cheek teeth and progressive complication of the crown patterns, increase in the size and complexity of the brain, etc.

*Eohippus* had four toes. From Lower Eocene to Upper Oligocene, the horses showed *Orthogenesis* or straight-line evolution. i.e.

*Eohippus* → *Orohippus* → *Epihippus* → *Mesohippus* → *Miohippus*
(L. Eocene) (M. Eocene) (U. Eocene) (L. Oligocene) (M&U.Oliocene)

The main line of evolution continued through *Parahippus* (L. Miocene) to *Merychippus* (U. Miocene), to *Pliohippus* (Pliocene) to *Eqqus* or modern horses (Pliestocene). *Merychippus* had three toes. In the Pliocene times the *Hipparian* evolved from the *Merychippus*. They were present in all continents except South America. They are typical of the Pliocene times. *Pliohippus* had only a single toe. *Hippidium* was a horse that evolved from *Pliohippus* and originated in South America.

## XI. Living Fossils

The term 'living fossil' was coined by Darwin who applied it to the East African Ginkgo tree. Living Fossils are relics of ancient stocks that continued long after their parent groups have vanished. They exhibit the phenomenon of *Bradytelic* evolution *i.e.*, in a declining lineage the survivors may often continue for a very long time, evolving very slowly or not at all. The sea-shell *Lingula* is a classic example of extremely slow evolution. Certainly the most famous 'living fossil' in the animal kingdom is the Coelacanth, *Latimeria Chalumnae*, which was not found until 1938. It is the only living member of a geologically very old group of fishes. The other living fossils among the fishes include the Australian Lung Fish (*Epiceratodus forsteri*), which is the oldest living form of lung fishes (*Diphnoi*). All these instances where the rate of evolution has halted are known as *Laggard*. The living fossils are also described as 'Panchronic forms' i.e. they appear to be characterised by a great morphological stability throughout geologic time.

Examples of Living Fossils include : (a) Vertebrates :Latimera, Lepisosteum, Sphenodon, Apteryx, Didelphis. Epiceratodus. etc. (b) Invertebrate : Peripatopsis, Anaspides, Limulus, Neopillina. Nucula, Lingula, Nautilus, etc. (c)Plants : Angiopteris, Dipteris, Cycas, Sequoia and Stylites.

## XII. Homology

*Homology* is the condition where two or more organisms possess organs which may look different at the first glance, but have a basic structural similarity which betrays a common origin. Thus, the man's hand, the bat's wing and the seal's flipper are all *homologues* and exhibit the phenomenon of *homology*.

## XIII. Homeomorphy

Homeomorphy is a condition where two or more organisms look similar, or have parts which look similar but have different origins, or relate to dissimilar arch types. Thus, the whale, the fish, and the Ichthyosaur all look similar - they all as a rule, are stream-lined, have two eyes and fin-shaped appendages. But they are homeomorphs because their internal structures are different from each other , this being due to the fact that the whale has descended from the primitive land mammals, the Icthyosaur from the primitive land reptiles, and the fish from primitive aquatic fish. Of course, homeomorphy is a purely relative concept as all land vertebrates are descended from fish but it usually related to more immediate ancestries than that. Homeomorphy may be considered as the result of Convergent evolution. The individuals exhibiting homeomorphy are called *homeomorphs*. They may be isochronous i.e. present at the same time or heterochronous i.e. present at different times. Homeomorphs may occur either within the same taxon or in different taxa.

## XIV. Some Laws of Evolution

**Dollo's Law :** It deals with the irreversibility of evolution. The disappearance of digits in Equids is an example.

**Cope's Law or (Deperet's Law) :** It deals with the increase in size observed in many lineages. Aspects opposite to the Cope's law : Decrease in the size of Equids and return of Ceratite sutures in Cretaceous Ammonoids.

**Bradytelic and Tachytelic Evolution** : The time of rapid evolutionary change of a new taxa or the adaptive radiation is described as *Tachytelic Evolution* (e.g. Spirifer). When, in a declining lineage the survivors continue for a very long time, evolving very slowly or not at all it is described as *Bradytelic Evolution* (e.g. Lingula). When the evolution is normal or average, it is described as *Horotelic*. Rhynconella exhibits both bradytelic and tachytelic evolution.

**Convergence :** Convergence occurs when organisms of different groups acquire similar morphologies due to identical modes of life. It is inverse of *adaptive specialization.* If closely related groups of organisms are involved, it called *Homology* e.g. Sharks, Icthyosaurs, Dolphins whereas, if distant groups are involved it is called *Analogy* e.g. Pterosaurs, Birds and Chiroptera.

**Parallelism** : When different lineages exploit the same adaptive possibilities, it is described as Parallelism.

**Cladogenesis** : It is the ramification or bursting forth of an evolutive line of descent.

*****

# 6

# Principles of Stratigraphy

Principle of order of superposition of Strata is the first fundamental generalisation of stratigraphy. In a sequence of layered rocks, the bottom-most layer is the oldest, while the topmost is the youngest. Exceptions—deposition in a cave, tectonically overturned beds.

Doctrine of uniformitarianism (Plutonism)—James Hutton—"Present is the Key to the Past".

**Geopetal.** Sander (1936)—refers to any feature that enables us to determine the relationship of top to the bottom at the time of rock formation.

**Correlation.** Is the process which determines the mutual time relations of local sequences. Lithologic or physical correlation is used in special circumstances when fossils fail. Bentonites, evaporites and glacial varves have been used for correlation.

Law of Faunal Succession—William Smith—Every formation encloses a fossil assemblage which is characteristic of it and is different from that of the underlying and overlying formations.

**Index Fossils.** Have wide geographic and narrow stratigraphic extent ; adaptable to wide range of environments ; and represented by abundant specimens. Organisms with complicated structure are good index fossils.

John Lehman was the first geologist to measure and record actual sections of stratified rock.

**Dual Stratigraphic Correlation.** In stratigraphy two types of units are considered : (*i*) Objective rock units subjected to tracing and mapping on the basis of observable criteria and (*ii*) groupings of strata differentiated with reference to their position in geologic time. Thus, the following categories of stratigraphic units are recognized :

I. Special divisions based on position in the geologic time including :

A. Time Units—segments of continuous geologic time (Eons, Eras, Periods, Epochs, Ages)

B. Time-Rock Units—Assemblages of strata representing deposition during distinct time units (Systems, Series, Stages).

II. Mappable assemblages of strata called rock units distinguished and identified by objectives physical criteria observable in the field (groups, formations, members, tongues, lentils).

**Biochron.** H.S. Williams (1901)—refers to the length of time defined by the range of fossil species.

**Teilchron.** W.J. Arkell (1933)—Time unit defined by the local range of a species.

**Hemera.** Buckmann (1893)—Time-equivalent of a zone. (=Epibole)

| *Group/Time rock units* | *Time units* |
|---|---|
| Group/Erathem | Era |
| System | Period |
| Series | Epoch |
| Stage | Age |

**Biostratigraphic Units**—Consists of strata that are defined and identified solely on the basis of their fossils without reference to any lithological characters. Wholly objective.

1, **Range Zones.** Simple biostratigraphic units ; consists of all beds included between the lowest stratigraphic occurrences of any fossil.

(*a*) **Biozone**—S.S. Buckmann (1902)—Consists of all strata deposited during the existence of the zonal organism, irrespective of the presence of fossils in it.

(*b*) **Teilzone.** Pompecks (1914)—It is defined by part of the total range, represented by local occurrences of a species.

2. **Assemblage zones.** Consists of strata characterized by the occurrence of two or more parts of zonal fossils, commonly species or genera, *e g*., overlap zones, negative association zones etc.

3. **Abundance Zone.** EPIBOLE—Denotes a BSZ consisting of strata deposited during the acme or climax, but not the entire existence of a particular organism.

# 7

# Petrology

## (a) IGNEOUS PETROLOGY

**Magmas**. Magma is defined as "naturally occurring mobile rock matter that consists in noteworthy part of a liquid phase having the composition of a silicate melt". Buffen was the first to suggest the existence of two primary magmas — granitic and basaltic. Daly and Brown suggested the occurrence of a single basaltic magma derived from the melting of the parent Periodotite mantle. Buddington suggested the existence of several primary magmas. The temperature of the magmas in the crust ranges from 700- 1200°C. Potassium and sodium forms the most active bases present in igneous magmas.

**Reaction Principle** : N.L. Bowen (1922) explained the crystallization of silicate magmas on the basis of reaction principle in rock building i.e., on the mutual interaction which takes place in a cooling magma between the solid and liquid phases. In a continuous reaction series there is a continuous change in the mineral composition with no concomitant change in its structure, whereas in a discontinuous reaction series there is a continuous change in the mineral composition as well as the crystal structure.

### Textures and Microstructures

Texture is defined as the intimate mutual relations of the crystalline mineral constituents and glassy matter in a rock of uniform aggregate. Each texture results from a particular set of geological conditions and their study reveals the geological history of a rock. It depends upon the crystallinity and granularity and fabric.

(a) ***Equigranular textures*** : In this type of texture the constituent minerals are of approximately the same size. When most of the minerals are anhedral, the texture is described as ***Allotriomorphic*** or Xenomorphic as in Aplites; when most grains are subhedral, it is called ***Hypidiomorphic*** or hypautomorphic or granitic and when most of the crystals are euhedral, the texture is described as ***Panidiomorphic*** or automorphic as in lamprophyres.

(b) ***Inequigranular textures*** : When in the same rock there are two dominating sizes, the texture is described as inequigranular or ***hiatal***. If there is a complete range in the size of grains between the ground mass and the phenocrysts, the texture is described as ***Seriate***.

1. ***Porphyritic texture*** — In this type of texture large and euhedral crystals known as phenocrysts or insets are set in a fine-grained glassy groundmass or mesostasis. It is most common in volcanic and hypabyssal rocks but it also occurs in some plutonic rocks. ***Varieties*** : Glomeroporphyritic (phenocrysts occur in clusters), Micrograniite (microcrystalline groundmass without phenocrysts), Pilotaxitic (matrix of feldspar microlites), Hyalopilitic (matrix is mainly of microcrystalline plagioclase laths and augite granules in flow pattern), Vitrophyric (matrix is glassy). The most common mode of origin is by a break in the physical conditions of a cooling magma. The growth of phenocrysts is also favoured when a magma contains high molecular concentration of a particular mineral.

2. ***Poikilitic texture*** — This is a reverse of porphyritic texture in which a number of small crystals of essential minerals are uniformly scattered through a large crystal of another mineral. The enclosed crystals are called Xenocrysts or

*Chadacrysts* and the host crystals are called *Oikocrysts*. These are very common in monzonites, syenites, peridotites and picrites

3. ***Ophitic Texture*** —— This is a particular case of poikilitic fabric in which the xenocrysts are feldspars and the host crystals are composed of pyroxene (augite). The feldspars are un-oriented and are idiomorphic while pyroxenes are allotriomorphic. This texture is typical of dolerites. When the enclosure of plagioclase laths by augite is partial due to increased grain size of laths, then the texture is described as ***sub-ophitic***. When the enclosed plagioclase shows marked parallelism, it is described as ***trachy-ophitic***.

4. ***Directive Textures*** —These are produced by flow in magmas during their crystallization. Trachytic and trachytoid textures are formed by the parallel arrangement of feldspars, the later being typical of certain syenites.

5. ***Intergrowth or Symplectic Textures*** — These may result from either eutectic or simultaneous crystallization or due to exsolution.

(a) *Graphic texture* : It is an intergrowth of quartz and feldspar in which quartz is disposed in wedge-shaped or prismatic areas intersecting at angles of 60o. At microscopic level, it is known as *micrographic texture*. Examples are seen in pegmatites and granophyres.

(b) *Granophyric texture* : In this type of texture, quartz occurs as irregular patches and shreds in the feldspar itself instead of occurring in spaces enclosed by it. All the small grains have the same orientation and appear as if they were parts of a second large crystal intergrowth within the first.

(c) *Perthite* : It is an intergrowth of two feldspar minerals in which a K-rich feldspar (orthoclase) contains strings, patches or lenticles of sodic plagioclase. If plagioclase contains similar inclusions of' orthoclase or microcline, it is described as ***Anti-Perthite***. Origin may be due to exsolution or replacement.

(d) *Myrmekite* : It is a cauliflower-like intergrowth between acid plagioclase and vermicular quartz. It is very common in charnockites. The origin of myrmekites is attributed to several reasons. Becke (1908) described the formation of myrmekite as similar to the phenomenon of albitization. Spencer (1945) suggested exsolution as the probable mode for the formation of myrmekites. However, the most widely accepted view is that myrmekite is formed due to the replacement of K-feldspar by a Na-rich plagioclase with the separation of excess quartz.

6. *Rapakivi texture* : It is characterized by large ovoids of K-feldspar surrounded by rims of plagioclase in a medium-grained, somewhat uneven ground mass of quartz, plagioclase, dark mica, hornblende and fluorite together with accessories. The general view of petrologists is that alkali feldspar crystallized early in these rocks, and thus had adequate time to develop large phenocrysts. The Rapakivi texture was originally described from Finnish granites.

7. ***Corona Structure (Synantectic structure)*** : It results when the early formed minerals of the discontinuous reaction series are surrounded by later members such as hypersthene, enstatite or amphibole. The term *'Reaction rim'* is used when the rims are of magmatic origin whereas the term *'Kelyphitic borders'* is applied when they are of secondary origin.

8. ***Xenolithic Structures*** : These are the inclusions of pre-existing rocks in an igneous rock. The xenoliths may show extensive reaction with the igneous rock and may have suffered partial assimilation. When the inclusions are genetically related to the enclosing rocks, these are described as *Cognate xenoliths*.

9. ***Orbicular structure*** : It consists of large spherical masses made up of concentric shells of different mineral composition which occur within certain

plutonic igneous rocks, mainly granites and diorites. Napoleonite and corsite are orbicular diorites.

10. ***Miarolitic Structure*** : This is a term used to describe the fabric of some plutonic rocks, especially granites, which have sparsely scattered, angular cavities into which idiomorphic or faceted crystals of quartz, rutile, tourmaline or fluorite project outward.

(c ) *Volcanic Textures* : (1) ***Spherulitic Structure*** : It results from the simultaneous crystallization of fibres with radiate arrangement about a common centre. Spherulites mainly occur in the acid volcanic, or hypabyssal rocks, and less abundantly, in basic lavas and intrusions. They represent rapid growths in a quickly cooling lava or glassy matter.

(2) ***Perlitic Structure*** : Most glassy rocks show minute, curved, sometimes partially concentric cracks due to shrinkage of the cooling glass, giving rise to a Perlitic structure.

(3) ***Flow Structure*** : Many volcanic rocks show a parallel or sub-parallel alignment of various elements in the fabric. This is due to movement or flow in the still liquid lave, and the feature is termed as flow structure.

(4) ***Variólitic Structure*** : It is a type of spherulitic texture often occurring in glassy basic rocks like trachylites, at the chilled margins of dykes or sills.

## Classification of Igneous Rocks

The classification of igneous rocks has been a subject of controversy, and even today there is little agreement concerning the principles or methods that should be employed in constructing classificatory schemes. Unlike the animal and plant types, rocks cannot be compared to species or genera.

The difficulties in the way of quantitative mineralogical composition are the presence of isomorphism among plagioclase feldspars which are found in most of the igneous rocks, the presence of zoning in the same minerals in many volcanic rocks, and lastly the aphanitic and glassy nature of the groundmass of many volcanic rocks render it impossible to determine the modal composition.

**Bases of Classification** : Igneous rocks can be classified on the bases of texture, silica percentage, minerology, colour and association. (A) ***Texture*** : The textural factor expresses the conditions under which the cooling of the magma took place. In other words, it is a reflection on the geological occurrence of rocks. On this basis igneous rocks are divided into two groups : plutonic and volcanic. The ***plutonic rocks*** are formed by deep-seated crystallization of magma with slow cooling under great pressure, and with the retention of volatile constituents. They exhibit coarse-grained holocrystalline textures. The ***volcanic rocks*** are formed at shallow depths under low pressure and high temperature conditions by rapid cooling of the magma and with the loss of volatile constituents. They exhibit merocrystalline or glassy textures. (B) ***Silica percentage or Chemical classification*** : In the past many petrologists, notably Loewinson and Lessing (1897), divided rocks into four classes : acid($SiO_2$ >66%), intermediate($SiO_2$ 66-52%), basic($SiO_2$ 52-45%) and ultrabasic($SiO_2$<45%), based upon silica percentage. The disadvantage of this method of division based upon silica percentage is that it brings together rocks of different mineral composition, which it is desired to separate, such as a granite and a syenite. This type of grouping has now been abandoned by many authors such as Hatch, Wells and Wells. (C) ***Mineralogical Classifications*** : The classification of rocks that are now followed are the qualitative or semi-qualitative mineralogical classifications. The use of quantitative mineralogical classifications, as proposed by Shand, Johannsen and others, is now increasing rapidly. The primary constituent minerals of a rock fall into categories — essential and accessory. Essential minerals are those whose presence is required for its definition. Accessory minerals are those whose presence is not essential to define a rock type. These include hornblende, biotite, muscovite, zircon, sphene and iron oxides. (D) ***Colour*** : The most obvious property of a rock is its colour which is reflected in the colours of its constituent minerals. The ratio between the mafic (dark) and felsic (light) groups of minerals is known as the colour index. This colour index is based on the percentage of dark minerals in a rock, as determined from a thin section. Based on the colour index, rocks are divided into four groups : leucocratic (mainly light-coloured minerals), mesocratic (mesotype), melanocratic (mainly dark minerals) and hypermelanic (exclusively dark minerals). The limits of these groups are given below:

| | **Johanssen** | **Shand** |
|---|---|---|
| Leucocratic | 0-5 | 0-30 |
| Mesotype | 5-50 | 30-60 |
| Melanocratic | 50-95 | 60-90 |
| Hypermelanic | 95-100 | 90-100 |

Development of various classificatory schemes : (A) ***Rosenbuch's Classification*** : H.Rosenbusch (1887) was the first person to classify rocks based on geological occurrence. His classification was based on a combination of texture, geological occurrence and mineral composition. Major divisions are : abyssal or plutonic; hypabyssal or dyke rocks; and effusive or volcanic rocks. (B) ***C.I.P.W. Classification*** : This is the most important of the chemical classifications and was proposed by four US scientists, W.Cross, J.P.Iddings, L.V.Pirsson and Henry Washington in 1903. These authors recognised the limitations of chemical composition and mineral composition as independent bases of the classification of igneous rocks. They, therefore, evolved a system in which the constituent minerals of an igneous rock are considered both chemically and quantitatively. In the CIPW system, the molecular amounts of each oxide are calculated, and then normative minerals are calculated according to a fixed plan. The minerals, however, are not exactly those which actually occur in the rocks but are calculated from the chemical composition in terms of a set of hypothetical minerals whose compositions are simple and invariable. These are designated as ***Norm.*** The volume percentage of minerals in a rock is described as the ***Mode***. These hypothetical mineral molecules are called standard mineral molecules. The norm may be divided into salic and femic groups. The salic group include minerals like quartz, corundum, zircon, orthoclase, albite, anorthite, leucite, nephelite, halite, etc. The femic group include minerals like acmite, diopside, wollastonite, hypersthene, olivine, magnetite, haematite, ilmenite, chromite, etc. This classification divides rocks into five groups on the basis of the ratio of salic to femic minerals : persalane (>7:1), dosalane (7:1 - 5:3), salfemane (5:3 - 3:5), dofemane (3:5 - 1:7) and perfemane (<1:7). Further subdivisions are made on the basis of the ratios of certain minerals or combination of oxides. (C) ***Shand's Classification*** : Shand (1947) classified rocks primarily on the silica saturation principle, according to which "certain minerals cannot exist in the presence of free silica. If silica were to be present, these minerals combine with it to give rise to different minerals". Shand first classified rocks as eurocrystalline and dyscrystalline. Next, according to the saturation principle rocks were grouped as follows: (i) ***Oversaturated Rocks***: These rocks contain free silica (quartz or tridymite) of primary origin. (ii) ***Saturated Rocks*** : These rocks do not contain either free silica or any unsaturated minerals but contain only saturated minerals viz; those that are compatible with excess silica under magmatic conditions like feldspars, pyroxenes, amphiboles, fayalite, spessartite, ilmenite, tourmaline, sphene, zircon, apatite and topaz. (iii) ***Undersaturated Rocks*** : These rocks consists either wholly or partly of unsaturated minerals like leucite, nepheline, sodalite, hauynite, noselite, cancrinite, analcite, melanite, pyrope, corundum, calcite and spinel. These may be (a) Non-Feldspathoidal — Alkalis are saturated but dark minerals are unsaturated e.g. olivine-gabbro, olivine-norite and melilite basalt; (b) Feldspathoidal — Alkalis are partly or wholly unsaturated. The dark minerals may be saturated (as in nepheline syenites) or unsaturated (as in nepheline-basalt and leucite-basanite).

Shand next considers the molecular ratio of alumia to those of soda, potash and lime. In feldspars, nepheline and leucite the ratio of $Al_2O_3$ to $Na_2O + K_2O +$ CaO is 1:1, hence any excess or deficiency of Al2O3 is reflected in the composition of the dark minerals. On this basis four groups are formed : (1) ***Peraluminian Rocks***: $Al_2O_3 > Na_2O + K_2O + CaO$. The excess alumina is used up in micas, tourmaline, topaz, corundum or garnet. These rocks are formed in the later stages of crystallization of acid hydrous magmas. (2) ***Metaluminian Rocks*** : $Na_2O + K_2O < Al_2O_3 > Na_2O + K_2O + CaO$. The available alumina after enabling the formation of alkali feldspars and anorthite, is used up in the formation of hornblende, epidote or melilite, or there will be pairs of aluminous and non-aluminous minerals in the same rock (viz; biotite with pyroxene; hornblende with olivine). These rocks contain minerals that are formed at low temperatures in the presence of water. (3) ***Subaluminian Rocks*** : $Al_2O_3 == Na_2O + K_2O + CaO$. In these rocks there is little or no excess of $Al_2O_3$ after the formation of feldspars and feldspathoids. The ferromagnesians are therefore non-aluminous, such as olivine or orthopyroxene or diopside or augite in which a small amount of $Al_2O_3$ replaces $SiO_2$. These rocks are formed at high temperatures from hot and comparatively anhydrous magmas. (4) ***Peralkaline Rocks*** : $Al_2O_3 < Na_2O + K_2O$ Here there is an excess of alkalies over alumina after the formation of the feldspars. The typical minerals are soda-rich such soda-amphiboles and soda-pyroxenes (riebeckite, aegirite, arfvedsonite, etc). Other characteristic minerals are those in which $Al_2O_3$ is substituted for by $TiO_2$ or $ZrO_2$. These are the products of last stages of crystallization of low-temperature alkaline magmas, such as the peralkaline granites and syenites.

## Nomenclature of Igneous Rocks

### *I. Plutonic rocks*

#### *(a) Granite-Granodiorite-Diorite*

Granite : Quartz (10-40% )+ Feldspars (K-feldpars > Plagioclase)
Granodiorite: Quartz+Feldspars (Na-Plagioclase > K-feldpars)
Diorite: dominant calcic plagioclase.

In the above series, from granite to diorite, the quartz content decreases, K-feldspar decreases and the plagioclase content increases. The plagioclase becomes increasingly calcic from left to right. Biotite is the dominant mafic mineral at the granite-end while hornblende is the dominant at the diorite-end.

Varieties : 1. Adamellite : K- feldspar = Plagioclase
2. Tonalites: Granites with Plagioclase >Orthoclase
3. Rockallite: Granites in which Quartz =Albite =Aegerine
4. Alaskites: Granites with Quartz and K-feldspar only

***Origin of Granites*** : The origin of granites is controversial ; some regard them as magmatic, others regard them as metamorphic. Modes of magmatic origin are magmatic differentiation and Anatexis. The criteria for this kind of origin are: chilled margins due to sudden cooling ; presence of xenoliths of the earlier country rocks ; sharp contacts; structural disharmony between the country rocks and granites; apophyses, tongues and dykes of granite in the country rock; etc. Granitization is the process of transformation of the pre-existing solid rocks into those of granitic character by ionic diffusion of fluids. The criteria for granitization are: gradational or diffuse contacts with country rocks; relict inclusions; foliation passes continuously into regional structure of the country rocks; dykes show no dilation ; presence of some metamorphic minerals, etc.

***Types of Granites*** : Chappell and White (1974) suggested that the granitoid rocks could be divided into two groups: S-type and I-type granitoids. The ***I-type granitic rocks*** are derived from igneous (or meta-igneous) sources and are typical of subduction zones in continental margin settings. They have high concentrations of Na and Ca, resulting in hornblende and sphene in the mode and diopside in the norm. Their initial Sr87/Sr86 ratios are <0.710. Molybdenum and copper deposits are generally associated with them. The ***S-type of granitic rocks*** are derived from a source which had been through a weathering cycle. They are typically found in catazonal migmatitic complexes. They are highly aluminous in composition and are characterized by absence of hornblende in the mode. Their initial Sr87/Sr86 ratios are >0.710. Many cassiterite deposits are associated with these type of granites. White (1979) also defined a third type of granitoid, the ***M-type granitoids***, which are presumably directly derived from the melting of subducted oceanic crust or the overlying mantle. Loiselle and Wones (1979) defined a fourth granitoid type, the ***A-type granitoids***, which derived from anorogenic alkaline magmas. These are typical of tensional (or non-compressive) environments in oceanic basins, along continental rift zones or within stable continental blocks. These rocks suites range in composition from quartz syenites, metaluminous granites and peralkaline granites. Mineralogically, these rocks are rich in fayalite, ferrohedenbergite, ferrohastinsite and annite and those of peralkaline compositions are enriched in alkali-amphiboles and sodic pyroxenes. Chemically, these granitoids are characterized by high alkali and low CaO contents, high FeO/MgO, high K2O/Na2O and show an abundance in halogens, HREE and large-ion lithophile elements (LILE). Their initial Sr87/Sr86 ratios fall between 0.703 and 0.712. Notable examples of A-type granitoids suites include, the White Mountain Magma Series, Malani igneous suite and Gabo and Mumbulla suites.

***(b) Syenites, Nepheline syenites and other alkaline rocks***

Syenites : alkali-feldspar is an essential mineral
Monzonite : Feldspar = plagioclase
Pullaskite : alkali-feldspar + nepheline
Nordmarkite : nepheline syenite with accessory quartz
Shonkinite: alkali-syenite with abundant mafics

Theralite : nepheline syenite in which plagioclase (labradorite) becomes dominant.

Teschenite : equivalent to theralite but with nepheline replaced by analcite.

Ijolite : nepheline syenite in which feldspars disappear

*(c) Gabbros, Anorthosites and Ultramafic rocks*

Gabbros : Mafelsic, Calcic plagioclase (labradorite)+augite or diallage, also hypersthene and olivine. Exhibits Coarse-grained hypidiomorphic texture; occurs as layered intrusions (cryptic layering).

Norite: gabbro in which hypersthene > clinopyroxene

Anorthosite : These are felsic, mono-minerallic containing labradorite only; the crystals being coarse-grained and antiperthitic . Anorthosites occur as vast intrusive bodies ranging in size from a few hundred feet across to great batholiths covering thousands of square miles. They are found in two strikingly different environments: (i) as vast sheets in the layered basic complexes and (ii) as major rock bodies of batholithic dimensions in Precambrian shield areas where they may outcrop over vast areas. Most anorthositic rocks exhibit a variety of reaction rims called Coronas or Kelyphitic rims.

Lherzolite : Ulramafic rock composed of olivine, orthopyroxene, clinopyroxene, garnet or sphene in decreasing order of abundance.

Hazburgite: composed of olivine and pyroxene

Dunite : monominerallic; nearly all olivine

Wehrlite: olivine and clinopyroxene

Websterite: Ortho and clino pyroxene rock.

## *II. Hybabyssal (dyke) rocks*

(a) Aplite : light-coloured, fine-grained, allotriomorphic rock consisting of quartz, alkali feldspars, muscovite, with or without fluorite and tourmaline.

(b) Dolerite (=Diabase) : Dyke rock composed of labradorite, augite and iron oxides. Characteristically exhibits ophitic texture.

(c) Lamprophyres : These are strongly porphyritic, melanocratic dyke rocks exhibiting panidiomorphic texture. Biotite, hornblende and pyroxenes are the essential minerals ; while, feldspars and feldspathoids are the accessories.

## Classification of Lamprophyres

| | *With Orthoclase* | *With Plagioclase* | *No Feldspars* |
|---|---|---|---|
| Biotite | Minette | Kersanite | Alnoite (with melilite) |
| Common hornblende | Vogesite | Spessartite Malchite | — |
| Barkevikite and/or Augite | — | Camptonite (with barkevikite) | Monchiquite (with analcite) |

(after Rosenbuch)

## III. Volcanic Rocks

*(a)* Rhyolite : lavas of granitic composition.

*(b)* Dacite : Plagioclase is greatly in excess of orthoclase. It is similar to andesite in texture and composition but differs from it by the abundance of quartz.

*(c)* Rhyodacite: intermediate between rhyolite and dacite.

*(d)* Obsidian : extreme glassy modification of rhyolite.

*(e)* Trachyte : volcanic rock of syenitic composition; rich in orthoclase or other alkali feldspars.

*(f)* Phonolite: volcanic equivalent of nepheline syenite or leucite syenite.

*(g)* Trachyandesite: volcanic equivalent of monzonite; in which alkali feldspars are more or less equal to soda-lime feldspars.

*(h)* Andesite: felsic volcanic rock; andesine is the predominant constituent.

*(i)* Basalts: Most widely distributed of all volcanic rocks; occur not only as lava flows but also as cone-sheets, dykes and other minor intrusions. The common texture is porphyritic and glomeroporphyritic with groundmass varying from holocrystalline to virtrophyric.

1. ***Alkali Olivine Basalts*** : These are characteristic of oceanic islands and the upper parts of submarine volcanic cones to about 1500-2000m below sea level. They also occur on the continents, in association with the tholeiitic flood basalts. They are typically unsaturated rocks in which olivine does not show any reaction relation with the liquid to form pyroxene. Olivine is strongly zoned while the pyroxene is diopsidic or titaniferous augite. Quartz is absent throughout.

2. ***Tholeiites*** : These are characteristic of most of the great plateau basalts found on each of the continents, over most of the deep ocean bottoms, being generated along spreading oceanic ridges (e.g. Mid-Atlantic Ridge), and on a few large oceanic islands (e.g. Hawaii). Their most striking feature is their vast extent and volume and generally uniform composition. All these rocks are essentially saturated in silica and contain little or no olivine. Such olivine as is present is unzoned and bears a reaction rim of Ca-poor pyroxene (hypersthene) on the phenocrysts or pigeonite in the groundmass, showing that it is out of equilibrium with the magma. Chemically they are on the average lower in $Na_2O$, $K_2O$ and $TiO_2$ and higher in $SiO_2$ than the alkali olivine basalts. It has been noticed by many workers that Na/K for oceanic tholeiites is greater than 10, whereas for those on the continents the ratio is generally greater than 3 but less than 6. Engel et al (1965) noted that oceanic tholeiites are also lower than alkali olivine basalts in elements usually considered characteristic of granitic rocks i.e. Ba, La, Nb, Sr, P, Pb, Th, U, Zr, etc and they are less oxidized ($Fe_2O_3/FeO$ is usually $<0.4$, versus $>0.4$ in alkali basalts). Miyashiro et al (1969b) have subdivided oceanic tholeiites into low- and high-alumina types; the order of crystallization in the former being : Olivine → Plagioclase → Clinopyroxene, and in the latter being : Plagioclase → Olivine → Clinopyroxene.

**Petrographic differences between Tholeiites and Alkali-olivine basalts**

| *Tholeiites* | *Alkali-olivine basalts* |
|---|---|
| 1. The tholeiitic magma is assumed to be the parent magma of the basaltandesite-rhyolite line of differentiation. | 1. The alkali-olivine basalt is supposed to be the parent magma of the olivine-basalt-trachy-andesite-trachyte; or olivine-basalt, hawaiitemugearite line of descent. |
| 2. Saturated | 2. Unsaturated |
| 3. Olivine may or may not be present; when present the quantity is small; olivine is magnesian and unzoned. | 3. Olivine is characteristically present; it is often strongly zoned with fayalite-rich margins. |
| 4. Pyroxene may be either a single subcalcic augite, or pigeonite, or hypersthene; pyroxenes of latter differentiates are enriched in iron. | 4. Pyroxene is a diopsidic or titaniferous augite |
| 5. Residual glass, when present, is quartzo-feldspathic in composition; it is often densely charged with dusty iron-ore and altered to chlorophaeite. | 5. Residual glass is often altered to chlorophaeite or is charged with iron-ore. It may also contain minute crystals of alkali feldspar. |
| 6. Zeolites, carbonates, chalcedony are common as amygdale fillings. | 6. Common zeolites are natrolite, analcite, Stillbite, etc. |

*3. **Spilites*** : Soda-rich basalt in which albite or albite-oligoclase is predominant. These strongly sodic lavas are prominently developed in the oceanic 'eugeosynclinal' environment with thick sections of graywacke and exhibit pillow structure. They are characteristically associated with ***Keratophyres***, which are highly-sodic, albite-rich andesites and dacites. The spilite-keratophyre associations are a necessary member of the well-known ***Steinmann Trinity*** — pillow lavas, serpentinite and bedded radiolarian cherts. The chemical features of the whole rock association are a high content of $Na_2O$, low $K_2O$ and MgO, and also low CaO and $Al_2O_3$.

4. Picrites : olivine-bearing basalts.

5. Tephrites : olivine-free basalts, containing both plagioclase and a feldspathoid.

6. Tachylite : glass of basaltic composition.

(j) *Carbonatites*: They are a group of rocks rich in Carbonates (50% or more) which are genetically related to alkaline magmas. They develop in the areas of crustal stability, rifting or block faulting or moderate folding and range in age from Proterozoic to Recent. Carbonatites are composed of mainly medium-grained carbonates but include very fine to coarse-grained varieties resembling marble. ***Sovite*** is nearly pure calcitic carbonate and has a maximum of 10% of other species. It is a medium or coarse-grained plutonic rock and has got its dyke equivalent. The dolomitic equivalent of sovite is known as ***Rauhaugite*** or ***beforsite.*** The carbonatites may also be composed of ankerite. The carbonatites are emplaced in various stages, which is typically depicted as: C1 → C2 → C3 → C4, where

C1 :: ***sovite*** intrusions dominant; emplaced in an envelope of explosively brecciated rocks; presence of calcite, apatite, pyrochlor, magnetite, biotite and aegirine-augite.

C2 :: ***alvikite*** is dominant; medium to fine-grained; shows marked flow banding.

C3 :: ferrocarbonatite; contains iron-bearing carbonate minerals, some rare-earths and radioactive minerals.

C4 :: late stage ***alvikites***; barren

***(k) The Peridotite-Serpentinite Association*** : Serpentinites consists essentially of mixtures of the different serpentine minerals like chrysotile, antigorite, lizardite and bastite in various proportions. They occur in two very different associations: (i) in the basal parts of layered complexes and (ii) as massive 'Alpine-type' ultramafic bodies in orogenic zones, and frequently bounded by thrust faults. The peridotites and serpentinite bodies of the Alpine-type occur as large sheets and lenses in folded geosynclinal sediments of orogenic belts. Characteristically, they occur with spilites and cherts where they make up the so-called ***Steinmann Trinity***, and with graywackes, gabbro and basalt as part of ***ophiolite*** complexes.

***(l) Ophiolite Complexes*** : An ophiolite complex is characteristic of volcanogenic troughs, consisting of clayey-sandy rocks and jasper, diabase outpourings and masses of gabbro and ultrabasic rocks altered to serpentinites. Most geological and geophysical evidences suggests that ophiolite complexes represent fragments of older oceanic crust that has been tectonically emplaced in continental orogenic belts. If so, ophiolites represent samples of oceanic crust and upper mantle produced at ancient oceanic ridges. A completely developed ophiolite complex may contain the following sequence of rock types :

Low-K tholeiitic flows; commonly pillowed; (Oldest)
Low-K tholeiitic sheeted diabase dyke complex;
Low-K tholeiitic gabbros and diabases, commonly exhibiting cumulus textures; and
Ultramafic rocks, usually exhibiting a metamorphic-tectonic fabric (Youngest)

The gabbro-ultramafic contact in ophiolite complexes is usually interpreted as a "Fossil Moho". The most common sediments overlying ophiolites are pelagic limestones and cherts.

## Rock Equivalents

| *Silica%* | *Plutonic* | *Hypabyssal* | *Volcanic* |
|---|---|---|---|
| Acid (66%) | Granite | Microgranite | Rhyolite (Liparite) |
| | Adamellite | - | Rhyodacite (Toscanite) |
| | Granodiorite | Microgranodiorite | Dacite |
| Intermediate (52-66%) | Syenite | Microsyenite | Trachyte |
| | Monzonite | - | Tracyandesite (Latite) |
| | Diorite | Microdiorite | Andesite |

*(Contd)*

| | | | |
|---|---|---|---|
| Mafic (44-52%) | Gabbro | Dolerite | Basalt |
| Ultramafic (44%) | Peridotite | Rare | Komatiite (rare) |

## (b) *SEDIMENTARY PETROLOGY*

### Classification of Sedimentary Rocks

Sedimentary rocks are broadly classified into two groups : (a) Detrital or Clastic or Terrigenous rocks and (b) Non-detrital or non-clastic rocks. The detrital rocks again consists of three sub-groups : (a) Rudaceous, (b) Arenaceous and (c) Argillaceous

a) **Rudaceous Rocks *(Psephites)* :** All coarser clastic sediments of a size 2 mm such as gravels, conglomerates and breccias are included in this group.

**Conglomerates :** Rudaceous deposits with more or less, rounded fragments are called Conglomerates. These are essentially composed of three components: detritals, matrix and cement. Detritals are allogenic; matrix is also allogenic but the cement is authigenic.

(1) **Epiclastic conglomerates** (also called "Terrigenous gravels" and "Extraformational conglomerates" ): These are formed by sedimentation processes on the continents and oceans i.e. by the breaking down of pre-existing source rocks outside the basin of deposition.

A. **Orthoconglomerates** : These are ordinary or normal conglomerates where the detritals are closely packed and the matrix less than 15% and sphericity and roundness is more. They are a result of a prolonged and multi-cyclic process of sedimentation. They are mature cong l omerates and indicate stable tectonic setting. They are again subdivided into two groups:

*(I)* ***Oligomictic or monogenetic conglomerates*** : These are simple, homogenous, conglomerates derived from one provenance and are indicative of long tectonic stability and always mark the basal Cambrian.

*(II)* ***Polymictic or polygenetic Conglomerates*** : These are varied in composition but the chief constituents are meta-stable rocks. The characteristic features of these rocks are their coarseness, large number of grades and the polymodal character. ***Fanglomerates*** are typically polymictic. They are a result of multicyclic sedimentation and indicate frequently disturbed tectonism.

**B. Paraconglomerates (Diamictites)** : They display a matrix of fine-grained materials in which the larger clasts are embedded or found as "float"; matrix is 15%; ill-sorted or unsorted ; polymodal size, but principal mode in fine grades. These are immature conglomerates which indicate an unstable tectonic environment. Glacial tillites and tilloids are included in this group.

**(2) Intraformational Conglomerates** : These are characterized by pebbles or intraclasts which are derived from the same formation. Tectonically they are indicative of minor oscillations in the basin or local folding or faulting.

(3) **Agglomerates** : These are formed by the consolidation of pyroclastics formed under the influence of volcanic eruptions.

(b) **Arenaceous Rocks *(Psammites)*** : These are the detrital rocks of the size range 0.10-2 mm. These are 'sand' grade deposits.

**Sandstones** : Sandstones are generally classified on the basis of presence or absence of matrix. In graywackes the matrix is >15% and the cement is absent; in arkose, the matrix is either absent or <15% but a chemical cement is present. In both the cases the silica percentage is <75. In orthoquartzites the silica % exceeds 95; matrix is absent or <15% and a chemical cement is present.

1. **Orthoquartzites** *(Quartz.Arenites):* These are sandstones made up of >95% detrital quartz and are cemented by silica. They are typically thin and blanket like deposits i.e., they have a great aerial extent in proportion to their thickness; sphericity and roundness are high; sorting is excellent and the over-all maturity is very high. Cross-bedding and ripple marks are the chief structures. Fossils are rare. They are a result of multicyclic sedimentation and are characteristic of cratonal and marginal miogeosynclinal sequences and mark periods of crustal stability e g., Paniam quartzites.

2. **Arkose and Sub-Arkose** : Feldspathic sandstone whose framework is composed of >25% feldspar particles and a small proportion of rock fragments is described as ***arkose***. Arkoses whose feldspar content varies between 10-25% are termed ***subarkose***. Feldspar is largely microcline. These are coarse-grained rocks with angular grains which are fair to ill sorted. They are characteristic of transitional environments i.e., hinge-line or "strand line". Arkose is most characteristically derived from material eroded rapidly from granitic uplands of high relief like shield areas. Arkoses are typical or zeugogeosynclines or 'yoked' basins

3. **Graywackes** : These are argillaceous sandstones containing about 15-50% of clastic feldspar, 30-40% of quartz, 5-10% of rock fragments with 5-10% of detrital chert and matrix (mostly chlorite) >20% . In terms of composition and texture, these are immature rocks with typically angular and poorly sorted grains and exhibit graded bedding. They are commonly associated with "'greenstones" and are the most typical sandstones of eugeosynclinal fold areas.

4. **Lithic Arenites** : In these, the rock fragments exceed feldspar particles. They are the most abundant of all sandstones. They exhibit "salt and-pepper" texture and are characteristic of post-orogenic molasse facies and originate particularly within miogeosynclines.

(c) **Limestones and Dolomites** : These rocks consists of at least 80% of the carbonates of Ca and Mg. Limestones are classified as calcrudites, calcarenites and calcilutites based on their grain size. ***Spergenites*** are calcarenites contain entirely of organic detritus. ***Encrinites*** are crinoidal coquinas. Folk divided limestones into two groups: *allochems* and *orthochems*. Allochems are not truly chemical, but partly chemical in origin and has four basic components: detritals, pellets, skeletal elements and oolites. Orthochems are truly chemical in origin. These are divided into micrites and sparites. *Micrite* is microcrystalline calcite and is <4.5*u* in diameter whereas *sparite* is sparry calcite >4.5 *u* in diameter. Micrite is the first generation calcite.

## Sedimentary Structures

## I. Current Structures

**1. Depositional structures :** Beach cusps, graded bedding, parallel lamination (parting lineation), sand waves (ripple-marks and cross-bedding), wave and swash marks.

**2. Erosional Structures** : Channels, obstacle scours, rill marks, scour marks.

**3. Tool Marks** : Bounce, brush, prod and skip marks; roll marks ; slide and groove marks ; straitions and grooves.

**II.** **Deformation Structures** : Dessication (mud cracks), eruption (sand volcanoes and spring pits), founder and load structures, Impact (spray, hail and rain pits), injections (neptunian dykes and sills), slumps (folds, faults and breccias).

**III.** **Chemical structures** : Cementation (sand crystals); crystallization (salt and ice), diffusion (colour banding), pressure solution (stylolites), replacement (nodules).

**IV.** **Biogenic structures** : Crawling trails, feeding trails, impressions, rootlets, etc.

**1. Cross-bedding** (Current, diagonal or false bedding) : A cross-bed is a single sedimentation unit consisting of internal laminae (foreset laminae) inclined to the principal surface of sedimentation. This sedimentation unit is separated from adjacent layers by a surface of erosion, non-deposition or abrupt change in character. In *planar cross-bedding* the bounding surfaces are planar and in *trough cross-bedding*, they are curved. In most cases, the cross- bedding is a result of the migration of small ripples and mega-ripples. In such cases the lee side laminae of the ripples form the foreset laminae of the cross-beds. Micro-deltas also contribute to the formation of cross-bedding. A simple cross-stratified layer exhibits scale, inclination and azimuth. *Scale* deals with the thickness of the unit. *Inclination* refers to the dip of the foreset laminations. The *azimuth* is the horizontal angle between meridian and the horizontal projection of the dip line of the foreset.

(a) **Herringbone Cross-bedding** : It is characterized by reversed foreset laminae in one coset. It is typical of tidal environments.

(b) **Wave-ripple Bedding** : It is produced when wave ripples migrate on a sediment surface. A characteristic feature of such ripples is the bundle-wise up-building of foreset laminae in a single unit. The foresets show offshoots that exhibit chevron structures.

(c) **Longitudinal Cross-bedding :** In this type of cross-bedding, the beds run parallel to the current direction due to the lateral shifting of tidal channels on the inter-tidal flats. These are characteristic of inter-tidal flat environments, especially mixed flats, and point bars of meandering and braided rivers.

(d) **Channel-fill Cross-Bedding** : It is a tough-shaped bedding formed by the filling up of small alluvial or erosional channels.

(e) **Flaser Bedding** : It is a ripple bedding in which mud streaks are preserved completely in troughs and partly on the crests. It is produced in environments in which conditions for deposition and preservation of sand are relatively more favourable than for the mud.

(f) **Lenticular Bedding** : It shows well-preserved sand lenses embedded within the muddy layers. It is produced under conditions which are more favourable for the deposition and preservation of mud than for sand.

(g) **Climbing Ripple Lamination** : These are the internal structure features formed in non- cohesive material from migration and simultaneous upward growth of ripples produced either by currents or waves. Its formation requires that abundant sediment is continuously available to a current or wave so that the ripples are built upward in an overlapping series rather than merely migrating in a forward direction.

**2. Graded Bedding** : Graded beds are sedimentation units characterized by a gradation in grain size, from coarse to fine, upward from the base to the top of a unit. It is apparently the result of relatively rapid pulsational sedimentation under unstable geologic environments. The rocks thus formed are called *turbidites*. It is particularly common in sandstones, especially graywackes. Graded beds, thus, commonly occur in thick sequences of *flysch* type of sediments. The sedimentation usually follows an orderly rhythmic cycle described as the *Bouma cycle*.

**3. Convolute Bedding** (Slip bedding): It is a structure showing marked crumpling or complicated folding of the laminae of a rather well-defined sedimentation unit. Convolute folds tend to die out toward both top and bottom of the bed. This bedding is common in turbidite beds, but they also occur in fluviatile, tidal flat and other sediments.

**4. Ripple Marks** : Ripple marks are one of the commonest features of sedimentary rocks, both in Recent and ancient sediments. Ripple marks are a group of wave-like depositional structures that may be aqueous or aeolian in origin. Most of them develop in shallow water or in sand dunes or in the flood plains of rivers. The *Current Ripples* are asymmetrical ripples formed by wind or water action. They have a gentle slope in the up-current direction and a steeper slope down-current. The *Oscillatory Ripples* are symmetrical ripples consisting of broad troughs that are convex downwards and of sharp crests that point upward. They form in bodies of standing water indicating a stable environment. Based on size, the ripples are classified as small (30-60 cm), mega (>60 cm) and giant (> 30 m). Tanner (1967) classified the ripples based on their *ripple index* i.e. the ratio of the horizontal projection of the stross side and the lee side. In general, ripples formed by wind origin have ripple index >15, whereas those formed by water action have ripple index <15. Based on shape, ripples are classified as straight-crested, undulatory, linguoid, rhomboid and lunate. *Linguoid ripples* are tongue-shaped ripples resembling mud flows. They are formed at higher stream velocities and face the direction of the current. The crest of the ripples is discontinuous and broken. *Lunate ripples* are essentially same as the linguoid ripples but they point up-current.

**5. Stromatolites** : These are organo-.sedimentary structures formed by the activity of blue green algae in the tidal zone. These are usually calcareous. In form they range from mound like forms to columnar, digitate and branching forms. Free rolling or mobile forms with concentrically structured bodies are known as ***Oncolites***. "*Weedia*" *crusts* are hemispherical or cabbage like forms ; *collenia* has a bean-shaped form while *cryptozoons* have concentric forms. Stromatolites range in age from about 3.5 b.y. in West Australia to modern. Stromatolites are excellent indicators of extremely shallow water and inter-tidal shore facies.

**6. Turbidites** : It is a sedimentary deposit resulting from the deposition of turbidity currents and are characterized by fining-upward cycle formed by a rhythmic alteration of relatively thin sandstones with thin shales. An ideal single turbidite sequence is made up of five units with specific sedimentary structures. This is known as the *Bouma sequence*. These units are: graded sandy beds, parallel lamination of sand, current ripple lamination, upper parallel lamination of fine sandy to silty clay and a top pelitic interval.

## Textural Features of Sedimentary Rocks

Texture deals with the size, shape and arrangement of component minerals of a rock. It is essentially the microgeometry of the rock. Three components give a rock its textural characteristics: clasts, matrix and cement. Clasts are the particles of the sediment while matrix is composed of finer material that fills the interstices among the particles. The cement is a post-depositional feature that binds clasts and matrix together. Sedimentary rocks are found with various combinations of clasts, matrix and cement.

Clasts only e.g. shale, mudstone

Clasts + Matrix e.g. graywacke, arkose, conglomerate

Clasts + Cement e.g. orthoquartzite, quartz-arenite

Clasts + Matrix + Cement e.g. sandstone

**Size of the grains** : The grain size of a detrital sediment is of considerable importance. It is a measure of the energy of the depositing medium and the energy of the basin of deposition. In general, coarser sediments are found in high-energy environments and finer sediments in low-energy environments. The grain size depends upon the mode of weathering, duration of transport and the texture of the parent rock. C.K.Wentworth (1922) introduced a *geometric grade* scale of size with the following broad grades : 1/2048 to 1/256 clay, 1/256 to 1/16 silt, 1/16 to 2 sand, 2 to 4 granule, 4 to 64 pebble, 64 to 256 cobble and >256 boulder. Subsequently, W.C.Krumbein (1938) introduced an arithmetic scale of phi (ø) units for grain-size measurement. The following are the ϕ-scale values for various grades: 0 to -1 granule; -1 to -5 pebble, -5 to -8 cobble, < -8 boulder. ***Sorting*** is a measure of the standard deviation, i.e. spread of the grain-size distribution. It is defined statistically as the extent to which the grains spread on either side of the average — the wider the spread, the poorer the sorting. ***Skewness*** is a measure of the symmetry of the distribution and is also a reflection of the depositional processes. Beach sands for example tend to have a negative skew while river sands are often positively skewed. ***Kurtosis*** is the difference in the peakedness of the two sediments.

## (c) METAMORPHIC PETROLOGY

Metamorphism essentially takes place in solid phase. It always involves the partial or complete recrystallization of original rock minerals. There is no order of crystallization. When the alteration of bulk chemical composition takes place, the metamorphism is described as allochemical ; if the bulk chemical composition remains unchanged it is called isochemical process. If rocks of lower metamorphic grade are produced from higher grades by decreasing P and T, it is known as diaphthoresis or retrograde metamorphism.

Factors Controlling metamorphism are (1) Temperature; (2) Pressure and (3) Chemically active fluids. *Temperature* : The temperature increases with depth in the earth. At a depth of a few miles, the temperature is so high that the rocks which are stable at the surface may be subjected to great changes because of the heat. This is referred as the geothermal energy. Heat is also generated due to igneous activity, friction between rock masses and radioactivity. The main effect of heat is to bring about recrystallization and therefore, it is of immense importance in identifying the assemblage of minerals in metamorphic rocks. *Pressure* : It also increases with depth and is a result of the super incumbent load. The *Hydrostatic or Uniform Pressure* acts on both liquids and solids and produces a decrease in volume (dilation) and an increase in the specific gravity of a specimen. It operates equally in all directions and generally increases with depth. The *Directed pressure or Stress* acts in one particular direction and it is one of the most powerful factors influencing the fabric of metamorphic rocks. It acts only on solids and decreases with depth, being active at shallow depths and practically zero at the centre. The effects range from distortion and rotation of minerals and mechanical fractures to complex minor folds. Stress also lowers the melting point of minerals and increases their solubility. Therefore, it is an important factor in promoting recrystallization. The resulting mineral grains grow perpendicular to the direction of the stress. *Chemically Active fluids* : Water, carbondioxide and boric and hydrofluoric acids play an important role in metamorphism. Using the above variables, metamorphism can be classified as:

**A. Thermal or Contact Metamorphism** : ( HT, LP, low strain and variable fluid pressure). The *Contact metamorphism* involves changes which take place in aureoles surrounding intrusive rocks at lower temperatures. *Pyrometamorphism* includes intense local changes at very high temperatures, affecting xenoliths immersed in basaltic lavas or the immediate contact of igneous plutons with country rocks. *Optalic or caustic metamorphism* denotes the indurating, burning and fritting effects produced by lavas and small dykes on the rocks. *Pneumatolytic metamorphism* involves changes taking place as a result of vapours and gases.

**B. Cataclastic or Kinetic or Dynamic metamorphism** : (High strain, variable pressure, variable temperature and high fluid pressure). It acts in the upper parts of the earth and leads to mechanical breakdown or rocks with no new mineral formation, except when heat produced by friction is involved. Its effects include granulation, flowage, distortion and rupture of mineral grains.

**C. Dynamothermal or Regional Metamorphism** : (HT, HP, variable strain, variable fluid pressure). It is the large-scale action of both heat and pressure which results in the production of a wide range of new minerals and a widespread development of *tectonites*. It is always associated with orogenies and igneous intrusions.

**D. Plutonic Metamorphism** : It takes place at great depths where the factor of stress is insignificant. The mineral transformations are complete but no new structures are formed.

## Products of Thermal Metamorphism

(a) Shale —Contact Met→ Hornfelses

Shale —Optalic Me→ Hornstones; Novaculite; Porcellanite; Lydian stone

(b) Sandstones →→ Quartzites

(c) (i) Pure limestones → Marble

(ii) Limestones with siliceous impurities → W ollastonite

(iii) Limestones with aluminous impurities → Anorthite

(iv) Dolomitic limestones → Periclase + CaCO3 (dedolomitisation)

$MgO+H_2O$ → $MgO. H_20$ (Brucite)

Brucite + CaCO3 -→ Brucite Marble (Predazzite and Pencatite)

(v) Dolomitic limestones with small amounts of silica → Forsterite marble

Forsterite marble + H2O → Serpentine Marble or Ophicalcite

(vi) Dolomitic limestones with intermediate amount of silica → Tremolite Marble

(vii) Dolomitic limestones with large amount of silica → Diopside

(viii) Dolomitic limestones with aluminous impurity → Spinel

(ix) Dolomitic limestones with alumina and silica → Forsterite and Spinel

**(d) Skarns :** These are lime silicate rocks rich in iron silicates and are formed due to pyrometamorphism. They exhibit granoblastic or porphyroblastic textures. Mineralogy is simple and includes diopside, hedenbergite, actinolite or wollastonite.

**(e) Tactites :** These are the rocks composed of ore mineral and silicates and are formed by thermal metamorphism of carbonate rocks. They are coarse-grained and exhibit xenoblastic textures. Typical ore minerals include sphalerite, galena, scheelite, bornite, molybdenite, etc and silicates such as tourmaline, zoisite, forsterite, etc.

## Products of Cataclastic Metamorphism

(i) Shale → Slates

(ii) Granites / Feldspasthic sandstones : Deformation and changes takes place in a sequence.

- (a) Some minerals exhibit optical anomalies e.g. quartz with undulose extinction; garnet showing birefringence
- (b) *Mortar structure* : Quartz is crushed but feldspars remain unaltered
- (c) *Augen structure* ; feldspars exhibit eye-like appearance
- (d) *Cataclasite* : feldspars are also crushed to give a pulverized form
- (e) *Flaser rocks* (augen gneisses) : It has minute, eye-shaped grains of orthoclase. They are formed when stress is accompanied by differential movement.
- (f) *Mylonite* : These are rocks which have been completely pulverized and rolled out by extreme differential movement; extreme forms are known as Pseudotachylites.
- (g) Crush breccia, crush conglomerates and *hartschiefer* (like ultramylonites) are the other products of cataclastic metamorphism.

## Products of Dynamothermal Metamorphism

Tectonites are widespread. It exhibits 3 types of deformation : blastic, plastic and elastic. The formation of "Snow Ball Garnets" is an example of elastic deformation which results in the change of form due to crushing.

(a) Effect on Olivine Gabbro (olivine, pyroxene and anorthite are the main mineral constituents)

Anorthite → Epidote + Feldspar (Saussiritisation)

Pyroxene → Chlorite → Calcite + Quartz (Chloritization)

→ Hornblende → Epidote + Quartz (Uralitization)

(b) Dolerite → Diabase

(c) Shale → Phyllite → Schist → Gneiss

(d) K-Feldspar → Sericite + Quartz (Sericitization)

(e) Amphibolites

## Products of Plutonic Metamorphism

These include many pyroxene-bearing rocks : granulites, charnockites, eclogites, leptynites, etc.

**Stress and Anti-Stress Minerals :** Harker divided the metamorphic minerals into two categories: Stress and Anti-Stress minerals. ***Stress minerals*** are those whose fields of stability on a P-T diagram are extended by the introduction of a non-hydrostatic stress. They have high index of crystal lattice and low molar volume. Restricted to schists and other deformed rocks. e.g. micas, chlorite, albite, epidote, zoisite, amphiboles, kyanite, talc, garnet, etc. ***Anti-stress Minerals*** are those whose fields of stability are reduced on the introduction of a non-hydrostatic stress. They have low packing index and high molar volume and are confined to undeformed rocks such as hornfelses. They are the products of thermal and contact metamorphism and favour the absorption of heat under formation. e.g. Anorthite, alkali feldspars, augite, hypersthene, olivine, nepheline, andalusite, sillimanite, spinel, scapolite, etc. Some petrologists have cast a doubt on the validity of this concept as many examples of 'stress' minerals occurring in 'unstressed' environments have been recorded.

Idioblastic minerals are those which develop their crystal form inspite of the resistance of a solid medium. Minerals with poorly-developed faces are called Xenoblastic.

**Idiobalstic (Crystalloblastic) Series** : Becke established the following sequence of perfection of idioblastic form in metamorphic minerals, which is called crystalloblastic or idioblastic series.

Magnetite, Sphene, Rutile, Ilmenite >> Perfectly idioblastic

Pyroxenes, Amphiboles, Wollastonite >> Less idioblastic

Micas, Chlorite, Calcite, Scapolite >> More xenoblastic than idioblastic

Quartz, plagioclase, Orthoclase, Microcline >> Perfectly Xenoblastic

## Metamorphic Textures and Structures

1. **Cataclastic Structure** : These are a result of the deformation of hard brittle rocks. There is almost no new mineral formation. Deformation is effected in various stages forming different products: Crush breccia → Cataclasite → Flaser rock → Mylonite

2. **Maculose Structure** : It is characterized by spotted appearance due to incipient crystallization of equidimensional minerals and porphyroblasts of strong minerals like andalusite, cordierite, biotite, chloritoid, etc. It develops under thermal metamorphic conditions belonging to the Hornfels facies. Typical of argillaceous rocks.

3. **Granulose Structure** : Developed largely in the rocks with granoblastic minerals such as quartz, feldspar, pyroxene, garnet, etc. It is characteristic of marbles, skarns and a few high-grade rocks such as Khondalites.

4. **Schistose Structure** : Foliated structure formed due to parallelism or linear minerals, due to cataclastic or dynamothermal metamorphism.

5. **Gneissose Structure** : It is basically a composite structure with alternation of foliate (schistose) and non-foliate (granulose) bands and lenticles. Common in quartzo-feldspathic rocks. If the foliated structure is dominated by prismatic minerals it is described as *Nematoblastic* and if it is dominated by tabular minerals it is described as *Lepidoblastic*.

6. **Mosaic or Saecharoidal Texture** : It is a texture having sugary appearance. It is commonly applied to marbles and quartzites.

7. **Helicitic Structure** : It refers to reverse S-shaped trails of inclusions in poikiloblastic crystals, especially garnets and staurolite, found in regionally metamorphosed rocks.

## Zones of Metamorphism

1. **Epizone** : It is the upper-most zone that is nearest to the earth's surface and is characterized by mechanical and chemical metamorphism and constituting

of hydrous silicates. Shearing stress is strong and hydrostatic pressure is low. The temperature is relatively low or moderate i.e.<300 0C. Cataclastic metamorphism is dominant. Typical minerals include: sericite, chlorite, paragonite, talc, brucite, zoisite, antigorite, albite etc.

2. **Mesozone** : It is an intermediate zone of chemical metamorphism. Pressure and temperature are higher than those of the epizone. Temperature ranges from 300-5000°C. Stress is very high or lacking Richest stress minerals are found in this zone. It is a zone of thermal or dynamothermal metamorphism. Typical minerals include : biotite, muscovite, staurolite, kyanite, anthophyllite, cummingtonite, grunerite, actinolite, calcite, brucite, almandine, zoisite, etc.

3. **Katazone** : It is the lower-most zone characterized by long continued chemical recrystallization often accompanied by deformation. The hydrostatic pressure and temperature are high while the stress is low. Temperature ranges from 500-7000C. Plutonic metamorphism is characteristic. Typical minerals include : biotite, K-feldspar sillimanite, enstatite, hypersthene, olivine, omphacite, hornblende, jadiete, garnet, spinel, periclase, etc.

**Barrowian zones of Progressive Regional Metamorphism** : Barrow (1912) mapped the Daldriadan Schist Field of the Scottish Highlands into a series of zones of progressive regional metamorphism based on mineralogical transformations in derivatives of argillaceous sediments and correlated them in a general way with increasing temperature and pressure. Each zone is defined by an index mineral, the first appearance of which marks the *Isograd* (or the line of equal grade of metamorphism) defining the outer limit of the zone. The zones established by Barrow are the same as those in other parts of the world, except minor departures due to variation in the chemical composition of parent rocks. The following is the sequence of Barrovian zones :

Chlorite zone — Brown Biotite zone— Almandine zone— Staurolite zone— — Kyanite zone— Sillimanite zone.

grade of metamorphism increases

⟶

**Riecke'e Principle** : When the stress is operating, solution takes place at the points of; maximum pressure with simultaneous or concurrent precipitation towards the region of minimum stress or pressure.

**Facies Concept** : Eskola (1915) proposed the term ***metamorphic facies*** to include any association of metamorphic rocks within which there is a correlation between the mineral composition and chemical composition. Fyfe and Turner (1966) defined metamorphic facies as " a set of metamorphic mineral assemblages, repeatedly associated in space and time, such that there is a constant and therefore predictable relation between mineral composition and chemical composition". Five metamorphic facies were originally defined by Eskola : Sanidinite (LP/HT); Hornfels (moderate Pr/HT); Amphibolite (moderate T/ moderate Pr) and Greenschist facies (LT/mod pr). Later on some more facies like the Granulite facies (HPIHT); and Glauchophane schist facies were identified.

(a) ***Greenschist Facies*** : It includes the common products of low-grade regional and dislocation metamorphism and comprises of the most voluminous and most widespread of the metamorphic rocks. They appear to develop under conditions of normal geothermal gradient in geosynclinal environments, major granitic batholiths and areas of Precambrian shields. Typical rocks in this facies include slate, phyllite and fine-grained schist from pelitic rocks,green schists from basalts and andesites, magnesian schists from peridotites, etc. Characteristic minerals include the green minerals chlorite, actinolite and epidote along with albite and quartz in pelitic and most other meta-sedimentary rocks. Talc and antigorite occurs in magnesian rocks.

(b) ***Amphibolite Facies*** : They occur in most orogenic belts and appear to develop under conditions of a normal or somewhat steepening geothermal gradients in the deeper levels where the temperatures range between 550-750 °C and pressures range between 4000-8000 bars. Typical rocks include micaceous schists and gneisses from pelitic rocks, quartzo-feldspathic schists and gneisses from feldspathic sandstones;, amphibolites from basalts and andesites, and calc-silicate rocks from calcareous sandstones and shales. The diagnostic mieral assemblages are hornblende + plagioclase + almandine in the higher grade rocks and hornblende + plagioclase + epidote in the lower grade rocks.

(c) ***Granulite Facies*** : These are the products of high-grade deep-seated regional metamorphism and are characterized by typical granulose fabric, in addition to the gneissic structure. It is also characterized by a wide range of composition derived from quartzo-feldspathic rocks. Most of the granulite facies complexes occur in the Precambrian Shield areas. Typical rock types include hypersthene granites (charnockites), hypersthene granodiorites, quartz-hypersthene gabbros, etc.

(d) ***Eclogite facies*** : Rocks of this facies occur in wide range of rocks of all ages from Precambrian to Tertiary. They have a pronounced tendency to occur in a number of situations: as inclusions in Kimberlites; as streaks and bands in dunites; as lenses in migmatite gneisses accompanied by amphibolitization ; as bands in amphibolites, mica schists, etc in the regions of alpine folding. Eclogites are dominantly composed of Na-rich clinopyroxene, Omphacite and Mg-rich garnet, Pyrope. Rutile is an accessory phase. Most eclogites are chemically equivalent to basalts, especially low-K oceanic tholeiites. Eclogites have 4 characteristic features : (1) distinctive mineralogy and exceptional density; (2) Restricted range of chemical composition corresponding approximately to basic igneous rocks ; (3) their high susceptibility to retrogressive metamorphism ; (4) their pronounced tendency to occur in a number of situations

(e) ***Blueschist Facies*** : These rocks develop under conditions of abnormally low geothermal gradients (100C/km) in eugeosynclinal environments along tectonically active continental margins. There they form tile high-pressure ocean-ward member of a common HP-LP pair of metamorphic belts. Most of these rocks are confined to the post-Palaeozoic orogenic belts. The diagnostic minerals are glaucophane, lawsonite, jadeite, soda- amphiboles, and stilpnomelane. The mineralogy suggests an origin involving HP/LT and depths of 15-20 kms where is the pressure is around 5-10 K bars and temperature is around 150-4000C.

(f) ***Sanidinite Facies*** : It belongs to the contact metamorphic facies and is characterized by maximum temperatures and minimum pressures. Its field of occurrence is restricted to xenoliths in lavas and dyke rocks, fragments in breccias, and local developments of contact rocks in the immediate vicinity of near-surface intrusions. The minerology is typically composed of high-temperature, anhydrous minerals like tridymite, montmorillonite, meliiite, larnite, rankinite, etc. Sanidine, often with a high content of soda, is a critical mineral of this facies. Glass is sometimes present due to partial or complete fusion.

(g) ***Zeolite Facies (Prehnite-Pumpellyite Facies)*** : The assemblages constituting the zeolite facies result wen chemically unstable rocks become completely converted to low-temperature assemblages rich in zeolites (especially alumontite) and duplicate the products of diagenesis. Thus, this facies is considered as a transition between diagenesis and regional metamorphism. They occur in the temperatures of 200-3000C and pressures of 2000-3000 bars and at depths ranging from 3000- 40,000 feet. Typical mineral assemblages include laumontite + albite + quartz, quartz + albite + pumpellyite and quartz + adularia + pumpellyite. Prehnite seems to be a characteristic mineral in rocks of this facies in some regions and pumpellyite (a Mg-epidote) is characteristic Ca-Al silicate of half-reconstituted schistose meta-graywackes.

**Charnockites** : Charnockite is defined as an orthopyroxene, quartz, feldspar rock, with or without garnet. They range in composition from basic to acidic and constitute a suite of rocks referred to as "Charnockite series". (a) *Acid Division* : It is representcd by charnockite, a hypersthene granite, having a sp. gr. of 2.67 and 75% silica. It also includes alaskites, birkremites, enderbites and hypersthene-quartz syenites. The type area is Magazine Hill (near Chennai). Microscopically, the rock is an even-grained aggregate of potash feldspar and quartz. 'Ihe potash feldspars, are microperthite, microcline-perthite and microcline. According to Tuttle, it suggests a magmatic origin for the rocks. Twinning is frequently the albite-ala type. Quartz is often highly granulated and is associated with myrmekite. Accessory minerals are apatite; zircon, titano-magnetite, ilmenite and pyrrhotite. The presence of pyrrhotite indicates a high temperature of consolidation. (b) *Intermediate Division* : These include the most abundant rocks in the suite. The rocks are characterized by an apparently composite structure and all the minerals of the acid and basic divisions are frequently found in the same specimen with a tendency for the coloured minerals. The average specific gravity is 2.77 and silica % is about 54. e.g. hyp-diorites, pyroxene granulites and granulitic migmatites. Type area is Shevroy Hills in the Salem district of Tamil Nadu. (c) ***Basic Division*** : These forms are considered equivalent to Norite. It is composed of hypersthene, augite, plagioclase and iron ores often with hornblende. They are generally granulitic in texture. They have a specific gravity of 3.03 (d) ***Ultrabasic Division*** : This division has rocks having a density of 3.73 and SiO2 % of 47-50. These are pyroxenites occurring as *'schlieren'* or as ramifying veins and dykes composed almost entirely of dark hypersthene (bronzite), constituting about 65% of the rock.

**Tourmalinization** : It is due to the ccmbined effect of $H_30$, B and F.

**Scapolitization** : It is due to combined effect of chlorine and carbondioxide.

**Fenitization** : It is the contact conversion of granite to a metasomatic rock of syenite composition (Fenite).

**Miscellaneous :** (1) Adinole is an albitized slate, (2) Greisen is a quartz, muscovite-topaz rock, connected with Fluorine metasomatism of granites, (3) Schorl is an aggregate of quartz and tourmaline: (4) Sericite is white mica and quartz ; (5) Pinite is composed of white mica and chlorite (6) Saussurite is composed of albite and epidote: (7) Cymatolite is composed of albite and muscovite (8) Leptite is a fine grained, granulose metamorphic rock composed of Quartz and feldspars with mafics.

# 8

# *Indian Geology*

## RELATIVE AGES

→ Age decreases from left to right

1. **Shales.** Binota Shales > Suket Shales = Auk Shales > Bijaigarh Shales > Panna Shales > Jhiri Shales > Ganugarh Shales > Sirbu Shales > Fenestella Shales > Productus Shales > Needle Shales (Talchir) > Olive Shales > Ironstone Shales (Damuda) > Daonella Shales > Monatis Shales > Kuti Shales > Raghavapuram Shales = Vemavaram Shales > Spiti Shales > Chikkim Shales.

2. **Limestones.** Vempalle limestone > Kazrahat limestone > Fawn limestone > Rohtas limestone = Narji limestone = Tirohan limestone > Nagode limestone > Mountain limestone (M. Permian) > Nodular limestone > Daonella limestone > Coral limestone (Triassic) = Kalapani limestone > Megalodon limestone = Kioto limestone > Chikkim limestone > Sylhet limestone.

3. **Sandstones.** Kaimur sandstone > purple sandstones > Jodhpur sandstones > Speckled sandstones (U. Carboniferous) > Calcareous sandstones (Permian/Spiti) > Chintalpudi sandstone = Kamthi sandstone > Golapilli sandstone = Budivada sandstone > Tirupati sandstone = Pavalur sandstone > Katrol sandstone > Giumal sandstone = Nimar sandstone > Guddalore sandstone = Rajahmundry sandstone.

4. **Quartzites.** Ranthambor quartzites > Gulcheru quartzites > Plateau quartzites > Pinnacled quartzites > Dhandraul quartzites > Muth quartzites.

5. **Beds.** Neobulus beds > Talchir boulder beds = Mandhali beds = Blaini beds > Umaria Marine beds > Gangamopteris beds > Zeewan beds > Halobia beds > Tropite beds > Juvavite beds > Kamthi beds > Macrocephalous beds > Sulcacutes beds (Jurassic/Spiti) > Ukra beds (Jurassic-Cretaceous) > Umia Plant beds (Jurassic-Cretaceous) > Lameta beds > Bagh beds > Warkala beds > Quilon beds = Baripada beds > Dwaraka beds.

6. **Volcanics.** (a) Dalma traps (Precambrian) > Cuddapah traps (Proterozoic) > Pirpanjal traps > Rajmahal traps (Jurassic) > Deccan traps (U. Cretaceous-Eocene) > Sylhet traps (Paleozoic- Eocene).

(b) Dalma traps > Dhanjori-Simlipal volcanics > Bijli Rhyolites > Pitepani volcanics > Sitagota volcanics > Mangikhuta volcanics > Cuddapah traps > Malani volcanics > Bhimtal volcanics > Panjal volcanics > Rajmahal traps (Jurassic-L. Cretaceous) > Deccan traps > Dras volcanics > Sylhet traps.

7. **Conglomerates.** Gulcheru conglomerate > Banganapalli conglomerate > Ralam conglomerate > Po Series conglomerate > Hemis conglomerate.

8. **Breccias.** Hornstone breccia > Tirohan breccia > Susnai breccia.

## SYNONYMS & EQUIVALENTS

Sulekere Series = Upper Dharwars; Hosur Series = Middle Dharwars; Complex Series = Lower Dharwars; Charnockites = Blue Mountain Gneiss = Nilgiri Gneiss; Clospet Granite = Bellary Gneiss = Hosur Gneiss; Granite Gneiss (Chota Nagpur) = Dome Gneiss = Bengal Gneiss; Khondalites = Bezwada Gneiss = Kailasa Gneiss; Older Greenstone Belt = Sargur Schist Complex; Newer Greenstone Belt = Dharwar Supergroup; Fundamental Gneiss = Peninsular Gneiss; Bhander limestones = Nagode limestones; Kioto limestone = Megalodon limestone; Great limestone = Riasi limestone = Jammu limestone (Precambrian); Barren Measures = Iron Stone Shales = Motur Stage = Kulti Formation; Ranigunj Formation = Bijori Formation; Arkasani Granophyre = Soda Granite; Talchir Tillites = Tannakki Boulder Beds (Kashmir-Hazara) = Mandhali Beds (Garhwal) = Blaini Boulder Beds (Simla); Kamthi Beds (U. Gondwanas) = Pali Beds = Himgir Beds = Bijori Beds = Athgarh Beds; Dongargarh Granite = Clospet Granite; Sausar Group = Chilpi Ghat Group; Indravati Group = Kurnool Group; Bap Formation = Talchir Tillites; Semri Group = Kurnool Group = Bhima Group = Palnad Group; Nagthat Series = Upper Vindhyans.

## IMPORTANT OROGENIES

Khairagarh Orogeny - 600 My; Singhbhum Orogeny - 850 My; Kolhan Orogeny - 950 My; Sausar Orogeny - 844-996 My; Satpura Orogeny - 1000 My; Sakoli Orogeny - 1335 My; Dhanjori Orogeny - 1500 My; Eastern Ghat Orogeny - 1600 My; Delhi Orogeny - 1600 My; Amgaon Orogeny - 2300 My; Iron Ore Orogeny - 2900 My.

## CHARACTERISTIC REGIONAL STRIKE DIRECTIONS

Aravalli & Delhi Cycles - NE-SW
Dharwar Cycle - NNW-SSE
Eastern Ghat Cycle - NE-SW
Mahanadi - ENE-WSW
Satpuras - ENE-WSW
Iron Ore Cycle - NNE-SSW
Chilpi Ghat Series (M.P.) - N-E
Upper Kaladgis - WNW-ESE
Pakhals - NW-SE
Spiti - NW-SE
Jurassic of Kutch - E-W
Gondwanas of Damodar, Narmada & Son Valleys - E-W

## MINERAL DEPOSITS ASSOCIATED WITH SOME IMPORTANT FORMATIONS

Iron Ore Series - iron
Sausar & Sakoli Series - manganese
Khondalites - graphite, manganese, hydrocarbons and some diamonds
Bijawer Series - diamonds (Golconda diamonds)
Gwalior Series - solfatoras (sulphurous hot-springs)
Vempalle Formation - barytes, steatite, asbestos
Narji Limestones - Paniam cement
Rohtas Limestones - Dalmia cement
Panna Shales - diamonds
Banganapalli Formation - diamonds
Bijaigarh Shales - pyrite, mica, coal (vitrain)
Mansar Formation (Sausar Group) - manganese
Nallamallai Group - lead and copper
Cheyair Group - specks of diamonds
Karewas - lignite
Bhimas - limestones with pyrite crystals

## IMPORTANT GONDWANA & SIWALIK FOSSILS

**Lower Gondwanas**

(a) **Plant Fossils.** Pteridosperms (Ferns), i.e., Gangamopteris and Glossopteris (U. Carboniferous - L. Triassic) = Gigantopteris of North America; Lepidodendron; Sigillaria.

Vertebraria (stem) in Talchir

Dadoxylon (fossil wood) in Ranigunj Stage (Damudar)

Pteridosperm leaves - Damudar Series

Schizoneura Gondwanensis - Panchet

Gondwanidium – Buriadia } Karharbari

Noeggerathiopsis – Paranocladus } Karharbari

Walkomiella, Pseudoctenis - Barakar

Cyclodendron, Eurydesma, Fenestella, Phyllotheca, Cyclopteris, Samoropsis, Taeniopteris

Buriada, Ottokaria, Arberia cladophlebis (Filicales), Sphenophyllum

Glossopteris showed maximum development in the Damudar

(b) **Vertebrates.** Stegocephalia, Gonioglyptus, Pachygonia (Amphibians) - Panchet;

Amblypterus - Fish - Panchet

Estheria - Crustacean - Panchet

Dicynodon, Ptychosiagum - Reptiles - Panchet

**Upper Gondwanas**

(a) **Plant Fossils.** Bennettitales - Cycad - Rajmahal

Homoxylon - fossil wood

Cladophlebis, Coniopteris, Pecopteris, Sphenopteris (Ferns)

Williamsonia - Cycad - Rajmahal

Dictyozamites - Pterophyllum

Plagiophyllum - Brachyphyllum, Ptilophyllum, Dicrodium, Otozamites, Arancorites, Elatocladus

(b) **Vertebrates.** Brachyops (Amphibian) & Estheria (Crustacean) - Mahadeva group - Mastodonasaurus indicus, Plesiosaurus (Reptile)

**Siwalik Fossils**

Siwalik fossils mostly comprise of Mammals, molluscs, ostracodes, charophytes, etc.

(a) **Upper Siwaliks.** Primates - Simla, Semnopithecus

Carnivores - Canis, Mellivora, Felis, Ursus

Elephants - Mastodon Sivalensis; Stegodon ganesa

Ungulates - Rhinocerus, Sus, Bos, Leptobos, Hippopotamus

(b) **Middle Siwaliks.** Primates - Paleaopithecus, Dryopithecus, Ramapithecus, Macacus

Carnivores - Felis, Lutra, Amphicyon

Rodents - Hystrix

Elephants - Dinotherium, Tetrabelodon, Stegodon clifti, Syconolophus

Ungulates - Listriodon, Hipparion, Teleoceras, Gazella

(c) **Lower Siwaliks.** Primates - Sivapithecus, Dryopithecus, Bramapithecus

Carnivores - Dissopsalis, Amphicyon

Proposidians - Dinotherium

Ungulates - Acerathium, Hyotherium, Brachyodus

Birds - Phalacrocorax, Pelecanus

Reptiles - Crocodilus, Gavialis, Rhamphosuchus (Crocodiles); Varanus (Lizard); Colossochelys atlas, Beilia, Trionnyx (Turtles); Ophiocephalus, Chrysichthys, Rita, Arius (Fishes)

**Characteristic Fossils of each Formation of the Siwalik Group**

| | |
|---|---|
| Basal Conglomerate Formation | Elephas, Equus, Buffelus, Camelus |
| Pinjor Formation | Hyppophys, Leptobos |
| Dhokpathan Formation | It is richest in fossils among all the Siwalik formations. Stegodon, Mastodon, Giraffoid, Sus, Several giraffes and short-jawed probosicideans, Syconolophus is characteristic. Bovidae appear first in Dhok Pathan |
| Nagri Formation | Hipparion, Sus, Ramapithecus, Brahmapithecus, No elephants |
| Chinji Formation | Macrotherium, Gaindatherium (Rhinocerus), Listriodon, Amphicyon; Hipparion; Dryopithecus, Sivapithecus; Ramapithecus (Primates) |
| Kamlial Formation | Aceratherium, Dinotherium, Trilophodon; Tetrabelodon, Hypoboops |

## IMPORTANT FORMATIONS/SYSTEMS/GROUPS/BEDS

1. **Dalma Traps.** These are the intercalations of contemporaneous lava flows, tuffs and agglomerates on top of the Iron Ore Series in the North Singhbhum. It is believed that the lava flows of the Dalma Traps reflects the existence of an island-arc system during the Archaen times.

Chota Nagpur Gneiss

Dalma & Dhanjori lavas

Dhanjori sandstones

– – – – – – – – – – – – UC – – – – – – – – – – –

Iron Ore Series

2. **Malani Igneous Suite.** It comprises of Lower Vindhyan rocks of Western Rajasthan composed of a group of rhyolitic lavas resting unconformably on Aravalli Schists. The lavas vary in acidity from rhyolites to quartz andesites and in which two types of granites are recognised : Hbl-Bt granite (Jalor granite) and Hbl-granite (Siwana granite). This igneous suite represents one of the world's largest examples of felsic magmatism.

Jodhpur Sandstone

– – – – – – – – – – – – UC – – – – – – – – – – –

Malani Igneous suite

Kirana Hills

Tosham Hills

– – – – – – – – – – – – UC – – – – – – – – – – –

Delhi Super Group

3. **The Pakhal Group.** These are the rocks comprising of conglomerate, arkose, sandstone, shale, dolomite with ferruginous quartzite and beds of hematite deposited in the Pakhal Basin (A.P.) and spreading over a length of 230 km. The Pakhal rocks are generally folded NW-SE and exhibit an increase in metamorphism from NW to SE.

There are two different opinions about the stratigraphic position of the Pakhals : (a) Pakhals are equivalents of Middle Dharwars and (b) Pakhals are equivalents of Cuddapahs. However, the generally accepted view is that the Pakhals are equivalents of the Cuddapahs.

M.S. Krishnan has classified the Pakhal Group into the following two stages :

| | |
|---|---|
| Pakhal Group | Albaka Stage — 3500 m |
| | Pakhal Stage — 2800 m |

However, Basumallick (1967), Srinivas (1979) and Dutt (1978) divided the Pakhals into the following three groups which are correlated with their corresponding formations of the Cuddapah Supergroup :

| | | |
|---|---|---|
| | ALBAKA SANDSTONE | = Srisailam Quartzite |
| | ------- Disconformity ------- | |
| MULUG GROUP | LAKNAVARAM PHYLLITES & QUARTZITES | = Cumbum Shales and Phyllites |
| | PATTIPALLI QUARTZITE | = Bairenkonda Quartzite |
| | -------- Disconformity ------- | |
| KOMARARAM GROUP | ENCHENCHERUVU SHALES | = Pullumpet Shales |
| | POLARAM SHALES | = Nagari Quartzite |
| | JAKARAM ARKOSE | |
| | ------- Disconformity ------ | |
| MALAMPALLI GROUP | PONDIKUNTA SHALES | = Vempalle Formation |
| | GUNJEDA DOLOMITIC LIMESTONES AND SHALES | |
| | BAYYARAM CONGLOMERATES, ARKOSE AND QUARTZITES | = Gulcheru Formation |

The Pakhal Group (south of Godavari River, A.P.)

**The Shernvala Outlier.** In the south of Yellandalapad, the Pakhals extend as a narrow strip called the 'Shernvala Outlier' in the N-S direction. It is exposed east of Khammam (A.P.). The lithology includes quartzofelspathic gneisses, granulites, charnockites, leptynites, khondalites, migmatites (Augen gneisses), greenschists, amphibolites, quartz-norite schists, pegmatite veins, etc.

**4. The Penganga Group.** These are the sedimentary formations comprising of grey and purple limestones with shale intercalations deposited in the northern area of the Godavari Basin in Andhra Pradesh. The Penganga rocks generally lie directly over the gneisses but in Adilabad district (A.P.), they show basal quartzites. In this area, the Upper Limestones of the Penganga Group show distinct sedimentary horizons of manganese ore with interlayered jasper. The rocks of the Penganga Group were subjected to folding during the mid-Kurnool tectonism. The Penganga Group has been correlated with rocks of the lower Kurnools.

**5. The Sullavai Group.** It consists of a 400 m thick deposition of sedimentary nature comprising of conglomerates, sandstones, quartzites and slates in the northern area of the Godavari basin, lying over the folded Penganga beds disconformably. The quartzites of the Sullavai Group have a great lithological similarity with the Pinnacled quartzites of the Kurnool Group.

| | | |
|---|---|---|
| Sullavai Group | Limestone<br>Red Mottled Sandstone | = Nandyal Shale Koilkunta limestone Panem quartzite & shale |
| | ------------ Disconformity ---------- | |
| Penganga Group | Penganga Upper Limestone<br>Penganga Shale<br>Penganga Lower Limestone<br>Pranhita Sandstone | = Auk shale = Narji Limestone = Banganapalli quartzite |

Geology of the rock exposures in the northern area of the Godavari basin (A.P.).

6. **The Kaladgi Group.** It rests unconformably upon the Peninsular Gneissic Complex and is unconformably overlaid by the Deccan traps of cretaceous-tertiary age. It comprises of 3-5 km thick succession of arenaceous, argillaceous and calcareous rocks. The sediments show a general decrease in coarseness from bottom to top representing at least two major cycles of deposition corresponding to the lower and the upper formations of the Kaladgi Group. Various types of Stromatolites (Collenia and Cryptozoan) and acritarchs are reported from Kaladgis.

DECCAN TRAPS

------------- UC -------------

| | |
|---|---|
| UPPER (3500′) (WNW–ESE) | Shale<br>Ferruginous Limestone<br>Basal Sandstone |
| LOWER (11000′) | Shales<br>Sandstones<br>Siliceous Limestones<br>Quartzites & Conglomerates |

----------- UC ---------

ARCHAEANS — Peninsular Gneissic Complex

The Kaladgi Group

7. **The Badami Group.** The Badami Group unconformably overlies the rocks of the Kaladgi Group and was established as a distinct post-Kaladgi lithounit by Vishwanath (1968). The age of the Badami Group is determined as Ordovician (500-430 MY) on the basis of its organic findings consisting of Chitinozoa, Scolecodontes and other microplanktons. The Badami Group has been correlated with the Bhima and Kurnool Groups.

| | |
|---|---|
| BADAMI GROUP | Gokak Quartz Arenite<br>Temple Quartz Arenite |

---------- UC ----------------

KALADGI GROUP

8. **The Indravati Group.** The Indravati Group is exposed in the Indravati river traversing westward across the Bastar depression. It comprises of basal quartzites and conglomerates followed by quartzites, shales, dolomites and limestones. It is considered as an equivalent of the Kurnool Group.

| | | |
|---|---|---|
| JAGDALPUR FORMATION (200-250 m) | U. Purple shale with quartzites Purple shale-limestone<br>Greenish grey & purple crypto-crystalline limestone<br>Purple shale with quartzite | = NANDYAL FORMATION |
| KANGER FORMATION (0-140 m) | Grey laminated limestone | = KOILKUNTLA LIMESTONE |
| TIRATGARH FORMATION (0-110 m) | Purple shale with quartzite intercalations<br>Subarkose & orthoquartzite<br>Basal Conglomerate | = PANEM FORMATION |

---------- UC ----------------

GRANITOIDS

The Indravati Group (after Dutt, 1963).

9. **The Bagh Beds.** These are the marine rocks of Cretaceous age occurring as inliers of older rocks surrounded by the younger Deccan Traps in the Narmada Valley. These rocks which overlie the Archaean gneisses and schists consist of

about a 30 m thick succession of largely unfossiliferous sandstones and conglomerates known as the Nimar Sandstone overlain by a 12 m thick sequence of fossiliferous, compact, white, argillaceous Nimar Limestone and a 10 m thick succession of Coralline Limestone.

| | LAMETA BEDS |
|---|---|
| BAGH BEDS | Coralline Limestone |
| | Nodular Limestone |
| | Nimar Sandstone |
| | UC |
| | PRECAMBRIAN |

10. **The Lameta Beds.** These are 6-30 m thick marine transgressed sediments occurring below the Deccan Traps and resting on various older formations such as the Archaeans, the Upper Gondwanas or the Bagh Beds and comprising of limestone with subordinate sandstones and clays. The Lameta Beds have yielded fossils of a large number of molluscs, terrestrial reptiles and fishes of which the following are important :

| | |
|---|---|
| Molluscs | Malania, Physa (Bullinus), Corbicula |
| Fishes | Lepidosteus, Pycnodus |
| Dinosaurs | Antarctosaurus, Titanosaurus indicus, Indosuchus raptoris, Lametosaurus, etc. |

The dinosaurs reached their highest development in India during the Lameta epoch. These fossil dinosaurs are closely allied to those of Brazil and Madagascar suggesting the existence of land bridges in the present Indian and Atlantic Oceans. The Lameta Beds which were deposited almost during the same time when the Bagh Beds were being deposited in the Narmada Valley are considered to be partly younger than the Bagh Beds. Fossil assemblages have suggested an upper Cretaceous age for the Lametas.

11. **Umaria Marine Beds.** It consists of a 10 ft thick marine assemblage comprising of shelly limestone and clay containing fossil shells of productus, spiriferina, reticularia and a few other gastropods. The fossil assemblage depicts highly individual characters of the species and is suggestive of a Lower Permian age for the beds. The Umaria Marine Beds overlie the Talchir Formation but passes upwards without any visible break into the overlying Karharbari Formation. The Umaria Marine Beds which can be correlated with the Eurydesma and Conularia Beds of the Salt Range and Gangamopteris Beds of Kashmir are suggestive of (a) marine transgression and (b) Warmer climates.

KARHARBARI FORMATION (200-300 ft) (L. Permian)

UC ---- Umaria Marine Beds (10 ft)

TALCHIR FORMATION (U. Carboniferous)

A part of the Lower Gondwana succession.

12. **Intertrappean Beds.** The volcanic eruptions of the Deccan Traps were not continuous but intermittent. Each successive lava flow was followed by long intervals of quiescence extending for several thousand years. The sedimentary rocks that were deposited on the irregularities of the surface of the traps during these intervals are known as Inter-Trappean Beds. These are lacustrine-fluviatile in origin and comprise of black, cherty rock resembling lyddite, stratified volcanic detritus, impure limestones and clays. The Inter-Trappean Beds contain rich faunal and floral assemblages which are very valuable in determining the geological history of the vast intervals of quiescence between successive lava flows. The important fossils contained in these Beds include : Palm (Palmoxylon), Chara (fresh-water

alga), Azolla (fern); and invertebrates such as Physa (Bullinus princepi, Lymnara, Unio, Melania, Natica, Pupa, etc. The vertebrate animals like fishes, frogs, tortoises, etc.

13. **Infratrappean Beds.** These are the beds occurring below the Deccan Traps in the Godavari Basin of Andhra Pradesh. These beds which are of lacustrine-fluviatile origin are composed of marine fossils such as nautilus, lamellibranchs and gastropods suggesting a Danian (Uppermost Cretaceous) age for them.

## TECTONIC ELEMENTS OF INDIA

1. **Great Boundary Fault.** It is a major fault of 500 miles length and 5000 feet throw which has brought the sandstones of the Bhander Series (Upper Vindhyans) in contact with the highly folded and foliated schists of Aravallis. It is roughly parallel with the course of River Chambal.

2. **Main Boundary Fault.** It comprises of a series of more or less parallel faults exhibiting similar tectonic and stratigraphic peculiarities and separates the Siwaliks from the earlier Tertiary and older rocks.

SIWALIKS MBT

F – – – – – – – – – – – – – – – – – – – – F

MURREE SERIES

3. **Syntaxial Bend.** At its western extremity, the Himalayan Orogenic Belt takes a sharp arcuate hair-pin bend in its normal orogenic trend of SE-NW and geological formations on both the arms of this bend exhibits a NW-SE strike. This bend, which is commonly referred as the Syntaxial Bend, depicts overthrusting of rocks directed from both sides towards the central axis. The formation of this syntaxial bed is believed to be the result of the rotational movement of the Indian plate during the Oligocene times.

4. **Narmada-Son Lineament.** It is generally a zone of crustal weakness (rift valley) that has often been activated in the geological past. It is reflected in a gravity anomaly map as a narrow belt of crustal subsistence along the Narmada Valley covered by 300 m of sediments in the east and 30-35 m in the centre. The rift is offset by a number of transverse tear faults. There is a line of carbonatite bodies and alkali syenites parallel to the rift valley. The faults which border the rift served as channels for the outpouring of Deccan lavas.

Origin - According to West and Auden, the lineament is a line of fundamental tectonic feature connected with some primary weakness of the Archaean crust, which has influenced the deposition and deformation of the Vindhyan and Gondwana rocks. The available geophysical data (Bouger, seismic and paleao-magnetic anomalies) and evidences from Plate Tectonics suggests that the Narmada-Son Lineament has originated due to Wrench Fault Tectonism with associated Warping and subduction.

5. **Owen Fracture Zone.** It lies in the western margin of the Indian Ocean and terminates against the sediment-filled Makran Subduction Zone. It has right-laterally offset the Carlsberg Ridge and continues southward, southwestward demarcating the faulted, straight coast of Malagasy Island. It is believed that the Chaman Fault System is the landward northerly expression of the Owen Fracture Zone.

6. **90° E Ridge.** This is a 4800 km long fracture or leaking transform fault running parallel to the island-arc chain of the Andaman and Nicobar. It extends northwards and joins with the Indo-Burman Border fault. Its northern end is buried under the flysch of the Ganga Basin. The northern part of the 90° E Ridge coincides

with the Sumatra-Java-Andaman Subduction Zone which is filled by sediments ot the Bengal Fan.

7. **Lakshadweep-Chagos Zone.** This transform fault occurs in the middle of the Indian Ocean and has offset the SE Indian Ridge from the Carlsberg Ridge. It separates the Chagos Islands from the Mauritian Plateau. It splits the Indian Ocean Plate into two major segments.

8. **Ganga Basin.** The Ganga Basin occupies 2,70,000 sq. km area in the North India. In the west, the Ganga Basin is defined by the Delhi Metasediments which form a subsurface ridge called the Delhi-Hardwar Ridge which is covered by alluvium. In the east the Ganga Basin is limited by another ridge of Satpura Metamorphics called the Monghyr-Saharsa Ridge. In the north it is limited by the Siwaliks which mark the Main Boundary Fault. In the southern region, it is demarcated by the Bundelkhand Gneissic Granites and Precambrian Sediments. The Ganga Basin is filled with alluvium of great thickness ranging in age from Pleistocene to Recent. Edward Suess recognised the Ganga Basin as a Fore Deep with reference to the Himalayas but Burrard felt that it is a Rift. Tectonically, the Ganga Basin represents a large-scale depression of Vindhyan Syneclise which shows a northward continuation of Vindhyan sedimentation.

**Summary of the Tectonic Elements of India**

| Tectonic Element | Examples |
|---|---|
| I. Intracratonic Basin or Parageosyncline | The Cuddapah Basin |
| (a) Yoked Basin or Zeugogeosyncline | Nallamallai Sub-basin<br>Srisailam Sub-basin |
| (b) Isolated Basin or Autogeosyncline | Papaghani Sub-basin<br>Kaladgi Basin |
| (c) Transverse Basin or Exogeosyncline | Siwaliks |
| (d) Taphrogeosyncline | Gondwana Basins |
| II. Grabens or Rift Valleys | Godavari Graben<br>Narmada Valley, Son Valley<br>Sabarmati Graben |
| III. Half-Grabens (taphrogenic basins) | Kurnool Sub-basin<br>Palnad Sub-basin |
| IV. Back-Arc Basin | Andaman & Nicobar Basin |
| V. Transform Faults | Owen Fracture Zone<br>90° Ridge<br>Chaman Fault System<br>Lakshadweep-Chagos Zone |
| VI. Mid-Oceanic Ridges | Carlsberg Ridge<br>Laccadive-Maldives Ridge |
| VII. Triple Junction | Godavari, Narmada and Son Rift Valleys form a Y-shaped Triple Junction. The area of this junction is now hidden under the Deccan Traps. |
| VIII. Aulacogen | Sabarmati-Kambhat graben |
| IX. Pericratonic Basin or Intracratonic rift basin | Krishna-Godavari Basin<br>Cauvery Basin |
| X. Horst or Block Mountains | Vindhya Mountains |

## Section II : Objective Questions

# 1

# *General Geology and Geotectonics*

1. The time taken for light from the sun to reach the earth is :
   a) 499.720 secs
   b) 499.012 secs
   c) 489.720 secs
   d) 489.012 secs

2. The most powerful radio object in the universe after the sun is/are :
   a) Black Holes
   b) Quasars
   c) Jupiter
   d) Venus

*3. Which of the following planets show the same "size-bearing" ?
   a) Earth-Mars
   b) Earth-Venus
   c) Mars-Venus
   d) Venus-Mercury

4. Which of the following possess both stellar and planetary characteristics ?
   a) Jupiter
   b) Venus
   c) Saturn
   d) Pluto

5. The Asteroid Belt-area between Mars and Jupiter-extends from :
   a) 2.2 to 3.6 A.U.

b) 2.2 to 4.8 A.U.

c) 3.6 to 4.8 A.U.

d) 4.8 to 6.4 A.U.

6. Olympus Mons, the largest known volcano in the Solar System is found on the surface of :

a) Earth

b) Moon

c) Mars

d) Jupiter

7. Match the following and choose the correct answer :

| *Theory* | *Propounder* |
|---|---|
| A. Big-Bang Theory | 1. Dr. Allan Sandage |
| B. Pulsating Universe Theory | 2. Abbe Georges Lemaitre |
| C. Steady State Theory | 3. Buffen |
| D Dynamic Encounter Theory | 4. Thomas Gold and Herman Bondi |

a) A—1 B—2 C—3 D—4

b) A—1 B—2 C—4 D—3

c) A—2 B—1 C—4 D—3

d) A—2 B—1 C--3 D—4

*8. The "Nebular Hypothesis" was proposed by :

a) Kant

b) Laplace

c) Kant and Laplace together

d) Kant and Laplace independently

9. Indentify the INCORRECT statement :

a) Nebular Hypothesis does not explain the principle of conservation of momentum

b) Fall of meteorites is best explained by the planetesimal hypothesis

c) Coplanar placement of planets is explained by the Tidal hypothesis

d) Schimdt's meteorite hypothesis does not explain the rotation of the planets and their distances from the sun

10. The spacing of planets is best explained by :

a) Dynamic Encounter theory

b) Planetesimal hypothesis

c) Nebular hypothesis

d) Tidal hypothesis

11. Which of the following is not 'Biparental' ?
   a) Tidal Hypothesis
   b) Planetesimal hypothesis
   c) Dynamic encounter theory
   d) Protoplanet hypothesis
12. Which of the following statements is NOT true of Schimdt's hypothesis of the evolution of Universe ?
   a) It correctly explained the rotation of planets and their distances from the sun
   b) It correctly explained the distribution of angular momentum between the planets and the sun
   c) It assumed that the earth was hot at the time of its formation
   d) None of the above
13. Which of the following planets has the least density ?
   a) Earth
   b) Mars
   c) Venus
   d) Saturn
   e) Pluto
14. The difference between the equatorial and polar diameters of the Earth is :
   a) 21 km
   b) 43 km
   c) 56 km
   d) 28 km
15. The Mohorovicic discontinuity is a :
   a) Compositional boundary
   b) Phase-change boundary
   c) Compositional as well as phase change boundary
   d) None of the above
16. Beneath the oceans, the MOHO lies at a depth of :
   a) 7–9 km
   b) 10–12 km
   c) 14–16 km
   d) 20–25 km
17. The temperature of the MOHO beneath the continents range from :
   a) 150°C–200°C
   b) 200°C–500°C
   c) 500°C–700°C
   d) 700°C–900°C

18. The earth's crust is thickest under :
    a) Shield areas
    b) Platforms
    c) Phanerozoic orogenic belts
    d) Archaen greenstone belts

19. The Inglemen discontinuity lies at a depth of : :
    a) 2900 km
    b) 3160 km
    c) 4980 km
    d) 5200 km

20. By area, which of the following occupies the largest % of the crust ?
    a) Continental crust
    b) Transitional crust
    c) Oceanic crust
    d) Oceanic and Transitional crust

21. In terms of seismicity, the oceanic crust is composed of three layers in which there is :
    a) A P-wave velocity overlap between layer 1 and layer 2
    b) A P-wave velocity overlap between layer 2 and layer 3
    c) A P-wave velocity overlap between layers 1, 2 and 3
    d) No P-wave velocity overlap between any of the three layers

22. The most important and drastic composition boundary within the Earth is :
    a) Conrad Discontinuity
    b) Moho-discontinuity
    c) Gutenberg Discontinuity
    d) Ingleman Discontinuity

23. Which of the following is a Second-order discontinuity ?
    a) Crust-Mantle discontinuity
    b) Mantle-Core discontinuity
    c) Upper-Mantle Lower Mantle discontinuity
    d) None of the above

24. Mohorovicic Discontinuity is marked by :
    a) Abrupt increase in the seismic wave velocities
    b) Abrupt decrease in the seismic wave velocities
    c) Gradual increase in the seismic wave velocities
    d) Gradual decrease in the seismic wave velocities

25. An important boundary occurs at a depth of 100 km which

marks a change in rheological properties rather than in composition. This is the boundary between :

a) Lower Crust-Upper Mantle
b) Upper Mantle-Lower Mantle
c) Lithosphere-Asthenosphere
d) Asthenosphere-Transition zone

26. When P waves are passed through the crust downwards, a sudden dip in P wave velocity is noticed at a depth of 100 km followed by an increase starting at about 200 km. This represents the presence of :

a) Mohorovicic discontinuity
b) Inversion surface
c) Low Velocity Zone
d) Transition Zone

27. The Low Velocity Zone apart from reduced seismic velocities is also characterized by :

a) Low heat flow
b) Low heat flow and high electrical conductivity
c) High heat flow and high electrical conductivity
d) High heat flow and low electrical conductivity

*28. The Low Velocity Zone (LVZ) is NOT poorly developed in the regions of :

a) Shield areas
b) Platforms
c) Marine Trenches
d) Mid-oceanic Ridges

29. The 'Mantle Transition Zone' at about 370—720 km depth is :

a) Overlain by the U. Mantle and underlain by the L. Mantle
b) Overlain by the L. Mantle and underlain by the Outer Core
c) Found between the Moho and the LVZ
d) Found between the Moho and the Gutenberg Layer

30. The seismic velocities attain their maximum value for the Earth in :

a) Upper Mantle
b) Lower Mantle
c) Outer Core
d) Inner Core

31. The linear magnetic anomalies occurring as parallel bands along many MORs are most probably caused by the rocks of :

a) Layer 1 of the oceanic crust

b) Layer 2 of the oceanic crust

c) Basaltic layer of the continental crust

d) Ultramafic rocks of the Mantle

32. The Mantle material does not contribute to the magnetic anomalies because :

a) It does not contain abundant metallic ions to induce a magnetic field

b) It contains only ultramafic rocks which are essentially non magnetic

c) It exhibits a more or less constant magnetic field

d) The temperatures are above the curie point of natural magnetic materials

33. The radial extent of the Core is :

a) 2886 km

b) 3486 km

c) 1227 km

d) 4113 km

34. The most abundant light element present in the core is :

a) either S or O

b) either S or Al

c) Potassium

d) Sodium

35. The composition of the Earth can be compared to that of some meteorites. It has been pointed out that the composition of the mantle is similar to——————and the composition of the core is similar to——————.

a) Chondrites/achondrites

b) Siderites/Chondrites

c) Chondrites/tektites

d) Chondrites/siderites

36. The topographic extremes of height and depth are respectively ———————at the Mount Everest and————————at the Challenger Deep :

a) 29, 140 m/36, 204 m

b) 11055 m/8840 m

c) 8840 m/11055 m

d) 6230 m/8430 m

*37. The most accurately known physical parameter in the deep Earth is :

a) Density

b) Elastic moduli

c) Seismic wave velocity

d) Gravity

*38. "Denser rocks generally have a higher velocity of seismic waves". Based on the above statement it can be predicted that :

a) Seismic velocities depend on the density of rocks and increases with an increase in density

b) Seismic velocities depend on depth, and as high density rocks are found at greater depths, the seismic waves have higher velocities

c) Seismic velocities depend on the value of elastic constants and increases with an increase in their value

d) None of the above

*39. The value of Rigidity modulus in the core is :

a) $1.0 \times 10^{12}$ dynes/cm$^2$

b) $0.5 \times 10^{12}$ dynes/cm$^2$

c) $15 \times 10^{12}$ dynes/cm$^2$

d) none of the above

40. Which of the following environments are NOT characterized by extensional stress regime ?

a) Oceanic ridges

b) Marginal sea basins

c) Continental rifts

d) Collision zones

41. Who is considered as the "Father of Volcanic geology" ?

a) Robert Dietz

b) Nicolas Desmarest

c) Vening Meinesz

d) James Hall

42. The First Geologist to distinguish and interpret different seismic waves was :

a) R.D. Oldham

b) Rayleigh

c) C.F. Ritcher

d) Robert Dietz

43. The oldest of all the scales devised to measure the intensity of seismic waves is :

a) Ritcher scale

b) Rossi-Forel Scale

c) Mercalli Scale

d) Beufort Scale

44. The layers of rock having intermediate P wave velocities between 7.2—7.7 km/sec are characteristic of :

a) Shield Areas

b) Ocean Basins

c) Precambrian Orogenic terrains

d) Continental Margins

*45. Which of the following exhibits maximum half-life period ?

a) $U^{238}$—$Pb^{206}$

b) $U^{235}$—$Pb^{207}$

c) $Rb^{87}$—$Sr^{87}$

d) $Sm^{147}$—$Nd^{143}$

46. Which of the following is used to distinguish the Connate water and underground water :

a) $C^{14}$

b) $K^{40}$—$Ar^{40}$

c) Tritium

d) Radium

47. Which one of the following is a non radioactive isotope of lead ?

a) $Pb^{204}$

b) $Pb^{206}$

c) $Pb^{207}$

d) $Pb^{208}$

48. Which of the following are most radioactive ?

a) Younger acidic rocks

b) Younger basic rocks

c) Older acidic rocks

d) Older basic rocks

49. Earthquakes whose depth of focus ranges between 300—700 km are known as :

a) Shallow focus earthquakes

b) Intermediate focus earthquakes

c) Deep focus earthquakes

d) Normal earthquakes

50. Over three-quarters of the energy released by the earthquakes is derived from :

a) Shallow-focus events

b) Intermediate-focus events

c) Deep-focus events

d) All of the above

51. Most of the earthquakes of the shallow focus range are caused due to :

o) Normal faulting

b) Gravity faulting

c) Reverse faulting

d) Thrust faulting

52. "Swarms" are minor earthquakes which are mostly of :

a) Shallow focus

b) Intermediate focus

c) Deep focus

d) A combination of the above

53. Recently thousands of small tremors or "Swarms" were felt in :

a) Anchan village (Gujrat)

b) Tambaran (T.N.)

c) Lahaul and Spiti area

d) NE Himalayan ranges

54 Which one of the following is a modified seismic scale having twelve divisions ?

a) Ritcher Scale

b) Rossi-Forel Scale

c) Mercalli Scale

d) Beufort Scale

55. A ring-shaped zone situated opposite the earthquake zone where seismic waves are deflected and not registered lies between the arc :

a) 103°–133°

b) 103°–143°

c) 123°–143°

d) 103°–123°

56. Low frequency, long wavelength seismic waves whose particle motion lies in a horizontal direction perpendicular to the direction of wave propagation are described as :

a) Push waves

b) Shear waves

c) Rayleigh waves

d) Love waves

57. The disaster sequence of the seismic waves is :

a) P>S>L

b) L>P>S

c) P>L>S

d) P<S<L

58. "At a certain level in the Earth's crust all columns of material having unit-cross sectional area must have the same mass". This statement refers to :

a) Base level of erosion

b) Principle of isostasy

c) Law of Conservation of Mass

d) Dirac's Principle

*59. Which of the following statements is NOT derived from Pratt's hypothesis of isostasy :

a) Compensation takes place by lateral density variations of the crust

b) The thickness of the crust varies by the top and the bottom

c) Lithosphere is synonymous with the crust

d) Mountains are underlain by anomalously low density materials

60. Pratt's hypothesis of isostasy does not take into account

i) Thickness of the crust

ii) Thickness of the substratum

iii) Density of the crust

iv) Density of the substratum

a) (i) only

b) (ii) only

c) (i) and (ii) only

d) (ii) and (iv)

61. According to Airy's Theory of Isostasy :

a) The density of the crust and the substratum is uniformly same throughout

b) The thickness of the crust and the substratum is uniformly same throughout

c) The crust and the substratum each have uniform but different densities

d) The crust has a constant thickness at the top whereas the substratum has a constant thickness at the bottom

62. According to the doctrine of Isostasy. a region of lower density shows :

1) Negative anomaly

2) Mass deficiency

3) Mass excess

4) Positive anomaly

a) 1 only
b) 2 only
c) 1 and 3 only
d) 2 and 4 only

63. Choose the correct answer from the choices given below .
I. Seismic refraction studies give the crustal structure and thickness but cannot give information on isostatic equilibrium
II. Seismic refraction studies give information about crustal structure, thickness as well as isostatic equilibrium
III. Gravity anomaly studies give information about isostatic equilibrium but it cannot give the geometry of the rock bodies
a) Only I is correct
b) Only II is correct
c) I and III are correct
d) II and III are correct

*64. Seismic crustal studies have shown that the oceans and continents are in broad isostatic equilibrium with each other mainly due to :
a) Lateral variation in the mean density of the crust
b) Lateral variation of density within the mantle
c) Variation in the thickness of the low density crust
d) A combination of the above three factors

65. The oldest areas on the surface of the earth are definitely :
a) Positive areas
b) Negative areas
c) Neutral areas
d) None of the above

66. The average Bouguer anomaly for the ocean as a whole is :
a) Positive
b) Negative
c) Zero
d) Unpredictable

67. Which of the following exhibits a negative gravity anomaly ?
a) Hawaiin Islands
b) Mid-Atlantic Ridge
c) Ryukyu & Kyushyu Arcs
d) Mariana's Trench

68. The part of the Ocean between 0—200 m depth is described as :

a) Continental shelf
b) Continental slope
c) Continental rise
d) Oceanic trough

69. In India, the broadest continental shelf is found in :
a) Gujrat
b) Maharashtra
c) Tamil Nadu
d) A.P.

70. The flattest portion of the Earth's surface is formed by :
a) Abyssal Hills
b) Abyssal Plains
c) Oceanic Rises
d) Oceanic Trenches
e) Sea Mounts

71. Tbe highest pressures in the earth's crust is encountered in :
a) Mid-Oceanic ridges
b) Oceanic rises
c) Oceanic trenches
d) Island arcs

72. Match the following and choose the correct answer from the choices given below :

| | |
|---|---|
| A. Mindanao Deep | 1. Indian Ocean |
| B. Planet Deep | 2. Atlantic Ocean |
| C. Milwaukee Deep | 3. Pacific Ocean |
| D. Nares Deep | |

a) A—3 B—2 C—2 D—1
b) A—2 B—2 C—3 D—1
c) A—3 B—1 C—2 D—2
d) A—2 B—1 C—3 D—3

73. Isolated fragments of continental crust occuring as shallow rises within the ocean basins are known as :
a) Guyots
b) Sea Mounts
c) Atolls
d) Microcontinents
e) Marginal Basins

74. Which of the following is NOT matched correctly ?
a) Sri Lanka———————Continental islands

b) West Indies——————Volcanic islands
c) Laccadive and Minicoy—————Coral islands
d) British Isles——————Diastrophic islands

75. The Mid-Atlantic Ridge System is an area of :
a) Shear Plate Boundary
b) Consuming Plate boundary
c) Accreting plate boundary
d) Stable plate boundary

76. The characteristic feature of the Mid-Atlantic Ridge system is its :
a) Topographic asymmetry
b) Topographic and thermal asymmetry
c) Thermal and magnetic asymmetry
d) Topographic, thermal and magnetic asymmetry

*77. Which of the following mid-oceanic ridges contain large regions of granitic layer ?
a) Carlsberg Ridge
b) Mid-Atlantic Ridge
c) Lomonosov Ridge
d) Eact Pacific Rise

78. The largest and the only outcrop of an oceanic ridge is :
a) Greenland
b) Iceland
c) Ryukyu Islands
d) Galapagos

79. The mid-oceanic ridges are characterized by :
a) High heat flow, deep-focus earthquakes and positive gravity anomalies
b) High heat flow, shallow-focus earthquakes and negative gravity anomalies
c) High heat flow, shallow-focus earthquakes and positive gravity anomalies
d) Low heat flow, deep-focus earthquakes and negative gravity anomalies

80. 'Dolphin Ridge' and 'Challenger Ridge' are parts of :
a) East Pacific Rise
b) Mid-Atlantic Ridge
c) Carlsberg Ridge
d) Lomonosov Ridge
e) Central Indian Ridge

*81. According to the Theory of Continental Drift, about 250 m. a. there existed a supercontinent, the Pangea, which contained all our present continents.

Which of the following statements about the Pangea is Incorrect ?

a) It consisted of North America and Eurasia in the North

b) It covered an area of about 150 million $Km^2$

c) Two thirds of its landmass was concentrated in the north

d) It began to split in the Mesozoic

82. One of the strongest evidences favouring the Continental Drift Theory was the distribution of a fresh water reptile Mesosaurus. This reptile was characteristic of :

a) Permian

b) Triassic

c) Jurassic

d) Cretaceous

83. The marsupials which are characteristic of Australia have clearly evolved in isolation since the early Tertiary. Before this period they were also found in the continent of :

a) North America

b) South America

c) Eurasia

d) Africa

84. Laurasia and Gondwanaland collided in the———to form a single continent Pangea :

a) Hercynian Orogeny

b) Caledonian Orogeny

c) Acadian Orogeny

d) Antler Orogeny

85. 'Isthamanian Links' represent :

a) Hereditary links established between two similar forms on different continents

b) Hereditary links established between two morphologically dissimilar forms found on the same continent

c) Land bridges helping unrestricted migrations of animals

d) Narrow strips of the Gondwanaland

e) Narrow strips of the Angaraland

86. The Paleomagnetic dip (I) is related to the latitude (L) of the locality as :

a) $\sin I = 2 \sin L$

b) $\tan I = \tan L$

c) $\tan I = 2 \tan L$

d) cos I=2 cos L

87 Curie Temperature is described as :

a) The critical temperature above which minerals lose their radioactive properties

b) The critical temperature at which radioactive minerals lose half their life

c) The critical temperature above which magnetic minerals lose their magnetism

d) The critical temperature at which molten materials become solidified

88. The supercontinent Pangea was supposed to have covered an area of :

a) $15\times10^7$ km$^2$

b) $1.5\times10^7$ km$^2$

c) $15\times10^6$ km$^2$

d) $1.5\times16^6$ km$^2$

89. The Tethys sea was located between :

a) North America and South America

b) North America and Eurasia

c) Eurasia and Africa

d) Antartica and Australia

90. The Northern rift developed in the Pangea in the early Mesozoic times opened up the :

a) Atlantic Ocean

b) Pacific Ocean

c) Indian Ocean

d) Arctic Ocean

91. According to Alfred Wegener, the movement of the continents was effected by :

a) Tidal forces of the sun and the moon

b) Pohlflucht forces

c) Both tidal forces and pohlflucht forces

d) Epeirogenetic oscillations of the crust during the sinking and formation of geosynclines

92. In the Gondwanaland, India was NOT adjacent to :

a) Africa

b) Australia

c) Antartica

d) South America

93. The Curie Temperature for iron is :

a) 350°C
b) 660°C
c) 760°C
d) 1400°C

94. Recent oceanographic studies have revealed that about 50% of the present sea floor (about 1/3rd of the earth's surface area) has been created during————————of geologic time :
a) 25%
b) 15%
c) 10%
d) 3%
e) 1.5%

95. The oceanic crust is not older than :
a) Permian
b) Triassic
c) Jurassic
d) Cretaceous

96. Which of the following has the youngest oceanic crust ?
a) Pacific ocean
b) Atlantic ocean
c) Arctic ocean
d) Indian ocean

*97. Which of the following oceans exhibits a more or less constant spreading of its floor ?
a) Indian Ocean
b) Pacific Ocean
c) South Atlantic Ocean
d) North Atlantic Ocean

*98. The fastest spreading of the sea-floor is exhibited by :
a) South Atlantic Ridge
b) North Atlantic Ridge
c) Central Indian Rigde
d) East Pacific Rise

99. Rocks formed at the ridge crests which are younger than 0.69 M.Y. are :
a) Normally magnetized
b) Reversely magnetized
c) Either normal or reversely magnetized
d) The information is insufficient to predict the nature of magnetization involved :

100. The oldest of the magnetic epochs in the Geo-Magnetic Time scale is :

a) Gauss reversed
b) Gilbert reversed
c) Gauss Normal
d) Gilbert Normal

*101. The Geologic Epoch in which we are now living can be depicted as :
a) Holocene/Normal epoch
b) Pliestocene/Normal epoch
c) Holocene/Reversed epoch
d) Pliestocene/Reversed epoch

102. In the Plate Tectonics concept the 'Plates' are made up of:
a) Continental Crust only
b) Oceanic crust only
c) Both continental crust and oceanic crust
d) Continental crust, oceanic crust and outer parts of the U. Mantle

103. The African Plate is composed of :
a) Continental Crust
b) Oceanic Crust
c) Combination of both continental as well as oceanic crust
d) None of the above

*104. Plates are essentially———————regions of the Earth :
a) Inert/aseismic
b) Active/mobile
c) Oldest/tectonic
d) Recent/active

105. Which of the following is/are characteristic of a subduction zone ?
I. Melting of old plates
II. Mountain building and upliftment
III. Generation of Earthquakes and volcanoes
a) I only
b) II only
c) I and III only
d) I, II and III

106. Sections along which plate material is neither created nor destroyed are described as :
a) Neutral Plate boundaries
b) Conservative Plate Boundaries
c) Parallel Plate boundaries

d) Transform Fault boundaries

e) All of the above

107. The most extensive modern continental volcanic mountain chain is :

a) Andes mountains

b) Himalayas

c) Alps

d) Zagros

108. A pattern of deep trench and island arc develops as a result of :

a) Ocean-continent collision

b) Ocean-ocean collision

c) Continent-continent collision

d) None of the above

*109. The movement of the Plates is caused by :

a) The convection currents in the solid mantle

b) A lateral gravitational pressure on the plates on either side by the ridge-push mechanism

c) The gravitational effect of cooler denser material in and around the sinking slab

d) The mantle drag, whereby lateral convective flow within the asthenosphere pulls the plate along

e) A variety of mechanisms causing plate motion

110. Which of the following is the correct sequence of the birth, growth and termination of a continental margin ?

a) Continental splitting ---→ birth of continental margin ---→ passive margin ---→ active margin ---→ collision mountain range

b) Birth of continental margin ---→ continental splitting ---→ passive margin ---→ active margin ---→ collision mountain range

c) Birth of continental margin ---→ continental splitting ---→ active margin ---→ passive margin ---→ collision mountain range

d) Continental splitting ---→ birth of continental margin ---→ active margin ---→ passive margin —— collision mountain range

*111. Passive margins are characterized by :

i) Aseismicity

ii) Lack of any considerable degree of subsidence

iii) Thick sedimentary sequences

iv) Their location bordering young and expanding oceans

a) (i) and (ii) are correct
b) (i), (ii) and (iv) are correct
c) (i), (iii) and (iv) are correct
d) (ii), (iii) and (iv) are correct

112. Which of the following is/are NOT the main point (s) of difference between Active Margins and Island Arc Systems ?

i) A deep trench occurs on the oceanward side of the Island Arc System only
ii) The convergent plate margin at Island Arc system separates two portions of the oceanic lithosphere
iii) An Island Arc System is separated from the mainland by a marginal sea whereas an active margin has a direct contact with the mainland
iv) Island Arcs are associated with andesite volcanism whereas Active Margins are associated with basaltic volcanism only

a) (i) only
b) (i) and (ii) only
c) (ii) and (iii) only
d) (i) and (iv) only

113. A typical rifted margin is morphologically subdivided into :
a) Continental shelf and continental slope
b) Continental slope and continental rise
c) Continental shelf, continental slope and continental rise
d) None of the above

114. A typical Active Margin has ;
a) A trench and island arc along its oceanic side
b) A trench and an island arc along its landward side
c) A trench along its oceanic side and an island arc along its landward side
d) An island arc on its oceanic side and a trench on the landward side

115. Island Arcs are characterized by :

I. Deep Focus earthquakes

II. Strong positive gravity anomalies

III. Fissure type of volcanic activity

IV. Crust of Transitional type

a) I, II, III and IV
b) I, III and IV
c) I, II and IV
d) II, III, and IV

116. The most characteristic rocks of the Island-Arc Systems are :

a) Andesites

b) Granodiorites

c) Blue Schists

d) Basalts

*117. Which of the following are considered to be the "Geologic Analogues" of the island-arc systems ?

a) Cores of Tertiary mountain belts

b) Precambrian Greenstone belts

c) Proterozoic sandstones and limestones

d) Lower Triassic red beds

e) Jurassic Lava flows

118. When a double arc occurs :

a) Both the inner and the outer arcs are volcanic

b) Both the inner and the outer arcs are non-volcanic

c) Only the outer one is volcanic

d) Only the inner one is volcanic

119. Which one the following denote the most probable main stages in the evolution of an early Precambrian volcanic-sedimentary assemblage ?

a) Basalt ---→ Dacite ---→ Rhyolite ---→ Andesite ---→ Sedimentary rocks ---→ Granitic intrusions

b) Basalt ---→ Andesite ---→ Rhyolite ---→ Dacite ---→ Granitic intrusion ---→ sedimentary rocks

c) Rhyolite ---→ Dacite ---→ Andesite ---→ Basalt ---→ Granitic intrusion ---→ sedimentary rocks

d) Basalt ---→ Andesite ---→ Dacite ---→ Rhyolite---→ Sedimentary rocks ---→ Granitic Intrusions

*120. The Afar Region of Ethopia is the best example of :

a) Aulacogen

b) Transform faulting

c) Triple Junction

d) Calc-alkaline volcanism

121. The only combination of a triple junction that is always stable is :

a) Ridge-Ridge-Ridge

b) Trench-Trench-Trench

c) Ridge-Ridge-Trench

d) Ridge-Trench-Fault

*122. Transform Faults boundaries are characterized by :

a) Shallow focus earthquakes of depths less than 15 km
b) Shallow focus earthquakes of depths less than 100 km
c) Deep focus earthquakes
d) Absence of earthquakes

123. Transform faults which retain a constant length as a function of time are known as :
a) Ridge-Ridge
b) Ridge-Trench
c) Trench-Trench
d) None of the above

124. The average thickness of the plates is about :
a) 70 km
b) 100 km
c) 120 km
d) 150 km

*125. Which of the following is NOT a closing ocean basin ?
a) Black sea
b) Red sea
c) Caspian sea
d) Mediterranean sea

*126. Choose the pair which is not correct :
a) J. Tuzo Wilson———Transform fault concept
b) F.J. Morgan ———Plate Tectonics concept
c) Dana ———Geosynclinal concept
d) Dutton ———Concept of Isostasy

127. In a double arc, the folded and thrust sediments are found in :
a) Outer arc only
b) Inner arc only
c) Both outer and inner arcs
d) None of the above

128. The Plate Boundary at the San Andreas Fault is an example of :
a) Constructive Plate Boundary
b) Consuming Plate Boundary
c) Convergent Plate Boundary
d) Conservative Plate Boundary

129. According to the Plate Tectonics theory

I. The Atlantic Ocean will disappear in about 50 M.Y

II. American continents may collide with Asia in about 200 M.Y.

III. African Rift will give birth to a new ocean

IV. India and Asia are converging on each other

a) I and II are correct

b) I and III are correct

c) I, III and IV are correct

d) II, III, and IV are correct

130. Some of the features associated with the Indian Ocean are given below. Choose the correct combinations by matching the two lists give below

| | |
|---|---|
| (i) Seychelles Bank | (a) Transform Fault |
| (ii) Owen Fracture Zone | (b) Seismic Ridge |
| (iii) Sunda and Bunda | (c) Aseismic Ridge |
| (iv) Maldive and Cocos | (d) Coral Islands |
| (v) 90° E Ridge | (e) Microcontinent |
| | (f) Island Arcs |
| | (g) Oceanic Trench |

a) (i)—(a) (ii)—(g) (iii)—(d) (iv)—(f) (v)—c)

b) (i)—(c) (ii)—(a) (iii)—(f) (iv)—(d) (v)—b)

c) (i)—(e) (ii)—(g) (iii)—(d) (iv)—(f) (v)—b)

d) (i)—(e) (ii)—(a) (iii)—(f) (iv)—(d) (v)—c)

131. Which of the following is a correct sequence ?

a) Sedimentation —→ Geosynclinal subsidence —→ Lateral compression —→ upliftment by orogeny

b) Geosynclinal subsidence —→ Sedimentation —→ Lateral compression —→ Orogenic upliftment

c) Geosynclinal subsidence —→ Lateral compression —→ Sedimentation —→ Orogenic upliftment

d) Lateral compression —→ Sedimentation —→ Geosynclinal subsidence —→ Orogenic upliftment

132. The rate of accumulation of sediments within a geosyncline is approximately———per million years

a) 10 m

b) 12 m

c) 13 m

d) 17 m

133. Secondary geosynclines formed within the cratons and not on active mobile belts are described as

a) Orthogeosynclines

b) Parageosynclines
c) Eugeosynclines
d) Miogeosynclines

134. The Extra-cratonic linear geosynclines characterized by abundant carbonate rocks and the absence of volcanic successions are known as :
a) Eugeosynclines
b) Miogeosynclines
c) Autogeosynclines
d) Exogeosynclines

135. A parageosyncline without marginal uplifts is described as
a) Yoked Basin
b) Transverse Basin
c) Isolated Basin
d) Taphrogeosyncline

136. Deltas are an example of
a) Exogeosynclines
b) Autogeosynclines
c) Zeugogeosynclines
d) Taphrogeosynclines

*137. Elongated zones of thick sediment sequences in which the rate of deposition is more than 13 m/MY but whose basements are not downwarped into synclines are called
a) Starved Basins
b) Geoclines
c) Pliomagmatic geosynclines
d) Taphrogeosynclines

138. Miogeosynclines differ from Eugeosynclines in the absence of
(i) Greywacke-shale association
(ii) Volcanic rocks
(iii) Nappes
(iv) Metamorphism and granitic intrusions
a) (i) and (iii) only
b) (i, (ii) and (iii) only
c) (ii, (iii) and (iv) only
d) All of the above

139. Sediments produced by the erosion of mountain ranges subsequent to an orogenic episode are described as
a) Flysch

b) Molasse
c) Preflysch
d) Geoclines

140. Black shales are major constituents of
a) Flysch sequences
b) Molasse sequences
c) Pre-Flysch sequences
d) Volcanic sequences

141. Rapid subsidence and slow deposition is a characteristic feature of
a) Eugeosynclines
b) Miogeosynclines
c) Shield areas
d) Starved basins
e) Cuspate Delta

142. The East African Rift system is an example of
a) Aulacogen
b) Triple Junction
c) Transform fault
d) Accreting Plate boundary

143. The 'Continental Terrace' includes
a) Continental shelf
b) Continental slope
c) Continental shelf and continental slope
d) Continental shelf, slope and rise

144. When an orogenic belt shows a sudden change in its trend as a result of impressed strain, it is described as a/an
a) Sphenochasm
b) Orocline
c) Plume
d) Suture line

145. The paired metamorphic belts associated with subduction complexes are composed of
a) Amphibolites/granulites
b) Amphibolites/blueschists
c) Blueschists/greenschists
d) Granulites/eclogites

146. Dietz has designated ---- as miogeocline and described it as an analogue of miogeosyncline ; and ---- a eugeocline and described it as an analogue of eugeosyncline :

a) Continental shelf/Continental slope
b) Continental slope/Continental shelf
c) Oceanic trench/Oceanic rise
d) Oceanic rise/Oceanic trench

147. Match the following and choose the correct answer :

| | |
|---|---|
| A. Papaghani sub-basin | (1) Autogeosyncline |
| B. Nallamalai sub-basin | (2) Taphrogenic basin |
| C. Palnad sub-basin | (3) Zeugogeosyncline |
| D. Kaladgi basin | (4) Isolated basin |

a) A—1 B—4 C—2 D—3
b) A—1 B—2 C—4 D—3
c) A—4 B—3 C—1 D—2
d) A—4 B—3 C—2 D—1

148. The most typical rock of 'Molasse' sequences is :
a) Orthoquartzite
b) Greywacke
c) Arkose
d) Lithic arenite

149. The molasse sequences reflect :
a) Reduced environments
b) Agitated marine enviroments
c) Oxygenated environments
d) Euxinic environments

150. The "Flysch" formations are :
a) Pretectonic
b) Syntectonic
c) Late tectonic
d) Post tectonic

151. The age of the Flysch sediments is :
a) 40—20 M.Y.
b) 40—80 M.Y.
c) 130—40 M.Y.
d) 60—120 M.Y.

152. Hawaiin volcanism belongs to which one of the following types :
a) MOR volcanism
b) Island arc volcanism
c) Geosynclinal volcanism
d) Intraplate volcanism

153. The Tholeiitic magmas of the MORs differs from the continental tholeiites in having :

I. High contents of $Al_2O_3$ and Cr

II. Low contents of $Na_2O$, $K_2O$ and $TiO_2$

III. Low contents of $SiO_2$

IV. Low contents of Rb, Sr, Ba, U, etc.

a) I and III are correct

b) II and IV are correct

c) I, II and III are correct

d) I, II and IV are correct

154. The 'Suture Line' marked by the presence of Ophiolite complexes in orogenic belts represent :

I. Fragments of older oceanic crust tectonically emplaced in continental orogenic belts

II A line along which two plates collide

III. The site of closer of former oceans

a) Only I is correct

b) I and II are correct

c) I and III are correct

d) I, II and III are correct

155. Which of the following is the correct sequence of rock units found in a completely developed Ophiolite complex ?

a) Ultramafic rocks——low-K tholeiitic gabbros———sheeted diabase dykes———pillow lavas

b) Ultramafic rocks———sheeted diabase dykes———low-K tholeiitic gabbros———pillow lavas

c) Low-K tholeiitic gabbros———sheeted dykes———ultramafic rocks———pillow lavas

d) Pillow lavas———sheeted dykes———low-K tholeiitic gabbros———ultramafic rocks

156. The gabbro ultramafic contact in Ophiolite complexes is usually interpreted as a/an

a) Orogen

b) Orocline

c) Sphenochasm

d) Fossil Moho

157. Isolated submarine volcanic hills having flat tops are known as :

a) Island arcs

b) Oceanic rises

c) Sea mounts

d) Guyots

158. Guyots give an indication of :

a) A rise in sea level

b) A sinking of volcanoes

c) Sinking of a large part of an ocean basin

d) Any of these

159. Match the following and choose the correct answer :

| | |
|---|---|
| A. Varsican orogeny | (1) M. Devonian-Carboniferous |
| B. Caledonian orogeny | (2) Cretaceous |
| C. Laramide orogeny | (3) U. Silurian-L. Devonian |
| D. Antler orogeny | (4) L. Permian |

a) A—4 B—3 C—2 D—1

b) A—1 B—2 C—3 D—4

c) A—4 B—2 C—3 D—1

d) A—1 B—3 C—2 D—4

160. Rapid upliftment of an area is indicated by the presence of :

a) Greywacke

b) Orthoquartzite

c) Arkose

d) Lithic Arenite

161. Slow vertical movements of the crust of wide a.eal extent are known as :

a) Thallasogenic movements

b) Orogenic movements

c) Epeirogenic movements

d) Eustatic movements

162. The epeirogenic movements are also described as 'Oscillatory' movements because :

a) The movements take place in a wave like manner

b) The upliftment is followed by subsidence and vice versa

c) They do not involve any great deal of folding or faulting

d) The outlines and dimensions of uplifted areas change continuously

163. Isokatabases are the lines connecting points of:

a) Equal subsidence

b) Equal uplift

c) Equal degree of tectonic activity

d) Equal degree of volcanic activity

164. Eperiogenesis is indicated by one or more of the following features :

I. Meandering of rivers and formation of oxbow lakes
II. Accumulation of sediments in river terraces
III. Changes in the shore lines
IV. Recession of the sea

a) I and III are correct
b) II and III are correct
c) II, III and IV are correct
d) I, II, III and IV are correct

165. The earliest continental crust appears to have bеen formed between :
a) 4.2—4.5 B.Y. ago
b) 4.0—3.8 B.Y. ago
c) 3.8—2.4 B.Y. ago
d) 2.4—1.2 B.Y. ago

166. The main event of the crustal evolution involving the development of fold belts and their stabilisation to produce young platforms took place between :
a) 2400—1600 M.Y. ago
b) 1600—800 M.Y. ago
c) 1600—240 M.Y. ago
d) 800—240 M.Y. ago

167. The pre-geologic stage of the formation of the Earth's crust ranged between :
a) 5000—4500 M.Y. ago
b) 4500—4000 M.Y. ago
c) 4000—3500 M.Y. ago
d) 3500—2000 M.Y. ago

168. The evolution of vertebrates took place around :
a) 880 M.Y. ago
b) 450 M.Y. ago
c) 425 M.Y. ago
d) 225 M.Y. ago

*169. The oldest rocks in the world are found in :
a) Australian craton
b) Indian craton
c) Antartica
d) Greenland

*170. The oldest minerals found on the surface of the Earth are the crystals of
a) Calcite

b) Quartz
c) Tourmaline
d) Zircon

171. The rocks of the Granulite facies found in high grade Archaen terrains are desci ibed as :
a) Supracrustal rocks
b) Infracrustal rocks
c) Intracrustal rocks
d) None of the above

172. The Archaen greenstone belts range in age from :
a) 3100—2800 M.Y.
b) 2800—2500 M.Y.
c) 2500—1600 M.Y.
d) 1600—800 M.Y.

*173. The Expanding Earth Hypothesis of Hilgenberg assumes that the Earth had a———density and———radius than the present :
a) Lower/longer
b) Higher/longer
c) Lower/shorter
d) Higher/shorter

174. The ideas similar to those of the Expanding Earth Hypothesis are shared by :
I. Sea-floor Spreading Hypothesis
II. Dirac's Principle of the expansion of universe
a) I only
b) II only
c) Both I and II
d) Neither I nor II

175. Which of the following is NOT a postulate of the Tetrahedral Theory of the Earth postulated by Lowethian Green ?
a) The Earth started as a hot body
b) It cooled and contracted and acquired a tetrahedral form
c) The continents were roughly equidistant from one another
d) The continents have periodically drifted and expanded to give rise to new ocean basins

176. A sediment filled trough formed by normal faulting in the early stage of a Mid-oceanic Ridge is known as
a) Aulacogen
b) Orogen
c) Suture zone
d) Samfrau

177. The Earth's axis of instantaneous rotation is affected by various wobbles, most of which is a free oscillation known as

a) Nutational wobble

b) Chandler wobble

c) Solar wobble

d) Gutenberg wobble

178. The period of the free oscillation described above provides evidence for

a) The fluid nature of the core

b) The solid nature of inner core

c) The radial extent of core-mantle discontinuity

d) The ultrabasic nature of the mantle

179. Given below are four principal energy transactions affecting the Earth. Arrange them in the order of increasing energy dissipation

I. Solar energy received and re-radiated by the Earth

II. Heat loss by geothermal processes

III. Energy lost by the Earth's rotation

IV. Energy released by earthquakes

a) III > II > IV > I

b) III > II > I > IV

c) I > II > III > IV

d) IV > III > II > I

180. Volcanic massive sulphide ore deposits generally show a progression of types associated with the evolution of an island-arc system. Which of the following is the correct sequence or appearance of various types ?

a) Besshi type—Cyprus type—Kuroko type

b) Cyprus type—Kuroko type—Besshi type

c) Cyprus type --Besshi type—Kuroko type

d) Kuroko type—Besshi type—Cyprus type

181. The ores of volcanic affiliations which occur in the form of Ophiolite in the ridge-rift environments are described as

a) Cyprus type

b) Besshi type

c) Kuroko type

d) Sekkoko type

182. The lead-bearing sulphide deposits which accompany the felsite volcanism of the final stage of the island-arc evolution are described

a) Cyprus type

b) Besshi type

c) Kuroko type

d) Sekkoko type

183. According to the most recent trends, geosynclines are classified as linear areas of subsidence in which sedimentation occurs. Six different categories are recognized according to different types of tectonic settings. The sedimentation in marginal sea basins is described as

a) Island-arc type Geosyncline

b) Red Sea type Geosyncline

c) Andean type Geosyncline

d) Japan Sea type Geosyncline

184. Which of the following types of geosynclines typically forms a miogeosynclinal-eugeosynclinal couple ?

a) Mediterranean type Geosyncline

b) Atlantic type Geosyncline

c) Island-arc type Geosyncline

d) Red Sea type Geosyncline

185. The Island-arc type Geosynclines characteristically exhibit

a) Plutonic-epigenetic affiliations

b) Plutonic-syngenetic affiliations

c) Volcanic-epigenetic affiliations

d) Volcanic-syngenetic affiliations

186. 'Bollide' is the term used to denote

a) Solid bodies of extraterrestrial matter that vapourize completely in the atmosphere

b) Solid lumps of extraterrestrial matter that lands on the surface of the earth

c) An exceptionally bright meteorite that explodes into several fragments

d) A meteorite of glassy composition

*187. A type of meteorites whose mineral aggregates almost duplicate the fabric and texture of some terrestrial rocks is

a) Chondrites

b) Achondrites

c) Siderites

d) Siderolites

188. Which of the following are the most abundant of all the meteorites ?

a) Siderites

b) Siderolites

c) Aerolites

d) Tektites

189. Which of the following groups of meteorites are considered to be a product of the original process by which the planets were made from the solar nebula ?

a) Chondrites.

b) Achondrites

c) Pallasites

d) Mesosiderites

190. The "Mergui Ridge" is a part of

a) The Bay of Bengal basin

b) The Arabian sea basin

c) The Andaman basin

d) The Mahanadi basin

191. The mountain ranges of "Queen Mand Land" are located in

a) The Arctic ocean

b) Antartica

c) South-West Africa

d) West pacific

*192. The term "Terrane" is of comparatively recent origin. It is connected with

a) Mid-oceanic ridges

b) Continental drift theory

c) Plate tectonics

d) Expansion of universe

e) Deep sea drilling programme

193. According to the theory of plate tectonics which of the following would probably happen in another 50 M.Y. ?

I. Australia will push on northward to come alongside Malaysia and collide with Asia

II. The Mediterranean sea would be converted into a series of inland lakes

III. Delhi will be shifted towards east by about 200 km to its present position

IV Los Angeles will go down the Aleutian trench and disappear into the mantle of the earth

a) I and IV

b) I, III and IV

c) II, III and IV

d) I, II, III and IV

194. The relict of the "Tethys of Suess" can be seen at the present time in

a) Mediterranean sea
b) Red sea
c) Arabian sea
d) Sea of Okhotshk
e) Phillipine sea

*195. Of all the continental plates of the world, which one has the unique distinction of having travelled the longest distance ?
a) Asian plate
b) Indian plate
c) American plate
d) African plate
e) Pacific plate

*196. When India, as a definite entity, was formed from the Gondwana complex, the earth was roughly ?
a) 4000 M.Y. old
b) 3500 M.Y. old
c) 3000 M.Y. old
d) 2500 M.Y. old

197. The northward drift of the Indian plate was most rapid during
a) Paleocene
b) Eocene
c) Middle miocene
d) Pliocene

*198. Which of the following served as a bridge across which faunal exchange took place in most of the Mesozoic ?
a) Siwalik gulf
b) Assam 'neck'
c) Palk strait
d) Irrawadi shelf
e) Chagos and Maldive Islands

199. The Siwalik River was dismembered during the Late Pliestocene (0.25 M. a) into two rivers which are :
a) Indus and Brahmaputra
b) Ganga and Saraswati
c) Ganga and Yamuna
d) Brahmaputra and Irrawadi

*200. Which one of the following does NOT give an indication that India was a part of Gondwanaland ?
a) Glossopteris and Gangamopteris fossils in the Raniganj coal seams

b) Occurrence of ancient arthopod, peripatus, in the humid jungles of Assam
c) Fossils of Ceratodus in the Upper Triassic Maleri beds of Godavari valley
d) Fossils of Nummulites in the Upper Himalayan paleogene successions

201. The Neotectonic movements of the Indian plate suggests that in about 50 M.Y. the Indian sub-continent will drift further towards :

a) North
b) South
c) East
d) West

202. The first movements of the Himalayan orogeny led to the emplacement of :

a) Raj Mahal Traps
b) Sylhet Traps
c) Molasse Facies
d) Dras volcanics

203. The Pliestocene Ice Ages coincided with the rise of

a) Tibetan Himalayas
b) Higher Himalayas
c) Outer Himalayas
d) Tethyan Himalayas

204. Which one of the following zones was presumably a geanticline ?

a) Outer Himalayan zone
b) Lesser Himalayan zone
c) Tethyan Himalayan zone
d) Central crystalline zone

205. A new and controversial theory that "the Himalayas were formed because of an upward flow of hot magma from beneath the earth's crust" and not because of the collision of Indian and Asian plates was put forward by :

a) Prof. Harry Hess
b) Prof. F.J. Morgan
c) Prof. F. Ahmad
d) Sarkar and Saha

206. One of the following basins has rock units which are develop-

ed by a second-cycle of sedimentation derived from the deposits of the first cycle sequcnces :

a) Cuddapah Basin

b) Bengal Basin

c) Kurnool Sub-Basin

d) Cauvery Basin

207. Match the following and choose the correct answer

| | |
|---|---|
| A. Gondwana Basin | 1. Yoked Basin |
| B. Andaman Basin | 2. Half-Graben |
| C. Kurnool Basin | 3. Back-Arc Basin |
| D. Srisailam Basin | 4. Graben |

a) A—4 B—3 C—2 D—1

b) A—3 B—1 C—4 D—2

c) A—4 B—3 C—1 D—2

d) A—2 B—3 C—4 D—1

208. Which of the following cities is NOT subsiding ?

a) Bombay

b) Madras

c) Calcutta

d) Srinagar

209. Choose the correct geologic sequence as we encounter while going from south to north of the Indian Peninsula :

a) The Indian Craton—The Sub-Himalayas—the Indo-Gangetic trough Main boundary thrust—the higher Himalayas—Main Central Thrust—the Tibetan Himalayas

b) The Indian Craton—the Indo-Gangetic trough—the sub Himalayas—main boundary thrust—the lower Himalayas—main central thrust—the higher Himalayas

c) The Indian Craton—the Indo-Gangetic trough—the lower Himalayas—main boundary thrust—the Sub-Himalayas—main central thrust—the higher Himalayas

d) None of the above

210. The Tectonic contact between the Peninsular India and the Himalayan folded belt is marked by :

a) the Ganga basin

b) the Krol Nappes

c) the Main Boundary Thrust

d) the Satpura Orogen

e) the Main Central Thrust

*211. The Ganga basin can be considered as a :

a) negative area lying on a large positive area
b) negative area lying on a large negative area
c) positive area lying on a large negative area
d) positive area lying on a large positive area

*212. The strongest of all the Himalayan orogenies is :
a) Karakoram orogeny
b) Kirthar orogeny
c) Sirmurian orogeny
d) Siwalik orogeny

*213. The Indo-Gangetic Alluvial plains took shape during :
a) Siwalik orogeny
b) Sirmurian orogeny
c) Kirthar orogeny
d) Karakoram orogeny

214. Most of the physiographic features of India were shaped during the———era :
a) Proterozoic
b) Paleaozoic
c) Mesozoic
d) Cenozoic

*215. In India, the main earthquake belt runs along the Himalayan ranges and South-West to the Rann of Kutch. Which of the following earthquakes did NOT occur in this belt ?
a) Kangra (1905)
b) Assam (1950)
c) Koyna (1967)
d) Himachal Pradesh (1975)

*216. A magnetic observatory is a place where the magnetic field changes are continuously monitored. India has magnetic observatories at :
a) Calcutta, Bombay, Delhi, Bangalore, Madras, and Hyderabad
b) Bombay, Delhi, Bangalore, Trivandrum, Ernakulum and Ahmedabad
c) Trombay, Hassan, Sriharikota, Allahabad, Gulmarg and Alibag
d) Bombay, Alibag, Kodaikanal, Hyderabad Gulmarg and Trivandrum

217. The Alpine-Himalayan seismic belt is characterized by :
a) shallow-focus earthquakes
b) intermediate-focus earthquakes

c) deep focus earthquakes

d) all of the above

218. Throughout their entire strike length, the **Himalayas show a** tectonic transport towards :

a) South

b) North

c) East

d) West

219. The youngest tectonic element **among the following is :**

a) Narmada-Son-lineament

b) Indus suture

c) Krol nappe

d) Ganga basin

*220. According to the most recent views, **the Narmada-son linea-**ment has originated due to :

a) **Normal fault tectonics**

b) **Wrench fault tectonics**

c) **Taphrogenic movements**

d) **Oscillatory tectonic movements**

221. The contact between the **Carlsberg ridge and Owen fracture** zone is an example of :

a) RTF triple junction

b) RRT triple junction

c) RRF triple junction

d) FFR triple junction

222. The Indian Ocean plate is split into **two major segments by :**

a) Quetta Chaman fault

b) Owen fracture zone

c) 90°E ridge

d) Lakshadweep-Chagos fault

223. The 'Aegean Arc' is the **island-arc system in :**

a) **Atlantic ocean**

b) **Pacific ocean**

c) **Indian ocean**

d) **Mediterranean sea**

224. Which of the following stress **regimes are favourable for the** eruption of tholeiitic **magma series ?**

a) Extensional

b) compressive

c) Minor extensional

d) Minor compressive

225. Which of the following tectonic settings does not contribute to the magma production ?

a) Constructive plate boundaries

b) Conservative plate boundaries

c) Consuming plate boundaries

d) Marginal sea basins

226. The boundary between the sialic masses of the continents and the basaltic floor of the ocean basin is described as :

a) Andesite line

b) continental slope

c) Benioff zone

d) Mohorovicic discontinuity

227. The science of 'Geodesy' deals with :

a) dating of the terrestrial rocks

b) measurements of the elevations and depressions on the Earth's surface

c) measurements of the dimensions of the Earth

d) recording of the changes undergone by the crust since its accretion

228. What is the name given to the single large ocean which existed before the beginning of the continental drift ?

a) Pangaea

b) Panthalasa

c) Samfrau

d) Tethys Sea

229. According to Alfred Wegener, the continents had drifted generally towards :

a) East

b) West

c) North

d) South

230. India was separated from the Antartica complex about 140 M.Y. ago. When did it join South Asia ?

a) 120 MY ago

b) 80 MY ago

c) 60 MY ago

d) 45 MY ago

231. Before the break-up of the Pangea and the drifting of the continents India was situated

a) north of the Equator

b) south of the Equator
c) in the centre of Pangea
d) at the southpole

232. The Sunda Trench runs parallel to the islands of :
a) Java
b) Sumatra
c) Maldives
d) Mauritius

*233. The Prince Edward Crozet Ridge is found in
a) Atlantic ocean
b) Indian ocean
c) Pacific ocean
d) Red sea

*234. Choose the correct answer
a) Andaman and Nicobar Islands———Diastrophic islands
b) Madagascar———Barrier Reef
c) Maldives———Continental island
d) Mauritius———Volcanic island

235. The Arabian Sea is divided into two parts by :
a) the Socotra-Chagos ridge
b) the Carlsberg ridge
c) the 90°E ridge
d) the Seychelles ridge
e) the St. Paul ridge

## GENERAL GEOLOGY (KEY)

| | | | |
|---|---|---|---|
| 1. a | 2. c | *3. b | 4. a |
| 5. a | 6. c | 7. c | *8. d |
| 9. d | 10. a | 11. d | 12. c |
| 13. d | 14. b | 15. c | 16. b |
| 17. c | 18. c | 19. d | 20. c |
| 21. d | 22. c | 23. c | 24. a |
| 25. c | 26. c | 27. b | *28. d |
| 29. a | *30. b | 31. b | 32. d |
| 33. b | 34. a | 35. d | 36. c |
| *37. c | *38. c | *39. d | 40. d |
| 41. b | 42. a | 43. b | 44. d |
| *45. d | 46. c | 47. a | 48. a |
| 49. c | 50. a | 51. d | 52. a |

53. a 54. c 55. b 56. d
57. d 58. b *59. b 60. b
61. c 62. c 63. c *64. c
65. a 66. a 67. d 68. a
69. a 70. b 71. c 72. c
73. d 74. b 75. c 76. d
*77. c 78. b 79. c 80. b
*81. c 82. a 83. b 84. a
85. c 86. c 87. c 88. a
89. c 90. a 91. c 92. d
93. c 94. e 95. c 96. b
*97. c *98. d 99. a 100. b
*101. a 102. d 103. c *104 a
105. d 106. e 107. a 108. b
*109. e 110. a *111. c 112. d
113. c 114. c 115. c 116. a
*117. b 118. d 119. d *120. c
121. a *122. a 123. a 124. b
125. b 126. c 127. a 128. d
129. d 130. d 131. b 132. c
133. b 134. b 135. c 136. a
*137. b 138. d 139. b 140. c
141. d 142 a 143. c 144. b
145. b 146. a 147. d 148. c
149. c 150. b 151. c 152. a
153. d 154. d 155. a 156. d
157. d 158. d 159. a 160. c
161. c 162. b 163. a 164. d
165. b 166. c 167. a 168. b
*169. d *170. d 171. a 172. b
*173. d 174. c 175. d 176. a
177. b 178. a 179. c 180. c
181. a 182. c 183. d 184. b
185. d 186. c *187. b 188. c
189. a 190 c 191. b *192. c
193. d 194. a *195. b *196. b
197. a *198. b 199. a *200. d
*201. c 202. d 203. c 204. d
205. c *206. c 207. a 208. d
209. b 210 a *211. b *212. c

| | | | |
|---|---|---|---|
| *213. a | 214. d | *215. c | *216. d |
| 217. b | 218. a | 219. d | *220. b |
| 221. d | 222. d | 223. d | 224. a |
| 225. b | 226. a | 227. c | 228. b |
| 229. b | 230. d | 231. b | 232. a |
| *233. b | *234. d | 235. b | |

## EXPLANATORY NOTES

*3. (b) : Radius in Km : Venus=6100, Earth=6371.

*8. (d) : The Nebular Hypothesis was developed by the German philosopher Kant (1755) and the French mathematician Laplace (1796) independentof each other.

*28. (d) : LVZ is not detected beneath Precambrian Shields. Also in the regions of negative free-air gravity anomalies, the LVZ is poorly developed or does not exists.

*30. (b) : The D' layer of the lower mantle has the maximum velocity of P and S waves (P wave 13.7 km/sec : S wave 7.3 km/sec). The P wave velocity in the outer core is 8–10 km/sec while in inner core it is 11 km/sec.

*37. (c) : Within 2% error.

*38. (c) : The seismic velocities bear a direct relationship to the elastic constants and increases with an increase in their value.

$$\therefore \quad V_p = \sqrt{\frac{\lambda + 4/3\mu}{\sigma}} \text{ and } V_s = \sqrt{M/\sigma}$$

where, $\sigma$=density, $\mu$=rigidity modulus and $\lambda$=Bulk's modulus.

*39. (d) : The value of rigidity modulus in the core is zero Because the S waves do not pass through the core, $V=0$ ; $\mu=0$ i.e. the null resistance to shear characteristic of liquids (in normal conditions).

*45. (d) : $Sm^{147}$ – $10^{19}$ years ; $U^{238}$ – $4.5 \times 10^9$ years ;
$U^{235}$ – $0.71 \times 10^9$ years ; $Rb^{87}$ – $59 \times 10^9$ years

*59. (b) : The thickness of the crust varies only by the top.

*64. (c) : Confirms the Airy's hypothesis

*77. (c) : Lomonosov Ridge in the Arctic Ocean has a crustal thickness ranging between 15–18 km. It rises on the Continental crust.

*81. (c) : The landmass was equally distributed in both the Hemispheres.

*97. (c) : Although data indicate that a constant spreading rate is unlikely in any ocean basin, the South Atlantic closely approaches constancy (~ 1·9 cm/yr) and is commonly chosen as a reference to extrapolate the time scale.

*98. (d) : East Pacific Rise —6 cm/yr/ridge flank ; North Atlantic — 1 cm/yr ; Central Indian Ridge—4 cm/yr.

*101. (a) : According to Geomagnetic Time Scale, the present epoch has normal polarity and is designated as 'Brunhes Normal'. The present geologic epoch is referred as 'Holocene'.

*104. (a) : The plates are the inert asiesmic regions of the earth. However, they are bounded by narrow mobile belts which are highly active and are characterized by violent seismic and volcanic activity.

*109. (e) : No one knows the exact mechanism for the plate motion, though every possibility has been suggested i.e. ridge-push ; slab-pull mantle-drag ; sliding, etc.

*111. (c) : Choice (ii) is in correct. Despite their aseismicity, passive margins undergo considerable subsidence during their evolution and are covered by thick sedimentary sequences which makes them prime target for hydrocarbon exploration.

*117. (b) : Island arcs contain thick assemblages of volcanic and sedimentary rocks which closely parallel the composition of Greenstone Belts of 3·5 – 2·5 b.y a;e.

*120. (c) : It is a Triple Junction where the African rift zone, the Red Sea rift and the Gulf of Aden rift meet at a common point.

*122. (a) : The depth characteristics of earthquakes is different at different plate boundaries. Thus,

Constructive Boundaries (e.g. MORs)—Shallow Focus (depth < 100 km)

Destructive Boundaries (e g. Island Arcs)—Deep focus

Conservative Boundaries (e.g. Transform faults)—Shallow focus (depth < 15 km)

*125. (b) : Red Sea represents an early stage in ocean development and is opening along a line of rift valleys. Its typical age is about 25 M.Y. Rest of the basins are over 200 M.Y. in age

*126. (c) : The Gecsynclinal concept was introduced by James Hall in 1859.

*137. (b) : The term "Geocline" was proposed by Robert Dietz. In a Geocline the basement and the overlying sediments does not seem to be downwarped into a syncline ; rather it appears to dip uniformly seaward at a low angle—more like monocline.

*169. (d) : The oldest known rocks belong to the Amitsoq gneisses

of Greenland (3800 M.a.).

*170. (d) : The oldest minerals found on the surface of the earth are Zircon crystals about 4200 M.Y. old, found in Australia.

*173. (d) : It assumed that originally the Earth had a higher density of 9·13 g/cm³ and not 5·52 g/m³ and shorter radius of 5430 km and not 6371 km, than the present.

*187. (b) : Achondrites are very similar to terrerial rocks of the basic type like basalts and dunites.

*185. (c) : Aerolites or stony meteorites constitute 92% of the total meteorites ; Next follows Siderites or iron meteorites with 6% of all the meteorites.

*192 (c) : It is a short form of "Tectono-Stratigraphic Terrane", a concept started in 1970s when dispute arose over the use of land in Alaska. It tends to examine Plate Tectonics in minute details in order to get a slow-motion picture of the growth and movement of plates and periods when geologic activity was high and low.

*195. (b) : It travelled about 9000 km in about 200 M.Y., moving at the rate of 4·5 cm/yr. Its almost rectangular shape and small size were ideal for the drift over such a long distance.

*196. (b) : India as a definite entity was born sometime during the Carboniferous era (340— 85 Ma) when the earth was roughly 3650—3300 M.Y. old.

*198. (b) : It was through this land route that India exchanged its fauna with that of Europe and Africa till the middle of Eocene Period (45 M.a ).

*200. (d) : Nummulites evolved only in the Eocene and Oligocene, hence, it is not of much significance. Peripatus occurs in India, Australia, Africa and South America ; similarly, Glossopteris and Gangamopteris are found in all the Gondwanalands ; Ceratodus is Australian lung fish. Thus, the last three indicate the required relationship.

*201. (c) : Since the tectonic cycle of the Indian Plate is not yet over, the subcontinent is supposed to drift further towards East in about 50 M.Y. from now. Delhi will then be 200 km east of its present position. An arc of islands will also appear along the west coast of India in the Arabian Sea.

*206. (c) : The Cuddapahs represent the first cycle of sediments derived from the Archaens ; whereas, the Kurnools represent the second cycle (recycled) sediments derived from the Archaens as well as the Cuddapahs.

*211. (b) : The Ganga Basin represents a large-scale depression on the Vindhyan syneclise

*212. (c) : It marked the third Himalayan upheaval during the M. Miocene time which was the pronounced of all the Himalayan orogenies.

*213. (a) : The pliocene to M. Pliestocene witnessed the fourth phase of the Himalayan upheaval (Siwalik Orogeny) which gave shape to the Indo-Gangetic Alluvial Plains. The Main Boundary Fault was also formed during this time.

*215. (c) : So far only the Koyna Earthquake of Maharashtra (1967) and the Idukki Earthquake of Kerala (1988) have occurred outside the seismic belt of India.

*216. (d) : Magnetic observatories of India are located at Bombay, Alibag, Kodikanal, Trivandrum, Hyderabad, Annamalainagar, Ettayapuram, Ujjain, Shillong, Jaipur, Sabhawala and Gulmarg The first magnetic observatory was constructed in 1826 in South Bombay. Later it was shifted to Alibag, about 30 km from Bombay in 1905.

*233. (b) : It is the extension of the South Madagascar Ridge which runs southward from Madagascar island in the Indian Ocean.

*234. (d) : Andaman and Nicobar Islands and Madagascar = Continental Islands; Maldives = Coral Islands; Mauritius = Volcanic Island.

# 2

# Physical Geology and Geomorphology

1. Although sodium and potassium have approximately equal concentrations in nature, the waters of the seas and oceans are relatively strongly enriched in sodium because
   a) the potash felspars have low stability relative to plagioclase felspar
   b) sodium is chemically more reactive than potassium
   c) potassium enters readily into newly formed clay minerals while sodium tends to remain in solution
   d) sodium is NOT particularly stable in the weathering environment and therefore passes into solution

2. The aluminium content of sedimentary rocks is usually very high because
   I. Aluminium is one of the most insoluble elements and does not readily go into solution
   II. Much of the aluminium is taken up by the clay minerals to form clay-rich rocks
   III. Formation of residual deposits leads to the concentration of aluminium in sedimentary rocks
   a) I and II are correct
   b) I and III are correct
   c) II and III are correct
   d) all of the above are correct

3. Physical weathering is NOT characteristic of
   a) Polar regions
   b) Hot deserts
   c) Cold deserts
   d) None of the above

4. Exfoliation is a form of
   a) Physical weathering
   b) Chemical weathering
   c) Biochemical weathering
   d) Mass wasting

5. Exfoliation is most characteristically found in
   a) Granites
   b) Spilites
   c) Arkoses

d) Orthoquartzites

6. Exfoliation, a dominant physical process, is also assisted by some chemical processes. The most important part is played by
   a) Hydration
   b) Oxidation
   c) Hydrolysis
   d) Carbonation

*7. Hydration of a salt is accompanied by one or more of the following
   (1) increase in volume
   (2) increase in porosity
   (3) decrease in density
   (4) decrease in volume
   a) 1 and 3 only
   b) 1, 2 and 3 only
   c) 2 and 4 only
   d) 2, 3 and 4 only

8. According to the Mineral Stability series of Goldich
   a) olivine weathers to pyroxene and pyroxene weathers to hornblende
   b) olivine, pyroxene and hornblende have equal susceptibility to weathering
   c) minerals rich in Si—O are least resistant to chemical weathering
   d) olivine is less stable than pyroxene and pyroxene is less stable than hornblende

9. Which one of the following places is more likely to be affected by chemical weathering ?
   a) Thar Desert
   b) Tibetan plateau
   c) Leh
   d) Siwalik

10. Mass Wasting may be described as
   a) Excessive loss of material by landslides
   b) Downslope movement of rock and soil under the direct influence of gravity
   c) Excessive loss of organically-rich top soil due to soil erosion
   d) Flowage of saturated debris in periglacial regions

11. Mass wasting or 'gravitative transfer' is NOT a result of
   a) vertical movements
   b) lateral movements

c) diagonal movements

d) spiral movements

12. Lateral movement of a series of blocks away from a centre is described as

a) Sliding

b) Gliding

c) Spreading

d) Cambering

e) Flow

13. The slow downslope movement of saturated rock debris which is NOT confined to definite channels is described as

a) Debris slide

b) Mud flow

c) Solifluction

d) Soil creep

14. Which of the following does not result from the lateral movements of rock debris ?

a) Spreading

b) Solifluction

c) Cambering

d) Sackung

15. Solifluction and Gelifluction were the important processes of mass wasting during

a) Archaean

b) Proterozoic

c) Miocene

d) Pliocene

e) Pliestocene

16. Mushroom-shaped structures having slender columns and wide tops resulting from the abrasive action of wind are described as

a) Ventifacts

b) Brazil nuts

c) Pedestal rocks

d) Draas

17. Which of the following features denote furrow deflation ?

a) Hohlwegs

b) Desert pavements

c) Blow outs

d) Playas

*18. Which of the following is NOT an erosional feature of wind ?

a) Zeugen
b) Yardang
c) Serir
d) Hohlweg

19. The aeolian features that have been noticed in the planet Mars are
a) Zeugens
b) Yardangs
c) Inselbergs
d) Barchans

20. Match the following and choose the correct answer

| | |
|---|---|
| (A) Sandy desert | (1) Reg |
| (B) Stony desert | (2) Koum |
| (C) Rocky desert | (3) Adyry |
| (D) Loess desert | (4) Hamada |
| | (5) Harras |

a) A-2 B-2 C-4 D-5
b) A-2 B-1 C-3 D-4
c) A-1 B-4 C-3 D-5
d) A-2 B-1 C-4 D-3

21. Special types of longitudinal dunes which resemble railway embankments are described as
a) Barchans
b) Parabolic dunes
c) Seifs
d) Dome dunes
e) Draas

22. The sand dunes migrate in the
a) leeward direction
b) windward direction
c) both leeward and windward directions
d) none of the above

23. Loess is a non-stratified, well-sorted aeolian deposit composed of———grade fragments
a) sand
b) silt
c) clay
d) pebble

24. Inselbergs characteristically consists of
a) isolated pillar like masses in desert regions
b) U-shaped troughs formed by winds

c) cup-shaped depressions formed by glacial erosion
d) sink-hole ponds formed by solution action in karst regions

25. Match the following and choose the correct answer

| | |
|---|---|
| (A) Bolson | (1) A series of coalescing alluvial fans on a piedmont slope |
| (B) Bajada | (2) A basin more or less rimmed by mountains |
| (C) Driekanter | (3) Non-stratified silt-grade aeolian deposit |
| (D) Adobe/Limon | (4) Angular rocks with polished surfaces |

a) A-1 B-2 C-3 D-4
b) A-2 B-1 C-4 D-3
c) A-1 B-2 C-4 D-3
d) A-2 B-1 C-3 D-4

26. Flat-topped hills or small mountains formed by stream action are called
a) Mesas
b) Buttes
c) Cuestas
d) Stream terraces

27. An erosional ridge having a long, gentle slope on one side and an abrupt slope on the other is known as
a) Gloup
b) Groyne
c) Cuesta
d) Hogback

28. Streams which flow in the opposite direction to the original consequent streams are described as
a) Obsequent
b) Subsequent
c) Insequent
d) Resequent

29. The deltaic structure is seen on a small-scale in
a) Cross-bedding
b) Gradded bedding
c) Convolute bedding
d) Ripple lamination

30. In a deltaic structure, the topset beds consists of
a) fine material
b) coarse material

c) coarse and fine material
d) none of the above

31. The Bird's foot digitate delta is characteristically exhibited by
a) River Rhine
b) River Nile
c) River Brahmaputra
d) River Mississippi

32. Which of the following rivers does not form a delta ?
a) Krishna
b) Godavari
c) Mahanadi
d) Narmada

33. Flat valleys are formed during the———stage of a river
a) initial
b) youth
c) mature
d) old

34. The erosive power of a river depends on its velocity. If the velocity is doubled, the erosive power increases by
a) two times
b) four times
c) six times
d) twelve times

35. An interlaced network of high sinuosity channels is a feature exhibited by
a) Meandering streams
b) Braided streams
c) Anastomosing streams
d) Incised meanders

36. Which of the following is the best example of lateral shifting of a braided stream ?
a) River Kosi
b) River Krishna
c) River Pamba
d) River Ganges

37. Fining upward sequences are characteristic of
a) Meandering streams
b) Braided streams
c) Anastomosing streams
d) All of the above

38. The drainage pattern which shows a lack of structural control to the stream-flow direction is described as
    a) Parallel
    b) Dendritic
    c) Rectangular
    d) Trellis

*39. Radial drainage patterns are associated with
    a) slopes of volcanoes
    b) strongly folded rocks
    c) uniformly dipping beds of varying rock hardness
    d) homogenous rock surfaces of uniform resistance

*40. 'Natural levee' is an example of
    a) Point-bar deposit
    b) Channel-fill deposit
    c) Flood plain deposit
    d) Flood basin deposit

*41. Which of the following is a feature exhibited by a river in its upper course ?
    a) Meanders
    b) River piracy
    c) Piedmont plains
    d) Sloughs

42. When water of a stream descends over a step-like structure it is described as a
    a) Cascade
    b) Rapid
    c) Waterfall
    d) Spring

43. The horse-shoe shaped lakes formed in the cut-off meanders of a river are known as
    a) Oxbow lakes
    b) Mort lakes
    c) Billabongs
    d) All of the above

44. Match the following and choose the correct answer

| | |
|---|---|
| (A) Mort lakes | (1) lakes formed in Corries |
| (B) Ribbon lakes | (2) lakes occupying floors of glaciated valleys |
| (C) Tarns | (3) lakes formed in arid regions |
| (D) Playas | (4) lakes formed in meanders of rivers |

a) A-4 B-1 C-2 D-3
b) A-4 B-2 C-1 D-3
c) A-3 B-1 C-2 D-4
d) A-3 B-2 C-1 D-4

45. The most distinguishing feature of lacustrine deposits is
a) deposition of coarser sediments near the margins and finer sediments in the central part
b) deposition of finer sediments near the margins and coarser sediments in the central part
c) progressive coarsening of sediments from the centre outwards
d) progressive fining of sediments from the centre outwards

46. Which of the following features denote the process of 'rejuvenation' of a river ?

I. Incised Meanders
II. River Terraces
III. Valley cut within a valley
IV. A slight break in slope and change in the shape of a valley

a) I and III only
b) II and IV only
c) I, II and III only
d) All of the above

47. The Knob and Kettle Topography is encountered in the regions of
a) glacial environment
b) fluvial environment
c) deltaic environment
d) arid environment

*48. 'Nunataks' are
a) very high peaks projecting above ice sheets
b) rough irregular surfaces of knobby hills and deep hollows
c) domed earth mounds caused by water in permafrost regions
d) stream meanders entrenched between steep and symmetrical sides

49. Low, rounded and abraded mounds on the glaciated valley flooors are known as
a) Pingos
b) Roches mountonees
c) Kame Terrace
d) Drumlins

50. In Roches mountonees the upstream and the downstream sides suffer two different types of erosive actions. These are respectively

a) Nivation/Calvation
b) Abrasion/Frost heaving
c) Abrasion/Plucking
d) Plucking/Abrasion

51. Which of the following shows features that are opposite to those of roches mountonees ?
a) Drumlins
b) Moulins
c) Pingos
d) Erratics

52. A rock-strewn area which is a relict of frost wedging is described as
a) Crag and Tail
b) Rock-glacier
c) Ice-Blister
d) Felsenmeer

53. A glacier advances forward when the conditions are such that
a) accumulation is greater than wastage
b) accumulation is lesser than wastage
c) accumulation is equal to wastage
d) none of the above

54. Which of the following statements about glaciers is NOT correct ?
a) the top portion moves faster than the bottom
b) the middle portion moves slower than the sides
c) the velocity increases with the thickness of ice
d) their velocity increases with temperature

55. A snow-line denotes
a) a line above which the snow fields are not permanent
b) a line on the mountains above which snow does not melt except in summers
c) a line on the mountains above which the snow does not melt even in summers
d) nonc of the above

56. The snowline in the Himalayas lies between
a) 10,000 ft – 15,000 ft
b) 15,000 ft – 17,000 ft
c) 20,000 ft – 25,000 ft
d) 25,000 ft – 30,000 ft

57. Identify the features related with various geomorphic pro-

cesses from the lists given below

| *Features* | *Process* |
|---|---|
| A. Draas | 1. Glacial |
| B. Osser | 2. Aeolian |
| C. Pepino | 3. Fluvio-glacial |
| D. Pingo | 4. Fluvial |
| E. Cataract | 5. Solution action |

a) A–2 B–5 C–3 D–1 E–4
b) A–2 B–3 C–5 D–1 E–4
c) A–4 B–1 C–5 D–3 E–2
d) A–4 B–5 C–1 D–3 E–2

58. Natural bridges are
a) Bridges fromed by the deposition of load from braided streams
b) Isolated bar-like deposits in desert regions
c) Bridges across two steeply inclined streams
d) Vertical or inclined hollows in limestone caves

59. The dripstones hanging from the top of the limestone caves are called
a) Stalactites
b) Stalagmites
c) Stylolites
d) Geodes

60. A depositional feature in limestone caves which develops in all possible directions is described as
a) Hornito
b) Helictite
c) Polje
d) Pepino

61. The karst topography shows a series of stages in its development. Which of the following features shows the correct sequence of karst formation ?
a) Uvalas ––→ Poljes/Hums ––→ Lapies/Dolines––→Ponors/Caverns
b) Lapies/Dolines ––→ Ponors/Caverns ––→ Uvalas ––→ Poljes/Hums
c) Lapies/Dolines ––→ Uvalas ––→ Ponors/Caverns ––→ Poljes/Hums
d) Poljes/Hums ––→ Uvalas ––→ Lapies ––→ Ponors

62. Which of the following are the ideal conditions for karst development ?
(A) Huge thickness of pure limestone with low permeability and high crushing strength

(B) A massive organised system of joints intersecting the country rocks

(C) Low mean annual rainfall

(D) Considerable uplift above sea level

a) A, B and C only

b) A and B only

c) A, B and D only

d) all of the above

63. Steep-sided valleys that develop on the floor of large limestone caves are known as

a) Natural tunnels

b) Blind valleys

c) Griekes

d) Haystacks

64. Rugged, grooved and pitted surfaces exposed in the limestone terranes of high relief are described as

a) Dolines

b) Lapies

c) Pepino

d) Karst windows

65. Many geomorphic environments are characterized by the formation of residual hills. Choose the right combination from the lists given below

| *Residual Hills* | *Geomorphic environment* |
|---|---|
| A. Mogotes | 1 Alluvial |
| B. Monadnocks | 2. Aeolian |
| C. Inselbergs | 3. Glacial |
| D. Roches mountonees | 4. Karst |

a) A – 4 B – 1 C – 2 D – 3

b) A – 2 B – 1 C – 4 D – 3

c) A – 4 B – 2 C – 3 D – 1

d) A – 2 B – 4 C – 3 D – 1

66. The deposition of amorphous silica layer by layer produces

a) Geodes

b) Comb structure

c) Liesegang banding

d) Cockade structure

67. When the water issued from the Underground sources is very small in quantity and does not have the ability to flow, it forms

a) Springs

b) Seepages
c) Quicksand
d) Geysers

68. A spring deposit of $CaCO_3$ is known as
a) Calc Tufa
b) Travertine
c) Sinter
d) Natron

69. An Archipelago may be described as
a) a series of pyramidal peaks forming a distinct pattern
b) a shallow lake surrounded by thick sedimentary deposits
c) an extended line of islands
d) a cluster of volcanic islands

70. Which of the following correctly designates the increasing degree of volcanic activity ?
a) Hawaiin —→ Strombolian —→ Plinian —→ Vesuvian —→ Pelean —→ Vulcanian
b) Hawaiin —→ Vulcanian —→ Strombolian —→ Vesuvian —→ Pelean —→ Plinian
c) Hawaiin —→ Strombolian —→ Vulcanian —→ Vesuvian —→ Plinian —→ Pelean
d) none of the above

71. Blasts of incandescent gases and lava flows known as "nuees ardentes" is a characteristic feature of———type of volcanic activity
a) Pelean
b) Plinian
c) Hawaiin
d) Vesuvian

72. Volcanic fissures or vents through which $CO_2$ vapours are emitted are called
a) Saffioni
b) Solfataras
c) Mofettes
d) Fumaroles

73. Calderas are
a) depressions on the top of volcanic cones
b) a chain of active volcanoes
c) craters of very large size
d) subsidiary volcanic cones near the vent

74. Steeply-sloping cones formed by the solidification of viscous lava are described as
a) Acid lava domes
b) Spines
c) Plug cones
d) all of the above

75. Alternate layers of pyroclastic materials and lavas are found in
a) Acid lava domes
b) Strato volcano
c) Shield cones
d) Cinder cones

76. Lapilli are the pyroclastic materials having the size range
a) less than 0·25 mm
b) 0.25—4 mm
c) 4—32 mm
d) more than 32 mm

77. Most of the world's great volcanoes are
a) Shield cones
b) Cinder cones
c) Plug cones
d) Composite cones

78. 'Mauna Lao' is
a) the tallest mountain in Europe
b) the tallest active volcano in the world
c) the largest extinct volcano in the world
d) the biggest volcanic cone in the world

79. Match the following and choose the correct answer.

| | |
|---|---|
| A. Steptoe | 1. very small lava flows |
| B. Hornitos | 2. isolated hill surrounded by lava flows |
| C. Tummuli | 3. bulging mounds on a lava surface with cracks in the crests |
| D. Saffioni | 4. volcanoes emitting boric acid vapours |

a) A—1 B—2 C—3 D—4
b) A—2 B—1 C—3 D—4
c) A—1 B—2 C—4 D—3
d) A—3 B—1 C—2 D—4

80. Which of the following is the deepest of all the abyssal oozes ?

a) Globigerina ooze
b) Diatom ooze
c) Radiolarian ooze
d) Pteropod ooze

81. **Which of the following is a good source of phosphorous ?**
a) Globigerina ooze
b) Diatom ooze
c) Radiolarian ooze
d) Pteropod ooze

82. **Coral Reefs are generally found in the latitudinal extensions of**
a) 20°N – 20°S
b) 30°N – 30°S
c) 45°N – 45°S
d) 60°N – 60°S

83. **Circular reefs enclosing a shallow body of water are called**
a) Lagoons
b) Atolls
c) Fringing reefs
d) Barrier reefs

84. **According to the 'Subsidence hypothesis' of Dana and Darwin**
a) a fringing reef forms a barrier reef which in turn forms an atoll
b) a barrier reef forms a fringing reef and an atoll
c) an atoll forms a fringing reef due to subsidence of volcanic cones
d) none of the above

85. **The Natron lake deposits are**
a) Sodium chlorides
b) Sodium sulphates
c) Sodium carbonates
d) Sodium sulphides

86. **The term 'Rock Glacier' denotes**
a) a hardened boulder clay
b) glacio-fluvial sediments scattered erratically on the country
c) flowage of eluvial deposits through valleys in the form of streams
d) huge blocks covered by large ice-sheets

87. **The sea water is cool in the deeper zones because**
a) energy is exchanged largely at the surface

b) heat is lost mostly at the surface

c) salinity increases with depth

d) temperature decreases with depth

88. The process in which the glacial wastage takes place by the double process of evaporation and melting is known as

a) Nivation

b) Calvation

c) Ablation

d) Plucking

89. Glaciers approximately occupy———% of the world's area

a) 10

b) 15

c) 18

d) 20

90. When an iceberg floats in the sea, only———of its mass is visible above the water

a) 1/4

b) 1/5

c) 1/8

d) 1/9

91. When the distributaries of a river are only few and the material deposited is fine-grained, it gives rise to a/an ?

a) Arcuate delta

b) Cuspate delta

c) Bird's foot delta

d) Elongated delta

92. Coastlines of submergence are produced by

I. Subsidence of land

II. Rise in sea level

a) I only

b) II only

c) both I and II

d) none of the above

93. Pick out the set that designate the features associated with coastlines of emergence

a) sea caves, blow holes, arches, stacks

b) sea caves, blow holes, benches, offshore bars

c) wave-cut benches, wave-built terraces, arches, stacks

d) wave-cut benches, wave-built terraces, offshore bars, spits

94. The coastline formed when the mountain ridges running parallel to the coast are submerged as termed

a) Fjord coastline

b) Ria coastline

c) Dalmatian coastline

d) Haff coastline

95. A beach of sand or gravel that connects two islands, or an island with the mainland is termed

a) Tombolo

b) Gloup

c) Bar

d) Groyne

96. Match the following and choose the correct answer

| | |
|---|---|
| A. Gloup | (i) coastline formed by partial submergence of a glaciated area of high relief |
| B. Hook | (ii) transverse barriers across the seashore to check coastal erosion |
| C. Groyne | (iii) a curved ridge formed by marine construction |
| D. Fjord | (iv) sea cave with holes on its roof |

a) A–(iv) B–(ii) C–(iii) D–(i)

b) A–(iv) B–(iii) C–(ii) D–(i)

c) A–(iii) B–(iv) C–(ii) D–(i)

d) A–(iii) B–(iv) C–(i) D–(ii)

97. Which of the following weathers most rapidly ?

a) Anorthite

b) Bytownite

c) Albite

d) Orthoclase

98. A soil type whose B-horizon or illuvial zone is dominantly composed of calcareous sediments may be

1. Pedalfer
2. Pedocal
3. Podzol
4. Cherozems

a) only 3 is correct

b) only 4 is correct

c) 1 and 3 are correct

d) 2 and 4 are correct

99. Soils which show good development of all its layers are

a) Podzols

b) Pedocals

c) Chernozems

d) Loam

100. The most important layer in a soil profile which facilitates geomorphic studies of a region is

a) Eluvial zone

b) Illuvial zone

c) Clay Pan

d) Weathered country rocks

101. Soils in which sand, clay and humus are found more or less in equal properties are calle.l

a) Loamy soils

b) Regur

c) Chernozem

d) Pedalfer

102. The chernozem soils characteristically exhibits

a) high humic acid content, high microbiological activity, high cation exchange capacity

b) low humic acid content ; low microbiological activity, low cation exchange capacity

c) low humic acid content, low microbiological activity, very high cation exchange capacity

d) none of the above

103. Granitic terranes are characterized by the presence of

a) Black cotton soils

b) Lateritic soils

c) Red loamy soils

d) Chernozems

104. The term 'Seat Earth' denotes

a) deposition of mud over plant debris

b) old soil penetrated by rootlets

c) accumulation of plant debris

d) preservatian of plant debris by cementation

105. Low lying lands where the water table has just reached the land surface are called

a) Swamps

b) Marshes

c) Lagoons

d) Oasis

106. Streams which flow in the same direction as the original consequents but at a lower topographic level are known as

a) Obsequent streams
b) Subsequent streams
c) Resequent streams
d) Insequent streams

107. Which of the following is NOT a feature associated with fluvial geomorphic cycle ?
a) Cascade
b) Bluff
c) Slough
d) Matterhorn

108. Cirques are formed by the process of
a) Nivation
b) Calvation
c) Ablation
d) Abrasion

109. Undulating topography of flat till plains formed by glacial boulder beds is described as
a) Knob and kettle topography
b) Basin and range topography
c) Swash and swale topography
d) None of the above

110. Areas which are found at the margins of ice-sheets are called
a) Permafrost regions
b) Periglacial areas
c) Sub-arctic regions
d) Englacial areas

111. Long sinuous ridges of sand deposited as top of the swamp deposits of a delta are called
a) Bayou
b) Chenier
c) Panplain
d) Hamada

112. Match the following and choose the correct answer

| | |
|---|---|
| A. Bourne | 1. Network of meandering ridges composed of stratified glacio-fluvial deposits |
| B. Talik | 2. Layer of frozen water on the top of land, ice or river |
| C. Naled | 3. Layer of rock with a temperature |

above 0°C and surrounded by frozen layers

D. Osser 4. A seasonal spring

a) A—4 B—2 C—3 D—1
b) A—1 B—3 C—2 D—4
c) A—4 B—3 C—2 D—1
d) A—2 B—3 C—1 D—4

113. Shield volcanoes are formed due to the accumulation of
a) Acidic lavas with low viscosity
b) Acidic lavas with high viscosity
c) Basic lavas with low viscosity
d) Basic lavas with high viscosity

114. When water from extraneous sources is heated within and cause explosions without blowing of incandescent materials, it is described as
a) Phreatic explosion
b) Fissure eruption
c) Crater eruption
d) Nuee's ardentees

115. The lithified and indurated ash formed when volcanic eruptions takes place under water or ice is termed
a) Agglomerate
b) Igdimbirite
c) Pumice
d) Palagonite

116. Seas which are nearly surrounded by land masses are called
a) Relict seas
b) Epeiric seas
c) Marginal seas
d) Shelf seas

117. Pick out the correct sequence of the ideal shore profile
a) Cliff——Bench——Beach——Terrace
b) Cliff——Beach——Bench——Terrace
c) Beach——Bench——Cliff——Terrace
d) Beach——Bench——Terrace——Cliff

118. Match the following and choose the correct answer.

A. Colluvium 1. Deposits formed by mass wasting processes

B. Proluvium 2. Sediments deposited by mudflows and other flash flows

C. Eluvium — 3. Residual products of the breakdown of primary rocks

D. Deluvium — 4. Residual products of the breakdown of mobile rocks

a) A—4 B—2 C—3 D—1
b) A—1 B—3 C—2 D—4
c) A—4 B—3 C—2 D—1
d) A—1 B—2 C—3 D—4

*119. Which of the following is NOT matched correctly
a) Ponors......................Swallow holes
b) Roche Mountonees.....................Sheep rocks
c) Kames Terraces.....................Island of Mountains
d) Erratics......................Rocking stones

120. Large whale-back dunes with extreme heights are described as
a) Dome dunes
b) Draas
c) Star dunes
d) Parabolic dunes

121. A desert consisting of extensive sheets of gravel and boulders is known as
a) Erg
b) Hamada
c) Koum
d) Reg

*122. Which of the following dunes are characterized by the presence of multiple slip faces ?
a) Barchans
b) Dome dunes
c) Blowouts
d) Star dunes

123. Most dunes migrate laterally but a few grow vertically. Those showing the later type of growth are called
a) Reverse dunes
b) Draas
c) Linear dunes
d) Star dunes

124. Match the following and choose the correct answer

A. Rhourd — 1. Circular bowl with a dune on the downwind side

B. Compound dunes — 2. Growth of two or more basic types

| | |
|---|---|
| | of dunes |
| C. Complex dunes | 3. Several dunes of the same type combined by overlapping or superposition |
| D. Blowout dune | 4. Large nodal star dune |

a) A—4 B—3 C—2 D—1

b) A—4 B—2 C—3 D—1

c) A—1 B—3 C—2 D—4

d) A—1 B—2 C—3 D—4

125. Which of the following dunes appear U-shaped in plan view ?

a) Barchans

b) Parabolic dunes

c) Seifs

d) Dome dunes

126. Which of the following aeolian processes is most important for the formation of yardangs ?

a) Attrition

b) Abration

c) Deflation

d) Transportation

127. The coalescence of two or more alluvial fans built by neighbouring streams gives rise to a/an

a) Alluvial cone

b) Accumulation terrace

c) Bahada

d) Pediment

128. In a meandering stream

a) The thickness of the stream channel remains constant

b) Erosion is prominent on the inside of the bend and deposition on the outside

c) Erosion is prominent on the outside of the bend and deposition on the inside

d) Fine-grained deposits are found on the convex side and coarsegrained on the concave

129. The land counterpart of a delta is a/an

a) Alluvial fan

b) Stream Terrace

c) Pediment

d) Natural levee

*130. A "Yazoo type" of river is one

which cuts through the overlying rocks and maintains its

course irrespective of surface character

b) which carries load more than its capacity

c) which shows a highly complieated interconnecting pattern of meandering and braided streams

d) which runs parallel to the main river for many miles before finally finding a place to enter it

131. The sudden collapse of vapour bubbles in the water of a stream is due to the process of

a) Solution

b) Cavitation

c) Corrosion

d) Corrasion

*132. 'Pater noster' lakes are found in

a) Fluvial environment

b) Semi-arid environment

c) Pluvial environment

d) Glacial environment

133. Which of the following is NOT a stratified glacial drift ?

a) Drumlins

b) Eskers

c) Kames

d) Varves

*134. Which of the following is correctly matched ?

a) Lahars............small round or oblong reefs with enclosed lagoons

b) Faros............mound-like deposits of $CaCO_3$ of organic origin

c) Sandur............valley fill deposits of braided streams

d) Bioherm........ ...mudflows formed by volcanic activity

*135. Peneplains are characterized by one or more of the following feature

i) Thick mantle of deeply weathered waste

ii) Truncation of rocks of varying resistance

iii) Drainage independent of the structure of underlying rocks

iv) Topographic unconfirmities or nick points

a) (i) and (ii) only

b) (i) and (iii) only

c) (i), (ii) and (iv) only

d) (ii), (iii) and (iv) only

136. A plain formed by the growth and joining of floodplains is

called

a) Pedeplain
b) Peneplain
c) Panplain
d) Etch plain

137. Which of the following stages designates the Geomorphic cycle of Davis ?
a) Rapid uplift——standstill period——degradation——peneplanation
b) Standstill period——rapid uplift——degradation——peneplanation
c) Peneplanation——rapid uplift——degradation——standstill period
d) None of the above

138. According to W.M. Davis, the evolution of landscape is a function of
a) Structure and time
b) Structure, stage and time
c) Structure, lithology and time
d) Structure, stage and process

139. According to Walter Penck's concept of geomorphic cycle
a) Rapid rise of land follows periods of standstill
b) Slow uplift of land follows periods of accelerated uplift
c) Slow uplift of land is followed by periods of accelerated uplift
d) Dynamic equilbrium ensues between erosion and deposition towards the end of the cycle

140. Landforms that bear the imprint of more than one-cycle of erosion are known as
a) Simple landforms
b) Compound landforms
c) Monocyclic landforms
d) Multicyclic landforms

141. Geomorphologists have estimated that probably 90% of the present landscapes are of
a) Pre-paleozoic age
b) Pre-mesozoic age
c) Post-tertiary age
d) Quaternary age

142. When a structural high is expressed as a topographic low the condition is described as
a) Topographic anomaly

b) Bouguer anomaly
c) Inversion of topography
d) Basin and range topogrophy

143. Which of the following features represents an inversion of topography
a) Homoclinal valleys
b) Homoclinal ridges
c) Anticlinal valleys
d) Synclinal valleys

144. The expression of topography is most striking in the case of
a) Faults
b) Folds
c) Joints
d) Domes

*145. The Basin and range topography is an effect of
a) Faulting
b) Folding
c) Jointing
d) All of the above

## PHYSICAL GEOLOGY AND GEOMORPHOLOGY (KEY)

| | | | |
|---|---|---|---|
| 1. c | 2. d | 3. d | 4. a |
| 5. a | 6. a | *7. b | 8. d |
| 9. d | 10. b | 11. d | 12. c |
| 13. b | 14. b | 15. e | 16. c |
| 17. a | *18. c | 19. b | 20. d |
| 21. c | 22. a | 23. b | 24. a |
| 25. b | 26. c | 27. c | 28. a |
| 29. a | 30. c | 31. d | 32 d |
| 33. d | 34. b | 35. c | 36. a |
| 37. a | 38. b | *39. a | 40. c |
| *41. b | 42. a | 43. d | 44. b |
| 45 a | 46. c | 47. a | *48. a |
| 49. b | 50. c | 51. a | 52. d |
| 53. a | 54. b | 55. c | 56. b |
| 57. b | 58. d | 59. a | 60. b |
| 61. b | 62. c | 63. b | 64. b |
| 65. a | 66. c | 67. b | 68. b |
| 69. d | 70. c | 71. a | 72. c |
| 73. c | 74. d | 75. b | 76. c |

| | | | |
|---|---|---|---|
| 77. d | 78. d | 79. b | 80. c |
| 81. d | 82. b | 83. b | 84. a |
| 85. c | 86. c | 87. a | 88. c |
| 89. a | 90. d | 91. c | 92. c |
| 93. d | 94. c | 95. a | 96. b |
| 97. a | 98. d | 99. a | 100. b |
| 101. a | 102. a | 103. c | 104. b |
| 105. a | 106. c | 107. d | 108. a |
| 109. c | 110. b | 111. b | 112. c |
| 113. c | 114. a | 115. d | 116. c |
| 117. b | 118. b | *119. c | 120. b |
| 121. d | *122. d | 123. d | 124. a |
| 125. b | 126. c | 127. c | 128. c |
| 129. a | *130. d | 131. b | *132. d |
| 133. a | *134. c | *135. c | 136. c |
| 137. a | 138. d | 139. c | 140. d |
| 141. c | 142. c | 143. c | 144. a |
| *145. a. | | | |

## EXPLANATORY NOTES

*7. (b) : Hydrated minerals tend to expand many time their original volume of anhydrous minerals. Therefore, they become more porous and attain a lower density. e.g.

| | *Density* | *Volume Change* (%) |
|---|---|---|
| $CaSO_4$ | 2.61 | 42.27 |
| $CaSO_4.2H_2O$ | 2.32 | |

*18. (c) : Serir is a stony desert composed of extensive sheets of gravel or boulders. It is also known as Reg or Harras.

*39. (a) : Radial drainage patterns are associated with domal uplifts, slopes of volcanoes, large basins and gulf depressions.

Choices (b) and (c) exhibit trellis pattern while (d) exhibits Dendritic pattern.

*41. (b) : The other three features i.e. (a), (c) and (d) are associated with the middle course of a river. Piedmont plains are formed by the coalescenece of two or more alluvial fans built by neighbouring streams. Sloughs are depressions upon the flood plains of meandering rivers and are excavated during flood times.

*48. (a) : (b) refers to knob and kettle topography, (c) refers to pingos and (d) refers to incised meanders.

*119. (c) : 'Island of Mountains' refers to inselbergs which are isolated pillar-like masses in desert regions with rounded tops and steep slopes.

*122. (d) : Star dunes have 3 or more slip faces

*130. (d) : It is named after the Yazoo River which parallels the Mississippi river for nearly 200 miles before finally joining it.

*132. (d) : These are the lakes formed in the 'rock basins' in a glaciated valley formed by the plucking action of glaciers.

*134. (c) : Lahars are mudflows formed by volcanic activity. Bioherms are mound like deposits of $CaCO_3$ of organic origin. Faros are small, round or oblong reefs with enclosed lagoons.

*135. (c) : The criteria used for identifying uplifted peneplains are : accordant interstream levels and summit areas ; topographic unconfirmities or nick points ; truncation of rocks of varying resistance ; drainage well adapted to the structure of the underlying rocks ; erosion remnants rising above it ; incised streams below it ; and deeply weathered soils upon its surface.

*145. (a) : It is found within the Cordillerean system of USA. Here a series of tilted fault-blocks uplifted in comparatively recent times have produced a number of N-S trending mountain ridges separated by depressions.

# 3

# *Crystallography*

1. In Crystallography the ratios of the three axes a, b, c, i.e a : b : c is known as axial ratios. While representing the axial ratios, the ———axis is arbitrarily taken as unity
   a) a-axis
   b) b-axis
   c) c-axis
   d) None of the above
2. The axial elements of the Orthorhombic system can be known fully only on determining
   a) c/a ratio
   b) a/b and c/b ratio
   c) a/b and c/b ratios and angle
   d) None of the above
3. A plane of symmetry is always
   a) Parallel to a possible edge of the crystal
   b) Parallel to a possible face of the crystal
   c) Perpendicular to a possible face of the crystal
   d) None of the above
4. A Fundamental or unit form has
   a) Maximum number of faces in a crystal
   b) Minimum number of faces in a crystal
   c) Faces each of which intersect the axes at unit lengths
   d) Faces each of which cuts all the axes at unequal lengths

*5. Which of the following are Orthogonal systems ?
   a) Cubic, Tetragonal, Orthorhombic
   b) Tetragonal, Hexagonal, Trigonal
   c) Orthorhombic, Monoclinic, Triclinic
   d) Tetragonal, Hexagonal, Orthorhombic

*6. A solid in which a face on one side is opposite a point on the other is definitely
   a) Centrosymmetric
   b) Non-centrosymmetric
   c) Enantiomorphous
   d) Hemihedral

7. There are 48 different forms in crystals. Of these, the Cubic system has
   a) 9 forms
   b) 15 forms
   c) 23 forms
   d) 33 forms

8. Any two faces of a form are called 'Congruent' if one face can be brought into the position of the other by
   a) Rotation
   b) Reflection
   c) Inversion
   d) All of the above

9. Which of the following systems has all closed forms ?
   a) Triclinic
   b) Monoclinic
   c) Trigonal
   d) Cubic

10. Axial planes are planes of symmetry which
    a) Bisect 'a' and 'c' axes
    b) Bisect 'b' and 'c' axes
    c) Enclose only one crystallographic axis
    d) Enclose two crystallographic axes

11. Of the 48 naturally occurring forms———are closed and ———are open
    a) 30, 18
    b) 18, 30
    c) 22, 26
    d) 26, 22

12. Which of the following is NOT a closed form
    a) Rhombohedron
    b) Scalenohedron
    c) Trapezohedron
    d) None of the above

13. A crystal form consisting of only one face is
    a) Sphenoid
    b) Basal pinacoid
    c) Pedion
    d) Dome

14. A 'Pyramid' is described as a form whose face always cuts

a) a-axis
b) b-axis
c) c-axis
d) All of the above

15. Domes may be considered as
a) Vertical prisms
b) Non-vertical prisms
c) Vertical pyramids
d) Non-vertical pyramids

16. The common feature between pyrite type and tetrahedrite type in the cubic system is
a) Both are centrosymmetric
b) Both are non-centrosymmetric
c) Both have same axial symmetry elements
d) Both have same number of symmetry planes

*17. Match the following and choose the correct answer

| | |
|---|---|
| A. Sphenoid | 1. Trigonal system |
| B. Rhombohedron | 2. Monoclinic system |
| C. Scalenohedron | 3. Tetragonal system |
| D. Dome | 4. Orthorhombic system |
| | 5. Hexagonal system |

a) A—5 B—1 C—1 D—2
b) A—2 B—1 C—3 D—4
c) A—4 B—5 C—1 D—3
d) A—2 B—1 C—3 D—5

18. Trapezohedron is a form that occurs in
a) Hexagonal system
b) Trigonal system
c) Tetragonal system
d) All of the above

*19. Scalenohedron is a form having
a) 8 faces
b) 12 faces
c) Either 8 or 12 faces
d) 24 faces

20. A rhombohedron has
a) Vertical symmetry planes
b) Horizontal symmetry planes
c) Diagonal symmetry planes

d) All of the above

21. Forms which have fixed Miller symbols are described as
a) Limiting forms
b) General forms
c) Special forms
d) Congruent forms

22. Which of the following combinations correctly designate limiting forms ?
a) 0001, $10\bar{1}1$, $11\bar{2}0$
b) 001, 101, 100
c) 100, 110, 111
d) 001, 111, 110

23. The limiting forms of Trisoctahedron are
a) Octahedron and cube
b) Rhombdodecahedron and cube
c) Rhombdodecahedron and octahedron
d) Positive and negative pyritohedra

24. Match the following and choose the correct answer

| *Crystal system* | *Characteristic symmetry* |
|---|---|
| A. Monoclinic | 1. 4 triads |
| B. Tetragonal | 2. 3 diads |
| C. Orthorhombic | 3. 1 tetrad |
| D. Cubic | 4. 1 diad |

a) A—1 B—3 C—2 D—4
b) A—3 B—1 C—4 D—2
c) A—4 B—3 C—2 D—1
d) A—2 B—3 C—4 D—1

*25. Which of the following pairs does NOT have the same number of classes in the system ?
a) Monoclinic—Orthorhombic
b) Cubic—Trigonal
c) Tetragonal—Hexagonal
d) Monoclinic—Triclinic

*26. In a 'general form' the faces lie
a) Parallel to symmetry axes
b) Parallel to symmetry planes
c) Perpendicular to symmetry axes
d) Perpendicular to symmetry planes
e) None of the above

27. "All forms are general forms" in
   a) Cubic system
   b) Hexagonal system
   c) Monoclinic system
   d) Triclinic system

28. Which of the following crystal systems has one asymmetric class ?
   a) Isometric
   b) Trigonal
   c) Monoclinic
   d) Triclinic

29. A hemiorthodome is also described as
   a) Horizontal pyramid
   b) Horizontal prism
   c) Vertical pyramid
   d) Vertical prism

30. The characteristic feature of a 'variable form' is
   a) Variation in the total number of faces
   b) Variation in the relative ratios of intercepts of a face on the crystallographic axes
   c) Variation in the symmetry elements
   d) Variation in the crystal habit

31. Which of the following statements is true of hemihedral forms ?
   (i) They belong to the same crystal class as the Holohedral forms
   (ii) They have the same indices as the Holohedral form
   (iii) They have half the number of faces as the holohedral form
   (iv) For each holohedral form there corresponds two hemihedral forms
   a) (i) and (iii) are correct
   b) (ii) and (iii) are correct
   c) (ii), (iii) and (iv) are correct
   d) (i), (ii), (ii) and (iv) are correct

32. Which of the following is NOT a hemihedral class ?
   a) Gyroidal
   b) Hexoctahedral
   c) Diploidal

d) Hextetrahedral

33. Pyritohedron is the hemihedral form of
   a) Trisoctahedron
   b) Tetrahexahedron
   c) Trapezohedron
   d) Deltoid-Dodecahedron

34. All holosymmetric classes are
   a) Centrosymmetric
   b) Non-centrosymmetric
   c) Enantiomorphic
   d) Hemimorphic

35. How many classes among the seven crystal systems does not possess a centre of symmetry ?
   a) 32
   b) 24
   c) 21
   d) 15

36. Which of the following statements is/are true of enantiomorphic crystals ?
   (i) In such crystals one face can be brought into the position of another by reflection or inversion
   (ii) They do not possess any symmetry elements
   (iii) They are incongruent
   (iv) They do not exist in the cubic system
   a) (i), (ii) and (iii) are correct
   b) (i) and (iii) are correct
   c) (ii), (iii) and (iv) are correct
   d) (ii) and (iv) are correct

37. Match the following and choose the correct answer

| | |
|---|---|
| A. Spinel | 1. Cubic |
| B. Idiocrase | 2. Tetragonal |
| C. Aragonite | 3. Orthorhombic |
| D. Hematite | 4. Trigonal |
| | 5. Monoclinic |

   a) A—1 B—2 C—3 D—4
   b) A—5 B—2 C—4 D—3
   c) A—1 B—2 C—5 D—3
   d) A—5 B—2 C—1 D—4

38. In the monoclinic system the diad axis is parallel to

a) X-axis
b) Y-axis
c) Z-axis
d) None of the above

39. The trigonal unit cell is designated to
a) $a=b\neq c$ ; $\alpha=\beta=90°$ $\gamma=120°$
b) $a=b=c$ ; $\alpha=\beta=\gamma\neq 120°$
c) $a=b=c$ : $\alpha=\beta=\gamma\neq 90°$
d) $a\neq b\neq c$ ; $\alpha=\beta=\gamma<120°>90°$

The line joining the centres of two opposite faces of a cube is a
a) Diagonal axis
b) Trigonal axis
c) Tetragonal axis
d) Hexagonal axis

41. In the triclinic system all the forms have
a) 2 faces
b) 4 faces
c) 6 faces
d) Some forms have 2 faces, others have 4 or 6 faces

42. Which of the following denotes a class that has no symmetry at all ?
a) Pinacoidal
b) Pedial
c) Domatic
d) Sphenoidal

43. The prismatic, domatic and sphenoidal classes are characteristic of
a) Trigonal system
b) Triclinic system
c) Monoclinic system
d) Orthorhombic system

44. Which of the following crystals does not have a centre of symmetry ?
a) Tetrahedrite
b) Tourmaline
c) Quartz
d) All of the above

45. Which of the following is NOT a general form ?
a) Hexoctahedron

b) Hextatrahedron

c) Ditrigonal pyramid

d) Hexagonal bipyramid

46. A face which is parallel to the vertical crystallographic axis is a

a Pyramidal face

b) Prism face

c) Basal pinacoid

d) Pedion

47. The relative abundance of a particular face is a function of

a) The number of molecules in the lattice structure

b) The number of atoms in the lattice structure

c) Mode of occurrence of the crystals

d) The degree of crystallization of the crystals

48. A 'corner' is formed by the intersection of a minimum number of

a) 2 faces

b) 3 faces

c) 4 faces

d) None of the above

49. When the parameters are 2a : infinity b : 1c, the miller indices would be

a) 200

b) 201

c) 102

d) 210

50. The most general expression for a crystal face in the Weiss notation is

a) Xa, Yb, Zc

b) na, mb pc

c) n/a, m/b, p/c

d) 1/a, 2/b, 3/c

51. The general symbol of hexagonal crystal is designated as (hkil). The sum of the first three axes h+k +i is equal to

a) 90°

b) 120°

c) >90° but <120°

d) Zero

52. The maximum number of zone axes are found in

a) Orthorhombic system
b) Cubic system
c) Tetragonal system
d) Monoclinic system

53. The cubic system is characterized by
a) Absence of singular axis
b) One singular axis
c) Three singular axes
d) One singular axis with innumerable singular axes normal to it

54. Which of the following has three axes of 4-fold symmetry ?
a) Barytes
b) Gypsum
c) Galena
d) Rutile

55. Variation in the habit of crystal changes
a) Crystal class
b) Crystal system
c) Symmetry elements
d) Crystal form

56. Zircon type may be designated as having
a) 3 planes, 5 axes and centre
b) 5 planes, 5 axes and centre
c) 5 planes, 3 axes and no centre
d) 5 planes, 5 axes and no centre

57. A brachypinacoid face is parallel to
a) a-axis
b) b-axis
c) b and c-axes
d) a and c-axes

58. The interfacial angles depend upon
a) Size of the crystal faces
b) Shape of the crystal faces
c) Form of the crystals
d) None of the above

59. The contact goniometer is useful in the case of
a) Large crystals with smooth faces
b) Small crystals with smooth faces
c) Large crystals with ill-polished faces

d) Small crystals with ill-polished faces

60. Match the axes of rotary inversion which corresponds with various operations involving rotary axes and/or mirror planes

| | | | |
|---|---|---|---|
| A. | $\bar{1}$ | 1. | Centre of symmetry |
| B. | $\bar{2}$ | 2. | Reflection plane |
| C. | $\bar{3}$ | 3. | A triad axis combined with a perpendicular mirror plane |
| D. | $\bar{6}$ | 4. | A triad axis combined with a centre |

a) A—1 B—2 C—3 D—4
b) A—1 B—2 C—4 D—3
c) A—2 B—1 C—3 D—4
d) A—2 B—1 C—4 D—3

61. In the Hermann-Mauguin symbols for the cubic system, the second position coincides with
a) a-axis
b) Normal to (111) face
c) b-axis
d) Normal to (110) face

62. Forms having unusually high indices are described as
a) Special forms
b) Incongruent forms
c) Viccinal forms
d) Irregular forms

63. The bravais cells are classified into seven groups based on
a) Symmetry of their shape
b) Difference in their shape
c) Difference in their angles
d) Mode of their origin

64. The maximum number of space lattices is found in
a) Orthorhombic system
b) Cubic system
c) Hexagonal system
d) Tetragonal system

65. Match the following and choose the correct answer

| | | | |
|---|---|---|---|
| A. | Bravais lattices | 1. | 230 |
| B. | Space groups | 2. | 14 |
| C. | Crystal classes | 3. | 5 |
| D. | Holoaxial classes | 4. | 32 |
| | | 5. | 11 |

a) A—2 B—5 C—4 D—3

b) A—2 B—1 C—4 D—5
c) A—2 B—1 C—4 D—3
d) A—5 B—1 C—4 D—3

66. In a primitive unit cell the points are found
a) Only at the centre
b) Only at the corners
c) Only at the edges
d) None of the above

67. Lattice energy is the
a) Energy released when one mole of a crystalline substance is broken into its ions
b) Energy released when one mole of a substance is crystallized from its ions
c) Energy absorbed by a crystalline substance when it is exposed to ultraviolet light
d) Energy required to convert an amorphous substance into a crystalline solid

68. Which of the following is wrongly matched ?

| *Crystal system* | *Cell shape* |
|---|---|
| a) Cubic | Cube |
| b) Tetragonal | Square prism |
| c) Orthorhombic | Rectangular parallelelepid |
| d) Hexagonal | Regular hexagon |

69. The bravais lattice of a sodium chloride structure is
a) Primitive cell
b) Body-centred cube
c) Face-centred cube
d) Base-centred cube

70. In the tetragonal system
a) The prism of the first order is bigger than the prism of the second order
b) The prism of the first order is smaller than the prism of the second order
c) Both the prisms are of equal size
d) None of the above

71. Tetrahedron is derived from octahedron by the development of
a) Adjacent octants

b) Alternate octants

c) Adjacent cubes

d) Alternate cubes

72. A pyritohedron is described as positive when its

a) Long axis is vertical

b) Long axis is horizontal

c) Long axis is inclined

d) None of the above

73. Parting is a property due to

a) Crystalline structure

b) Crystal Defects

c) Twinning

d) Molecular disorders

74. The most common types of twins encountered in nature are

a) Contact twins

b) Penetration twins

c) Growth twins

d) Polysynthetic twins

75. An axis of———symmetry in an individual crystal can serve as a twinning axis

a) Two fold

b) Three fold

c) Four fold

d) Six fold

76. Twinning that originates right from the time of crystallization is termed

a) Paragenetic

b) Metagenetic

c) Diagenetic

d) Polygenetic

77. If the twin planes in a repeated twin does not remain parallel it is described as

a) Polysynthetic twinning

b) Lamellar twinning

c) Cyclic twinning

d) Complex twinning

78. A type of twinning in which the individual immitates the symmetry of a higher rank is described as

a) Mimetic twinning

b) Supplimentary twinning

c) Nuclear twinning
d) Tartan twinning

79. A crystal twinned on the lattice plane is known as
a) Rotation twin
b) Reflexion twin
c) Inversion twin
d) Contact twin

80. In reflexion twins
a) Twin axis is parallel to zone axis
b) Composition plane is parallel to twin axis
c) Composition plane is parallel to twin plane
d) Composition plane is parallel to zone axis

81. Match the following and choose the correct answer

| | |
|---|---|
| A. Geneculated twins | 1. Staurolite |
| B. Cruciform twins | 2. Rutile |
| C. Butterfly twins | 3. Gypsum |
| D. Arrow head twins | 4. Calcite |

a) A—2 B—3 C—4 D—1
b) A—2 B—4 C—1 D—3
c) A—2 B—3 C—1 D—4
d) A—2 B—1 C—4 D—3

82. Which of the following is correctly matched ?

| *Twin* | *Twin plane* | *Composition plane* |
|---|---|---|
| a) Carlsbad | 100 | 100 |
| b) Carlsbad | 100 | 010 |
| c) Baveno | 021 | 110 |
| d) Skew | 032 | 032 |

83. Twinning in quartz does NOT take place on which of the following laws
a) Dauphine law
b) Brazil law
c) Japan law
d) None of the above

84. If a quartz crystal is twinned by the two parts of the same hand i.e., right and right or left and left, it is described as
a) Optical twins
b) Electrical twins
c) Japan twins
b) Saw-horse twins

85. In the 'Skew' twins the twin plane is a

a) Dome
b) Bipyramid
c) Prism
d) Side pinacoid

86. Which of the following is wrongly matched ?

| Twin | C.P. |
|---|---|
| a) Ala A | 001 |
| b) Ala B | 010 |
| c) Acline A | 001 |
| d) Carlsbad A | 100 |

87. The 'Pericline' twins of plagioclases are often associated with
a) Skew twins
b) Carlsbad twins
c) Manebach twins
d) Albite twins

88. The type of twinning in which two individuals having less symmetry than the normal class interpenetrate so as to produce a crystal having outwardly the holohedral symmetry of the system is known as
a) Multiple twinning
b) Growth twinning
c) Supplementary twinning
d) Compound twinning

89. A crystal of quartz is capable of exhibiting
a) Piezoelectricity
b) Pyroelectricity
c) Both of the above
d) None of the above

*90. Crystals of which of the following classes are neither piezo-electric nor pyroelectric
a) Tetartoidal
b) Diploidal
c) Sphenoidal
d) Pedial

91. The law for the reflection of X-ray from a crystal of Sylvite (KCI) is given by the equation
a) $n=2d \sin \theta$
b) $n\lambda=2 \sin \theta$
c) $2n=d\lambda \sin \theta$
d) $n\lambda=2d \sin \theta$

*92. X-ray methods cannot subdivide the classes belonging to

a) Cubic system

b) Tetragonal system

c) Hexagonal system

d) Orthorhombic system

93. The accurate and meaningful portrayal of 3-dimensional objects on two-dimensional paper is described as

a) Projection

b) Tracing

c) Contouring

d) Air photography

94 The simplest of all the crystal projections is

a) Spherical projections

b) Stereographic projection

c) Gnomonic projection

d) None of the above

95. In the stereographic projection the axis normal to the plane of projection is

a) a-axis

b) b-axis

c) c-axis

d) None of the above

96. The characteristic feature of Gnomonic projection is

a) All great circles plot as arcs of a circle

b) All great circles plot as straight lines

c) The angular relationship between two or more planes is preserved

d) None of the above

97. In an orthographic projection the plane of projection is

a) parallel to projection lines

b) perpendicular to projection lines

c) inclined to projection lines

d) none of the above

98. A 'Screw axis' is obtained by combining the symmetry operations of

a) Reflection and translation

b) Rotation and reflection

c) Rotation and translation

d) None of the above

99. Match the following and choose the correct answer

| | | | |
|---|---|---|---|
| A. | Hexahedron | 1. | Pyritohedron |
| B. | Icositetrahedron | 2. | Diploid |
| C. | Didodecahedron | 3. | Trapezohedron |
| D. | Pentagonal dodecahedron | 4. | Cube |

a) A–1 B–2 C–3 D–4
b) A–4 B–3 C–1 D–2
c) A–1 B–3 C–2 D–4
d) A–4 B–3 C–2 D–1

100. [331] and [111] are
a) face symbols
b) form symbols
c) zone symbols
d) none of the above

101. Which of the following faces lie in the zone $\bar{1}10$ –011 – 101 – $1\bar{1}0$ ?
a) 010
b) 001
c) $\bar{1}\bar{1}1$
d) 121

*102. The characteristic symmetry of the triclinic system is
a) Monad axis
b) Diad axis
c) Triad axis
d) None of the above

103. The only symmetry that a triclinic crystal can display is
a) Centre of symmetry
b) Planes of symmetry
c) Axes of symmetry
d) None of the above

104. "Anorthic System" is the synonym for
a) Cubic system
b) Orthorhombic system
c) Triclinic system
d) Monoclinic system

105. A zone denotes a set of faces which meets in a
a) series of parallel edges
b) series of parallel reflection planes
c) series of alternate edges
d) none of the above

106. The different forms in a crystal class are distinguished on the basis of
   a) axial ratios
   b) crystallographic notation
   c) symmetry elements
   d) interfacial angles

107. Which of the following classes is both hemimorphic as well as enantiomorphic ?
   a) Gyroidal
   b) Pedial
   c) Sphenoidal
   d) Pinacoidal

108. Which of the following is NOT an enantiomorphic form ?
   a) Pedion
   b) Tetartoid
   c) Trapezohedron
   d) Hexahedron

109. 'Scalenohedron' occurs in
   a) Trigonal system
   b) Tetragonal system
   c) Both 'a' and 'b'
   d) None of the above

110. 'Dome' is characteristic of
   a) Orthorhombic and monoclinic
   b) Monoclinic and triclinic
   c) Orthorhombic and triclinic
   d) Orthorhombic and trigonal

111. Two non-parallel faces symmetrical with respect to an axis of 2-fold symmetry constitute
   a) Pinacoid
   b) Sphenoid
   c) Dome
   d) Pedion

*112. Arrange the crystal systems in the decreasing order of abundance of known crystals
   a) Cubic > Tetragonal> Hexagonal>Orthorhombic>Monoclinic
   b) Monoclinic > Orthorhombic > Triclinic > Cubic>Tetragonal
   c) Orthorhombic>Cubic > Tetragonal > Hexagonal>Mono-

clinic

d) Cubic>Orthorhombic>Monoclinic > Hexagonal > Tetragonal

113. Arrange the crystal systems in the order of decreasing importance of twinning

a) Monoclinic > Orthorhombic > Cubic>Trigonal > Tetragonal

b) Monoclinic>Triclinic>Orthorhombic>Hexagonal>Cubic

c) Cubic>Hexagonal > Monoclinic > Triclinic>Orthorhombic

d) Triclinic>Monoclinic>Orthorhombic>Hexagonal>Cubic

114 Match the following and choose the correct answer

| | | |
|---|---|---|
| A. | α-Quartz | 1. Hexagonal |
| B. | β-Quartz | 2. Trigonal |
| C. | Tridymite | 3. Tetragonal |
| D. | Cristobalite | 4. Monoclinic |

a) A–1 B–2 C–3 D–4

b) A–1 B–2 C–4 D–3

c) A–2 B–1 C–4 D–3

d) A–2 B–1 C–3 D–4

115. A pattern produced by reflection combined with translation parallel to the reflecting plane is called

a) Glide plane

b) Screw axis

c) Mirror plane

d) Inversion axis

116. Optical activity is the positive evidence of

a) Hemimorphism

b) Hemihedrism

c) Enantiomorphism

d) Tetartohedrism

*117. Which of the following classes is optically active ?

a) Pinacoidal (Triclinic)

b) Tetartoidal (Cubic)

c) Prismatic (Monoclinic)

d) Rhombohedral (Trigonal)

*118. In a crystal that has no symmetry (like the pedial class of triclinic system) every form consists of

a) only one face

b) only two faces

c) only three faces

d) either one, two or three faces

119. Match the following and choose the correct answer

| | | | |
|---|---|---|---|
| A. | Cube | 1. | Quadrilateral |
| B. | Rhombohedron | 2. | Equilateral triangle |
| C. | Trapezohedron | 3. | Square |
| D. | Rhombdodecahedron | 4. | Rhomb |

a) A–3 B–4 C–1 D–2
b) A–3 B–4 C–2 D–1
c) A–3 B–2 C–1 D–4
d) A–3 B–2 C–4 D–1

120. Which of the following combinations is incorrect ?
a) Optical twins——–Quartz
b) Cyclic twins——–Aragonite
c) Glide twins——–Calcite
d) Tartan twins——–Pyrite

121. When a crystal growth takes place from a supersaturated solution, the resulting crystals are known as
a) Hopper crystals
b) Microlites
c) Viccinal forms
d) Spiral crystal

122. In the Hopper crystals
a) The centres of faces are more developed than the edges
b) The edges are more developed than the centres of faces
c) The faces, edges and corners are more less equally developed
d) None of the above

123. In the Pericline law, the composition plane is a
a) Cubic section
b) Octahedral section
c) Rhombic section
d) Prismatic section

124. "Saw-Horse" twins are exhibited by
a) Chiastolite
b) Gypsum
c) Fluorite
d) Staurolite

125. Galena crystallises in the form of

a) Cubes and octahedra
b) Trapezohedra
c) Cubes and rhombohedra
d) Rhombdodecahedra

126. Which of the following forms corresponds to the 'Scalenohedron' of the calcite type ?
a) Trigonal pyramid of tourmaline type
b) Ditrigonal pyramid of tourmaline type
c) Trigonal trapezohedron of quartz type
d) Trigonal pyramid of quartz type

127. Match of the following and choose the correct answer

| | *Mineral* | *Crystal system* |
|---|---|---|
| A. | Haematite | 1. Cubic |
| B. | Olivine | 2. Monoclinic |
| C. | Wollastonite | 3. Triclinic |
| D. | Epidote | 4. Orthorhombic |
| | | 5. Trigonal |

a) A–1 B–4 C–2 D–3
b) A–4 B–2 C–5 D–3
c) A–5 B–4 C–3 D–2
d) A–5 B–2 C–3 D–1

128. Enantiomorphic crystal forms possess
a) Inversion and alternating axes only
b) Rotary axes of symmetry only
c) Planes of symmetry only
d) Centre of symmetry only

129. Piezoelectricity is characteristic of
a) Holosymmetric classes
b) Merosymmetric classes
c) Centrosymmetry classes
d) Non-Centrosymmetric classes

130. Which of the following classes, though non-centrosymmetric, do not exhibit the phenomenon of Piezo- and Pyro-electricity ?
a) Gyroidal class (432)
b) Tetartoidal class (23)
c) Rhombic disphenoidal (222)
d) Sphenoidal (2)

131. Select the form (s) which have two faces

(i) Pinacoids (ii) Domes
(iii) Sphenoids (iv) Pedion

a) (i) only
b) (i) and (iii) only
c) (i), (ii) and (iii) only
d) all of the above

*132. The symmetry of a crystal depends upon
a) shape of the faces
b) size of the faces
c) distribution of angular elements
d) crystallographic notation

133. The crystal class having two mutually perpendicular symmetry planes can be designated as
a) mm
b) 2/mm
c) 222
d) 2/m

134. The gyroidal class of the cubic system can be represented as
a) 23
b) $2/m^3$
c) $\bar{4}3$ m
d) 432

135. Which of the following classes possess a centre of symmetry ?
a) 4
b) $\bar{4}$
c) $\bar{6}$
d) $\bar{3}$

136. In the face-centred cubic cell every point is surrounded by ———nearest neighbours
a) 6
b) 8
c) 12
d) 24

137. The base-centred arrangement of unit cell is found in
a) Cubic crystals
b) Tetragonal crystals
c) Orthorhombic crystals
d) Triclinic crystals

138. In microcline, the twinning takes place on

a) Albite+Pericline laws
b) Albite+Carlsbad laws
c) Manebach+Acline laws
d) Albite+Ala A laws

139. The "Herringbone structure" of augite is due to polysynthetic twins on
a) (100)+(010)
b) (010)+(001)
c) (100)+(001)
d) (100)+(111)

140. In the Penine law, the composition face is
a) 001
b) 010
c) 100
d) 032

*141. Find the odd man out
a) $1A_2$
b) $1A_3$
c) $1A_4$
d) $1A_5$

142. Which one of the following is not a factor affecting the habit of growth of a particular substance into a crystal form ?
a) temperature of crystallization
b) presence of impurities
c) nature of the medium in which crystallization takes place
d) density of the crystallizing substance

143. Crystal faces showing solution cavities of definite shapes are described as
a) Striations
b) Etch figures
c) Erosion markings
d) Internal imperfections

144. Which of the following crystal faces always meet only one crystallographic axis and remains parallel to the other two ?
a) Pinacoid face
b) Prism face
c) Pyramid face
d) Pedion

145. As the orthopinacoid is not a plane of symmetry in the normal class of the monoclinic system———

a) the orthodome has only 4 faces
b) the clinodome has only 4 faces
c) the clinodome has only 2 faces
d) the orthodome has only 2 faces

146. Barrel-shaped crystals are formed by

a) Siderite
b) Calcite
c) Corundum
d) Topaz

147. Which of the following crystal forms have the least number of faces ?

a) Pyritohedron
b) Scalenohedron
c) Rhombohedron
d) Octahedron

148. The Trigonal pyran id of III order is the———form of dihexagonal bipyramid

a) Hemihedral
b) Tetartohedral
c) Ogdohedral
d) None of above

149. Which of the following crystal systems does not possess tetartohedral classes ?

a) Orthorhombic, Monoclinic and Triclinic
b) Tetragonal and Hexagonal
c) Orthorhombic and Monoclinic
d) Monoclinic and Triclinic

150. The 'rhombic prism' or the prism with four rhomb-shaped faces occur in

a) Tertagonal system
b) Trigonal system
c) Orthorhombic system
d) Hexagonal system

## CRYSTALLOGRAPHY (KEY)

| | | | |
|---|---|---|---|
| 1. b | 2. b | 3. c | 4. c |
| *5. a | *6. b | 7. b | 8. a |
| 9. d | 10. d | 11. a | 12. d |
| 13. c | 14. c | 15. b | 16. c |

*17. b  18. d  19. c  20. a
21. a  22. c  23. c  24. c
*25. d  *26. e  27. d  28. d
29. b  30. b  31. c  32. b
33. b  34. a  35. c  36. b
37. a  38. b  39. c  40. c
41. a  42. b  43. c  44. d
45. c  46. b  47. b  48. b
49. c  50. b  51. d  52. a
53. a  54. c  55. d  56. b
57. d  58. d  59. c  60. b
61. b  62. c  63. a  64. a
65. c  66. b  67. b  68. d
69. c  70. b  71. b  72. a
73. c  74. b  75. b  76. a
77. c  78. a  79. b  80. c
81. d  82. b  83. d  84. b
85. b  86 d  87. d  88. c
89. c  *90. b  91. d  *92. d
93. a  94. a  95. c  96. b
97. b  9‹. c  99. 3  100. c
101. d  *102. a  103. a  104. c
105. a  106. b  107. c  108. d
109. c  110. a  111. b  *112. b
113. a  114. c  115. a  116. c
*117. b  *118. a  119. c  120. d
121. a  122. b  123. c  124. d
125. a  126. b  127. c  128. b
129. d  130. a  131. c  *132. c
133. a  134. d  135. d  136. c
137. c  138. a  139. c  140. a
*141. a  142. d  143. b  144. a
145. d  146. c  147. c  148. c
149. a  150. c.

## EXPLANATORY NOTES

*5. (a) : In all these system $\alpha=\beta=\delta=90°$

*6. (b) : Like a Tetrahedron, in which the centre is absent

*17. (b) : Scalenohedron also crystallizes in Tetragonal system

*19. (c) : The Scalenohedron of tetragonal system has 8 faces

while that of trigonal system has 12 faces

*25. (d) : Monoclinic system has 3 classes (Prismatic, Domatic and Sphenoidal), whereas, triclinic system has 2 classes (pedial and pinacoidal) ; Cubic and trigonal systems have 5 classes each ; Tetragonal and Hexagonal systems have 7 classes each and the Orthorhombic sytem has 3 classes.

*26. (e) : A general form has no special relationship to any symmetry element, it intersects all three crystallographic axes and has indices of entirely general type (hkl), where, h, k and l may have any non-zero integral values.

*90. (b) : Diploidal class is centrosymmetric. Piezoelectricity and pyroelectricity is possible only in non-centrosymmetric classes.

*92. (d) : This is because X-ray methods cannot distinguish between pairs of classes in which one is non-centrosymmetrical and the other can be derived from it by adding a centre of inversion. This is true if attention is paid only to the direction of diffracted X-rays.

*102. (a) : The symmetry in triclinic system is designated as $\bar{1}$ (i.e. only a centre of symmetry) in the case of pinacoidal class and as 1 (i.e. absence of any symmetry )in the case of Pedial class.

*112. (b) : 50% of the known crystals belong to Monoclinic, 25% to Orthorhombic, 15% to Triclinic and the rest in the order given.

*117. (b) : This class is enantiomorphic

*118. (a) : Because there is no repetition at all

*132. (c) : The symmetry of a crystal is expressed in the repetition of the elements of its boundary i.e. faces, edges and apices. The symmetry is dependent on the type of space lattice and unit-cell involved in its construction.

*141. (d) : A five or ten-fold symmetry does not appear to exist in crystals. This is so because it is not possible to fill the entire crystal space in a periodic arrangement.

# 4

# *Mineralogy*

1. A mineral may be defined as
   a) A heterogenous solid body which occurs natuıally
   b) An inorganically formed substance having a definite chemical composition and occurring naturally
   c) A heterogenous solid body characterized by an orderly arrangement of atoms
   d) None of the above

2. The simplest of all the silicate structures is that of
   a) Orthosilicates
   b) Sorosilicates
   c) Metasilicates
   d) Ionosilicates

3. Epidote exhibits
   a) Nesosilicate structure
   b) Sorosilicate structure
   c) A combination of Neso- and Sorosilicate structures
   d) A three dimensional framework

4. Minerals exhibiting 'Metasilicate' structure are characterized by
   a) Tetrahedra independent of one another
   b) Tetrahedra forming closed units having a ring-like structure
   c) Tetrahedra occurring in pairs and the sharing of one oxygen between two silicons
   d) Tetrahedra forming a three dimensional network

5. Which of the following shows a 6-membered ring structure ?
   a) Benitoiite
   b) Tourmaline
   c) Beryl
   d) Axinite

*6. The maximum amount of substituion of Al for Si is found in
   a) Orthoclase
   b) Microcline
   c) Albite
   d) Anorthite

7. In a three dimensional network
   a) One oxygen is shared between two tetrahedra
   b) Two oxygens of each tetrahedron are shared
   c) Three oxygens of each tetrahedron are shared
   d) All oxygen atoms of each tetrahedron are shared

8. In sorosilicate structures, the neutralization of charges takes place by
   (1) Sharing of one oxygen atom between two adjacent tetrahedra
   (2) Presence of cations such as $Ca^{+2}$ or $Mg^{+2}$
   (3) Substitution of some Si by Al
   (4) Presence of —OH group in the structure
   a) Only 1 is correct
   b) Only 2 is correct
   c) 1 and 2 are correct
   d) 1 and 3 are correct
   e) All of the above

9. Prismatic habits are characteristic of
   a) Nesosilicates
   b) Ionosilicates
   c) Metasilicates
   d) Phyllosilicates

10. Which of the following is a non-diagnostic property ?
    a) Colour
    b) Streak
    c) Luster
    d) Hardness

11. Minerals which exhibit constant and characteristic colouration are known as
    a) Pseudochromatic
    b) Allochromatic
    c) Idiochromatic
    d) Parachromatic

*12. Cobalt imparts blue colour to a mineral when it is surrounded by
    a) Six oxygen atoms
    b) Four oxygen atoms
    c) Three oxygen atoms
    d) Two oxygen atoms

13. The purple colour of Amethyst is due to the presence of

a) Titanium
b) Manganese
c) Iron
d) Chromium

14. 'Play of colour' is a property resulting from
a) Low dispersion
b) High dispersion
c) Low interference
d) High interference

15. The phenomenon in which the surfacial covering of thin layers of a mineral exhibit different colours in different directions is described as
a) Iridescence
b) Opalescence
c) Change of colours
d) Pleochroism

16. 'Luster' is defined as the appearance of a mineral in
a) White light
b) Monochromatic light
c) Incident light
d) Reflected light

17. The adamentine luster is characteristic of minerals with R.I. value ranging between
a) 1.3 and 1.9
b) 1.9 and 2.6
c) 2.6 and 3.0
d) 3.0 and 4.8

18. The adamentine luster is exhibited by
(1) Diamond
(2) Zircon
(3) Cassiterite
(4) Cinnabar
a) 1 Only
b) 1 and 3 only
c) 1, 2 and 3 only
d) All of the above

19. Which of the following shows a decreasing intensity of reflectivity of minerals ?
a) Splendent———glistening———shining———glimmering ———dull

b) Splendent———shining——— glimmering ——glistening ———dull

c) Splendent———shining———glistening———glimmering ———dull

d) Shining———splending———glistening——— glimmering ———dull

20. Diaphaneity is the property related to
   a) Transmission of light
   b) Dispersion of light
   c) Reflection of light
   d) Refraction of light

21. The 'Sclerometer' is an instrument used to determine the——— of minerals
   a) Specific gravity
   b) Refractive index
   c) Chemical composition
   d) Hardness

22. Hardness of a mineral is defined as
   a) The ability of a mineral to scratch other substances
   b) The strength of a mineral to withstand compression
   c) The resistance of a mineral to abrasion on its surface
   d) All of the above

23. Hardness of minerals increases with a/an
   a) Increase in valency
   b) Increase in bond strength
   c) Increase in the density of packing of atoms
   d) All of the above

24. The Moh's Scale of hardness is a/an
   a) Arithmatic scale
   b) Geometric scale
   c) Logarithmic scale
   d) Arbitrary scale

25. When a mineral under test is scratched by quartz, and itself scratches orthoclase, its hardness must lie between
   a) 5 and 6
   b) 6 and 7
   c) 7 and 8
   d) 4 and 5

26. The hardest substance known in the world is

a) Boron trichloride
b) Boron triphosphate
c) Boron nitrate
d) Boron nitride

27. Which of the following is a non-vectorial property ?
a) Hardness
b) Specific gravity
c) Electricity
d) Cleavage

28. Specific gravity of a mineral is the ratio of its weight to that of an equal volume of water at
a) 0°C
b) 4°C
c) 100°C
d) 273°C

29. A pycnometer is used to find the specific gravity of
a) Bigger pieces of minerals with even surfaces
b) Bigger pieces of minerals or liquids
c) Small pieces of minerals or liquids
d) Solids of any size as well as liquids

30. Arrange the following minerals in the correct order of Sp. gr.
a) Diamond > celestine > barite > anglesite
b) Diamond > barite > celestine > anglesite
c) Diamond < barite < celestine < anglesite
d) Diamond < celestine < barite < anglesite

31. Quartz, tridymite and cristobalite have different values of specific gravity because
a) Their crystal systems are different
b) Their chemical compositions are different
c) Their stability fields are different
d) Their atomic structures are different

32. Cleavage is a————property
a) Non-directional and non-diagnostic
b) Directional and diagnostic
c) Non-directional and diagnostic
d) Directional and non-diagnostic

33. Fluorspar possess
a) Cubic cleavage
b) Octahedral cleavages

c) Rhombohedral cleavage
d) None of the above

*34. Which of the following is matched correctly ?
a) Diamond-----Cubic cleavage
b) Halite-----Octahedral cleavage
c) Calcite-----Rhombdodecahedral cleavage
d) Pyroxene-----Prismatic cleavage

35. In which of the following crystal systems each cleavage direction is parallel to a different form ?
a) Isometric
b) Hexagonal
c) Monoclinic
d) Triclinic

36. Which one of the following shows a bluish-white fluorescent colours due to the replacement of $W^{+6}$ by $MO^{+6}$ ?
a) Diamond
b) Fluorite
c) Calcite
d) Scheelite

37. Triboluminence is the phenomenon of emission of light due to
a) The application of heat
b) The application of electricity
c) The application of pressure
d) Incandescence

38. When diamond is exposed to sun, it exhibits ?
a) Phosphorescence
b) Fluorescence
c) Opalescence
d) Play of colours

39. Minerals which are attracted by a magnetic field are described as
a) Ferrimagnetic
b) Ferromagnetic
c) Paramagnetic
d) Diamagnetic

40. QUARTZ is
a) Strongly magnetic
b) Weekly magnetic

c) Non-magnetic

d) None of the above

41. Which of the following minerals is described as 'magnetic pyrites'?

a) Iron pyrites

b) Chalcopyrite

c) Pyrrhotite

d) Magnetite

42. According to Von Kobell's fusibility scale the most infusible mineral is

a) Almandine

b) Stibnite

c) Enstatite

d) Orthoclase

43. Match of following and choose the correct answer

| *Property* | *Mineral* |
|---|---|
| A. Taste | (1) Galena |
| B. Malleability | (2) Mica |
| C Combustibility | (3) Halite |
| D. Elasticity | (4) Sulphur |

a) A—1 B—2 C—3 D—4

b) A—3 B—4 C—1 D—2

c) A—4 B—3 C—2 D—1

d) A—3 B—1 C—4 D—2

44. Which of the following minerals does NOT exhibit needle-like crystals?

a) Wavellite

b) Sillimanite

c) Tourmaline

d) Rutile

45. Match the following and choose the correct answer

| *Form* | *Mineral* |
|---|---|
| A. Tuberose | 1. Flossferri |
| B. Stellate | 2. Wavellite |
| C. Scaly | 3. Tridymite |
| D. Lamellar | 4. Wollastonite |

a) A—1 B—2 C—3 D—4

b) A—1 B—3 C—2 D—4

c) A—4 B—2 C—3 D—1

d) A—4 B—3 C—2 D—1

46. In the double chain silicates or 'ribbon structures'

a) Two tetrahedra share one oxygen ion

b) Each tetrahedron share two oxygen ions

c) Tetrahedra share alternately two and three oxygen ions

d) Each sheet of tetrahedra share three oxygens

*47. Which of the following pairs of minerals belong to the same group or family ?

a) Chrysolite——Chrysotile

b) Coccolite——Crocidolite

c) Cossyrite——Crossite

d) Francolite——Franklinite

48. When anions and cations of similar size and in same numbers crystallizes in similar type of structure, the phenomenon is known as

a) Isomorphism

b) Polymorphism

c) Diadochy

d) Eutectic crystallization

*49. Under ordinary conditions which of the following would NOT be ideal isomorphs ?

a) Cerrusite—strontonite

b) Calcite—aragonite

c) Magnesite—siderite

d) Calcite—magnesite

50. Most calcium and magnesium compounds are NOT isomorphous because

a) They differ in chemical activity

b) Their ionic charges are different

c) Their ionic radii are different

d) They belong to different groups in the periodic table

51 Different polymorphs of the same substance exhibit

a) Different physical properties but identical chemical properties

b) Different chemical properties but identical physical properties

c) Different physical and chemical properties

d) Identical physical and chemical properties

52. When polymorphism occurs in elements it is described as
 a) Allotropy
 b) Enantiotropy
 c) Monotropy
 d) Paramorphism

53. The transformation Quartz $\xrightarrow{867°C}$ Tridymite designates
 a) Enantiotropy
 b) Monotropy
 c) Solid solution series
 d) Dimorphism

54. A type of polymorphism in which the two polymorphs differ only in the stacking of identical two dimensional sheets is described as
 a) Order-disorder
 b) Polytypism
 c) Paramorphism
 d) Displacive transformation

55. The minerals tantalite ($FeTa_2O_6$) and brookite ($TiO_2$) are examples of
 a) Isomorphs
 b) Pseudomorphs
 c) Paramorphs
 d) Polymorphs

56. Crystals exhibiting 'false form' are said to be pseudomorphs. This is a result of
 (1) Substitution of one mineral by the other
 (2) Filling up of the original cavities by the infiltration of a solution
 (3) Deposition of a coating of one mineral on the crystals of the other
 (4) Gradual chemical alteration
 a) 1 and 2 are correct
 b) 1, 2 and 3 are correct
 c) 2 and 4 are correct
 d) All of the above

57. The change in molecular constitution without a change of chemical composition is referred as
 a) Paramorphism

b) Polytypism

c) Polymorphism

d) Hemihedrism

*58. Which of the following is the best example of 'defect' or 'omission solid solution' ?

a) Olivine

b) Plagioclase

c) Pyrrhotite

d) Amphiboles

59. In the following transformation : Marcasite$\longrightarrow$Pyrite

a) Pyrite is unstable whereas marcasite is stable

b) Both pyrite and marcasite are stable

c) Both pyrite and marcasite are unstable

d) Pyrite can again revert back to marcasite

e) Pyrite can revert to marcasite only if its structure is first completely destroyed by melting and other processes

60. DIADOCHY is described as

a) The simultaneous occupation of a lattice position by similar elements

b) The simultaneous crystallization of two different substances

c) The ability of two different elements to exhibit same physico-chemical properties

d) The ability of different elements to occupy the same lattice position in a crystal

61. Which of the following features are typical of metamict minerals ?

I. These are non-crystalline pseudomorphs after the original crystalline material

II. They are isotropic

III. They are formed when the crystal structure breaks down by nuclear radiation

IV. They are generally compounds of weak acids and weak bases

a) I and III only

b) II and IV only

c) I, II and III only

d) All of the above

*62 The transformation $\alpha$-quartz$\xrightarrow{573°C}$$\beta$-quartz involves

a) Slight increase in volume
b) Slight decrease in volume
c) Great increase in volume
d) Great decrease in volume
e) No change in volume

63. Any two given paramorphs have the same
a) Crystal system
b) Crystal class
c) Chemical composition
d) Structure

64. Anhydrite changes to gypsum by
a) Loss of chemical components
b) Gain of chemical components
c) The exchange of chemical components
d) Neither the loss nor the gain of chemical components

65. When a mineral shows a smooth surface when cut by a knife, it is described as
a) Brittle
b) Flexible
c) Sectile
d) Ductile
e) Malleable

66. Sulphur exhibits
a) Waxy luster
b) Pearly luster
c) Silky luster
d) Metallic luster

67. Ring silicate structures are characteristic of minerals crystallising in
a) Hexagonal system
b) Hexagonal and trigonal systems
c) Hexagonal, tetragonal and orthorhombic systems
d) Hexagonal, orthorhombic and monoclinic systems

*68. Which of the following type of ring silicate structures are unlikely to occur in nature ?
a) 2-membered ring
b) 3-membered ring
c) 4-membered ring
d) 6-membered ring

69. A mineral appearing green in colour, having white streak, hardness of 7.5—8.0 and sp. gr. of 2.7 must be
   a) Olivine
   b) Chlorite
   c) Nephelene
   d) Beryl

70. Match the following and choose the correct answer

| | |
|---|---|
| A. Sphalerite | (1) Cleavage in one direction |
| B. Fluorite | (2) Cleavage in two directions |
| C. Dolomite | (3) Cleavage in three directions |
| D. Hornblende | (4) Cleavage in four directions |
| E. Topaz | (5) Cleavage in six directions |

   a) A—2 B—4 C—3 D—1 E—5
   b) A—3 B—1 C—4 D—2 E—5
   c) A—5 B—4 C—3 D—2 E—1
   d) A—5 B—3 C—4 D—2 E—1

71. The mineral Kyanite exhibits a variation in its hardness in different directions. The highest value is found parallel to
   a) c-axis
   b) b-axis
   c) a-axis
   d) b and c-axis

*72. In the two carbonate minerals calcite ($CaCO_3$) and magnesite ($MgCO_3$)
   a) Calcite is harder than magnesite
   b) Magnesite is harder than calcite
   c) The hardness of calcite equals that of magnesite
   d) None of the above

73. Meigen's test is used to distinguish
   a) Tridymite and cristobalite
   b) Marcasite and pyrite
   c) Aragonite and calcite
   d) Actinolite and trimolite

74. The word 'glance' as in 'lead glance' is a miner's term for minerals possessing
   a) Low reflectivity
   b) High reflectivity
   c) Complete transparency
   d) Partial transparency

75. In which of the following minerals Al is exclusively found in 4-fold coordination ?

a) Garnet

b) Hornblende

c) Augite

d) Biotite

*76. Choose the correct statement from the following

a) Calcite is harder than aragonite

b) Aragonite has lower specific gravity than that of calcite

c) Calcite and aragonite crystallises in the rhombohedral system

d) None of the above

77. "Floss-Ferri" is

a) Stalactitic coralloid variety of calcite

b) Stalactitic coralloid variety of aragonite

c) Concretionary anhydrite

d) Transparent, crystalline form of gypsum

78. "Cairngorm stone" is the name given to

a) Milky quartz

b) Smoky quartz

c) Rose quartz

d) Rock crystal

79. The mineral Piedmonite is a manganese-rich

a) Zeolite

b) Felspathoid

c) Epidote

d) Pyroxenoid

80. The mineral leucite does not occur ih

a) Undersaturated lavas

b) Sedimentary rocks

c) Volcanic rocks

d) Plutonic rocks

81. The rarest of all the garnets is

a) Pyrope

b) Almandine

c) Grossularite

d) Uvarovite

82. The garnet having the maximum density is

a) Almandine

b) Spessartite

c) Pyrope

d) Glossularite

83. The individuals of which of the following groups of minerals show a number of polymorphs ?

a) Mica

b) Garnet

c) Olivine

d) Pyroxene

84. "Aesterism" is characteristic of minerals crystallizing in

a) Monoclinic system

b) Triclinic system

c) Hexagonal system

d) Orthorhombic system

85. Which of the following felspathoid minerals is found in crystalline limestones ?

a) Nepheline

b) Cancrinite

c) Leucite

d) Lazurite

86. Leucoxene is a variety of

a) Tourmaline

b) Talc

c) Topaz

d) Sphene

87. 'Spinel' is

a) Mg-Al silicate

b) Mg-Al oxide

c) Mg, Fe, Al-silicate

d) Mg, Fe, Al-oxide

88. Malacon (Crytolite) is an alteration product of

a) Zircon

b) Staurolite

c) Sphene

d) Tourmaline

*89. Find the odd man out

a) Dickite

b) Halloysite
c) Illite
d) Fuchsite

90. Ophicalcite is a rock made up of
a) Serpentine and calcite
b) Serpentine and chlorine
c) Calcite and magnetite
d) Calcite and dolomite

91. Which of the following is a mixture of white mica and chlorite ?
a) Sericite
b) Pinite
c) Saussurite
d) Greisen

92. Micas are classified as 'dioctahedral' when the number of Y-ions i.e. Al, Mg, Fe, etc. is
a) 3
b) 4
c) 6
d) 8

93. In mica, the greatest hardness in the (001) plane is found in the direction parallel to
a) X-axis
b) Y-axis
c) Z-axis
d) None of the above

94. Which of the following forms of mica is a characteristic product of fluorine metasomatism or griesenization ?
a) Muscovite
b) Biotite
c) Phlogopite
d) Lepidolite

95. In Margarite, a form of mica, the cleavage flakes are
a) Flexible and elastic
b) Flexible but not elastic
c) Neither flexible nor elastic
d) None of the above

96. As the crystallization of an olivine-rich melt proceeds, the

Mg/Fe ratio

a) Increases

b) Decreases

c) First increases than decreases

d) First decreases than increases

e) Remains constant

97. Which of the following is the correct order of crystallization of the olivine minerals from a liquid of a given composition ?

a) Fosterite-Olivine-Fayalite

b) Fayalite-Olivine-Forsterite

c) Olivine-Forsterite-Fayalite

d) Forsterite-Fayalite-Olivine

98. Which of the following pairs is incompatible with free silica ?

a) Forsterite-Fayalite

b) Fayalite-Olivine

c) Forsterite-Olivine

d) All of the above

99. Which of the following is the correct decreasing order of the specific gravity of the olivine minerals ?

a) Olivine-Fayalite-Forsterite

b) Forsterite-Olivine-Fayalite

c) Fosterite-Fayalite-Olivine

d) Fayalite-Olivine-Forsterite

*100. The melting point of Forsterite is more than that of Fayalite because

a) The unit cell of fayalite has larger dimensions than that of forsterite

b) Forsterite is the first olivine mineral to crystallise from a liquid melt

c) The Fe—O bonds are weaker than Mg—O bonds

d) Fayalite is saturated whereas forsterite is unsaturated

101. The felspathoid mineral 'Nepheline' is isomorphous with

a) High quartz

b) High tridymite

c) High cristobalite

d) High leucite

102. The essential features of basaltic hornblendes are their

a) low $Fe^{+3}$ ; $Fe^{+2}$ ratio and low hydroxyl content

b) high $Fe^{+3}$ : $Fe^{+2}$ ratio and high hydroxyl content

c) low $Fe^{+3}$ : $Fe^{+2}$ ratio and high hydroxyl content
d) high $Fe^{+3}$ : $Fe^{+2}$ ratio and low hydroxyl content

03. Which of the following minerals does NOT occur as detrital grains in sands and sandstones ?
a) Quartz
b) Zircon
c) Garnet
d) Muscovite
e) None of the above

104. In which of the following minerals the frameworks has a Si : Al ratio of 2 : 1?
a) Quartz
b) Orthoclase
c) Analcite
d) Nepheline

105. SCHORL is an aggregate of
a) tourmaline and quartz
b) tourmaline and calcite
c) tourmaline and corundum
d) tourmaline and emery

106. Emery is a rock that consists of finely granular
a) Beryl and Quartz
b) Beryl and Magnetite
c) Corundum and Quartz
d) Corundum and Magnetite

107. Leucite is dimorphous. Above 650°C it exhibits
a) Isometric symmetry
b) Tetragonal symmetry
c) Hexagonal symmetry
d) Orthorhombic symmetry

*108. Which of the following pairs of minerals crystallizes in the same crystal system ?
a) Calcite-Aragonite
b) Orthoclase-Microcline
c) Gypsum-Anhydrite
d) All of the above
e) None of the above

109. Which of the following is the most ubiquitous of all the micas ?

a) Muscovite
b) Biotite
c) Phlogopite
d) Zinwaldite

110. Find the odd man out
a) Clintonite
b) Xanthophyllite
c) Margarite
d) Zinwaldite

111. When the aluminium of muscovite is replaced by $Cr^{+3}$ it gives rise to a green-coloured variety called ?
a) Fuchsite
b) Roscoelite
c) Phengite
d) Rubellite

112. In the structure of olivine, the layers are parallel to
a) Basal pinacoid
b) Side pinacoid
c) Front pinacoid
d) Pyramid

112. In the structure of olivine
a) $Mg^{+2}$ ions are located at the centre symmetry and $Fe^{+2}$ at the reflection planes
b) $Fe^{+2}$ ions are located at the centres of symmetry and Mg at the reflection planes
c) half of the $Fe^{+2}$/Mg are located at the centres of symmetry and half at the reflection planes
d) there is no specific arrangement of $Fe^{+2}$/Mg

114. Which of the following is/are the alteration products of olivine ?
(i) Chlorite (ii) Chlorophaeite
(iii) Serpentine (iv) Bowlingite
a) (iii) only
b) (i) and (iii) only
c) (i), (ii) and (iii) only
d) all of the above

115. Olivine alters to serpentine on the addition of
a) $H_2O$
b) $SiO_2$

c) $H_2O$ and $SiO_2$
d) $H_2O$, $SiO_2$ and $CO_2$

116. The mineral series Hortonite-Knebellite-Tephroite is an extension of the isomorphous series of
a) Olivine minerals
b) Plagioclase felspars
c) Orthopyroxenes
d) Clinopyroxenes

117. The mineral "Monticellite" is isomorphous with olivine. Its main metallic constituents are
a) Mn/Mg
b) Mn/Fe
c) Mg/Fe
d) Ca/Mg
e) Ca/Mg/Fe

118. Which of the following conditions are unfavourable for the formation of olivine ?
a) Magmatic
b) Metamorphic
c) Sedimentary
d) None of the above

119. The formula $(Si_2Al)O_6$ as in Analcite $Na(Si.Al)O_6.H_2O$ results where
a) $\frac{1}{4}$ of the Si atoms in a continuous three dimensional lattice are replaced by Al atoms
b) $\frac{1}{3}$ of the Si atoms are replaced by Al
c) $\frac{1}{2}$ of the Si atoms are replaced by Al
d) 1/5 of the Si atoms are replaced by Al

*120. Which of the following polymorphic transformations of silica shows the least change in volume ?
a) β-Quartz———α-Quartz
b) α-Quartz———α-Tridymite
c) α-Tridymite———α-Cristobalite
d) α-Cristobalite———Liquid silica

121. Match the following and choose the correct answer

| *Mineral* | *Variety* |
|---|---|
| (A) Fluorspar | (1) Schungite |
| (B) Graphite | (2) Blue John |
| (C) Riebeckite | (3) Crocidolite |
| (D) Chalcedony | (4) Bastite |

(E) Hypersthene (5) Plasma

a) A—2 B—1 C—3 D—5 E—4
b) A—2 B—3 C--1 D—5 E—4
c) A—1 B—4 C—3 D—5 E—2
d) A—3 B—4 C—1 D—5 E—2

122. 'Collophane' is a
a) cryptocrystalline variety of fluorite
b) cryptocrystalline variety of apatite
c) amorphous variety of fluorite
d) amorphous variety of apatite

123. 'Melanite' is a variety of
a) Andradite garnet
b) Almandine garnet
c) Uvarovite garnet
d) Spessartite garnet

124. The main difference between graphite and diamond is
a) composition
b) hardness
c) density
d) crystal structure

125. Lithophone is obtained by mixing
a) $ZnSO_4 + BaSO_4$
b) $ZnS + BaSO_4$
c) $ZnS + BaCO_3$
d) ZNO + BaO

126. Which of the following is matched correctly ?
a) Adularia———Amphibole
b) Edenite———Mica
c) Morion———Quartz
d) Kalsilite———Epidote

*127. Which of the following does NOT contain chromium ?
a) Uvarovite
b) Fuchsite
c) Picotite
d) Withamite

128. Some of the rare minerals are given below. Match these minerals with the groups to which they belong.

(A) Schorlomite (1) Pyroxene

(B) Hortonite (2) Garnet

(C) Kataphorite (3) Olivine

(D) Johannsenite (4) Amphibole

a) A—3 B—2 C—4 D—1
b) A—3 B—2 C—1 D—4
c) A—2 B—3 C—4 D—1
d) A—2 B—4 C—3 D—1

129. Which of the following is an oxide of titanium and iron ?
a) Rutile
b) Anatase
c) Brookite
d) Ilmenite

130. Martitization generally takes place when
a) Pyrite is oxidized to limonite
b) Hematite is oxidized to Goethite
c) Magnetite is oxidized to Hematite
d) Pyrite is oxidized to Hematite

131. Match the following and choose the correct answer

| | |
|---|---|
| (A) Paacock ore | (1) Kyanite |
| (B) Kidney ore | (2) Bornite |
| (C) Mispickel | (3) Arsenopyrite |
| (D) Blende | (4) Hematite |
| (E) Disthene | (5) Sphalerite |

a) A—2 B—4 C—3 D—5 E—1
b) A—2 B—4 C—1 D—3 E—5
c) A—4 B—2 C—1 D—5 E—3
d) A—4 B—2 C—3 D—5 E—1

132. Which of the following pairs is incorrect ?
a) Magnetite——— $FeO.Fe_2O_3$
b) Chromite——— $FeO.Cr_2O_3$
c) Borax——— $Na_2B_4O_{7\cdot5}H_2O$
d) Beryl——— $Be_3Al_2Si_6O_{18}$

133. Identify the Triclinic amphibole from the following
a) Grunerite
b) Arfvedsonite
c) Cossyrite
d) Tremolite

134. Chiastolite and viridine are varieties of

a) Andalusite
b) Sillimanite
c) Zircon
d) Tourmaline

*135, Which of the following is matched incorrectly ?
a) Calcite———Icelandspar
b) Gypsum———Satinspar
c) Hypersthene———Schillarspar
d) Dolomite———Dog-toothspar

136. Tourmaline has many varieties exhibiting various beautiful colours. Each variety is rich in a particular element. Match the following varieties with their characteristic colours and element

| | |
|---|---|
| A. Schrol | (1) Li-rich pink variety |
| B. Dravite | (2) Mg-rich brown variety |
| C. Rubellite | (3) Fe-rich black variety |
| D. Achroite | (4) K-rich white variety |

a) A—3 B—1 C—2 D—4
b) A—3 B—2 C—1 D—4
c) A—3 B—4 C—2 D—1
d) A—2 B—3 C—1 D—4

137. Which of the following is true about the compostion of Quartz ?
a) Si=46.7% O=53.3%
b) Si=53.3% O=48.3%
c) Si=48.3% O=51.7%
d) Si=51.7% O=48.3%

138. The Si : Al ratios in the tectosilicates is/are
a) 1 : 1
b) 3 : 1
e) 1 : 1 or 3 : 1
d) 1 : 2
e) either 1 : 1, 1 : 2 or 3 :1

139. Which of the following forms of silica is pseudo-isometric ?
a) Quartz
b) Tridymite
c) Cristobalite
d) Coesite

140. The only truly amorphous form of silica is
a) Chert

b) Chalcedony
c) Opal
d) Lechatelierite

141. Which of the following forms of silica indicates the highest pressures ?
a) Tridymite
b) Cristobalite
c) Coesite
d) Stishovite

142. Chrystobalite is stable between the temperatures of
a) 573°C–867°C
b) 867°–1470°C
c) 1470°–1713°C
d) 1713°C–1820°C

143. The maximum stability limit for Tridymite is
a) T-1470°C Pr-3000 kg/cm²
b) T-1470°C Pr-5000 kg/cm²
c) T-837°C Pr-3000 kg/cm²
d) T-867°C Pr-5000 kg/cm²

144. The high-temperature form of Tridymite inverts to the low temperature form at a temperature of
a) 573°C
b) 867°C
c) 117°C
d) 200–268°C

145. AT NTP Quartz is always present as
a) low quartz
b) high quartz
c) smoky quartz
d) opal

*146. Which of the following forms of silica crystallizes in the Monoclinic system ?
a) Milky quartz
b) Rose quartz
c) Rock crystal
d) Averturine

147. The purest and the most transparent form of quartz is
a) Milky quartz
b) Rose quartz

c) Rock crystal
d) Averturine

148. The form of silica that is specially characteristic of the Mantle Transition zone is
a) Stishovite
b) Coesite
c) Lussatite
d) Lechatelerite

*149. The high and low temperature forms of quartz exhibit
a) Hexagonal symmetry
b) Trigonal symmetry
c) Hexagonal and trigonal symmetry respectively
d) Trigonal and hexagonal symmetry respectively

150. A cryptocrystalline variety of quartz that is banded in alternate light and dark layers is
a) Chalcedony
b) Agate
c) Onyx
d) Cat's eye

151. A low temperature variety of potash feldspar showing an opalescent play of colours is
a) Adularia
b) Moonstone
c) Sanidine
d) Peristerite

152. "Saussurite" a mixture of albite and epidote, is a common alteration product of
a) potash feldspars
b) Sodic feldspars
c) Lime feldspars
d) Barium feldspars

*153. When Al and Si occupy strictly defined positions in the crystal lattice, then the resulting crystal exhibits
a) Low symmetry
b) High symmetry
c) Irregular symmetry
d) Random growth

154. Which of the following shows the correct order of symmetry in the feldspars ?
a) Sanidine>orthoclase>microcline
b) Orthoclase>microcline>sanidine

c) Orthoclase>sanidine>microcline
d) Microcline>orthoclase>sanidine

155. The feldspar mineral sanidine is characteristic of
a) Potash-rich volcanic rocks
b) Soda-rich volcanic rocks
c) Monzonites and granodiorites
d) Granites and granitic pegmatites

156. Match the following and choose the correct answer

| *Mineral* | *Mode of occurrence* |
|---|---|
| A. Adularia | 1. Contact metamorphosed limestones |
| B. Bytownite | 2. Gabbroic igneous rocks |
| C. Microcline | 3. Veins |
| D. Anorthite | 4. Pegmatites |

a) A–4 B–3 C–2 D–1
b) A–3 B–4 C–2 D–1
c) A–3 B–4 C–1 D–2
d) A–3 B–2 C–4 D–1

157. The solid solution between albite and anorthite in plagioclases can be described by the following ionic substitutions
a) $2Na^+ \longrightarrow Ca^{2+}$
b) $Na^+ + Si^{4+} \longrightarrow Ca^{2+} + Al^{3+}$
c) $Ca^{2+} + Si^{+4} \longrightarrow 2Na^+ + Al^{+3}$
d) $Na^+ + Al^{+3} \longrightarrow Si^{+4}$

158. The most fundamental and common form of pyroxenes is
a) Basal pinacoid
b) Front pinacoid
c) Side pinacoid
d) Prism
e) Pyramid

159. The pyroxene that is very characteristic of the HT-HP eclogite facies is
a) Pigeonite
b) Omphacite
c) Spondumene
d) Acmite

160. Omphacite can be considered as a solid-solution of
a) Jadiete-Diopside-Hedenbergite
b) Jadiete-Augite-Acmite
c) Jadiete-Augite-Pigeonite

d) Enstatite-Hypersthene-Orthoferrosilite

161. The length of the unit cell in pyroxenes is
a) 4.8Å
b) 5.3Å
c) 6.7Å
d) 7.3Å

162. In pyroxenes the fundamental form has interfacial angles of
a) 56° and 124°
b) 87° and 93°
c) 56° and 16°
d) 87° and 48°

163. Enstatite is a typical mineral of
a) Greenschist facies
b) Glaucophane schist facies
c) Granulite facies
d) Eclogite facies

164. Diopside characteristically exhibits
a) Orthorhombic symmetry
b) Monoclinic symmetry
c) Triclinic symmetry
d) Hexagonal symmetry

165. The end members of the orthopyroxene series are
a) Enstatite-Orthoferrosilite
b) Enstatite-Hypersthene
c) Diopside-Augite
d) Diopside-Aegirine

166. The pyroxene mineral Jadiete is intermediate in composition between
a) Enstatite and hypersthene
b) Diopside and hedenbergite
c) Albite and anorthite
d) Albite and nepheline

167. In the structure of augite, Al occurs in 4-fold coordination when it substitutes for
a) $Na^+$
b) $Mg^{+2}$
c) $Fe^{+2}$
d) $Si^{+4}$

168. Which of the following is NOT matched correctly ?

a) Anigmatite———amphibole
b) Analcite———zeolite
c) Sodalite———pyroxenoid
d) Sanidine———feldspar

169. Staurolite crystallizes in monoclinic system but its forms are
a) Pseudo-hexagonal
b) Pseudo-orthorhombic
c) Pseudo-triclinic
d) Pseudo-tetragonal

170. 'Microcline' is the general name given for
a) Monoclinic potash feldspar
b) Orthorhombic potash feldspar
c) Triclinic potash feldspar
d) Monoclinic sodic feldspar

171. Which of the following is an orthorhombic amphibole ?
a) Cummingtonite
b) Grunerite
c) Anthophyllite
d) Tremolite

172. According to the classification of orthopyroxenes by Poldvaart and Haas, which of the following best represents the composition of hypersthene ?

(En=enstatite)

a) $En_{90}-En_{70}$
b) $En_{70}-En_{50}$
c) $En_{50}-En_{30}$
d) $En_{30}-En_{10}$

173. Which of the following shows 'reverse zoning' ?

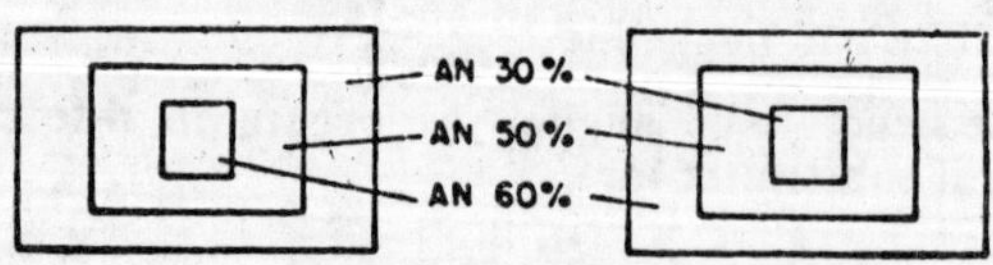


a) I only
b) II only
c) both I and II
d) none of the above

174. Uralite is a pseudomorphous ambhibole formed on replacing pyroxene. It has

a) Amphibole form
b) Pyroxene cleavages
c) Amphibole form but pyroxene cleavages
d) Pyroxene form but amphibole cleavages

175. The Alumina : Silica ratio in the anhydrous aluminium silicates is

a) 1 : 1
b) 1 : 2
c) 1 : 3
d) 2 : 1

176. When there is a progressive increase in Na content from the core to the rim of a plagioclase crystal, the zoning is described as

a) Normal zoning
b) Reversed zoning
c) Oscillatory zoning
d) Random zoning

177. 'Anorthoclase' can be designated as

a) $KAlSi_3O_8$
b) $NaAlSi_3O_8$
c) $(K, Na) AlSi_3O_8$
d) $(Na, K) AlSi_3O_8$

178. Which of the following exhibits the 'base exchange' property ?

a) Sodalite
b) Natrolite
c) Scapolite
d) Cancrinite

*179. Which of the following can be regarded as 'double salts' of plagioclase and NaCl or $CaCO_3$ ?

a) Scapolites
b) Zeolites
c) Chloritoids
d) Epidotes

180. 'Scorlomite' is

a) a variety of tourmaline
b) a variety of apatite

c) a Ti-rich variety of andradite
d) a Ti-rich variety of almandine

## MINERALOGY (KEY)

| | | | |
|---|---|---|---|
| 1. b | 2. a | 3. c | 4. b |
| 5. c | *6. d | 7. d | 8. c |
| 9. b | 10. a | 11. c | 12. b |
| 13. a | 14. b | 15. c | 16. d |
| 17. b | 18. c | 19. c | 20. a |
| 21. d | 22. c | 23. d | 24. d |
| 25. b | 26. d | 27. b | 28. b |
| 29. c | 30. d | 31. d | 32. b |
| 33. b | *34. d | 35. d | 36. d |
| 37. c | 38. a | 39. c | 40. c |
| 41. c | 42. c | 43. d | 44. a |
| 45. a | *46. a | *47. c | 48. a |
| *49. d | 50. c | 51. a | 52. a |
| 53. a | 54. b | 55. a | 56. d |
| 57. a | *58. c | 59. e | 60. d |
| 61. d | *62. a | 63. c | 64. b |
| 65. c | 66. a | 67. b | *68. a |
| 69. d | 70. c | 71. a | *72. b |
| 73. c | 74. b | 75. a | *76 d |
| 77. b | 78. b | 79. c | 80. d |
| 81. d | 82. a | 83. a | 84. c |
| 85. d | 86. d | 87. b | 88. a |
| *89. d | 90. a | 91. b | 92. b |
| 93. b | 94. a | 95. c | 96. b |
| 97. a | 98. b | 99. d | *100. c |
| 101. b | 102. d | 103. e | 104. c |
| 105. a | 106. d | 107. b | 108. e |
| 109. b | 110. d | 111. a | 112. c |
| 113. c | 114. d | *115. c | 116. a |
| 117. d | 118. c | 119. b | *120. d |
| 121. a | 122. b | 123. a | 124. d |
| 125. b | 126. c | *127. d | 128. c |
| 129. d | *130. c | 131. a | 132. c |
| 133. c | 134. a | *135. d | 136. b |
| 137. a | 138. c | 139. c | 140. d |
| 141. d | 142. c | 143. a | 144. c |

| | | | |
|---|---|---|---|
| 145. a | 146. c | 147. c | 148. a |
| *149. c | 150. c | 151. b | 152. c |
| *153. a | 154. a | 155. a | 156. d |
| 157. b | 158. d | 159. b | 160. b |
| 161. b | 162. b | 163. c | 164. b |
| 165. a | 166. d | 167. d | 168. c |
| 169. b | 170. c | 171. c | 172. b |
| 173. a | 174. d | 175. a | 176. a |
| 177. d | 178. b | *179. a | 180. c |

## EXPLANATORY ANSWERS

*6. (d) : A maximum of 50% substitution occurs.

*12. (b) : The environment in a crystal lattice can result in the same element being responsible for two different colours ; Cobalt imparts pink colour to a mineral when it is surrounded by 6 oxygen ions and blue colur when it is surrounded by 4 oxygen ions.

*34. (d) : Diamond——Octahedral cleavage ; Halite——cubic cleavage Calcite——rhombohedral cleavage

*47. (c) : Both of them crystallizes in the Tetragonal system

*49. (d) In isomorphism similarity in ionic sizes are more important than chemical similarity. Thus, calcium ($Ca^{+2}$=0.99 Å) and magnesium ($Mg^{+2}$=0.66Å) compounds are not isomorphous. All others have similar ionic radii.

*58. (c) : Pyrrhotite most often does not occur as pure FeS ; there are always more S ions than Fe ions in the structure. Electrostatic neutrality is maintained since some of the Fe ions are ferric.

*62. (a) : There is an increase of about 2.4% in volume

*68. (a) : A two-membered ring is unlikely because of the necessary close approach of two highly charged silicon ions.

*72. (b) : Hardness increases with a decrease in ionic radius. Thus, calcium with an ionic radius 0.99 Å has a hardness of 2.5—3.0 but magnesium with 0.66 Å has a hardness of 4.0—4.5

*76. (d) : Aragonite is harder than calcite and has higher sp. gr. Calcite crystallizes in rhombohedral system while aragonite crystallizes in orthorhombic system.

*89. (d) : Fuchsite is a mica while the rest are clay minerals

*100. (c) : The radius of $Fe^{+2}$ (0.74 Å) is larger than that of $Mg^{+2}$ (0.66 Å). Consequently the bonds are weaker in Fayalite and it has a lower melting temperature (1250°C) than that of Forsterite (1890 °C).

*108. (d) : Calcite——rhombohedral : aragonite——orthorhombic ; Orthoclase——monoclinic ; Microcline——Triclinic ; Gypsum——monoclinic ; Anhydrite——orthorhombic ;

*115. (d) : $\underset{\text{Olivine}}{3Mg_2SiO_4} + SiO_2\ H_2O \longrightarrow \underset{\text{Serpentine}}{2Mg_3Si_2O_5(OH)_4}$

*120. (d) : (a) =+2.4% ; (b) =+12.7%
(c) =+4.7% (d) =+0.4%

*127. (d) : Withamite is a Ca-Al-Fe silicate

*135. (d) : Dogtoothspar is a variety of calcite

*153. (a) : The greater the ordering the lower is the symmetry

*179. (a) : The scapolites resemble the plagioclase feldspars in composition and structure but have interspersed anions so that they can be described as 'double salts' of plagioclase and NaCl or $CaCO_3$. Common scapolite is an isomorphous series of two end members, namely :

Marialite : 3 $(NaAlSi_3O_8)$. NaCl cf. 3 Al bite + NaCl

Meionite : 3 $(CaAl_2Si_2O_8)$. $CaCO_3$ cf. 3 Anorthite + $CaCO_3$

# 5

# *Optical Mineralogy*

1. The wave theory of light was proposed by
   a) Newton
   b) Huygens
   c) Maxwell
   d) All of the above

2. Which of the following theories of light could explain all the properties of light ?
   a) Corpuscular theory
   b) Wave theory
   c) Electromagnetic theory
   d) Quantum theory

3. The surface determined at a given instant by all the parts of a system of waves travelling along the same direction and in the same phase is called
   a) Electric vector
   b) Wave length
   c) Wave front
   d) Wave surface

4. According to Brewster's law, polarization is maximum when the direction of reflection and refraction are
   a) 30° apart
   b) 60° apart
   c) 90° apart
   d) Parallel

5. Which of the following properties is NOT observed under ordinary light ?
   a) Colour
   b) Inclusions
   c) R.I.
   d) Pleochroism

6. Which of the following properties is observed under convergent light ?
   a) Pleochroism
   b) Isotropic and anisotropic character

c) Twinning
d) Optic sign

7. Match the following and choose the correct answer

| *Property* | *Nature of light* |
|---|---|
| A. Cleavage | 1. Plane polarized light |
| B. Twinkling | 2. Ordinary light |
| C. Twinning | 3. Crossed nicols |
| D. Dispersion | 4. Convergent light |

a) A—2 B—1 C—3 D—4
b) A—1 B—2 C—3 D—4
c) A—1 B—3 C—2 D—4
d) A—2 B—1 C—4 D—3

8. Sphalerite and sodalite show cleavage in
a) 6 directions
b) 4 directions
c) 3 directions
d) None of the above

9. In VIBGYOR, as we move from left to right
a) R I. decreases and velocity increases
b) R.I. increases and velocity decreases
c) Both R.I. and velocity decreases
d) Both R.I. and velocity increases

10. Leucite shows
a) Star-like inclusions
b) Curved inclusions
c) Radial inclusions
d) Straight inclusions

11. Calcite has maximum birefringence. This is best reflected in which of the following properties ?
a) Pleochroism
b) Twinning
c) Anisotropism
d) Twinkling

12. Extinction angle is the angle between
a) The light position and the dark position of a mineral
b) The zero of the circular scale and the dark position
c) The vibration direction of the lower and the upper nicols
d) A vibration plane and a crystallographic direction in a mineral section

13. Pleochroism and pleochroic halos are observed by using
    a) Polarizer only
    b) Analyzer only
    c) Both polarizer and the analyzer
    d) Polarizer, analyzer and the iris diaphragm

14. A Uniaxial mineral is said to be positive when
    a) The velocity of E.R. is more than that of OR
    b) The R.I. of E.R. is less than that of OR
    c) The R.I. of E.R. is more than that of OR
    d) None of the above

15. In Quartz
    a) E=1.553 Å W=1·658 Å
    b) E=1.553 Å W=1.486 Å
    c) E=1.553 Å W=1.544 Å
    d) E=1.486 Å W=1.658 Å

16. Match the following and choose the correct answer

| | |
|---|---|
| A. Wave front (isotropic) | 1. Ellipse |
| B. Wave surface (isotropic) | 2. Sphere |
| C. Wave front (anisotropic) | 3. Ellipsoidal |
| D. Wave surface (anisotropic) | 4. Circle |

    a) A—4 B—2 C—3 D—1
    b) A—2 B--4 C—3 D—1
    c) A—4 B—2 C—1 D—3
    d) A—2 B—4 C—1 D—3

17. In the construction of a Nicol Prism a long cleavage rhombohedron of Icelandspar (variety of Calcite) is used. The rhomb is chosen in such a way that its
    a) Length=thickness
    b) Length=2 (thickness)
    c) Length=3 (thickness)
    d) Length=4 (thickness)

18. The light ray that comes out of the Nicol prism is
    a) OR which vibrates parallel to the short diagonal of the rhomb face at the end of the Nicol
    b) ER which vibrates parallel to the short diagonal of the rhomb face at the end of the Nicol
    c) OR which vibrates parallel to the long diagonal of the rhomb face at the end of the Nicol
    d) ER which vibrates parallel to the long diagonal of the rhomb face at the end of the Nicol

19. The refractive index of the Canada Balsam is
    a) 1.50
    b) 1.54
    c) 1.486
    d) 1.65

20. Which of the following is used for the identification of polished sections of metallic ores ?
    a) Compound microscope
    b) Polarizing microscope
    c) Reflecting microscope
    d) Binocular microscope

21. A condenser is used to facilitate the change from
    a) Parallel to convergent light
    b) Parallel to divergent light
    c) Convergent to parallel light
    d) Divergent to parallel light

22. The Apochromatic objectives are made of
    a) Glass components only
    b) Fluorite components only
    c) A combination of glass and fluorite components
    d) Glass, fluorite and a little borax

23. The degree to which the objective can bring out the fine structural details of an object is described as its
    a) Magnifying power
    b) Resolving power
    c) Initial Magnification
    d) Dispersive powes

24. For photomicrography, the ocular used is
    a) Positive huygens type
    b) Positive ramsden type
    c) Negative huygens type
    d) Negative ramsdens type

25. In the ocular of the huygen's type the focal length of the eyepiece is
    a) One-half the focal length of the objective
    b) One-third the focal length of the objective
    c) Thrice the focal length of the objective
    d) Twice the focal length of the objective

26. The planochromatic objectives correct
    a) Chromatic aberration only
    b) Spherical aberration only
    c) Both chromatic and spherical aberrations
    d) None of the above
27. The optical tube length divided by focal length gives
    a) Initial magnification
    b) Numerical aperture
    c) Angular aperture
    d) Free-working distance
28. The Magnifying power of a microscope
    a) Increases with the tube length
    b) Decreases with the tube length
    c) Increases with the focal length
    d) None of the above
29. The free working distance is small in the case of
    a) High power objectives
    b) Medium power objectives
    c) Low power objectives
    d) Apochromatic objectives
30. Amici-Bertrand lens is inserted in the tube of the microscope between
    a) Ocular and Analyzer
    b) Objective and analyzer
    c) Analyzer and polarizer
    d) Objective and polarizer
31. A microscope is said to be 'centered' when
    a) The axis of rotation coincides with the centre of the field of view
    b) The vibration directions of the polars coincides with the cross-wires in the oculars
    c) The polarizer and the analyzer are mutually perpendicular
    d) A ray of light passes from the polarizer to the analyzer without any deviation
32. Orthoscopic illumination is caused by
    a) A bundle of approximately parallel light rays
    b) Strongly convergent light
    c) Strongly divergent light
    d) Light rays of a single wavelength

33. For carrying out conoscopic observation of minerals
   a) Crossed polars are used
   b) Crossed polars and Bertland lens are used
   c) Crossed polars, Bertland lens and Condensing lens are used
   d) None of the above

34. In the Becke Test, the lowering of the objective results in the movement of the Becke lines into the field of
   a) lower R.I. medium
   b) higher R.I. medium
   c) any of the above
   d) none of the above

35. Becke lines are observed best by using
   a) high power objectives and closing down the substage iris diaphragm
   b) high power objectives and opening up the substage iris diaphragm
   c) low power objectives and closing down the substage iris diaphragm
   d) low power objective and opening up the substage iris diaphragm

36. Oblique illumination (Shadow Method) is produced by
   a) slow insertion of a cardboard into the light path below a polarizer
   b) partial insertion of an accessory into the slot until the metal frame intercepts the light
   c) partial insertion of the analyzer until its frame intercepts light
   d) All of the above

37. In the substage mirror of microscope, the concave side is generally used for
   a) low magnification
   b) medium magnification
   c) high magnification
   d) any of the above

38. When the velocity of light is same in all the directions, the wave surface formed would be a/an
   a) Ellipse
   b) Sphere

c) any surface other than a sphere
d) any surface other than an ellipse

39. In the case of Negative Uniaxial crystals, the wave front is a/an
a) Circle
b) Ellipse
c) Circle inside an ellipse
d) Ellipse inside a circle

40. Tourmaline shows polarization due to
a) Absorption
b) Reflection
c) Refraction
d) Dispersion

41. Elliptical polarization occurs when the path difference is
a) $\frac{1}{2}\lambda$ or $\lambda$
b) $\frac{1}{4}\lambda$, $3/4\lambda$, or $5/4\lambda$
c) zero
d) other than 0, $\frac{1}{4}\lambda$ $\frac{1}{2}\lambda$ 3/4 $1\lambda \ldots n+\frac{1}{2}\lambda$

42. Polaroids show polarization by
a) Reflection
b) Repeated refraction
c) Absorption
d) Dispersion

43. The indicatrix of a uniaxial positive crystal is a/an
a) Sphere
b) Ellipsoid
c) Oblate spheroid
d) Prolate spheroid

44. The indicatrix is described as 'triaxial ellipsoid' for crystals belonging to
a) Monoclinic and Triclinic systems
b) Orthorhombic, Monoclinic and Triclinic system
c) Hexagonal and Trigonal systems
d) Orthorhombic, Hexagonal and Trigonal systems

45. A section cut perpendicular to optic axis through the indicatrix centre of a uniaxial crystal is described as
a) Principal ellipse
b) Principal section
c) Circular section
d) Random section

46. Two interfering waves are said to be "in phase" when their path difference is
   a) $1\lambda$ $2\lambda$ $3\lambda ... n\lambda$
   b) $\frac{1}{2}\lambda$ $3/2\lambda$ $5/2\lambda ... \frac{2n+1}{2}\lambda$
   c) any distance other than $n\lambda$ or $2n+\frac{1}{2}\lambda$
   d) none of the above
47. Destructive interference occurs when is
   a) $1\lambda$ $2\lambda$ $3\lambda ... n\lambda$
   b) $\frac{1}{2}\lambda$ $3/2\lambda$ $5/2\lambda .. \frac{2n+1}{2}\lambda$
   c) any distance other than n or $2n+\frac{1}{2}$
   d) none of the above
48. In a constructive interference, the resultant wave has
   a) same wavelength and same amplitude as the parent
   b) same wavelength and twice the amplitude as the parent
   c) twice the wavelength and same amplitude as the parent
   d) twice the wavelength but half the amplitude of the parent
49. Maximum intensity for the red light occurs when the phase difference is
   a) 0
   b) $\frac{1}{2}\lambda$
   c) $1\lambda$
   d) $3/2\lambda$
50. Maximum intensity of interference colours occurs when the value of retardation is equal to
   a) $n\lambda$
   b) $(2n+1)\lambda$
   c) $\frac{(2n+1)}{2}\lambda$
   d) zero
51. In an interference figure the isogyres are
   a) areas of equal vibration direction
   b) areas of identical interference colours
   c) points along which ER and OR have same velocity
   d) none of the above
52. The 'Melatope' of an interference figure coincides with
   a) the point of emergence of the optic axis
   b) direction perpendicular to the optic axis
   c) direction 45° to the optic axis

d) none of the above

53. Isochromes are more in number when the
    a) birefringence is high and the sections are thin
    b) birefringence is high and the sections are thick
    c) birefringence is low and the sections are thin
    d) birefringence is low and the sections are thick

54. Off-centred optic axis figures are obtained when the optic axis is———to the plane of the microscope stage
    a) Parallel
    b) Perpendicular
    c) neither 'a' nor 'b'
    d) either 'a' or 'b'

55. The interference colours with retardation less than 550 mu belongs to
    a) First order
    b) Second order
    c) Third order
    d) Fourth order

56. 'Sensitive Tint' is the interference colour corresponding to
    a) 560 mu
    b) 470 mu
    c) 410 mu
    d) 700 mu

57. In the Newton's colour chart the diagonal lines represent
    a) Colour only
    b) Birefringence only
    c) Colour and Birefringence
    d) Thickness

58. In the Michel-Levy colour chart the interference colours are given by
    a) horizontal lines
    b) vertical lines
    c) diagonal lines
    d) all of the above

59. Flash figure is the interference figure of
    a) Uniaxial minerals
    b) Biaxial minerals
    c) Isotropic minerals
    d) Colloidal mineral

60. Interference colours depends upon
   a) Phase difference
   b) Thickness of the slice
   c) Birefringence
   d) All of the above

61. Anamolous interference colours are exhibited by
   a) Zoisite
   b) Leucite
   c) Garnet
   d) Topaz

*62. In a Uniaxial cross interference figure
   a) the E and W vibration directions are radial
   b) the E and W vibration directions are concentric
   c) the W vibration directions are radial and the E directions are concentric
   d) the E vibration directions are radial and the W directions are concentric

*63. In biaxial optic axis figures
   a) Isogyres remain in the field of view when the stage is rotated
   b) An isogyre remains centred in the field of view during rotation
   c) Isogyres move slowly out of field of view when the stage is rotated
   d) Isogyres move quickly out of the field of view when the stage is rotated

64. If the 2 V of a mineral is 90° it can be described as
   a) positive
   b) negative
   c) either positive or negative
   d) neither positive nor negative

65. A mineral shows the following interference figure (see figure). It can be interpreted as a

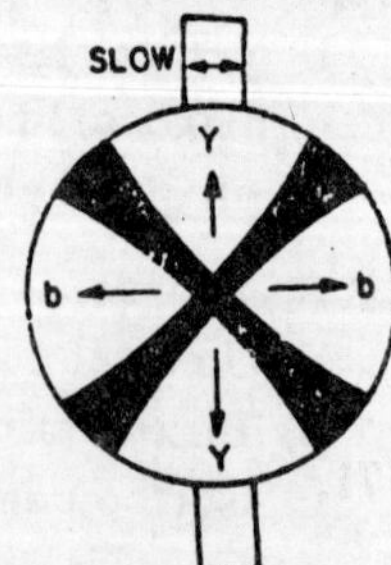


   a) Uniaxial positive figure
   b) Uniaxial negative figure
   c) Biaxial positive figure
   d) Biaxial negative figure

66. Which of the following sections gives the lowest interference colours ?

a) A section cut parallel to the c-axis of a uniaxial mineral

b) A section parallel to X optical direction in a biaxial mineral

c) A section parallel to Z optical direction in a biaxial mineral

d) A section cut parallel to one of the circular sections of the biaxial indicatrix

67. Match the following and choose the correct answer

| *(Light)* | *(Wavelength)* |
|---|---|
| (A) Red | (1) 560 mu |
| (B) Yellow | (2) 470 mu |
| (C) Blue | (3) 410 mu |
| (D) Violet | (4) 700 mu |

a) A—1 B—2 C—3 D—4

b) A—4 B—2 C—3 D—1

c) A—4 B—1 C—2 D—3

d) A—4 B—2 C—1 D—3

68. As the microscope stage is rotated, the polarization colours changes in

a) colour only

b) intensity only

c) both colour and intensity

d) none of the above

69. With an increase in the birefringence the intensity of interference colours

a) increases

b) decreases

c) increases or decreases

d) none of the above

70. Retardation is denoted as

a) $\Delta = \frac{(N-n)}{t}$

b) $t = \frac{(N-n)}{\Delta}$

c) $\Delta = t(N-n)$

d) $\Delta = t(n-N)$

71. The birefringence of Quartz is

a) 0.009

b) 0.010
c) 0.021
d) 0.172

**72. A Quartz wedge is always cut**
a) parallel to b-axis
b) perpendicular to b-axis
c) parallel to c-axis
d) perpendicular to c-axis

**73. Which of the following is designated as a ¼ wavelength plate ?**
a) Quartz wedge
b) Gypsum plate
c) Mica plate
d) Bertland ocular

**74. Exact extinction positions can be determined by**
a) Quartz wedge
b) Selenite plate
c) Mica plate
d) Wright biquartz wedge

**75. Which of the following is used to measure retardation ?**
a) Bertland lens
b) Condenser
c) Berek compensator
d) Iris diaphragm

**76. A Berek Compensator is made up of**
a) Calcite plate
b) Selenite plate
c) Mica plate
d) Satinspar plate

**77. In the Selenite plate, the cleavage present is**
a) basal pinacoidal
b) orthopinacoidal
c) clinopinacoidal
d) prismatic

**78. The thickness of the thin sections generally used under the polarizing microscope is**
a) 0.03 mm
b) 0·04 mm
c) 0.3 mm

d) 0.4 mm

79. In a Uniaxial negative mineral, the vertical crystalographic axis is
   a) fast
   b) slow
   c) slow or fast
   d) none of the above

80. A dichoiscope is an instrument used for observing
   a) Polarization colours
   b) Pleochroism
   c) Solar spectrum
   d) Twinkling effects in crystals

81. Minerals of which of the following crystal systems exhibit the phenomenon of dichroism ?
   a) Cubic system
   b) Triclinic system
   c) Tetragonal system
   d) Orthorhombic system

82. In the Hexagonal system, vertical sections give
   a) Straight extinction
   b) Oblique extinction
   c) Symmetrical extinction
   d) None of the above

83. In the Monoclinic system, the straight extinction is obtained in sections cut along
   a) Orthopinacoid
   b) Clinopinacoid
   c) Basal pinacoid
   d) Orthopinacoid and basal pinacoid
   e) Clinopinacoid and basal pinacoid

84. In the Triclinic system, oblique extinction is given by
   a) Basal pinocoidal sections
   b) Orthopinacoidal sections
   c) Clinopinacoidal sections
   d) Any of the above
   e) None of the above

85. Twinking is an optical phenomenon in which a mineral fragment shows variations in ———as the stage is rotated
   a) Relief

b) Refractive index
c) Quantity and Quality of colour
d) Reflectivity

86. Refractometers are instruments used to determine the R.I. of
a) small crystals
b) big crystals
c) liquids
d) liquids or solids

87. The Universal Stage helps in the determination of
a) Optic orientation
b) 2V
c) Optic sign
d) All of the above
e) None of the above

*88. In uniaxial minerals, the section cut with the c-axis perpendicular to mineral section gives
a) birefringence
b) partial birefringence
c) zero partial birefringence
d) none of the above

*89. In the biaxial minerals crystallizing in the Monoclinic system
a) the ab plane contains the X and Z optical directions
b) a//x and c//z
c) b-axis is parallel to the optical direction Y
d) all of the above are correct

90. In biaxial minerals, the maximum double refraction and the birefringence of a mineral is given by a section cut parallel to
a) X and Z of the indicatrix
b) X and Y of the indicatrix
c) Y and Z of the indicatrix
d) either of the two circular sections of the indicatrix

91. In which of the following crystal systems, the axes of the indicatrix coincides with the crystallographic axes ?
a) Hexagonal
b) Orthorhombic
c) Monoclinic
d) Triclinic

92. For practical purposes, a biaxial crystal can be considered as positive when

a) $(n\gamma-n\beta)>(n\beta-n\alpha)$
b) $(n\gamma-n\beta)<(n\beta-n\alpha)$
c) $(n\alpha-n\beta)>(n\beta-n\gamma)$
d) $(n\alpha-n\beta)>(n\beta-n\gamma)$

*93. In biaxial positive crystals
a) $2V\gamma<90°$
b) $2V\gamma>90°$
c) $2V\alpha<90°$
d) $2V\alpha=90°$

*94. Which of the following indicatrix can be obtained if $2V=180°$ ?
a) Uniaxial positive
b) Uniaxial negative
c) Biaxial positive
d) Biaxial negative

*95. In uniaxial positive crystals the length of the unique axis of revolution (the optic axis direction) is directly proportional to
a) The ordinary R.I.
b) The extraordinary R.I.
c) The minimum value of extraordinary R.I.
d) The maximum value of extraordinary R.I.

96. The dispersive power of a liquid is given as
a) $\frac{nF-nC}{nD-1}$
b) $\frac{nD-1}{nF-nC}$
c) $\frac{nC-nF}{nD-1}$
d) $\frac{nD-1}{nC-nF}$

97. Which of the following has the least dispersive power ?
a) Fluorite
b) Diamond
c) Calcite
d) Quartz

98. The wavelength of the F-Fraunhofer line is
a) 4308Å
b) 4861Å
c) 5893Å
d) 6870Å

99. When isometric crystals such as boracite and analcite show more or less pronounced double refraction it is described as

a) Chatoyancy

b) Refringence

c) Optical anamoly

d) Optical disharmony

100. The "bird's eye effect" caused due to the speckled appearance of interference colours resulting from strong birefringence is found in

a) Micas

b) Garnets

c) Olivines

d) Pyroxenes

101. Microcline shows 'cross-hatched' appearance between crossed nicols due to

a) Unidirectional twinning on Albite law

b) Undirectional twinning on Pericline law

c) Twinning in two directions on the Albite law

d) Twinning in two directions : one on the Albite law and the other on the Pericline law

102. Augite exhibites 'Hourglass structure' due to— —and 'Herring-bone structure' due to— —respectively

a) Twinkling/twinning

b) Wavy extinction/polysynthetic twinning

c) Polysynthetic twinning/wavy extinction

d) Alteration/dimorphism

103. Which of the following is NOT a diagnostic optical property of quartz ?

a) Absence of cleavage and alteration

b) Its transparency and lack of colour

c) Absence of shagreen surface or relief

d) Presence of easily distinguishable twins under the microscope

104. Optically, quartz can be distiugnished from Orthoclase by its

(i) Optically positive character

(ii) Absence of cleavage

(iii) Presence of inclusions

(iv) Higher refractive index

(v) Absence of twinning

a) (i), (ii) and (iii) only

b) (i), (ii) and (v) only
c) (i), (ii) and (iv) only
d) (i), (ii), (iv) and (v) only

105. Orthoclase is characterized by
a) Large optic axial angle
b) Low birefringence
c) Simple twins
d) Optically positive character
e) All of the above

106. Sanidine is distinguished from orthoclase by its
a) Smaller optic axial angle
b) Optically positive character
c) Higher relief
d) Absence of twinning

107. Which of the following is NOT true of the optical properties of Olivine ?
a) Very high relief
b) Presence of irregular fractures
c) Weak birefringence
d) Alteration

108. Orthopyroxenes are described as positive when
a) $FeSiO_3 < 15\%$
b) $FeSiO_3 > 15\%$
c) $FeSiO_3 = MnSiO_3$
d) None of the above

109. The Orthopyroxenes are commonly zoned and have
a) Mg-rich cores and Fe-rich margins
b) Fe-rich cores and Mg-rich margins
c) Mg-rich cores and Si-rich margins
d) Si-rich cores and Mg-rich margins

110. Acmite is characterized by
(i) Very low extinction angles
(ii) Strong pleochroism
(iii) Optical negative character
(iv) Strong tendency for alteration

a) (i) and (ii) only
b) (ii) and (iii) only
c) (i), (ii) and (iii) only

d) (i), (ii) and (iv) only

111. The pyroxene having an optical negative character among the following is
a) Enstatite
b) Hypersthene
c) Diopside
d) Augite

112. The cross-section of an Augite crystal is
a) 4-sided
b) 6-sided
c) 8-sided
d) None of the above

113. The 2V value of leucite is
a) 23°
b) 45°
c) 60°
d) None of the above

114. In Hornblendes the cleavage directions are at angles of
a) 56° and 124°
b) 54° and 126°
c) 87° and 103°
d) 70° and 120°

115. The biaxial positive variety of chlorite is
a) Clinochlore
b) Pennite
c) Pseudophite
d) Ripidolite

116. A biaxial mineral is said to be positive if
a) Z is the acute bisectrix
b) X is the acute bisectrix
c) E>W
d) W>E

117. In the mineral biotite, the absorption formula is
a) X>Y>Z
b) Z>Y>X
c) X>Z>Y
c) Z>X>Y

118. Chiastolite, a variety of the mineral andalusite, shows

inclusions which appear cross-shaped when viewed under a microscope. These inclusions contain

a) Zircon
b) Mica
c) Graphite
d) Quartz

119. The 'Scapolites' are
a) Isotropic
b) Uniaxial
c) Biaxial positive
d) Biaxial negative

120. Kyanite is distinguished from Sillimanite and Andalusite by its
a) Strong pleochroism
b) Strong birefringence
c) Large optic axial angle
d) Oblique extinction

121. An isotropic substance has.
a) One optic axis
b) Two optic axes
c) No optic axes
d) Infinite optic axes

122. A mineral with adamentine lustre shows the brilliant appearance because of
a) High interference
b) Low interference
c) High dispersion
d) Low dispersion

123. Optically clear fluorite is in demand for microscope lenses of high magnification because of its
a) Low dispersion
b) High dispersion
c) Low interference
d) High interference

124. The phenomenon of double refraction was discovered by
a) W. Nicol
b) Erasmus Bartholinus
c) Wahlstrom
d) Miller

125. The ordinary or extraordinary character of light is determined by

a) Selenite plate
b) Mica plate
c) Tourmaline plate
d) Bertland ocular

## OPTICAL MINERALOGY (KEY)

| | | | |
|---|---|---|---|
| 1. b | 2. d | 3. c | 4. d |
| 5. d | 6. d | 7. a | 8. a |
| 9. a | 10. c | 11. d | 12. d |
| 13. a | 14. c | 15. c | 16. c |
| 17. b | 18. b | 19. b | 20. c |
| 21. a | 22. c | 23 b | 24. b |
| 25. b | 26. c | 27. a | 28. a |
| 29. a | 30. a | 31. a | 32. a |
| 33. c | 34. a | 35. a | 36. d |
| 37. c | 38. b | 39. d | 40. a |
| 41. d | 42. c | 43. d | 44 b |
| 45. c | 46, a | 47. b | 48. b |
| 49. b | 50. c | 51. a | 52. a |
| 53. b | 54. c | 55. a | 56. a |
| 57. b | 58. b | 59. a | 60. d |
| 61. a | *62. d | *63. b | 64. d |
| 65. b | 66. d | 67. c | 68. b |
| 69. a | 70. c | 61. a | 72. d |
| 73. b | 74. d | 75. c | 76. a |
| 77. c | 78. a | 79. a | 80. b |
| 81. c | 82. a | 83. d | 84. d |
| 85. a | 86. d | 87. d | 88. c |
| *89. c | *90. a | 91. b | 92. a |
| *93. a | *94. b | *95. d | 96. b |
| 97. a | 98. a | 99. c | 100. a |
| 101. d | 102. b | 103. d | 104. d |
| 105. e | 106. a | 107. c | 108. a |
| 109. a | 110. c | 111. b | 112. c |
| 113. d | 114. a | 115. a | 116. a |
| 117. b | 118. c | *119. b | 120. d |
| 121. d | 122. a | 123. a | 124. b |
| 125. c | | | |

## EXPLANATORY ANSWERS

*62. (d) :

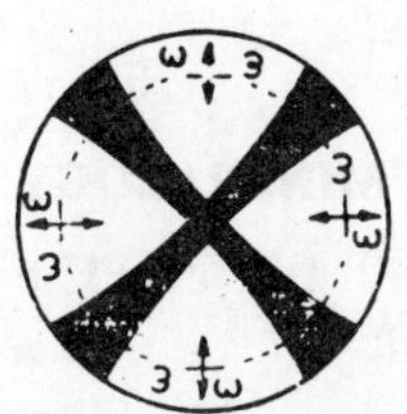

*63. (b) : The first case i.e. (a) arises in the case of acute bisectrix figure with small to moderate 2V ; the third case i.e. (c) is attained in the acute and obtuse bisectrix figures with moderate to large 2V and the last case i.e. (d) occurs in obtuse bisectrix figures with small to moderate 2V and in the uniaxial flash figures.

*88. (c) : In this special case, light can only vibrate in the circular section of the indicatrix and all possible vibration directions represent equal refractive indices. As a result, light is not doubly refracted and passes through the mineral with the vibration direction of the polarizer unchanged. Inserting the analyzer caused total absorption of light and the section will appear black in the common position of the rotation of the microscope stage. In other words, partial birefringence of this section is zero.

*89. (c) : In monoclinic minerals the ac plane contains 2 optical directions, commonly X and Z, but neither a nor c is parallel to either optical direction. b is parallel to the other optical direction, commonly Y.

*93. (a) : Biaxial minerals are positive when X is the acute bisectrix or when $2V\gamma < 90°$ (hence $2V\alpha > 90°$, since $2V\gamma + 2V\alpha = 180°$).

*94. (b) : As $2V\gamma$ approaches 180°, the two optic axes are nearly coincident so that in the limits. ($2V\gamma = 180°$) a negative indicatrix is obtained.

*95. (d) : This gives the indicatrix a shape of prolate spheroid in contrast to the oblate spheroid of negative crystals in which the length of the axis of revolution is directly proportional to the minimum value of extraordinary R.I. The circular sections in both positive and negative crystal indicatrix is directly proportional to the ordinary R.I.

*119. (b) : Scapolites crystallizes in the tetragonal system.

# 6

# *Geochemistry*

1. All the elements and their isotopes are believed to have synthesized from the nuclei of
   a) Hydrogen
   b) Helium
   c) Oxygen
   d) Nitrogen
2. Choose the correct statement among the following
   a) Gaseous element are most abundant in the universe
   b) Elements in the solid state are most abundant in the universe
   c) Elements in the gaseous and solid state are equally abundant
   d) None of the above
3. Which of the following is the correct order of elemental abundance in the universe ?
   a) $H_2 > He > Ne > O > N_2$
   b) $H_2 > He > O > N_2 > Ne$
   c) $H_2 > He > O > Ne > N_2$
   d) $H_2 > He > N_2 > O > Ne$
4. Of all the known elements those that occur naturally are about
   a) 92
   b) 96
   c) 102
   d) 106

*5. Of all the naturally occurring elements which one has not been detected in the universe ?
   a) Technetium
   b) Tantallum
   c) Osmium
   d) Promethium

6. According to the Oddo-Harkins Rule
   a) Elements of even atomic number are more abundant than those of odd atomic number
   b) Elements of odd atomic number are more abundant than those of even atomic number
   c) Elements of even and odd atomic numbers are equally abundant in the universe

d) None of the above

7. Which of the following have least Electro-Negativity (E.N.) ?
   a) Lithophiles
   b) Chalcophiles
   c) Siderophiles
   d) Atmophiles

*8. A mineral that is unknown on the earth but found on the moon is
   a) Promethium
   b) Armacolite
   c) Tantallum
   d) Teschenium

9. The crust consists of
   a) About 95% of silicate minerals
   b) About 75% of silicate minerals
   c) Over 50% of quartz
   d) Over 50% of non-silicates

10. Which of the following groups is characterized by the presence of metallic bonds ?
   a) Lithophiles
   b) Chalcophiles
   c) Siderophiles
   d) Atmophiles

11. Elements with completely full outermost shells are grouped under
   a) Lithophiles
   b) Chalcophiles
   c) Siderophiles
   d) Atmophiles

12. Comouflage is a type of ionic-atomic substitution which takes place in two atoms having
   a) Same charge and similar radius
   b) Same charge and lower radius
   c) Same charge and higher radius
   d) Lower charge and similar radius
   e) Higher charge and similar radius

13. Consider the following ionic-substitutions ;

$$Mg^{+2} \rightleftharpoons Li^{+} \; ; \qquad Ca^{+2} \rightleftharpoons Sr^{+}$$

   These are examples of
   a) Comouflage

b) Capture
c) Admittance
d) Diadochy

14. Silicon is the second most abundant element in the earth's crust after oxygen. It occupies——% of the total volume of the crust
a) 27%
b) 21%
c) 4%
d) 1%

15. The most abundant minerals in the earth's crust belongs to
a) Quartz
b) Pyroxenes
c) Plagioclase feldspars
d) Potash feldspars

16. Which is the most correct sequence of the elemental abundance in the earth ?
a) O>Si>Mg>Fe>Al
b) O>Si>Al>Mg>Fe
c) Fe>O>Si>Mg>Al
d) Fe>O>Si>Al>Mg

*17. The order of abundance of elements on the moon is
a) O>Si>Mg>Fe>Ca
b) O>Si>Al>Mg>Ca
c) O>Si>Mg>Al>Fe
d) O>Si>Fe>Al>Ca

18. 'Isobars' are different elements with
a) same at. wt. but different at. nos.
b) different at. wts. but same at. no.
c) different at. wts. and different at. nos.
d) same number of neutrons but different number of protons in the nucleus

19. 'Isotones' are characterized by
a) same number of neutrons in the nucleus
b) same number of neutrons but different number of protons
c) same number of neutrons, but different number of protons and different atomic weights
d) different number of neutrons and protons but nearly same atomic number

*20. Nuclei of which of the following combinations are the most abundant ?

a) even neutron number and even proton number

b) even neutron number and odd proton number

c) odd neutron number and even proton number

d) odd neutron number and odd proton number

21. In the ultrabasic model of Clark and Ringwood, the upper mantle is believed to be composed of 'Pyrolite'. Pyrolite corresponds in composition to a mixture of

a) one part of basalt to three parts of granodiorite

b) one part of basalt to three parts of dunite

c) basalt, rhyolite and eclogite

d) granites, granodiorites and ultramafics

22. The mineralogical composition of the mantle pyrolite varies as a function of

a) Temperature only

b) Pressure only

c) Both temperature and pressure

d) None of the above

23. Rubidium is usually found dispersed in

a) K-rich minerals

b) Ca-rich minerals

c) Na-rich minerals

d) Al-rich minerals

24. The term 'clarke' is used to define

a) The average % of an element within the earth

b) the average % of an element within the earth's crust.

c) $SiO_2/Al_2O_3$ ratio of an element in the earth's crust

d) The average volume % of an element within the earth

25. Which of the following layers of the earth can be described as 'Oxysphere' ?

a) Biosphere

b) Atmosphere

c) Asthenosphere

d) Lithosphere

e) Barysphere

26. Elements having a strong affinity for silicates are known as

a) Siderophiles

b) Chalcophiles

c) Lithophiles

d) Atmophiles

27. The most important source of radiogenic heat production in the earth just after its accretion was

a) $U^{238}$

b) $U^{236}$

c) $Th^{232}$

d) $U^{235}$

28. The interval between the formation of the earth and the formation of the oldest rocks is approximately

a) 1000 years

b) 100,000 years

c) $1 \times 10^6$ years

d) $1 \times 10^9$ years

*29. The 'Lemberg's Test' is used to distinguish

a) Calcite and dolomite

b) Calcite and aragonite

c) Olivine and serpentine

d) Limestone and marble

*30. Which of the following is NOT true of ionic or heteropolar bonding ?

a) It is generally formed between elements from opposite sides of the noble elements

b) The ionic bonds formed between two similar elements are strongest

c) Minerals with ionic bonding have high M.Ps and are mechanically hard and strong

d) No molecules are formed by ionic bonding

*31. On the basis of E.N. values the bonding within the most common coordination units ($SiO_4$) should be

a) 100% ionic

b) 100% covalent

c) 50% ionic and 50% covalent

d) Coordinate covalent

32. By volume, oxygen occupies about---- of the continental crust

a) 94%

b) 63%

c) 47%

d) 27%

33. When the centres of the three anions surrounding a cation lie

in one plane, the coordination polyhedra has the shape of

a) Triangle

b) Tetrahedron

c) Octahedron

d) Cube

34. Two ions can replace each other in a crystal structure only if the difference between their ionic radii does not exceed

a) 5%

b) 10%

c) 15%

d) 20%

*35. Consider the two ions $Ca^{+2}$ and $Na^{+1}$. When feldspars are formed from a melt

a) The minerals with $Ca^{+2}$ appear first

b) The minerals with $Na^{+1}$ appear first

c) Both $Ca^{+2}$ and $Na^{+1}$ appear simultaneously

d) there is no specific order of incorporation of these two ions in the crystal structure

36. Stishovite, a high pressure polymorph of quartz, exhibits

a) Tetrahedral structure

b) Octahedral structure

c) Hexagonal closed packed structure

d) Cubic structure

37. The inter-atomic linkage that is responsible for the coherence of inert gases when they are condensed to solids at low temperatures is

a) Ionic bonds

b) Covalent bonds

c) Coordinate covalent bonds

d) Vander Waal's bonds

38. Which of the following has the least ionic radius ?

a) $Na^{+}$

b) $Mg^{+2}$

c) $Al^{+3}$

d) $Si^{+4}$

39. The most important factor governing the atomic substitution is

a) Ionic size

b) Charge of the ion

c) Valency of the atom

d) E.N. differences between atoms

40. Colloidal particles range in size between
   a) $10-10^{-3}$
   b) $10^{-3}-10^{-6}$
   c) $10^{-6}-10^{-9}$
   d) $10^{-9}-10^{-12}$

41. Which of the following represents a solid-liquid colloidal system ?
   a) Sol
   b) Fog
   c) Smoke
   d) Emulsion

42. β-decay of $K^{40}$ leads to the formation of
   a) $Ar^{40}$
   b) $Ca^{40}$
   c) $K^{39}$
   d) $Ar^{39}$

43. Natural uranium is
   a) A mixture of $U^{238}$, $U^{236}$ and $U^{235}$
   b) A mixture of $U^{238}$ and $Th^{232}$
   c) $U^{238}$ with a coating of $U^{235}$
   d) None of the above

44. Rainwater is distinguished from groundwater by
   a) Tritium method
   b) Berylium method
   c) Rb-Sr method
   d) Radio carbon method

45. In the tectosilicates, the $(AlSiO_4)$ unit is found in
   a) Potash feldspars
   b) Plagioclase feldspars
   c) Leucite
   d) Nepheline

## GEOCHEMISTRY (KEY)

| | | | |
|---|---|---|---|
| 1. a | 2. a | 3. c | 4. a |
| *5. d | 6. a | 7. a | *8. b |
| 9. a | 10. c | 11. d | 12. a |
| 13. c | 14. d | 15. c | 16. c |
| 17. a | 18. a | 19. c | *20. a |
| *21. b | 22. c | 23. a | 24. b |

| | | | |
|---|---|---|---|
| 25. d | 26. c | 27. d | 28. d |
| *29. a | *30. b | *31. c | 32. a |
| 33. a | 34. c | 35. a | 36. d |
| 37. d | 38. d | 39. a | 40. b |
| 41. a | 42. b | 43. c | 44. a |
| 45. d | | | |

## EXPLANATORY ANSWERS

*5. (d) : Promethium (At. No. 61) has not been detected in the universe while technetium (At. No. 43) is absent from the earth but it has been detected in stars.

*8. (b) : Armacolite (Ca, Mg) $Ti_2O_5$ is named after the three Apollo II astronauts Armstrong, Aldrin and Collins.

*20 (a) : This is due to the greatest nuclear binding energy for even N-even Z nuclei.

*29. (a) : Calcite boiled for 15 minutes in a solution of Al or ferric chloride and logwood is stained pink, while dolomite is not.

*30. (b) : Ionic bonding does not occur between two similar elements.

*31. (c) : The E.N. values of Si (1.81) and O (3.5) indicate that the bonding in the $(SiO_4)$ tetrahedra is about 50% covalent and 50% ionic. However, the composition and structures of the common silicate minerals can be understood by regarding them as ionic structures.

*35. (a) : This is according to the Rule-3 of the ionic substitution i.e., when two ions have the same radius but different electrical charges, the one with the greatest charge tends to be incorporated first.

# 7

# *Structural Geology*

1. Structural Geology deals with matters in the
   a) Solid state
   b) liquid state
   c) Gaseous state
   d) All of the above
2. The 'true dip' of a bed is considered to be a
   a) Vector quantity
   b) Scalar quantity
   c) either vector or scalar quantity
   d) Neither vector nor scalar quantity
3. The dip of a horizontal bed striking E-W is measured in the
   a) Horizontal plane
   b) Vertical plane
   c) Axial plane
   d) Inclined plane
4. The amount of dip measured along the strike direction of a vertical bed is always
   a) 90°
   b) 60°
   c) 45°
   d) zero
5. The maximum amount of dip of a bed is measured along N 40°E. The strike direction of the bed is
   a) N—S
   b) E—W
   c) N 50°W
   d) S 50°W
6. Isogons are defined as the points joining beds of
   a) Equal thickness
   b) Equal dip amount
   c) Same age
   d) Same strike direction
   e) Equal elevation above the sea level

7. In a strata composed of horizontal beds, one outcropping at higher contours is
   a) Less resistant than the lower ones
   b) More resistant than the lower ones
   c) Older than the lower ones
   d) Younger than the lower ones

8. In the case of dipping beds, progressively younger beds occur along the direction of dip if
   a) Dip>Slope
   b) Dip<Slope
   c) Dip=Slope
   d) None of the above

9. The behaviour of perfectly elastic bodies is governed by
   a) Hooke's law
   b) Hilt's law
   c) Lambert's law
   d) Bode's law

10. Hooke's law is applicable in the case of
    a) Stretching
    b) Bending
    c) Twisting
    d) All of the above

*11. Liquids and gases possess
    a) Bulk modulus of elasticity
    b) Young's modulus of elasticity
    c) Both 'a' and 'b'
    d) Rigidity modulus of elasticity

12. Compressibility can be described as the reciprocal of
    a) Bulk modulus
    b) Young's modulus
    c) Rigidity modulus
    d) None of the above

13. Poisson's ratio is the ratio between
    a) Tangential stress and shearing strain
    b) Transverse pressure and axial strain
    c) Hydrostatic pressure and volume strain
    d) None of the above

14. The average value of Poisson's ratio for rocks is
    a) 0.25
    b) 0.50
    c) 1.00
    d) 1 50

15. Substances that undergo a large plastic deformation before rupture are called
    a) Brittle
    b) Ductile
    c) Amorphous
    d) Malleable

16. When the strain is recoverable but is also time dependent, the deformation is known as
    a) Elastic
    b) Anelastic
    c) Plastic
    d) Ductile

17. Direct stress acts———to the surface of the body
    a) Parallel
    b) Inclined
    c) Normal
    d) None of the above

18. In the common geologic practice a compressive stress is considered to be
    a) Positive
    b) Negative
    c) Either positive or negative
    d) Neither positive nor negative

*19. If the orientations of the principal stress axes X, Y and Z have changed during the deformation, the process is known as
    a) Pure shear
    b) Simple shear
    c) Axial strain
    d) Triaxial stress

20. In a general stress (Triaxial stress) system
    a) The three principal stressess are equal
    b) the three principal stresses are unequal
    c) The three principal stresses are unequal and have non-zero values

d) None of the above

21. The inter-limb angles of folds give an indication of the
   a) Genesis of folds
   b) Morphology of folds
   c) Intensity of deformation
   d) Type of deformation

22. Which of the following are the folds with parallel arrangement of limbs ?
   a) Isoclinal folds
   b) Inverted folds
   c) Reclined folds
   d) Symmetrical folds

23. In 'Flexures' the inter-limb angle lies between
   a) 180°—120°
   b) 120°—70°
   c) 70°—30°
   d) 30°—0°

24. Fold axis lies in the direction
   a) Parallel to hinges
   b) Normal to hinges
   c) Inclined to hinges
   d) None of the above

25. A fold profile is taken in the direction
   a) Parallel to fold axis
   b) Perpendicular to fold axis
   c) 45° to fold axis
   d) 60° to fold axis

26. A fold which is convex upwards and having younger rocks in its core may be described as a/an
   a) Anticline
   b) Antiform
   c) Anticlinorium
   d) Synform

27. A synform may be described as a/an
   a) Downward facing syncline
   b) Downward facing anticline
   c) Upward facing syncline
   d) Upward facing anticline

28. Folds having sharp hinges and smooth sides are termed as
    a) Concertina folds
    b) Concentric folds
    c) Cuspate folds
    d) Chevron folds

29. Folds which exhibit a constant profile are termed as
    a) Curvilinear folds
    b) Concentric folds
    c) Concertina folds
    d) Cylindrical folds

30. Match the following and choose the correct answer

| | |
|---|---|
| A. Conjugate folds | 1. Folds developed in many stages |
| B. Polyclinal folds | 2. Folds in which axial surfaces dip towards each other |
| C. Fan folds | 3. Folds with variable axial surfaces |
| D. Cross folds | 4. Folds in which both the limbs are overturned |

    a) A—2 B—1 C—4 D—3
    b) A—4 B—3 C—2 D—1
    c) A—1 B—2 C—3 D—4
    d) A—2 B—3 C—4 D—1

31. Choose the correct statement among the following
    a) Parallel folds maintain their shapes both upwards and downwards
    b) Parallel folds maintain their shape upwards only
    c) Parallel folds maintain their shape downwards only
    d) Parallel folds do not maintain their shape either upwards or downwards

32. In the parallel folds
    a) Anticlines and synclines become sharper with depth
    b) Anticlines and synclines become broader with depth
    c) Synclines become sharper with depth
    d) Anticlines become sharper with depth

33. Thickening and thinning of beds at crests and troughs is found in
    a) Open folds
    b) Similar folds
    c) Parallel folds
    d) Concentric folds

34. The plan views of some of the common folds are given below. Match these with the folds to which they belong

A. 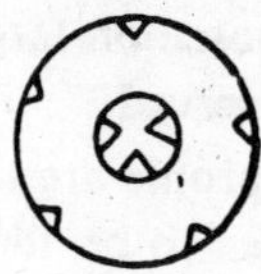 1. Dome

B.  2. Basin

C.  3. Anticline

D. 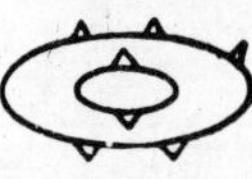 4. Syncline

a) A—1 B—2 C—3 D—4
b) A—2 B—1 C—3 D—4
c) A—2 B—1 C—4 D—3
d) A—1 B—2 C—4 D—3

35. In which of the following folds the limbs dip towards each other ?

a) Anticlines
b) Antiforms
c) Anticlinal fanfolds
d) Synclinal fanfolds

36. Choose the non-tectonic folds (s) from the following

(i) Decollement
(ii) Supratenuous fold
(iii) Generative fold
(iv) Brachymorphic fold

a) (i) only
b) (i) and (ii) only
c) (ii) and (iii) only
d) (i), (ii) and (iv) only

37. In supratenuous folds the beds are

a) Thin at the limbs but thickened at the crests
b) Thin at the crests but thickened at the limbs
c) Uniformly thick at the crests and the troughs

d) None of the above

38. A complex fold is one in which

a) Smaller folds are superimposed on larger folds
b) Hinges are numerous and curved
c) Axis is folded
d) The structure is too complex to be identified

39. The diagram given below represents

a) Non-plunging anticline
b) Non-plunging syncline
c) Doubly plunging syncline
d) Doubly plunging anticline

40. The difference between folds and faults is that

a) Folds are two dimensional while faults are three dimensional
b) Folds are three dimensional while faults are two dimensional
c) Folds are three dimensional structures while faults are planar discontinuities
d) Folds are associated with sedimentary rocks while faults are found in igneous rocks

41. Faults are generated by

(1) Extension
(2) Shear
(3) Compression
(4) Torsion

a) 1 only
b) 1 and 3 only
c) 1, 2 and 3 only
d) 3 and 4 only

42. Faults are divided into parallel faults, enechelon faults, peripheral faults and radial faults based on

a) Fault pattern
b) Rake of the net slip
c) Dip amount
d) Apparent movement of the blocks

43. If the rake of the net slip of a fault is 90° the fault may be a
   a) Strike-slip fault
   b) Dip-slip fault
   c) Diagonal-slip fault
   d) Tear fault

44. Hade of a fault is
   a) 90°+ dip
   b) 90°−dip
   c) Plunge+rake
   d) Plunge+dip

45. Tear faults are a variety of
   a) Strike faults
   b) Dip faults
   c) Strike-slip faults
   d) Dip-slip faults

46. In which type of faults the stratigraphic throw is measurable in many thousands of feet
   a) Overthrust faults
   b) Underthrust faults
   c) Gravity faults
   d) Step faults

47. Which of the following are NOT low-angle faults ?
   a) Overthrusts
   b) Uuderthrusts
   c) Cylindrical faults
   d) Detachment faults

48. In the case of normal faults, the hade is towards
   a) Downthrown side
   b) Upthrown side
   c) Either downthrown or upthrown side
   d) None of the above

49 The main feature of thrust faults is
   a) The footwall has moved down in relation to the hanging wall
   b) The footwall has moved up in relation to the hanging wall
   c) Shortening of the crustal rocks involved in the faulting
   d) Very large stratigraphic throw

50. Lag faults are
   a) Low angle faults with normal displacement

b) Low angle faults with reverse displacement
c) High angle normal faults
d) High angle reverse faults

51. Gravity faults are formed under
a) Compressive stress regime
b) Tensile stress regime
c) Shear stress regime
d) Effects of torsional forces

52. Schuppen structures are associated with
a) Normal faulting
b) Reverse faulting
c) Thrust faulting
d) Recumbent folding

53. Thrust faults differ from reverse faults in
a) The relative position of hanging wall and footwall
b) The amounts of dips involved in the two faults
c) The nature of tectonic forces responsible for faulting
d) All of the above

54. Overthrusts can be described as
a) Low angle th-ust faults
b) High angle reverse faults
c) High angle reverse faults with large net slip
d) Low angle thrust faults with large net slip

55. The greatest principal stress axis is vertical in
a) Normal faults
b) Reverse faults
c) Thrust faults
d) Strike-slip faults

56. The stress characteristic of Wrench faults is that
a) $\sigma_1$ is horizonal and $\sigma_3$ is vertical
b) $\sigma_1$ is horizontal and $\sigma_2$ is vertical
c) $\sigma_1$ is vertical and $\sigma_3$ is horizontal
d) $\sigma_1$ is vertical and $\sigma_2$ is horizontal

57. Splay faults are
a) Faults which display arc like outcrop on level surface
b) Short faults that overlap each other
c) Normal or reverse faults with more or less parallel strikes
d) A complex arrangement of branching faults

58. Match the following and choose the correct answer

| | | |
|---|---|---|
| A. Grabens | (1) | A series of parallel normal faults |
| B. Klippen | (2) | A zone of relative displacement |
| C. Step fault | (3) | Erosional remnant of a nappe |
| D. Triangle zone | (4) | A structure formed due to back-thrusting |
| | (5) | Miniature rift valley |

a) A—5 B—3 C—1 D—2
b) A—4 B—3 C—1 D—2
c) A—5 B—3 C—1 D—4
d) A—5 B—2 C—4 D—3

59. Which of the following pairs is incorrect ?
a) Triangular facets———normal faults
b) Offset streams————strike-slip faults
c) Plumose markings———thrust faults
d) Palaeontologic break———disconfirmity

60. Fluccan or Pug consists of
a) Large blocks caught along faults
b) Broken fragments in a dark matrix of ultramicroscopic grains
c) Minute glassy spherulites
d) Fine powder found along fault surfaces

61. Joints are encountered in
a) Sedimentary rocks only
b) Igenous rocks only
c) Metamorphic rocks only
d) All of the above

62. The distance between any two joints may be described as
a) Slip cleavage
b) Crenulation cleavage
c) Fracture cleavage
d) Bedding fissility

63. Match the following and choose the correct answer

| *Type of joints* | *Genesis* |
|---|---|
| A. Extension joints | 1. Residual stresses |
| B. Sheeting | 2. Tensional forces |
| C. Conjugate joints | 3. Tectonic stresses |
| D. Columnar joints | 4. Shearing |

a) A—2 B—1 C—4 D—3

b) A—3 B—4 C—1 D—2
c) A—3 B—1 C—4 D—2
d) A—2 B—4 C—1 D—3

64. Joints which are developed perpendicular to the fold axes are called
a) Extension joints
b) Released joints
c) Tension joints
d) Shear joints

65. Cross joints are
a) Columnar joints produced due to the inter-section of two or more sets of vertical joints
b) Cracks formed due to prolonged exposure of clayey sediments to the atmosphere
c) Joints formed perpendicular to column axes due to contraction
d) Joints formed parallel to the fold axes

66. Select the suitable alternatives from the choices given below

| | |
|---|---|
| (1) Cleat | (A) Three sets of mutually perpendicular joints found in granites |
| (2) Mural joints | (B) Joints that are parallel in plan view and cut across other joints |
| (3) Sheeting | (C) Joints parallel to lineation |
| (4) Systematic joints | (D) Ruptures similar to joints found in artificial openings |
| | (E) Jointing found in coal |

a) 1—E 2—A 3—D 4—B
b) 1—E 2—A 3—D 4—C
c) 1—E 2—A 3—C 4—B
d) 1—D 2—B 3—C 4—A

67. Very small ridges and depressions on the surface of joints are described as
a) Slickensides
b) Feather joints
c) Plumose markings
d) Piedmont scarps
e) Striations

68 Match the following and choose the correct answer

(A)  (1) Vertical foliation

(B)  90° (2) Vertical joint

(C) (3) Vertical cleavage

(D) (4) Vertical lineation

a) A—1 B—2 C—3 D—4
b) A—2 B—1 C—4 D—3
c) A—3 B—2 C—4 D—1
d) A—2 B—3 C—1 D—4

69. When the two sets of beds separated by an unconfirmity are NOT parallel to each other, it is described as a/an
a) Angular unconfirmity
b) Disconfirmity
c) Non-confirmity
d) Heterolithic unconfirmity

70. Which of the following statements is true of unconfirmities ?
a) Unconfirmities are three dimensional, non-directional features
b) Unconfirmities are two dimensional, directional features
c) Unconfirmities are three dimensional, directional features
d) Unconfirmities are two dimensional, non-directional features

71. An unconfirmity which marks an abrupt change in the velocity of seismic waves is termed
a) Angular unconfirmity
b) First order unconfirmity
c) Second order unconfirmity
d) Zero order unconfirmity

72. Short breaks in sedimentation associated with slow vertical movements of the earth's crust is known as
a) Blended unconfirmities
b) Heterolithic unconfirmities
c) Pinch outs
d) Diastems

73. Intrusive igneous contracts are indicative of a/an
a) Angular unconfirmity
b) Disconfirmity
c) Non-confirmity
d) Blended unconfirmity

74. Disconfirmities are most easily distinguished with the help of
a) Erosional and non-depositional surfaces
b) Variation in the dip amounts between two sets of beds

c) A significant break in paleontologic record

d) Intrusive relationship of older igneous rocks with younger sedimentary formations

75. An unconfirmity contact can be distinguished from faulted contact by means of one or more of the following criteria

(1) In an UC contact older rocks always strike into the younger rocks

(2) A faulted contact shows features such as breccia, gouge, etc , whereas UCs show conglomerates at their contacts

(3) Contacts between older and younger rocks show divergence of dip in the case of a fault

(4) Unconfirmities are three dimensional whereas faults are two dimensional

a) 1 and 2 only

b) 1, 2 and 3 only

c) 2 and 3 only

d) 1, 2 and 4 only

76. A limited area of an exposure in which older rocks are surrounded by younger rocks is known as an

a) Overlap

b) Offlap

c) Outlier

d) Inlier

77. Synclinal hill is an example of

a) Outlier

b) Inlier

c) Overstep

d) Overlap

78. When the lower beds of older series extend further than the younger ones it is described as an

a) Overlap

b) Offlap

c) Onlap

d) Overstep

*79. Identify the fault given in the diagram below

a) Normal fault

b) Reverse fault

c) Thrust fault

d) Lag fault

*80. Which of the following statements are true of a 'Geologic window' structure or 'Fenster' ?

I. In it older rocks are surrounded by younger rocks

II. These are erosional depressions in large thrust sheets or nappes

III. The older rocks crop cut through the windows

IV. Windows occur in normal stratigraphic sequences

a) I and III only
b) II and IV only
c) I, II and III only
d) II, III and IV only

81. Pitch and plunge of a fold coincides when the folds are
a) symmetrical
b) asymmetrical
c) disharmonic
d) isoclinal

82. An anticlinal bend is a fold in which the dip of the bed
a) increases suddenly
b) decreases suddenly
c) first increases and then decreases
d) remains more or less constant

83. A fold which increases in size as it is traced upwards along the axial plane is termed
a) Growth fold
b) Reclined fold
c) Ptygmatic fold
d) Generative fold
e) Glitbrett fold

84. Folds having multiple hinges are called
a) Polyclinal folds
b) Box folds
c) Concentric folds
d) Curvilinear folds

85. A Pericline is a fold in which
a) the axial plane is curved
b) crests are thick while the limbs are thin
c) the diameter increases along its axial plane
d) the amplitude decreases regularly to zero in both the directions

86. A set of folds with smaller wavelengths occuring superimposed on those with larger wavelength are described as
a) Disharmonic folds

b) Enechelon folds
c) Parasitic folds
d) Digitations

87. A 'Culmination' may be described as a combination of
a) two anticlines of different sets of folding
b) two synclines of different sets of folding
c) an anticline and a syncline
d) a thrust fault and a recumbent fold

88. In Polyclinal folds the axial planes are
a) horizontal
b) vertical
c) inclined
d) parallel
e) randomly oriented

89. If the Plunge of a fold reverses its direction within a given limit the fold is described as
a) Doubly plunging fold
b) Non-plunging fold
c) Reclined fold
d) Linear fold

90. A sudden flattening of dip without change in its direction results in

| | |
|---|---|
| (i) Anticlinal bend | (ii) Synclinal bend |
| (iii) Structural terrace | (iv) Monocline |

a) (i) and (ii) only
b) (i) and (iii) only
c) (ii) and (iv) only
d) (ii) and (iii) only

91. Match the following and choose the correct answer

| | |
|---|---|
| (A) Imbricate zone | (1) a thrust sequence in which new thrusts develop in footwall |
| (B) Depression | (2) folds with angular hinges and planar limbs |
| (C) Accordian folds | (3) combination of an anticline and a syncline of different sets of folding |
| (D) Piggyback | (4) stacking of many thrust sheets |

a) A—4 B—3 C—2 D—1
b) A—1 B—3 C—2 D—4
c) A—1 B—4 C—2 D—3

d) A—4 B—2 C—3 D—1

92. Thrust faults in which the hanging wall is the active element are called

a) Underthrusts
b) Overthrusts
c) Upthrusts
d) Oversteps

93. An 'Overstep' is a thrust sequence in which new thrusts develop

a) in the footwall only
b) in the hanging wall only
c) both in the footwall and the hanging wall
d) alternately in the footwall and the hanging wall

94. In strike-slip faults the sense of simple shear movement implies

a) horizontal contraction of blocks parallel to $\sigma_1$ and vertical extension parallel to $\sigma_3$
b) horizontal extension of blocks parallel to $\sigma_1$ and vertical extension parallel to $\sigma_3$
c) horizontal extension of blocks parallel to $\sigma_3$ and vertical extension parallel to $\sigma_1$
d) horizontal contraction of blocks parallel to $\sigma_1$ and horizontal extension parallel to $\sigma_3$

95. Transverse strike-slip faults are known as

a) Transform faults
b) Tear faults
c) Transcurrent faults
d) Lag faults

96. The horizontal displacement of the fractured ends of a rock stratum is termed as

a) Rake
b) Throw
c) Heave
d) Plunge

97. Relative motion along a fault may be deciphered by means of

a) Fault gouge
b) Drags
c) Mylonite
d) Horses and slices

98. Mylonite forms when
   a) confining pressures are very low
   b) confining pressures are moderate
   c) confining pressures are very high
   d) none of the above

99. A Conjugate system of joints consists of two sets of joints
   a) at right angles to one another and produced by the same stress system
   b) parallel to each other and produced by the same stress system
   c) which are mutually perpendicular and produced by two different stress regimes
   d) none of the above

100. Feather joints are formed due to
   a) Tectonic stresses
   b) Residual stresses
   c) Shearing
   d) Tensional forces

101. Which of the following show the effects of tensional forces ?
   (1) Columnar joints (2) Cross joints
   (3) Mudcracks (4) Bedding joints
   a) 1 and 3 only
   b) 1 and 2 only
   c) 1, 2 and 3 only
   d) all of the above

102. In Flexure folding
   a) both the competents and incompetent beds thicken and thin
   b) the incompetent beds thicken and thin
   c) the competent beds thicken and thin
   d) none of the above

103. Differential movement along closely spaced fractures results in
   a) Flexure folds
   b) Shear folds
   c) Flow folds
   d) Glitbrett folds

104. The majority of folds encountered in nature are
   a) Flexure folds
   b) Shear folds
   c) Flow folds

d) A combination of flexure, shear and flow folds

105. In which of the following folds an axial plane may not be present ?

a) Monoclines
b) Anticlines
c) Antiforms
d) Flow folds
e) Ptygmatic folds

*106. A homocline in which the dip continuously decreases is termed a/an

a) Plane homocline
b) Antihomocline
c) Synhomocline
d) Monocline

107. Folds of oval shape in which the ratio of the length and the width is less than 3 are called

a) Linear folds
b) Brachymorphic folds
c) Concentric folds
d) Cleavage folds

108. In dome-shaped folds the ratio of the length (a) and width (b) i.e. a/b is

a) less than 3
b) more than 3
c) approximately equal to unity
d) infinity

109. The property of the rocks whereby they break along approximately parallel surfaces is termed

a) Bedding fissility
b) Rock cleavage
c) Lineation
d) Foliation

110. The term "Continuous cleavage" includes

a) Slaty cleavage and Schistosity
b) Slaty cleavage and Fracture cleavage
c) Fracture cleavage and Crenulation cleavage
d) Axial Plane cleavage and Bedding cleavage

*111. Slaty cleavage can be distinguished from Fracture cleavage on the basis of one or more of the following criteria

(i) slaty cleavage depends on the parallel arrangement of mineral constitutents whereas fracture cleavage does not

(ii) slaty cleavage is penetrative while fracture cleavage is non-penetrative

(iii) fracture cleavage is a shear phenomenon whereas slaty cleavage is not

(iv) fracture cleavage is normal to the axial planes whereas slaty cleavage is parallel to axial planes

a) (i), (ii) and (iv) only
b) (i), (ii) and (iii) only
c) (i) and (iii) only
d) (ii) and (iv) only

112. The term "Spaced cleavage" includes
a) Fracture cleavage+Slaty cleavage
b) Fracture cleavage+Flow cleavage
c) Fracture cleavage+Schistosity
d) Fracture cleavage+Shear cleavage+Slip cleavage

113. Cleavage or schistosity that is parallel to the bedding is a result of

(i) isoclinal folding
(ii) load metamorphism
(iii) mimetic recrystallization
(iv) flow parallel to bedding

a) (i) and (iv) only
b) (i), (ii) and (iv) only
c) (i), (iii) and (iv) only
d) all of the above

114. Major folds can be delineated by the cleavage-bedding relationship. If the cleavage dips more gently than the bedding then the folds are inferred as
a) Symmetrical folds
b) Asymmetrical folds
c) Overturned folds
d) Isoclinal folds

115. When sausage-shaped bodies of one rock layer are sandwiched between layers of different rock types, the structure is described as a
a) Boudinage
b) Mullion
c) Mineral lineation
d) Crenulation

116. Mullions are formed under

a) Compressive stress regime

b) Tensile stress regime

c) Shearing

d) All of the above

117. A Shear zone may be defined as

a) a brittle zone between two undeformed rocks

b) a zone of ductile deformation between two undeformed rocks

c) a zone of relative displacement between two deformed rocks

d) none of the above

118. Pure shear can be denoted as

a) $\sigma_1 \neq \sigma_2 \neq \sigma_3$

b) $\sigma_2 = \sigma_1 + \sigma_3 = 0$

c) $\frac{\sigma_1 + \sigma_2 + \sigma_3}{3}$

d) $\sigma_2 = \sigma_3 - \sigma_1$

*119. In the case of pure shear the relationship between principal strain axes and the principal stress axes is such that

a) $X//\sigma_1$, $Y//\sigma_2$ and $Z//\sigma_3$

b) $X//\sigma_3$, $Y//\sigma_2$ and $Z//\sigma_1$

c) only X corresponds with the orientation of $\sigma_3$

d) only Y corresponds with the orientation of $\sigma_2$

*120. For strongly linear fabrics the relationship between the principal strain axes is such that

a) $X > Y > Z$

b) $X \approx Y > Z$

c) $X > Y \approx Z$

d) $X = Y = Z$

*121. A planar flattening fabric, such as slaty cleavage, schistosity or gneissosity lies in——— plane of the strain ellipsoid

a) XY

b) XZ

c) YZ

d) Any of the above

122. The simplest relationship between shear stress $\sigma$ and normal stress $\tau$ at failure is given by the formula $\tau = C + \mu\sigma$ where C and $\mu$ are constants. This relationship is described as

a) Mohr equation
b) Mohr failure envelope
c) Coulomb failure criterion
d) Griffith failure creterion

123. A material which basically obeys the viscous law but which behaves elastically for stresses of short duration is termed
a) Viscoelastic
b) Elastoviscous
c) Plastic
d) Ductile

124. A typical creep curve or strain-time diagram for long time periods is composed of distinct stages each of which corresponds to a different type of rock deformation. The correct sequence of rock deformation is
a) Elastic ⟶ Viscoelastic ⟶ Viscous ⟶ Rupture
b) Viscous ⟶ Viscoelastic ⟶ Elastic ⟶ Rupture
c) Viscous ⟶ Anelastic ⟶ Accelerated viscous ⟶ Rupture
d) Viscoelastic ⟶ Viscous ⟶ Accelerated Viscous ⟶ Rupture

125. The effects of confining pressure on rock deformation is to
(i) decrease the volume of rocks
(ii) increase the density
(iii) decrease the elastic limit
(iv) increase the ultimate strength
a) (i) and (ii) only
b) (i) and (iii) only
c) (ii), (iii) and (iv) only
d) (i), (ii) and (iv) only

126. The presence of solution in the pore spaces of rocks creates a pore pressure. Its role in rock deformation is
(1) It counters the direct pressure between the adjoining grains
(2) It promotes mineralogical reactions particularly at elevated temperatures
(3) It increases ductility
(4) It decreases the ultimate strength of the rocks
a) 1 only
b) 1 and 2 only
c) 1, 2 and 3 only
d) all of the above

127. The statistical study of the recognition and description of the geometrical properties of planar and linear discontinuities of rocks is described as

a) Dynamic analysis

b) Mechanical analysis

c) Kinematic analysis

d) Petrofabric analysis

128. Which of the following secondary structures is a combination of planar and linear fabric components ?

a) Joints

b) Faults

c) Fold axes

d) Cleavage and Schistosity

*129. All fabric components are referred in space to three coordinates or fabric axes which are mutually perpendicular. Which of these corresponds to the direction of tectonic transport ?

a) a-axis

b) b-axis

c) c-axis

d) none of the above

*130. Which of the following groups of minerals are most commonly studied in the petrofabric analysis ?

a) Tetragonal+Hexagonal+Trigonal

b) Tetragonal+Hexagonal+Orthorhombic

c) Orthorhombic+Monoclinic+Triclinic

d) Cubic+Tetragonal+Hexagonal

131. Which of the following exhibits a spherical symmetry of fabric ?

a) Spherulites

b) Geodes

c) Ooliths

d) All of the above

e) None of the above

132. The petrofabric diagram of rock samples collected from an area which had been subjected to repeated phases of deformation shows

a) Axial symmetry of fabric

b) Orthorhombic symmetry

c) Monoclinic symmetry

d) Triclinic symmetry

133. Identify the symmetry in the petrofabric diagram given below

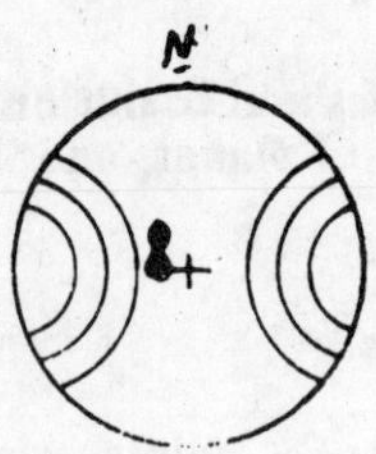


a) Axial symmetry
b) Orthorhombic symmetry
c) Monoclinic symmetry
d) Triclinic symmetry

134. A B-tectonite exhibits
a) Penetrative linear fabric
b) Penetrative planar fabric
c) Non-penetrative linear fabric
d) Non-penetrative planar fabric

135. The vertical sectional view of an area along a definite line is known as a/an
a) Isolith
b) Outcrop
c) Contour
d) Profile

136. The intersection of contour lines with the bedding plane indicates the presence of a/an
a) Vertical bed
b) Horizontal bed
c) Dipping bed
d) None of the above

137. If the spacing between the various contour lines is very narrow it indicates the presence of a
a) Steeply dipping terrain
b) Gently dipping terrain
c) Almost horizontal terrain
d) None of the above

138. On level surface, elliptical outcrops are commonly exhibited by
a) Non-plunging folds
b) Plunging folds
c) Doubly plunging synclines
d) Doubly plunging anticlines

139. In a geologic map domes and basins are represented by
   a) Concentric ellipses
   b) Concentric circles
   c) Parabolas
   d) Hyperbolas

140. In a slope between any two points when one points is visible from the other it is described as a
   a) Convex slope
   b) Concave slope
   c) Scarp slope
   d) uniform slope

141. The simplest example of a contour line is a/an
   a) Orocline
   b) Coastline
   c) Continental slope
   d) Continental shelf

142. Higher contours always enclose the lower ones in the case of a
   a) Hill
   b) Valley
   b) Basin
   c) Spur

143. Identify the geomorphic features from their typical contour forms

| *Geomorphic feature* | *Contour form* |
|---|---|
| (1) River valley | (A) |
| (2) Plateau | (B) |
| (3) Ridge | (C) |
|  | (D) |

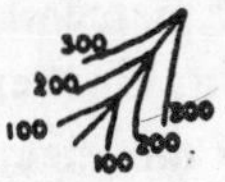

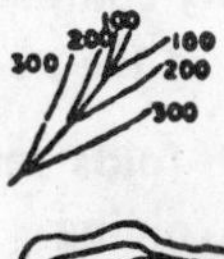

   a) 1—A 2—B 3—C

b) 1—A 2—C 3—D

c) 1—B 2—C 3—D

d) 1—A 2—D 3—C

144. Contour lines cannot cross or touch each other except in the case of

a) Escarpments

b) Scarp slope

c) Waterfall

d) Plateau

145. A clinometer compass is used to measure

a) Strike direction

b) Strike and dip directions

c) Strike direction, dip direction and dip amount

d) None of the above

146. When drilling is employed to infer the structure of the beds the following cannot be determined

a) Strike

b) Dip

c) Both strike and dip

d) None of the above

147. Pi diagrams and beta diagrams are used for analyzing

a) Folds

b) Faults

c) Folds and faults

d) Joints

148. Diapiric or piercement folding results from

a) Horizontal movements in an initially competent fold

b) Vertical movements in an initially competent fold

c) A combination of horizontal and vertical movements in an initially competent fold

d) None of the above

149. Gleitbrett folds result from

a) Shear folding

b) Flow folding

c) Flexure folding

d) All of the above

150. Horizontal shear faults that are commonly found at right angles to the mid-oceanic ridges and often appear to offset them are called

a) Tear faults
b) Transcurrent faults
c) Transform faults
d) Thrust faults

## STRUCTURAL GEOLOGY (KEY)

| | | | |
|---|---|---|---|
| 1. d | 2. a | 3. b | 4. d |
| 5. c | 6. b | 7. c | 8. a |
| 9. a | 10. d | *11. a | 12. a |
| 13. b | 14. a | 15. b | 16. b |
| 17. c | 18. a | *19. b | 20. c |
| 21. c | 22. a | 23. a | 24. a |
| 25. b | 26. a | 27. b | 28. c |
| 29. d | 30. d | 31. d | 32. d |
| 33. b | 34. c | 35. c | 36. b |
| 37. b | 38. c | 39. a | 40. c |
| 41. b | 42. a | 43. b | 44. b |
| 45. c | 46. a | 47. c | 48. a |
| 49. c | 50. a | 51. b | 52. c |
| 53. b | 54. d | 55. a | 56. b |
| 57. d | 58. c | 59. c | 60. d |
| 61. d | 62. c | 63. c | 64. a |
| 65. c | 66. a | 67. c | 68. d |
| 69. a | 70. a | 71. b | 72. d |
| 73. c | 74. c | 75. b | 76. d |
| 77. a | 78. b | *79. d | *80. b |
| 81. a | 82. a | 83. d | 84. a |
| 85. d | 86. c | 87. a | 88. c |
| 89. a | 90. d | 91. a | 92. b |
| 93. b | 94. d | 95. c | 96. c |
| 97. b | 98. c | 99. a | 100. c |
| 101. c | 102. b | 103. b | 104. d |
| 105. a | *106. c | 107. b | 108. c |
| 109. d | 110. a | *111. b | 112. d |
| 113. d | 114. c | 115. a | 116. a |
| 117. b | 188. b | 119. b | 120. c |
| *121. a | 122. c | 123. b | 124. d |
| 125. d | 126. d | 127. d | 128. c |
| *129. a | *130. a | 131. d | 132. d |
| 133. d | 134. b | 135. d | 136. a |
| 137. a | 138. c | 139. b | 140. b |

141. b 142. c 143. b 144. c
145. c 146. a 147. a 148. c
149. a 150. c.

## EXPLANATORY NOTES

*11. (a) : Since liquids and gases can permanently withstand only a hydrostatic pressure, they possess only bulk modulus of elasticity. On the other hand, solids possess all the three moduli of elasticity

*19. (b) : If the orientations of the principal strain axes X, Y and Z have not changed during the deformation, the strain is described as irrotational and the process is known as 'Pure Shear'. Where a change has occurred, the strain is described as rotational and the process is known as 'Sample shear'.

*79. (b) : Faults dipping less than 45° or low-angle faults, are distinguished from high-angle dip-slip faults. If the sense of movements is reverse they are termed thrusts and if the sense of movement is normal, they are termed 'lags'.

*80. (b) : In a geological window structure older rocks generally surround younger ones and it is the younger rocks that crop out through the fenster.

*111. (b) : Both slaty cleavage and fracture cleavage are roughly parallel to the axial plane of folds. Slaty cleavage converge downwards to the anticlinal core while fracture cleavage converge upwards towards the top of an anticline.

*120. (c) : In general the three principal strain axes X, Y and Z are unequal and $X>Y>Z$. Some special cases are :

(i) Axially symmetric extension $(X>Y=Z)$ which involves uniform extension in the X direction and equal shortening in all direction at right angles to it.

(ii) Axially symmetric shortening $(X=Y=Z)$ involves uniform shortening in the Z direction and equal extension in all directions at right angles to it.

(iii) Plane strain $(X>Y=1>Z)$—here the Y axis has unit length, X is extended and Z is shortened. Thus, plane strain is a special type of triaxial ellipsoid.

*121. (a) : This enables the orientation of Z axis to be found immediately.

*129. (a) : The a-axis is the direction of tectonic transport i.e., in which the beds slide past one another. It coincides with the slip direction on an S-surface, or with maximum elongation of particular rock particles. The b-axis is the axis of internal or external rotation or the intersection of bedding and slip planes. It coincides with the fold axes.

*130. (a) : Mostly uniaxial minerals are studied in petrofabric analysis. Biotite and muscovite are pseudo-hexagonal. Olivine is found more frequently in fracture zones, thus it is studied even though it crystallizes in the orthorhombic system.

# 8

# Paleontology

1. 'Taphonomy' is the science dealing with
   a) Study of the conditions of burial of fossils
   b) Reconstruction of paleo-environments by means of fossils
   c) Modes of preservation of fossils
   d) Study of fossil pores and spores
2. Which of the following modes of preservation of fossils is rare?
   a) Preservation of the entire organism
   b) Preservation of hard parts
   c) Preservation of original form
   d) Alteration of hard parts
3. Jelly fishes and worms are mostly preserved by
   a) Silicification
   b) Carbonization
   c) Petrifaction
   d) Pyritization
4. 'Petrifaction' is a type of fossilization where
   a) Original form is preserved
   b) Original form and structure and preserved
   c) Entire organism is preserved
   d) Only the hard parts are preserved
5. Thc exoskeleton of sponges is made up of
   a) Silica
   b) Calcite
   c) Aragonite
   d) Calcium phosphate
6. Fossil fecal pellets of ancient animals are described as
   a) Gastroliths
   b) Coproliths
   c) Beekite rings
   d) Pseudo-fossils
7. The most favourable environment for the preservation of fossils is
   a) Terrestrial
   b) Lacustrine
   c) Fluvial
   d) Marine

8. The term *'Living fossil'* refers to
   a) a species which was thought to be extinct but which is in fact living
   b) a species which has acquired the characters of its predecessors
   c) a species which has retained all the primitive characters of its ancestors and has survived much longer than its counterparts
   d) a species which is at its acme of evolution

*9. Living fossils exhibit the phenomenon of
   a) Bradytelic evolution
   b) Orthogenesis
   c) Palingenesis
   d) Tachygenesis

10. Which of the following is NOT a living fossil ?
   a) *Lingula*
   b) Nucula
   c) Nautilus
   d) Nummulites

11. Which of the following can be considered as a 'Living Fossil' among the mammals ?
   a) *Ginkgo*
   b) Opossum
   c) Neopilina
   d) *Sphenodon*

*12. Superficial similarities in the morphology of members of different Phyla is described as
   a) Homology
   b) Homeomorphy
   c) Homomorphy
   d) Paedomorphosis

13. The passing of a feature of the adult stage of the ancestor into the ontogenies of descendents is described as
   a) Hypomorphosis
   b) Paedomorphosis
   c) Neoteny
   d) Gerontomorphosis

14. 'Neoteny' is described as
   a) The retention of larval characters in the adult form
   b) The hereditary characters acquired by the off-springs from their ancestors
   c) Mutation- induced in an individual due to environmental effects
   d) The regeneration of fragile parts in an individual

15. The time of rapid evolutionary change of a new taxa or the adaptive radiation is described as
   a) Tachytelic evolution

b) Bradytelic evolution
c) Orthogenesis
d) Paragenesis

*16. The pattern of evolution in which an ancestral stock frequently gives rise to short-lived groups which very often replace each other successively is described as
a) Iterative evolution
b) Bradytelic evolution
c) Convergent evolution
d) Adaptive radiation

17. Fossils of stratigraphic interest are generally provided by
a) Bradytelic evolution
b) Tachytelic evolution
c) Convergent evolution
d) Adaptive radiation

18. Homeomorphy is an example of
a) Convergent evolution
b) Divergent evolution
c) Parallel evolution
d) Adaptive specialization

19. The Dollo's law deals with
a) Irreversibility of evolution
b) Increase in the size of morphoogical parts with evolution
c) Decrease in the size of morphological parts with evolution
d) Reversibility of evolution

20. According to the Theory of Recapitulation
a) The 'Present is the key to the Past'
b) 'The Function creates the Organ'
c) 'Natural selection follows the Struggle for Existence'
d) 'Ontogeny recapitulates Phylogeny'

21. Man's hand, bat's wing and seal's flipper are all examples of
a) Analogous organs
b) Homologous organs
c) Vestigial organs
d) Homeomorphs

22. The study of Trace fossils is known as
a) Paleo-ecology
b) Synecology
c) Ichthyology
d) Ichnology

23. 'Conchology' is the study of
a) Fossil remains of fishes
b) Fossil shells of marine organisms

c) Various types of molluscs
d) Various types of mollusc shells

24. The fauna which comprises exclusively of active swimmers is described as
a) Plankton
b) Nekton
c) Benthic
d) Pelagic

25. Marine animals which are confined to waters of normal salinity are called
a) Stenohaline
b) Eu.yhaline
c) Polyhaline
d) None of the above

26. 'Biostrome' means
a) Any sheet-like mass of inorganic materials derived from limestones
b) Any sheet-like mass of organic materials derived from limestones
c) A sedimentary rock containing skeletal remains of plants
d) A sedimentary rock containing skeletal remains of animals

27. Who among the following is considered as the 'Father of Invertebrate Paleontology'?
a) Charles Lyell
b) William Smith
c) James Hutton
d) Lamarck

28. The first Vertebrates appeared during
a) Cambrian
b) Ordovician
c) Silurian
d) Devonian

29. Match the following and choose the correct answer

| | |
|---|---|
| A) First land plants | 1. Triassic |
| B) Origin of Amphibians | 2. Carboniferous |
| C) Origin of Reptiles | 3. Silurian |
| D) Rise of Dinosaurs | 4. Devonian |
| | 5. Cambrian |

a) A-5 B-3 C-2 D-1 b) A-3 B-4 C-2 D-1
c) A-5 B-4 C-2 D-1 d) A-1 B-5 C-4 D-2

30. The Prokaryotes are characterised by
i) Absence of nuclei in the cells
ii) DNA not arranged in chromosomes
iii) Chromosomes bound by nuclei and organelles

a) i only
b) ii only
c) i & ii only
d) All of the above

31. The earliest occurrence of *Stromatolites* is reported from
a) Pilbara, Western Australia (3.4 B.Y.)
b) Bulawayan Group, Zimbabwe (2.9- 3.2 B.Y.)
c) Vaimpalli dolomites (1400-1600 M.Y.)
d) Kheinjua formation, Semri Group (900-600 M.Y.)

32. Which of the following represents the earliest known sexually reproducing organisms?
a) Bitler Spring Formation, Central Australia
b) Fig Trees Series, Swaziland
c) Pilbara, Western Australia
d) Bulawayan Group, Zimbabwe

33. The earliest record of microfossils in the Indian Shield is the "doubtful organic remains" reported by Slater in 1899 from
a) Limestones of the Kunchanahalli in the Shimoga Belt
b) Dodguni cherts of Chitradurga Belt
c) G.R.Formation, Chitradurga Schist Belt
d) Kaladgi Formation

34. The age of the Badami Group has been fixed as Ordovician (500-430 M.Y.) on the basis of
a) Acritarchs (*Leiosphaeridium* & *Rugocytis*)
b) Chitinozoa and Scolecodontes
c) Coccoid-type of micro-biota
a) Silicified Oomicrites

35. Amoeba belongs to
a) Phylum Protozoa
b) Phylum Coelentrata
c) Phylum Arthopoda
d) Phylum Echinodermata

36. The *Foraminifers* belong to the class
a) Sarcodina
b) Sporozoa
c) Flagelleta
d) Mastigophora

37. Which of the following represents the most primitive of all the Foramanifers ?
a) *Saccamina*
b) Astorhiza
c) Hyperamina
d) *Allogromia*

38. In some Foraminifers the shell is provided with perforations in order to facilitate
    a) Breathing of the organism
    b) Passage of pseudopodia
    c) Disposal of organic wastes
    d) Exchange of constituents between the body fluids and water
39. The shells of Foraminifers are made up of
    a) Calcium carbonate
    b) Arenaceous sediments
    c) Chitin
    d) Any of the above
    e) None of the above
40. In which of the following foraminifers, the shell consists of numerous chambers arranged in the form of concentric rings?
    a) *Textularia* b) Globigerina
    c) Orbitolites d) Nummulites
41. In *Globegerina* the chambers of the shell are arranged in the form of
    a) Planospiral b) Helicoid spiral
    c) Conical spiral d) Irregular spiral
42. Many foraminifers are dimorphous i.e. a single species exists in two different forms. The main difference between the two forms is
    a) Size of the proloculus of the first chamber
    b) Size of the animal as a whole
    c) Composition of the shell
    d) Nature of coiling of the chambers
43. *Nummulites* were the abundant rock-building Foraminifers during
    a) Silurian b) Devonian
    c) Carboniferous d) Eocene
44. Some Foraminifers with hyaline shells exhibit the principle of *Lipopalingenesis* during the development of their shells. This denotes
    a) Hastening in development
    b) Skipping of Stages
    c) Repetition of Stages
    d) Imitation of lower forms by the higher forms
45. *Fusulina* is a Foraminifer with arenaceous shell. Its porcellaneous form is described as
    a) *Archaias* b) Orbitolites
    c) Monalysidium d) Fasciolites
46. Which is the correct sequence of the evolution of Foraminifers?
    a) Calcareous foraminifers à Arenaceous foraminifers à Chitinous foraminifers
    b) Calcareous foraminifers à Chitinous foraminifers à Arenaceous foraminifers

c) Chitinous foraminifers à Arenaceous foraminifersà Calcareous foraminifers
d) Arenaceous foraminifersà Calcareous foraminifersà Chitinous foraminifers

47. Which of the following phyla consists of organisms whose grade of organisation is considered to be intermediate between that of Protózoans and Metazoans ?
a) Phylum Porifera
b) Phylum Echinodermata
c) Phylum Coelentrata
d) Phylum Bryozoa

*48. The simplest grade of organisation in sponges is described as
a) Sycon b) Leucon
c) Rhagon d) Ascon

49. *Heliolites* and *Syringopora* are classified under
a) Complex corals b) Simple corals
c) Tabulate corals d) Rugose corals

50. The skeleton of an entire coral colony is described as
a) Corallum b) Corallite
c) Collumella d) Rhabdosome

51. A compound coral in which the corallites are parallel or sub-parallel is described as
a) Dendroid b) Fasciculate
c) Massive d) Basaltiform

52. The distribution of reef-building corals depends upon
i) Depth
ii) Temperature
iii) Light
iv) Salinity
a) (i) and (ii) only b) (i), (ii) and (iv) only
c) (i), (ii) and (iii) only d) All of the above

53. Fossils of which of the following groups of animals are mostly found preserved in black carbonaceous shales ?
a) Nummulites b) Echinoids
c) Graptolites d) Trilobites

54. Graptolites belong to the class
a) Hydrozoa
b) Actinozoa
c) Scyphozoa
d) Ctenophora

55. The earliest forms of Graptolites were found from the rocks belonging to
a) Lower Cambrian
b) Upper Cambrian

c) Ordovician
d) Silurian

56. In Graptolites. the skeleton of a simple colony is known as———and that of a complex colony is known as ———
a) Rhabdosome/Stock
b) Rhabdosome/Nema
c) Rhabdosome/Synrhabdosome
d) Synrhabdosome/Rhabdosome

57. Graptolites exhibit
a) Radial symmetry
b) Bilateral symmetry
c) Biradial symmetry
d) Radio-bilateral symmetry

58. The thread-like prolongations of graptolites are called
a) Sicula
b) Stipes
c) Virgula
d) Polypary

59. Graptolites were found in abundance in
a) Black carbonaceous shales
b) Cherty limestones
c) Black sandstones
d) Rose-pink granites

60. Which of the following forms of Graptolites does NOT belong to the Cambrian period?
a) *Tetragraptus, Bryograptus, Monograptus*
b) Monograptus, Cyrtograptus, Rastrites
c) Didymograptus, Dichograptus, Diplograptus
d) Tetragraptus, Clonograptus, Bryograptus

61. In Graptolites, the hard parts are made up of
a) Silica
b) Calcium Carbonate
c) Chitin
d) Calc-Arenite

62. Which of the following types of corals are found as fossils?
a) Zoantharian corals
b) Antipatharian corals
c) Medreiporian corals
d) None of the above

63. The Nepionic stage of the development of Corals involves
i) development of *axial septum* and *calyx*
ii) formation of *counter* and *cardinal septa* by the break-up of axial septum

iii) formation of *alar* and *counter lateral septa*
iv) appearance of two pairs of septa on the cardinal side
v) complex development of septa on the cardinal side
a) i, iii & iv only b) i, ii, iv & v only
c) ii, iv & v only d) ii and iii only

64. The Corals start exhibiting marked radial symmetry in the
a) Embryonic stage b) Nepionic Stage
c) Neanic Stage d) Adult Stage

65. The optimum temperature under which the reef-building corals thrive ranges from
a) 27°C and 29°C b) 22°C and 27°C
c) 29°C and 33°C d) 14°C and 22°C

66. Which of the following corals is used for decorations?
a) *Alcyonacea* b) Sertularia
c) Pennatula d) *Alcyonium*

*67. Which of the following corals is referred as "Dead man's fingers"?
a) *Alcyonium* b) Gorgonia
c) Leptogorgia d) Pennatula

68. Which of the following are described as "Sea Urchins"?
a) Diatoms b) Conodonts
c) Ichthyodonts d) Echinoids

69. The earliest known Echinoid is
a) *Echinocory* b) Miocidaris
c) Clypeastrina d) *Bothriocidaris*

70. Echinoids evolved during
a) Cambrian b) Ordovician
c) Silurian d) Devonian

71. Choose the correct statement from the following.
a) Irregular echinoids were derived from regular echinoids
b) Regular echinoids were derived from irregular echinoids
c) Irregular and regular echinoids had separate lines of descent
d) None of the above

72. The Endocyclic echinoids belong to
a) Regularia b) Irregularia
c) Holectypina d) Spatangina

73 . In Echinoids the mouth is surrounded by a membrane called
a) Peristome b) Periproct
c) Madreporite d) Aristotle's Lantern

74. The periproct of echinoderms is encircled by a ring of plates known as oculars and genitals, of which
a) Oculars are hexagonal and genitals are triangular
b) Oculars are triangular and genitals are hexagonal
c) Both the oculars and genitals are hexagonal

d) Both the oculars and genitals are triangular

75. In echinoids the corona is made up of
   a) 5 ambulacral and 5 interambulacral plates
   b) 5 ambulacral and 5 interambulacral areas
   c) 10 ambulacral and 10 interambulacral areas
   d) None of the above

76. 'Biplication' is a feature commonly exhibited by
   a) Brachiopods b) Gasteropods
   c) Pelecypods d) Cephalopods

77. The shell of a Brachiopod consists of
   a) Two equal but inequilateral valves
   b) Two unequal but equilateral valves
   c) Two equal and equilateral valves
   d) None of the above

78. The vertical section of a Brachiopod valve consists of
   a) Three layers with no vertical canals
   b) Three layers with two vertical canals
   c) Two layers with three vertical canals
   d) Two layers with no vertical canals

79. In the Brachiopods belonging to *'Pygocaulia'*, the pedicle is formed from
   a) Head segment b) Ventral mantle lobes
   c) Tail segment d) Any of the above

80. In Brachiopods locomotory muscles are
   a) Present b) Absent
   c) May or may not be present d) None of the above

81. In which of the following genera of Brachiopods, the delthyrium is completely closed by a single large plate
   a) Protremata b) Atremata
   c) Teleotremata d) Neotremata

82. The Trilobite fauna is restricted to
   a) Lower Paleozoic b) Upper Paleozoic
   c) Mesozoic d) Cenozoic

83. The body of Trilobites is divided into
   a) Two parts
   b) Three parts
   c) Two parts by means of a furrow
   d) Three parts by means of two furrows

84. In Trilobites, the most primitive suture line is described as
   a) Hypoparian
   b) Opisthoparian
   c) Proparian
   d) Gonatoparian

85. When the suture line from the anterior region of a Trilobite touches the posterior side it is described as
    a) Protoparian
    b) Opisthoparian
    c) Proparian
    d) Gonatoparian
86. Identify the type of facial sutures of a Trilobite cephalon in the diagram given below.
    a) Proparian
    b) Gonatoparian
    c) Opisthoparian
    d) Hypoparian
87. The lateral and posterior borders of Trilobites meet to form an angle known as
    a) Glabellar angle
    b) Genal angle
    c) Spire angle
    d) Angle of divergence
88. The genal angles are NOT pointed or produced into spines in
    a) *Paradoxides*
    b) Deiphon
    c) Trinucleus
    d) Calymeme
89. Arrange the following in the decreasing order of age.
    a) *Olenus - Paradoxides-Ollenelus-Calymene*
    b) Ollenelus-Paradoxides- Olenus - Calymene
    c) Paradoxides-Olenus—Ollenelus-Calymene
    d) Calymene-Ollenelus-Olenus-Paradoxides
90. The Lamellibranchs are also described as
    a) Pelecypoda
    b) Acephala
    c) Didisoconch
    d) All of the above
91. The earliest of the Pelecypods belong to
    a) Lower Ordovician
    b) Lower Devonian
    c) Upper Carboniferous
    d) Upper cretaceous
92. In Lamellibranchs, thė two leaf-like projections present on either side of the mouth are called
    a) Lunules b) Divaricators
    c) Labial palps d) Escutcheon

93. In Lamellibranchs, the valves are closed by —— and opened by ——
    a) Divaricators/ Adductors
    b) Adductors/Divaricators
    c) Adductors/Ligament
    d) Ligament/ Adductors
94. The hinge line in Lamellibranchs is composed of
    a) Teeth on the ventral valve and sockets on the dorsal valve
    b) Teeth on the dorsal valve and sockets on the ventral valve
    c) Alternate teeth and sockets in the dorsal and the ventral valves
    d) None of the above
95. The embryonic membrane of the Pelecypod larva is described as
    a) Prodisoconch
    b) Protoconch
    c) Protegulum
    d) Zooea
96. Match the following and choose the correct answer.

| | |
|---|---|
| A) Brachiopoda | 1. *Calceola* |
| B) Lamellibranchia | 2. *Bellerophon* |
| C) Gasteropoda | 3. *Crania* |
| D) Coelentrata | 4. *Spondylus* |

    a) A-3 B-4 C-2 D-l
    b) A-3 B-4 C-1 D-2
    c) A-4 B-3 C-2 D-l
    d) A-4 B-3 C-l D-2
97. Which of the following is NOT matched correctly ?
    a) *Montivaltia*——Coral
    b) *Anomia*—Brachiopod
    c) *Helix*—— Gasteropod
    d) *Textularia*—Foraminifer
98. Gasteropods exhibit
    a) Bilateral symmetry b) Radial symmetry
    c) Biradial symmetry d) Asymmetry
99. Gasteropods having conical shape and long, acute spire are described as
    a) Globular b) Turreted
    c) Turbinate d) Convolute
100. Which of the following Gasteropods is known as 'Sea lemon' ? .
    a) *Dorbis* b) Pila
    c) Trochus d) *Conularia*
101. Which of the following Gasteropods consists of spines and exhibit dextral type of coiling ?
    a) *Murex* b) Physa
    c) Voluta d) *Turritela*

102. The changes is the shell shape and coiling in Gasteropods has resulted due to a
   a) Free-swimming habit
   b) Bottom-crawling habit
   c) Fixed mode of life
   d) Burrowing mode of life
103. The largest number of Gasteropods belong to
   a) Protogasteropoda
   b) Prosobranchia
   c) Opisthobranchia
   d) Pulmonata
104. Which of the following Molluscs are entirely marine ?
   a) Pelecypods
   b) Gasteropods
   c) Cephalopods
   d) None of the above
105. The most gigantic of all the Molluscs were
   a) Pelecypods b) Gasteropods
   c) Nautiloids d) Ammonites
106. The part of the Ammonoids which resembles the operculum of the Gasteropods is called
   a) Apical plate b) Aptichus
   c) Miopta d) Camerae
107. The correct sequence of the evolution of cephalopod sutures is
   a) Nautiloid—Goniatite—Ammonite—Ceratite
   b) Nautiloid-—Ceratite-—Goniatite-—Ammonite
   c) Nautiloid—Goniatite—Ceratite-—Ammonite
   d) Goniatite—Ceratite—Nautiloid—Ammonite
108. Identify the type of suture lines shown in the diagram below

   a) Agoniatite b) Goniatite
   c) Ceratite d) Ammonite
109. The fastest moving species among the invertebrates belong to
   a) Pelecypods b) Gasteropods
   c) Cephalopods d) Brachiopods
110. Which of the following belongs to the order 'Dibranchia' of the sub-phylum Cephalopoda ?
   a) *Orthoceras* b) Nautilus
   c) Bacculites d) *Belemnites*

111. 'Squids' are the present day representatives of
   a) Brachiopods
   b) Cephalopods
   c) Gasteropods
   d) Lamellibranchs

112. "Ink Gland" is present in
   a) Monobranchiate cephalopods
   b) Dibranchiate cephalopods
   c) Tribranchiate cephalopods
   d) Tetrabranchiate cephalopods

113. Match the following and choose the correct answer ?

| *Group* | *Larvae* |
|---|---|
| (A) Trilobites | (1) Planula |
| (B) Corals | (2) Veliger |
| (C) Gasteropods | (3) Nauplius |
| (D) Cephalopods | (4) Trochophore |

   a) A-3 B-2C-4 D-l
   b) A-l B-3 C-4 D-2
   a) A-3 B-4 C-1 D-2
   b) A-3 B-1 C-4 D-Z

114. Lamellibranchs with two unequal adductors are described as
   a) Isomaria   b) Anisomyaria
   c) Monomyaria   d) Dimyaria

115. Monomyarian lamellibranchs are generally characterized by
   a) Presence of a well-developed anterior adductor
   b) Absence of an anterior adductor
   c) Absence of a posterior adductor
   d) None of the above

116. Lamellibranchs adapted themselves to the burrowing mode of life by one or more of the following modifications
   (i) The foot became tongue-shaped
   (ii) Antero-ventral portion of the shell became elongated
   (iii) Shells became compressed and/or cylindrical
   a) (i) only
   b) (ii) only
   c) (i) and (ii) only
   d) (i), (ii) and (iii)

117. Which of the following exhibits a byssal attachment ?
   a) *Mytilus*
   b) Pinna
   c) *Arca*
   d) All of the above
   e) None of the above

118. The lamellibranchs belonging to the *'Cavicole type'* are characterized by
   a) Burrowing mode of life
   b) Free swimming mode of life
   c) Fixing by cementation
   d) Byssal mode of attachment

119. In Lamellibranchia
   a) Pallial sinus is invariably present
   b) Pallial sinus is always present towards the posterior end
   c) Pallial sinus is always present towards the anterior end
   d) Pallial sinus is generally absent

120. When the ligament of a Pelecypod lies wholly in front of the beaks the condition is described as?
   a) Amphidetic
   b) Opisthodetic
   c) Prosodetic
   d) None of the above

121. When the two valves are tightly closed, the ligament is under
   a) Compressive stress
   b) Tensional stress
   c) Shearing stress
   d) Torsion

122. The earliest type of dentition in the bivalves is described as
   a) Ctenodont
   b) Actinodont
   c) Asthenodont
   d) Dysodont

123. Obsolete teeth present in some burrowing and boring bivalves are described as
   a) Pantodont
   b) Cyclodont
   c) Asthenodont
   d) Anamalodont

124. Match the following and choose the correct answer

| *Bivalve* | *Dentition* |
|---|---|
| (A) *Trigonia* | (1) Pantodont |
| (B) *Spondylus* | (2) Dysodont |
| (C) *Mytilus* | (3) Isodont |
| (D) *Allodesma* | (4) Schizodont |

   a) A-4 B-3 C-2 D-1
   b) A-4 B-2 C-1 D-3
   c) A-3 B-4 C-1 D-2
   d) A-2 8-3 C-4 D-1

*125. The dental formula of Lamellibranchs can be designated as

a) $\frac{1\text{-}3ab\text{-}5ab}{2ab\text{-}4ab}$

b) $\frac{2ab\text{-}4ab}{1\text{-}3ab\text{-}5ab}$

c) $\frac{1\text{-}2ab\text{-}3ab}{4ab\text{-}5ab}$

d) $\frac{4ab\text{-}5ab}{1\text{-}2ab\text{-}3ab}$

126. Which of the following exhibits wing-like extensions called *'Ears'* in their shells ?

a) *Pecten*
b) Spondylus
c) *Productus*
d) None of the above
e) All of the above

127. A brachiopod that shows both accelerations and decelerations in development is

a) *Lingula* b) *Spirifer*
c) *Rhynconella* d) *Obolella*

128. In the brachiopods belonging to Articulata, the pedicle is formed by the juxtaposition of two notches. The one found to the brachial valve is known as

a) Delhyrium b) Deltidium
c) Notothrium d) Chilidium

129. In which of the following genus the pedicleis confined only to the ventral valve

a) *Lingula* b) *Linguella*
c) *Discina* d) *Orthis*

130. The Brachiopod '*Syringohyris*' belongs to

a) Ordovician b) Silurian
c) Devonian d) Carboniferous

131. Brachiopods having long and straight hinge line which coincides with the hinge axis are said to possess

a) Dentulous shells b) Edentulous shells
c) Strophic shells d) Non-strophic shells

132. 'BILBITES' are

a) Ancestors of Trilobites
b) Trail-like markings of Trilobites
c) Temporary resting traces or burrows of Trilobites
d) The most dominant type of Graptolites during Ordovician

133. Which of the following Trilobites is/are blind ?

a) *Agnostos*

b) Microdiscus
c) *Acidaspis*
d) All of the above
e) None of the above

134. The most ancient kind of eyes in Trilobites are known as
a) Holochroal b) Schizochroal
c) Achroal d) None of the above

135. The photo-sensitive elements in the eye of Trilobites are described as
a) Ommatidia b) Rhabdom
c) Corneal lens d) Microvilli

136. The eyes are unusually large in
a) *Cyclopyge* b) Harpes
c) Olenellus d) Ampyx

137. Trilobites consists of a series of movable thoracic segments, which range in number from
a) 2 to 32 b) 2 to 42
c) 4 to 36 d) 2 to 38

138. In Trilobites the segments of the thorax and the pygidium can be differentiated by
a) their shape and size
b) their movable and immovable character
c) the nature of the furrows
d) the total number of segments

139. The central convex lobe of the cephalon of Trilobites is known as
a) Glabella b) Occipital furrow
c) Free cheek d) Neck, furrow

140. The head shield of trilobites is composed of —— pairs of appendages
a) 3 b) 4
c) 5 d) 6

141. Which of the following is not matched correctly ?
a) *Olenellus*——Lower Cambrian
b) *Paradoxides*—Middle Cambrian
c) *Calymene*—Upper Cambrian
d) *Trinucleus*—Ordovician

142. The correct order of the larval development of Trilobites is
a) Protaspid stage—Meraspid stage—Holaspid stage
b) Meraspid stage— —Protaspid stage——Holaspid stage
c) Protaspid stage—Holaspid stage—Meraspid stage
d) Holaspid stage—Meraspid stage—Protaspid stage

143. Match the following and choose the correct answer

| | |
|---|---|
| A. *Syringopora* | 1. Brachiopod |
| B. *Holaster* | 2. Coral |
| C. *Cidaris* | 3. Regular echinoid |

D. *Productus* — 4. Irregular echinoid
5. Lamellibranch

a) A-2 B-3 C-4 D-l
b) A-2 B-3 C-4 D-5
c) A-2 B-4 C-3 D-l
d) A-5 B-3 C-4 D-l

144. 'Megalodon' belongs to the phylum
a) Brachiopoda b) Pelecypoda
c) Gasteropoda d) Cephalopoda

145. '*Scallops*' is the name given to the species of
a) *Pecten* b) Spirifer
c) Globigerina d) *Nautilus*

146. Which of the following shows sinistral type of coiling ?
a) *Murex* b) Conus
c) Physa d) Bellerophon

147. Each coil in a shell of Gasteropod
a) increases towards the base
b) decreases towards the base
c) increases towards the apex
d) is uniform throughout the shell

148. The intersection of siphuncle and septa in Gasteropods gives rise to
a) Septal tube b) Siphuncle tube
c) Septal neck d) None of the above

149. The most characteristic type of coiling in Ammonoid shells is
a) Helicoid spiral b) Conical spiral
c) Planispiral d) Discoidal

150. Ammonoid shells that have become straight as a result of uncoiling are described as
a) Baculiticones b) Bactriticone
c) Ammoniticone d) Dactylioceracone

*151. Which of the following Ammonoids shows a tendency to uncoil?
a) *Dactilioceras* b) Tornoceras
c) Rhabdoceras d) Gomphoceras

152. The typical Ammonite of the Paleozoic was
a) *Goniatites* b) Ceratites
c) Dactilioceras d) Tornoceras

153. The Goniatitic suture of the Paleozoic is closely similar to that of the earlier stage of Dactilioceras belonging to the Mesozoic. This reflects the principal of
a) Palingenesis b) Tachygenesis
c) Palingenesis and Tachygenesis d) Orthogenesis

154. The correct sequence of the successive stages in the development of Ammonoids is

a) Nepionic stage—Ephebic stage—Neanic stage—Gerontic stage
b) Neanic stage—Nepionic stage—Ephebic stage—Gerontic stage
c) Ephebic stage—Nepionic stage—Neanic stage—Gerontic stage
d) Nepionic stage—Neanic stage—Ephebic stage—Gerontic stage

155. The Ammonoids became extinct during
a) Upper Carboniferous b) Late cretaceous
c) Upper Cretaceous d) Eocene

*156. Which of the following groups is described as 'Head-footed' ?
a) Pelecypoda
b) Gasteropoda
c) Brachiopoda
d) Cephalopoda

157. Given below are some index fossils. Which of these are NOT matched correctly ?
a) *Pentamerus*— Silurian
b) *Nummulites*—Eocene
c) *Tornoceras*——Triassic
d) *Glossopteris*—Permian

158. Match the following and choose the correct answer.

| | | |
|---|---|---|
| A. | *Cardita beaumonti* | 1. Cephalopod |
| B. | *Rhynconella griesbachi* | 2. Gasteropod |
| C. | *Conularia* | 3. Brachiopod |
| D. | *Octoceras* | 4. Pelecypod |

a) A-1 B-3 C-4 D-2
b) A-4 B-3 C-2 D-1
c) A-l B-3 C-2 D-4
d) A-4 B-Z C-3 D-l

159. The evolution of Graptolites was marked by a sequence of environmental adaptation. The correct order of evolution of graptolites is
a) Benthonic —Epiplanktonic—Plancktonic
b) Epiplanktonic— Planktonic- —Benthonic
c) Planktonic—Epiplanktonic—Benthonic
d) Benthonic- - - Planktonic—- Epiplanktonic

*160. Which of the following evolutionary changes took place from Regular to Irregular Echinoids?
1. Radial symmetry changed to bilateral symmetry
2. The shell changed its shape from circular to heart- shaped
3. The anus moved forward and the mouth moved backward
4. Ambulacral areas became petaloid

a) 1 and 3 are correct
b) 1, 2 and 3 are correct
c) 1, 2 and 4 are correct
d) All of the above are correct

161. In some Echinoids such as Salena, an extra plate is found in addition to the oculars and genitals. This is described as
   a) Anal plate
   b) Sur-anal plate
   c) Ambulacral plate
   d) Compound plate

162. *Aristotle's lantern* is an organ used for
   a) Respiration
   b) Reproduction
   c) Excretion
   d) Mastication

163. The Aristotle's lantern is composed of
   a) 5 teeth and 40 calcareous pieces
   b) 40 teeth and 5 calcareous pieces
   c) 40 calcareous pieces only
   d) None of the above

*164. Aristotle's Lantern is absent in
   a) Regularia b) Holectypina
   c) Clypeastrina d) Spatangina

165. The skeletons of which of the following Phyla are purely calcareous
   a) Pelecypoda b) Gasteropoda
   c) Brachiopoda d) Bryozoa

166. The brachial skeleton is definitely present in
   a) Atremata b) Neotremata
   c) Protremata d) Teleotremata

167. In which of the following genera, the brachial skeleton is represented by two short curved processes called the Crura ?
   a) *Terebratula* b) Rhynconella
   c) Spirifer d) Lingula

168. The brachiopod '*Lingula*' is included under the order Atremata because
   a) The pedicle is confined to the ventral valve
   b) The pedicle is confined to the dorsal valve
   c) The pedicle is shared by both the valves
   d) The brachial skeleton is absent

169. The littoral genera of brachiopods does not include
   a) *Lingula* b) Discina
   c) Discinisca d) Atrypa

170. A spoon-shaped plate which serves for the attachment of muscles in the Brachiopods is described as
   a) Radula b) Pedicle
   c) Spondylium d) Foramen

171. Spondylium is characteristic of Brachiopods belonging to the order
   a) Protremata b) Teleotremata

c) Atremata d) Neotremata

172. Brachiopod shells with a triangular notch (the delthyrium) are described as
   a) Holoperipheral
   b) Hemiperipheral
   c) Mixoperipheral
   d) None of the above

173. The valves of the inarticulates are—in shape
   a) Biconvex b) Biconcave
   c) Conical d) Circular

174. Which of the following statements is true ?
   a) The articulates are more diverse but less abundant than the inarticulates
   b) The inarticulates are more diverse but less abundant than the articulates
   c) The inarticulates are much more diverse and abundant than the articulates
   d) The articulates are much more diverse and abundant than the inarticulates

175. Choose the correct pair from the following.
   a) Sulcacutes beds........ Belemnites
   b) Eurydesma zone....... Gasteropod
   c) Otoceras zone........... Pelecypod
   d) Megalodon limestone............Cephalopod

176. The most primitive form of ligament in Lamellibranchs
   a) extended wholly in front of the umbo or beak
   b) extended wholly behind the umbo or beak
   c) extended in front of as well as behind the umbo or beak
   d) none of the above

*177. When the valves of a Lamellibranch are tightly closed, the ligament and the resilium are under
   a) tension and compression respectively
   b) compression and tension respectively
   c) tension
   d) compression

*178. Which of the following Lamelibranchs does NOT belong to the same Order?
   a) *Ostrea, Gryphaea, Exogyra*
   b) Cardita, Chama, Venus
   c) Trigonia, Schizodus, Unio
   d) *Pecten, Spondylus, Pinna*

179. The earliest known Gasteropod was
   a) *Helcionella*

b) Pleurotomaria
c) Trochus
d) Littorina

180. The bottom-crawling Gasteropods were characterised by
i) Planispiral coiling
ii) Helicoid Spiral coiling
iii) Well-developed Umbilicus
iv) Absence of Umbilicus
a) i & iii only
b) i & iv only
c) ii & iii only
d) ii & iv only

181. Match the following and choose the correct answer.

| *Formation* | *Typical Gasteropod* |
|---|---|
| A. Productus Limestone, Salt Range (Permian) | 1. *Bellerophon* |
| B. Bagh Beds (U. Cretaceous) | 2. *Turritella* |
| C. Surma Series, Assam (U. Oligocene - L. Miocene) | 3. *Natica* |
| D. Cuddalore Sandstone (Miocene) | 4. *Conus* |

a) A-1 B-2 C-3 D-4
b) A-1 B-3 C-2 D-4
c) A-4 B-1 C-3 D-2
d) A-4 B-3 C-1 D-2

182. The peak period for the Goniatite suture in Ammonoids was
a) Lower Carboniferous
b) Middle Carboniferous
c) Upper Carboniferous
d) Permian

183. Which of the following does NOT exhibit the correct evolutionary change among the different groups of Echinoids?
a) Bothriocidaroidea à Miocidaris
b) Cidaroidsà Miocidaris
c) Ectobranchiates àIrregularia
d) Cidaroids à Ectobranchiates

184. Which of the following represents the evolutionary changes that are brought about in an Echinoid shell due to burrowing mode of life?
1) Mouth tends to move towards the anterior margin.
2) Shell becomes elongate and bilaterally symmetrical.
3) Aristotle's Lantern and Perignathic Girdle disappear
4) Ambulacral areas are formed of simple plates.
a) 1 and 3 are correct
b) 1, 2 and 3 are correct

c) 1, 2 and 4 are correct
d) All of the above are correct

185. The earliest known bird belongs to
a) Upper Jurassic
b) Lower Cretaceous
c) Lower Jurassic
d) Upper Cretaceous

186. The Solenhofen limestone had well-preserved fossils of
a) Archeaofavosites
b) Archaeoptrix
c) Jamoytis
d) Ceratodus

187. "The primitive birds bad toothed beaks, but the teeth were lost , in the advanced stages of development". This statement is
a) True
b) False
c) Cant say
d) The information is insufficient to come to a definitive conclusion.

188. The largest terrestrial carnivore that has ever lived on Earth was
a) Gorgosaurus
b) Brontosaurus
c) Tyrannosaurus
d) Ornitholestes

189. Which of the following reptiles adapted themselves for flight ?
a) Ichthyosaurs
b) Pterosaurs
c) Theriodonts
d) Thecodonts

190. The most characteristic feature of the reptile 'Cynognathus' is
a) Presence of swimming paddles
b) Streamlined body
c) Mammal-like form
d) Presence of a heavy armour on the body

191. Which of the following are described as the "horned dinosaurs" ?
a) Ankylosaurs b) Stegosaurs
c) Sauropods d) Ceratopsians

192. In India the dinosaurs attained their highest development during the deposition of
a) Bagh Beds
b) Lameta Beds
c) Umaria Marine Beds
d) Bap Beds

193. According to a recent theory proposed by the Chinese scientists, the sudden disappearance of dinosaurs about 60 m.a. was due to

a) Arsenic poisoning
b) Extra-terrestrial catastrophes
c) Sudden and drastic changes in the climate
d) The growing competition with other reptiles which caused excessive food shortages

*194. Match the following and choose the correct answer.

| | |
|---|---|
| A. *Dicynodon* | 1. Fish |
| B. *Estheria* | 2. Amphibian . |
| C. *Stegocephalia* | 3. Crustacean |
| D. *Amblypteris* | 4. Reptile |

a) A-2 B-l C-3 D-4
b) A-4 B-3 C-2 D-l
c) A-2 B-4 C-3 D-l
d) A-l B~3 C-2 D-4

195. The Perrisodactyls consists of

a) Even-toed ungulates
b) Uneven-toed ungulates
c) Both even and uneven-toed ungulates
d) None of the above

196. The earliest representative of the horses was

a) Orohippus
b) Epihippus
c) Pliohippus
d) Eohippus

197. The horse *Hyracotherium* possessed

a) Four toes
b) Three toes
c) Two functional toes
d) A single functional toe

198. Horses are very good examples of 'straight-line evolution' i.e. t they have evolved with little deviation with time. This phenomenon is known as

a) Orthogenesis  b) Paragenesis
c) Tachygenesis  d) Ontogeny

*199. The first species of horses which showed divergence into several branches was

a) Eohippus  b) Miohippus
c) Merychippus  d) Pliohippus

200. The "Hipparian Faunas" were characteristic of

a) Eocene  b) Miocene
c) Pliocene  d) Pliestoceoe

201. It is believed that the average rate of evolution of one genus of horses was approximately

a) 5.5 M.Y. b) 6.5 M.Y.
c) 7.5 M.Y. d) 10 M.Y.

202. The earliest evidence of elephants was reported from

a) North America
b) South America
c) Africa
d) Egypt
e) India

203. The Mastodons originated in

a) Eocene b) Oligocene
c) Miocene d) Pliocene

204. The immediate ancestors of the mammoths were

a) Stegodons b) Mastodons
c) Laxodonts d) Phiomia

*205. The dental apparatus of mammoths consisted of

a) Molars with ridges but without cement
b) Molars with ridges completely submerged in cement
c) Molars with cement encasing only the lower half of the ridges of the enamel
d) None of the above

206. The mammoths lived in all continents except

a) Africa b) Asia
c) North America d) South America

*207. The Hominids are characterized by

a) Brachycephaly. b) Platycephaly
c) Dolichocephaly d) Acephaly

208. The earliest man-like primate recovered in 1930s from the Siwalik Hills is named as

a) Dryopithecus b) Ramapithecus
c) Sivapithecus d) Laetolil man

209. The *'Ramapithecus'* was discovered by

a) G.E. Lewis b) Prof. Dart
c) Dr. Broom d) Richard Leaky

*210. The 'Homo Erectus' or the erect man was the first specie to achieve what is called the Homo Erectus Complex. This consists of

i) Achieving pre-language
ii) Developing a shuffling gait
iii) Beginning to lose their hair
iv) Venturing out into open savannas

a) (i), (ii) (iii) and (iv)
b) (i), (ii) and (iv) only

c) (i), (iii) and (iv) only
d) (ii), (iii) and (iv) only

211. The origin of man is believed to have taken place in
a) Europe
b) North America
c) Africa
d) Central Asia
e) Arctic region

*212. The Cro Magnon Man was discovered in
a) Africa
b) Asia
c) France
d) Germany

213. The advantages of 'microfossils' over 'megafossils' in biostratigraphy and paleogeological reconstructions is/are
1) They can be analysed up to the species level
2) They have facies independence
3) They are restricted vertically but widespread geographically
4) They have small size
a) 1 only
b) 1 and 2 only
c) 1, 2 and 3 only
d) All of the above

214. The dominant forms of microfossils in the abyssal region are
a) Diatoms
b) Ostracods
c) Foraminifers
d) Radiolarians

215. '*Vertebraria indica*' is
a) a plant fossil belonging to Conifers
b) a plant fossil belonging to Cycads
c) the stem of Glossopteris
d) the stem of Gangamopteris

216. The floral species '*Williamsonia*' is a
a) Cycad
b) Conifer
c) Fern
d) Bryophyta

217. Identify the floral species given in the diagram below

a) *Elatocladus*
b) Siggillaria
c) Nilsonia
d) Sphenophyllum

218. The commonest Mesozoic compression-impression fossils belong to
a) Cycadales
b) Coniferales
c) Lycopsida
d) Sphenopsida

*219. Which of the following does NOT belong to the same group ?

a) *Ptillophyllum* b) Williamsonia

c) Otozamites d) *Sigillaria*

220. The Blue-green algae belong to

a) Cyabophytes b) Pteridophytes

c) Lycophytes d) Pterophytes

221. Gangamopteris is distinguished from Glossopteris by

a) Presence of mid-rib

b) Presence of venation

c) Presence of spines

d) None of the above

222. The term "Glorified Reptiles" is used for

a) Birds b) Dinosaurs

c) Ichthyosaurs d) Sauropods

223. Which of the following can be regarded as a "Connecting Link" between the birds and reptiles?

a) Archaeornis b) Hesperornis

c) Duck-billed Platypus d) Emu

224. Which of the following can be regarded as a "Connecting Link" between the amphibians and reptiles?

a) Archaeopterix b) Duck-billed Platypus

c) Seymouria d) Limnoscelis

225. Which one of the following can be considered as the most primitive of all reptiles?

a) Petrolacosaurus b) Limnoscelis

c) Hylonomus d) Captorhinus

226. The first Dinosaurs appeared in

a) Early Triassic

b) Late Triassic

c) Early Jurassic

d) Late Jurassic

e) Cretaceous

227. Which of the following dinosaurs was small as a chicken?

a) Compsognathus b) Utahraptor

c) Struthiomimus d) Diplophosaurus

228. Which one was NOT a dinosaur?

a) Tyrannosaurus b) Pachycephalosaurus

c) Megalosaurus d) Ichtyosaurus

229. Which was the largest dinosaur to walk on the land?

a) Ultrasaurus b) Tyrannosaurus

c) Diplodocus d) Brachiosaurus

230. Which one was a dinosaur?

a) Pterosaur b) Elasmosaurus

c) Megalosaurus d) Ichthyosaurus

231. Which one had its body covered with thorns?
a) Tornosaurus b) Plesiosaurus
c) Deinonychus d) Acanthophosis

232. Which was a crested, duck-billed dinosaur?
a) Psittacosaurus b) Corythosaurus
c) Velociraptor d) Utahraptor

233. Which one was the most deadly meat-eater that ever lived?
a) Pteronodon b) Tyrannosaurus Rex
c) Dilophosaurus d) Brachiosaurus

234. Which of the following had a sharp spike on each of the thumbs?
a) Iguanadon b) Diplodocus
c) Orthnithomimus d) Archaeopteryx

235. Which of the following was the largest flying creature that ever lived?
a) Quetzacoatlus b) Archaeopteryx
c) Flying Albatross d) Rhamphoryncus

236. Which one was the "Plated Lizard"?
a) Sordes b) Platosaurus
c) Spinosaurus d) Stegosaurus

237. What was a Theropod?
a) Two-legged, meat-eating dinosaur with bird-like body
b) Four-legged, plant eating dinosaur
c) Four-legged, fish eating dinosaur
d) Two-legged, plant-eating dinosaur

238. Which one of the following was a gigantic turtle?
a) Anatosaurus b) Triceratops
c) Archelon d) Maisaura

239. Match the following and choose the correct answer.

A. "Egg-thief" 1. Corythosaurus
B. "Helmet Lizard" 2. Panoplosaurus
C. "Armoured lizard" 3. Oviraptor
D. "Spiked Lizard" 4. Ankylosaurus

a) A-3 B-1 C-4 D-2 b) A-3 B-1 C-2 D-4
c) A-4 B-1 C-3 D-2 d) A-1 B-4 C-2 D-3

240. Which of the following is called the "Age of Mammals"?
a) Paleozoic
b) Mesozoic
c) Cenozoic
d) None of the above

241. The time of slow evolutionary change is described as
a) Bradytelic Evolution
b) Horotelic Evolution

c) Tachytelic Evolution
d) Panchronism

242. The return of Ceratite sutures in the Cretaceous Ammonoids can be explained by
a) Dollo's Law
b) Cope's Law
c) Inverse of Dollo's Law
d) Inverse of Cope's Law

243. The disappearance of digits in Equids can be explained by
a) Dollo's Law
b) Cope's Law
c) Deperet's Law
d) None of the above

244. The ramification or 'bursting' forth of an evolutive line of descent is described as
a) Panchronism
b) Parallelism
c) Cladogenesis
d) Convergence

245. According to the concept of 'Lamarckism"
a) The 'Present is the key to the Past'
b) 'The Function creates the Organ'
c) 'Natural selection follows the struggle for existence'
d) 'Ontogeny recapitulates Phylogeny'

246. Which of the following is a mammal?
a) Psilotum b) Opossum
c) Sphenodon d) Ginkgo

247. Which of the following is considered as the central ancestor of the Chordates?
a) Amphioxus b) Jamoytius
c) Pteraspida d) Anaspida

248. The 'Cartilaginous fishes' belong to the class
a) Placodermi b) Chondrichthyes
c) Osteichthyes d) Ostracodermi

249. The first appearance of jaws in fishes took place in
a) Cambrian
b) Ordovician
c) Silurian
d) Devonian

250 The most primitive of all the fins is described as
a) Homocercal
b) Heterocercal
c) Leptocercal
d) Protocercal

# PALEONTOLOGY (KEY)

| | | | | | | | | | |
|---|---|---|---|---|---|---|---|---|---|
| 1-a | 2-a | 3-b | 4-b | 5-a | 6-a | 7-d | 8-c | *9-a | 10-d |
| 11-b | *12-c | 13-d | 14-a | 15-a | *16-a | 17-b | 18-c | 19-a | 20-d |
| 21-b | 22-c | 23-d | 24-b | 25-a | 26-b | 27-d | 28-b | 29-b | 30-c |
| 31-a | 32-a | 33-a | 34-b | 35-a | 36-a | 37-d | 38-b | 39-d | 40-c |
| 41-d | 42-a | 43-d | 44-b | 45-d | 46-c | 47-a | *48-d | 49-c | 50-a |
| 51-d | 52-d | 53-c | 54-a | 55-b | 56-c | 57-b | 58-c | 59-a | 60-d |
| 61-c | 62-c | 63-d | 64-b | 65-a | 66-b | *67-a | 68-d | 69-d | 70-b |
| 71-a | 72-a | 73-a | 74-b | 75-b | 76-a | 77-b | 78-b | 79-c | 80-b |
| 81-a | 82-a | 83-d | 84-a | 85-b | 86-b | 87-b | 88-d | 89-b | 90-d |
| 91-a | 92-c | 93-c | 94-c | 95-a | 96-a | 97-b | 98-d | 99-b | 100-a |
| 101-a | 102-b | 103-c | 104-c | 105-c | 106-b | 107-c | 108-a | 109-c | 110-d |
| 111-d | 112-b | 113-d | 114-b | 115-b | 116-d | 117-d | 118-a | 119-b | 120-c |
| 121-b | 122-a | 123-c | 124-a | *125-a | 126-e | 127-c | 128-c | *129-c | 130-d |
| *131-c | *132-b | 133-d | 134-a | 135-b | 136-a | 137-b | 138-b | 139-a | 140-c |
| 141-c | 142-a | 143-c | 144-b | 145-a | 146-c | 147-a | 148-c | 149-c | 150-a |
| *151-c | 152-a | 153-c | 154-d | 155-c | *156-d | 157-c | 158-b | 159-a | *160-d |
| 161-b | 162-b | 163-a | *164-d | 165-c | 166-d | 167-b | 168-c | 169-d | 170-c |
| 171-a | 172-c | 173-c | 174-d | 175-a | 176-d | *177-a | *178-d | 179-a | 180-c |
| 181-a | 182-c | 183-b | 184-d | 185-a | 186-b | 187-a | 188-c | 189-b | 190-c |
| 191-d | 192-b | 193-a | 194-b | 195-b | 196-d | 197-a | 198-a | 199-c | 200-c |
| 201-c | 202-d | 203-b | 204-a | 205-a | 206-d | 207-a | 208-b | 209-a | 210-a |
| 211-d | *212-c | 213-d | 214-d | 215-c | 216-a | 217-d | 218-a | *219-d | 220-a |
| 221-d | 222-a | 223-a | 224-c | 225-b | 226-b | 227-a | 228-d | 229-d | 230-c |
| 231-d | 232-b | 233-b | 234-a | 235-a | 236-d | 237-c | 238-c | 239-b | 240-c |
| 241-a | 242-d | 243-a | 244-c | 245-b | 246-b | 247-a | 248-b | 249-d | 250-d |

# EXPLANATORY ANSWERS

*9. (a) : Living fossils exhibit the phenomenon of Bradytelic evolution i.e. in a declining lineage the survivors may often continue for a very long time, evolving very slowly or not at all. All instances where the rate of evolution has halted are known as Laggard.

*12. (c) : Homeomorphy denotes morphological similarities shown by genetically distinct members of the same phylum. e.g., the whale, the fish and the ichthyosaurs. Homology is the condition where two or more organisms possess organs which may look different at the first glance, but have a basic structural similarity. e.g., man's hand and bat's wing. Paedomorphosis is the retention or larval characters in adult forms.

*16. (a) : Ammonoids exhibit this pattern of evolution.

*48. (d) : There are three grades of organisations in Sponges. In the simplest grade i.e., ascon, the sac-like body is merely a single chamber lined with choanocytes. Sycon sponges have a number of grouped ascon-like chambers with a central opening, but the vast majority of sponges are of leucon (rhagon) type in which sycon-like elements open into a large central cavity.

59. (a): Alcyonium is referred as "Dead man's fingers", Gorgonia is referred as "Sea Fan", Leptogorgia is referred as "Sea Whip" and Pennatula is referred as "Sea Pen".

*125. (a): The system of notation for the hinge teeth was devised by the French paleontologists Bernard and Munier-Chalmas. The odd numbers are used for the right and even numbers for the left value. The letter (a) follows the numbers of anterior teeth and the letter (b) follows those for the posterior teeth. The sockets are represented by a dash.

*131. (c) : *Spiriferida* and *Orthida* of the Paleozoic are the Brachiopods which possess strophic shells. Others such as *Terebratula* and *Rhynonella* possess non-strophic shells in which the hinge axis passes through the teeth and sockets which are its only fulcra The shell is curved and not coincident with the hinge-line.

*132. (b): Bilobites are ribbon-like markings with paired oblique chevron structures. These are the trail-like markings always occurring at the interfaces of sandy and shaly layers in the rock. Trilobite burrows or temporary resting traces of trilobites are called Rusophycus.

*151 (c): The Ceratoids show some retrogressive changes. In Rhobdoceras of Triassic, the tendency to uncoil is set in an early stage of development and practically the whole shell had assumed a straight condition. The suture line is reverted to a simple Goniatitic condition.

*156. (d): The class name Cephalopoda means "head-footed" and refers to the modification of the foot into many arms or tentacles which are symmetrically arranged in a crown about the head.

*160 (d) In the regular echinoids the mouth and anus were in the centre on opposite sides of the test and in the irregular echinoids, the anus moved backwards and the mouth forwards.

*164 (d): This is due to the burrowing mode of life.

*177 (a): Resilium is a triangular pad similar to the ligament and resides in a central pit along and below the hinge line of a Lamellibranch shell. It is an organ of compression i.e. It is elastic chiefly under compression stress, in contrast to the ligament which is an organ of tension. Thus, the ligament is under tensional stress when the valves are tightly closed, whereas the resilium is compressed under similar conditions.

*178 (d) : Lamellibranchia is divided into three main orders: Taxodonta, Anisomyaria and Eulamellibranchia. Nucula, Arca and Ctenodonta are examples of Taxodonta. The Anisomyaria are divided into five sub-orders: Mytilacea (e.g. Mytilus, Lithophaga), Pteriacea (e.g. Pinna, Pteria), Pectinacea (e.g. Spondylus, Pecten), Anomiaceae (e.g. Anomia) and Ostraceae (e.g. Ostrea, Gryphaea, Exogyra). Eulamellibranchia is divided into four sub-orders: Schizodonta (e.g. Trigonia, Schizodus, Unio, Cardina), Heterodonta (e.g. Cardita, Chama, Diceras, Hippurites, Venus), Desmodonta(e.g.Mya, Pholas, Teredo, Pleuromya) and Palaeoconcha (e.g. Cardiola, Cardiomorpha).

*212 (c) : It was discovered in 1863 by Lartet at Cro Magnon in France.

*219 (d) : Sigillaria is an arborescent lycopod while the others belongs to cycadales.

****

# 9

# *Principles of Stratigraphy*

1. **The litho-stratigraphic classification was first adopted in**
   a) Central Africa
   b) Western Europe
   c) USA
   d) UK
2. **The proterozoic formations were first described from the shield area of**
   a) Canda
   b) Brazil
   c) India
   d) South America
3. **Which of the following systems was established later than the others in the group ?**
   a) Cambrian
   b) Ordovician
   c) Silurian
   d) Devonian
4. **The stages Geogian, Acadian and Potsdamian belongs to**
   a) Cambrian
   b) Ordovician
   c) Silurian
   d) Devonian
5. **The Neogene period includes**
   a) Paleocene and Eocene
   b) Paleocene, Eocene and Oligocene
   c) Oligocene, Miocene and Pliocene
   d) Miocene and Pliocene
6. **The transitional formations between the U. Cretaceous and Palaeocene known as the "Passage beds" are represented worldwide by the**
   a) Maastrichtian stage
   b) Danian stage
   c) Montain stage
   d) Landenian stage

7. 'Permian' is named after 'Perm' a province in
   a) USA
   b) USSR
   c) UK
   d) FRG

8. The type locality for triassic is
   a) Canada
   b) England
   c) Russia
   d) Germany

9. All branches of stratigraphy are based on
   a) Theoretical knowledge
   b) Laboratory procedures
   c) Experimental analyses
   d) Field studies

10. The first fundamental generalisation of stratigraphy is
    a) Order of Superposition
    b) Principle of uniformatarianism
    c) Law of Faunal succession
    d) Stratigraphic analysis

11. If the deposition takes place in a cave
    a) It follows the law of superposition
    b) It does not follow the law of superposition
    c) It may or may not follow the law of superposition
    d) None of the above

12. Which of the following is a tectonic criteria for superposition ?
    a) Graded bedding
    b) Cross stratification
    c) Index fossils
    d) Slaty cleavage

13. Which of the following in the strongest criteria for superposition ?
    a) Graded bedding
    b) Cross bedding
    c) Convolute bedding
    d) Index fossils
    e) Drag folds

14. The correlation of precambrian terranes is mostly based on
    a) Lithology
    b) Fossils
    c) Crustal upliftment
    d) Eustatic changes

15. Which of the following has been used for lithological correlations ?
    a) Kaolin
    b) Fuller's earth
    c) Bentonites
    d) Ball clays

16. According to the LAW OF FAUNAL Succession
    a) The fossil content of each formation is non-diagnostic
    b) The fossil assemblage of each formation is the same as that of the overlying and the underlying formations
    c) The fossil assemblage of each formations is very distinctive and characteristic of it
    d) The fossils in a formation are arranged successively according to their age

17. The chief characteristics of a good index fossil are
    (1) presence of complicated structures
    (2) Presence of simple structures
    (3) Ability to float or swin
    (4) Tendency to remain stationary or attached
    a) 1 and 4 are correct
    b) 1 and 3 are correct
    c) 2 and 3 are correct
    d) 2 and 4 are correct

19. A community of interrelated organisms inhabiting an area is described as
    a) Biotype
    b) Thenatocoenose
    c) Biocenose
    d) Thenotype

20. The concept of 'biostratigraphy' was introduced by
    a) W.J. Arkell
    b) S.S. Buckman
    c) L. Dollo
    d) William Smith

21. A biostratigraphic zone consisting of a strata deposited during the climax of an organism is called

a) Biozone
b) Assemblage zone
c) Range zone
d) Epibole

22. A zone consisting of all the strata deposited during the existence of a zonal organism is called
a) Biozone
b) Overlap zone
c) Abundance zone
d) Barren zone

23. Who is required as the 'Father of Stratigraphy' ?
a) John Lehman
b) Charles Lyell
c) Werner
d) James Hutton

24. The first geologist to measure and record actual sections of stratified rocks was
a) Werner
b) James Hutton
c) William Smith
d) John Lehmann

25. The greatest modern advance in the principles of correlation i.e., correlation by means of Eustatic changes was made by
a) Suess
b) John Strachey
c) Gottlob
d) M.F. Maury

26. The concept of Werner that all rocks were formed in water is described as
a) Plutonism
b) Hydrolism
c) Neptunism
d) Hydrostasis

27. Mappable assemblages of strata are called
a) Time units
b) Time-rock units
c) Rock units
d) None of the above

28. Which of the following is NOT a time-rock unit ?
a) System

b) Series
c) Epoch
d) Stage

29. Which of the following is the principal unit of geologic time ?
a) Era
b) Epoch
c) Eon
d) Period

30. Geologic time is divided at the beginning of the Cambrian period into two
a) Erathems
b) Eons
c) Eras
d) Epochs

31. Which of the following is a fundamental rock unit ?
a) Group
b) Formation
c) Member
d) Lentil

32. The world 'Formation' denotes similiarity of
a) Lithologic composition
b) Genesis
c) Time of deposition
d) All of the above
e) None of the above

33. Wedge-like lithologically identifiable rock units are called
a) Tongues
b) Lentils
c) Members
d) Strata

34. The time-rock unit 'Stage' corresponds with———of the time units
a) Epoch
b) Age
c) Chron
d) Era

35. A zone is a
a) Lithostratigraphic unit
b) Chronostratigraphic unit
c) Litho-chronostratigraphic unit

d) Biostratigraphic unit

36. Match the following and choose the correct answer

| | |
|---|---|
| A. Group | 1. Epoch |
| B. System | 2. Age |
| C, Series | 3. Era |
| D. Stage | 4. Period |

a) A–4 B–3 C–4 D–2
b) A–4 B–3 C–2 D–1
c) A–3 B–4 C–1 D–2
d) A–3 B–4 C–2 D–1

37. In the stratigraphic nomenclature which of the following consists of only two words
a) Formations
b) Members
c) Periods
d) Systems

38. Which of the following is designated as 'Psychozoic' era ?
a) Quaternary
b) Tertiary
c) Secondary
d) Primary

39. The Cryptozoic Eon is designated as
a) The age of no life
b) The age of well-displayed life
c) The age of hidden life
d) The age of modern life

40. The first land plants appeared in
a) Ordovician
b) Silurian
c) Devonian
d) Carboniferous

41. Amphibians originated in
a) Silurian
b) Devonian
c) Carboniferous
d) Cambrian

42. First leaves and flowers appeared during
a) Early cretaceous

b) Late cretaceous
c) Early jurassic
d) Late jurassic

43. The Alps Mountains originated during
a) Eocene age
b) Oligocene age
c) Miocene age
d) Pliocene age

44. Which of the following is known as the 'Age of Cycads' ?
a) Triassic
b) Jurarssic
c) Cretaceous
d) Tertiary

45. The Mesozoic group existed about——ago
a) 120 MY
b) 160 MY
c) 150 MY
d) 280 MY

46. The most widely used data in seismic stratigraphy is
a) P-wave reflections
b) S-wave reflections
c) L-wave reflections
d) All of the above
e) None of the above

47. The simplest of all the biostratigraphic units is
a) Assemblage zone
b) Abundance zone
c) Range zone
d) Epibole

48. Choose the correct combination from the following
a) Eons, Eras, Period, Epoch, age
b) System, Period, series stage
c) Group, Formation, Member
d) Biozone, Biochron, Teilzone, Teilchron

49. Which of the following is the oldest magnetic epoch ?
a) Brunhes normal
b) Matayama reversed
c) Gauss normal
d) Gilbert reversed

50. Which of the following is the correct sequence of the Indian Geological time scale ?
   a) Archaen–Purana–Aryan–Dravidian groups
   b) Dravidian–Archaen–Purana–Aryan groups
   c) Archaen–Purana–Dravidian–Aryan groups
   d) Archaen–Aryan–Purana–Dravidian groups

## PRINCIPLES OF STRATIGRAPHY (KEY)

| | | | |
|---|---|---|---|
| 1. b | 2. a | 3. b | 4. a |
| 5. c | 6. b | 7. b | 8. d |
| 9. d | 10. a | 11. b | 12. d |
| 13. a | 14. a | 15. c | 16. c |
| 17. a | 18. b | 19. c | 20. c |
| 21. d | 22. a | 23. b | 24. d |
| 25 a | 26. c | 27. c | 28. c |
| 29. d | 30. b | 31. b | 32. d |
| 33. b | 34. b | 35. d | 36. c |
| 37. b | 38. a | 39. c | 40. b |
| 41. b | 42. b | 43. c | 44. b |
| 45. c | 46. a | 47. c | 48. a |
| 49. d | 50. c | | |

# 10

# *Indian Geology*

1. The two-fold classification of the Precambrian into the lower Archaen and the upper Proterozoic Groups was first introduced in
   a) Canada
   b) Western Australia
   c) South America
   d) Northern Europe

2. The Main Unconfirmity separating the Archaen and Proterozoic successions occur at
   a) 3200 M.Y.
   b) 2750 M.Y.
   c) 2500 M.Y.
   d) 1600 M.Y.

3. Which one of the following marks the boundary between the Archaen and the Proterozoic formations in India ?
   a) Untala Granite
   b) Berach Granite
   c) Erinpura Granite
   d) Idar Granite

4. The Dharwar schists have been named as
   a) Older Greenstone belt
   b) Newer Greenstone belt
   c) Sargur Schist Complex
   d) Fundamental Gneissic Complex

5. The Keewatians of Canada are correlated with
   a) The Sargur Schist Complex
   b) The Dharwar Supergroup
   c) The Cuddapah Supergroup
   d) The Vindhyan Supergroup

6. The type area for the Sausar Group is
   a) Western Rajasthan
   b) Nagpur-Bhandara
   c) Singhbhum (Bihar)
   d) Chindwara (M.P.)

7. The emplacement of Erinpura Granite marks the close of
   a) Dharwar orogeny
   b) Delhi orogeny
   c) Eastern Ghat orogeny
   d) Satpura orogeny

8. The Erinpura Granite is a
   a) Hornblende granite
   b) Biotite granite
   c) Hornblende-biotite granite
   d) None of the above

9. Among the following, the oldest rocks belong to
   a) Sargur Schist Complex
   b) Banded Gneissic Complex
   c) Older Metamorphic Group
   d) Iron Ore Group

10. The unmetamorphosed facies of the Aravallis is termed as
    a) Binota shales
    b) Ranthambor formation
    c) Kolhan series
    d) Khardeola grits

11. Match the following and choose the correct answer

| | |
|---|---|
| (A) Kolhan orogeny | (1) 1660 M.Y. |
| (B) Satpura orogeny | (2) 1600 M.Y |
| (C) Eastern Ghat orogeny | (3) 1000 M.Y. |
| (D) Delhi orogeny | (4) 950 M.Y. |

    a) A—4 B—1 C—2 D—3
    b) A—3 B—1 C—2 D—4
    c) A—4 B—3 C—2 D—1
    d) A—3 B—2 C—4 D—1

12. The Granite gneiss of the Chotanagpur area is named as
    a) Bezwada Gneiss
    b) Kailasa Gneiss
    c) Hosur Gneiss
    d) Dome Gneiss

13. Which of the following represents the transitional phase from the Azoic to Phanerozoic eons of the earth's history ?
    a) Riphean
    b) Vendian
    c) Georgian

d) Acadian

14. Which of the following belongs to the lower Purana formations of the northern Peninsula ?
   a) Pakhal Group
   b) Kaladgi Group
   c) Chattisgarh Group
   d) Kolhan Group

15. The Cuddapah Basin is
   a) Linear
   b) Curved
   c) Çurvo-linear
   d) Concavo-convex

16. In the Cuddapah Supergroup there occurs an unconfirmity between any two successive series. The largest occurs between
   a) Papaghani-Cheyair
   b) Cheyair-Nallamalai
   c) Nallamalai-Kistna
   d) Kistna-Kurnool

17. In the Cuddapah Basin, the highest lateral extent is shown by
   a) Kistna Group
   b) Nallamalai Group
   c) Cheyair Group
   d) Papaghani Group

18. In the Cuddapah Basin the effects of metamorphism is seen to increase from
   a) North to South
   b) South to North
   c) East to West
   d) West to East

19. The total thickness of the Cuddapah Supergroup is
   a) 10,000 ft.
   b) 20,000 ft.
   c) 30,000 ft.
   d) 40,000 ft.

20. The lower and upper parts of the Delhi Supergroup are separated from each other by an intervening succession known as.
   a) Rialo series
   b) Susnai breccia
   c) Kushalgarh limestone
   d) Hornstone breccia

21. Dalma traps are the intercalations of lava flows, tuffs and conglomerates found on the top of
   a) Sausar-Sakoli series
   b) Iron-ore series
   c) Rialo series
   d) Kolhan series

*22. Arrange the following in the decreasing order of their age
   I. Dras volcanics — II. Panjal volcanics
   III. Malani volcanics — IV. Pitepani volcanics
   a) I, II, III, IV
   b) IV, III, II, I
   c) I, III, II, IV
   d) II, III, IV, I

23. Match of the following and choose the correct answer

| | |
|---|---|
| (A) Needle shales | (1) Talchir stage |
| (B) Khardeola grits | (2) Cambrian of Salt Range |
| (C) Fawn limestone | (3) Semri series |
| (D) Purple sandstone | (4) Gwalior series |

   a) A–3 B–4 C–2 D–1
   b) A–1 B–2 C–3 D–4
   c) A–3 B–2 C–4 D–1
   d) A–1 B–4 C–3 D–2

24. One of the world's largest examples of felsic magmatism during the upper Proterozoic times is evident in
   a) Pitepaṇi volcanics
   b) Malani volcanics
   c) Dhanjori lavas
   d) Khewra traps

25. Which of the following groups of rocks have been subjected to intense folding ?
   a) Semri Group
   b) Kaimur Group
   c) Rewa Group
   d) Bhander Group

*26. The 'Nagode Limestone' belongs to
   a) Kaimur Group
   b) Nallamalai Group
   c) Kolhan Group
   d) Bhander Group

27. The Bhander Group of the Vindhyans is
    a) Argillaceous
    b) Arenaceous
    c) Calcareous
    d) Partly arenaceous, partly calcareous

28. In the Semri Group, well-preserved stromatolites are found in
    a) Suket shales
    b) Bleaching shales
    c) Fawn limestone
    d) Kajrahat limestone

29. The most common rocks of the Vindhyan Supergroup are
    a) Limestones
    b) Sandstones
    c) Shales
    d) Gneisses

30. The succession : Mallampalli/Komamaram/Mulug/Albaka Sandstone belongs to
    a) Pakhal Group
    b) Sullavai Group
    c) Bheema Group
    d) Mesozoic of W. Rajasthan

*31. The Albaka Sandstone of the Pakhals is correlated w.th
    a) Bairenkonda quartzites of the Cuddapahs
    b) Srisailam quartzites of the Cuddapahs
    c) Pinnacled quartzites of the Kurnools
    d) Dhandraul quartzites of the U. Vindhyans

32. In the Pranhita Valley, rocks similar to the Pakhals are described as
    a) Gollapilli beds
    b) Penganga beds
    c) Sullavai Group
    d) Indravati Group

33. Which of the following correlations is/are correct ?
    (i) Sullavai Quartzites ——— Pinnacled Quartzites of Kurnools
    (ii) Pranhita Sandstone of Pengangas ——— Banganapalli Quartzites of Kurnool Group
    (iii) Dogra slates ——— Dharwar schists
    (iv) Bhander Sandstone ——— Purple Sandstone of Salt Range

a) (i), (ii) and (iv) only
b) (i), (ii) and (iii) only
c) (ii) and (iv) only
d) (i) and (iv) only

34. The 'Shahabad Limestone' belongs to
a) Penganga Group
b) Sullavai Group
c) Bhima Group
d) Badami Group

35. Revised geological mapping of the Kaladgi Basin has led to the separation of a part of the Kaladgi succession into
a) Indravati Group
b) Badami Group
c) Martoli Formation
d) Morar Formation

36. On the basis of new advanced organic findings composed of Chitinozoa and other microplanktons, the Badami Group has been elevated to the rank of
a) Cambrian
b) Ordovician
c) Silurian
d) Devonian

37. Which of the following groups contain well preserved stromatolites ?
a) Subathu
b) Kasauli.
c) Shali
d) Jutogh

38. In the Nepal-Sikkim area, the gneissose Archaen rocks are known as
a) Salkhala series
b) Vaikrita system
c) Daling series
d) Darjeeling series

39. The Archaen rocks of Gujarat are known as
a) Sonawani series
b) Sausar series
c) Champaner series
d) Jutogh and Chail series

40. In India, the Precambrian-Cambrian boundary can be located in

(1) Lolab formation (2) Ralam conglomerate
(3) Kunzamla formation (4) Bhander Group

a) 1 only
b) 2 only
c) 1, 2 and 3 only
d) All of the above

41. The Syringothyris Limestone belongs to
a) Early Carboniferous
b) Middle Carboniferous
c) Upper Carboniferous
d) Permian

42. The 'Nagthat formation' is of
a) Archaen age
b) Proterozoic age
c) Paleozoic age
d) Mesozoic age

43. Which of the following represents the correct order of cyclic sedimentation in the Gondwanas ?
a) Coal, shale sandstone
b) Coal, sandstone, shale
c) Sandstone, shale, coal
d) Shale, sandstone, coal

44. The Middle Gondwanas were formed under
a) Warm and humid climate
b) Warm and dry climate
c) Cold and humid climate
d) Cold and dry climate

45. The rocks belonging to Gondwanas are of
a) Fluviatile origin
b) Lacustrine origin
c) Marine origin
d) Glacial origin

46. The Glossopteris flora reached its highest stage of development during the deposition of
a) Damuda formation
b) Barakar formation
c) Ranigunj formation
d) Panchet formation

47. Which of the following groups is richest in the plant fossils ?

a) Damuda formation
b) Barakar formation
c) Ranigunj formation
d) Rajmahal formation

48. Which of the following is NOT a Gondwana fossil ?
a) Schizoneura
b) Pecopteris
c) Wilsonia
d) Williamsonia

49. The Glossopteris flora existed during the time span of
a) Permian to Triassic
b) Permian to Jurassic
c) L Carboniferous to U. Triassic
d) U. Carboniferous to L. Triassic

50. Identify the Panchet fossils among the following
a) Schizoneura gondwanensis, Glossopeteris indica, Noeggerathiopsis hislopi
b) Schizoneura gondwanensis, Glossopteris indica, Gonioglyptus, Pachygonia
c) Glossopteris, Gangamopteris, Angiopteridium, Thinnfeldia
d) Dicrodium hughesi, Sphenopteris, Dadoxylon

51. Choose the correct order in which the following plants appear in the Gondwona sequence
a) Glossopteris——→Schizoneura——→Dicrodium——→ Williamsonia
b) Glossopteris——→Schizoneura——→Williamsonia——→ Dicrodium
c) Glossopteris——→Williamsonia——→Schizoneura——→ Dicrodium
d) Schizoneura——→Glossopteris——→Dicrodium——→ Williamsonia

52. The earliest evidence of glacial action in India is obtained from the scratched pebbles of
a) Kaldurga conglomerate
b) Bap and Badhaura formation
c) Talchir tillites
d) Blaini beds

53. The equivalents of the Talchir tillites in the Kashmir Hazara area are known as
a) Mandhali beds

b) Blaini beds
c) Tannaki beds
d) Umia beds

54. The stratigraphic position of the Umaria marine beds is
a) Post-Talchir and pre-Damuda
b) Post-Damuda and pre-Panchet
c) Post-Panchet and pre-Mahadeva
d) None of the above

55. Which of the following rock successions show evidence of glacial erosion in its lower parts ?
a) Zewan formation
b) Muth quartzite
c) Haimanta group
d) Agglomeratic slate series

56. A world-wide phase of mountain building which led to the emergence of several mountain chains during the Ordovician and Silurian times is described as
a) Varsican orogeny
b) Antler orogeny
c) Caledonian orogeny
d) Alpine orogeny

57. The Garbyang formation of Kumaon belongs to
a) Precambrian
b) Cambrian
c) Ordovician
d) Carboniferous

58. Which of the following successions is conspicuously absent from the Indian Peninsula ?
a) Lower Paleozoic
b) Upper Paleozoic
c) Lower Mesozoic
d) Cenozoic

59. Which of the following successions contains plant fossils of Gondwana affinity ?
(1) Agglomeratic Slate series
(2) Gangamopteris Beds
(3) Infra Krol
(4) Subathu formation
a) 1 and 2 only

b) 2 and 3 only
c) 1, 2 and 3 only
d) 2, 3 and 4 only

60. Match the following and choose the correct answer

(A) Blaini beds — (1) U. Carboniferous
(B) Bivalve beds — (2) Triassic
(C) Sulcacutes beds — (3) Jurassic
(D) Umia plant beds — (4) E. Cretaceous

a) A—1 B—3 C—2 D—4
b) A—4 B—2 C—3 D—1
c) A—4 B—3 C—3 D—1
d) A—1 B—2 C—3 D—4

61. Choose the correct succession among the following

a) Panjal volcanics
Agglomeratic slates
Passage beds
Fenestella shales
Syringothyris limestone

b) Panjal volcanics
Agglomeratic slates
Fenestella shales
Passage beds
Syringothyris limestone

c) Panjal volcanics
Passage beds
Agglomeratic slates
Syringothyris limestone
Fenestella shales

d) Panjal volcanics
Agglomeratic slates
Passage beds
Syringothyris limestone
Fenestella shales

62. Which of the following formations is devoid of fossils ?
a) Blaini
b) Infra Krol
c) Krol
d) Tal

63. The Punjab Salt Range is the type area for
a) Cambrian
b) Ordovician
c) Triassic
d) Eocene

64. 'Hormuz Series' of Iran is homotaxial with
a) Talchir tillites
b) Saline series
c) Zeewan beds
d) Bagh beds

65. The sedimentary origin for the saline series was proposed by
a) R.D. Oldham
b) C.S. Middlemiss

c) E.H. Pascoe
d) Christie

66. The Saline series is intruded by volcanic rocks belonging to
a) Sylhet traps
b) Panjal traps
c) Deccan traps
d) Khewra traps

67. The succession of rock formations exposed in the Spiti valley range in age from
a) Precambrian to Recent
q) Precambrian to Cretaceous
c) Cambrian to Recent
d) Cambrian to Cretaceous

68. The Kanawar system of the Spiti valley consists of rock formations deposited during
a) Lower to middle Carboniferous
b) Upper Carboniferous to Permian
c) Permian to Triassic
d) Triassic to Jurassic

69. The Permian system in the Spiti synclinorium is described as
a) Kuling system
b) Lilang system
c) Haimanta system
d) Po series

70. The Cambrian succession of Spiti valley is named as 'Haimanta system' because of
a) Its location in a town of that name
b) Its location along the River Haimanta
c) Its occurrence in high show-capped peaks
d) Its most prominent fossils belonging to the Phylum Bryozoa

71. 'Parahio Series' is the name given to
a) Lower Haimantas
b) Middle Haimantas
c) Upper Haimantas
d) Lower and middle Haimantas

72. Conglomcrates of "Autoclastic" origin are found among the Slates of
a) Cambrian of Spiti
b) U. Carboniferous of Spiti
c) Proterozoic of Western Rajasthan
d) Garbyang formation

73. "Productus limestone series" is the name given to——— succession of India

a) L. Permian
b) M. Permian
c) U. Permian
d) L. Triassic

74. In India, the triassic system of rocks are found as

a) Marine facies
b) Marine geosynclinal facies
c) Marine transgressed
d) Fluvio-Lacustrine facies

75. In India, deposition of rocks took place in diverse type of environments during

a) Triassic
b) Jurassic
c) Cretaceous
d) Quaternary

76. The Lameta beds of Jabalpur represent

a) Marine facies
b) Marine transgressed facies
c) Volcanic facies
d) Fluvio-lacustrine facies

77. The Permian-Triassic boundary in the Gondwana sequence of India lies

a) Below the Panchet
b) Above the Panchet
c) In the middle of the Panchet
d) Between the Panchet and the Mahadva

*78. Which of the following statements is/are true of the Kioto Limestone of the Central Himalayas ?

(i) The lower part is dominantly composed of Pelecypods
(ii) The upper part is known as Sulcacutes beds
(iii) The entire limestone succession shows well-preserved fossils belonging to pelecypods, gasteropods and ammonites
(iv) Its age is U. Triassic to M. Jurassic

a) (i) and (iii) only
b) (i), (ii) and (iii) only
c) (ii) and (iv) only
d) (i), (ii) and (iv) only

79. 'Chikkim limestone' and 'Chikkim shale' belongs to

a) The Eocene rocks of Kashmir

b) The Precambrians of Sikkim

c) The Cretaceous rocks of spiti valley

d) The Jurassic rocks of spiti valley

80. Which of the following is the correct sequence of the Mesozoic of the Kutch Basin ?

| a) | b) | c) | d) |
|---|---|---|---|
| Bhuj | Umia | Bhuj | Umia |
| Umia | Katrol | Umia | Katrol |
| Katrol | Chari | Katrol | Bhuj |
| Chari | Patcham | Patcham | Chari |
| Patcham | Bhuj | Chari | Patcham |

81. Bhuj formation is the name given to

a) Ukra beds

b) Umaria marine beds

c) Umia Ammonite beds

d) Umia plant beds

82. During the Jurassic period a marine transgression took place from

a) East

b) West

c) Southeast

d) Southwest

83. The youngest succession of rocks in the Kutch basin is known as

a) Bhuj formation

b) Umia formation

c) Porbander formation

d) Madh formation

84. The Cenomain Transgression occurred during

a) Triassic period

b) Jurassic period

c) Cretaceous period

d) Miocene period

85. Which of the following successions show evidence of marine transgressions ?

(i) U.Gondwana deposits of East coast

(ii) U. Jurassic of Kutch

(iii) Cretaceous of Tiruchinpolly

(iv) Bagh beds of Narmada valley

a) (i), (ii) and (iii) only

b) (i), (ii) and (iv) only

c) (ii), (iii) and (iv) only

d) All of the above

86. Match the following and choose the correct answer

| | |
|---|---|
| A. Indian Geological time scale | 1. B. Rama Rao |
| B. Three-fold division of Dharwars | 2. T. H. Holland |
| C. Two-fold classification of Gondwanas | 3. C. S. Middlemiss |
| D. Hypogene intrusive origin of Saline series | 4. E. Vredenburg |

a) A—2 B—1 C—4 D—3

b) A—1 B—2 C—4 D—3

c) A—2 B—1 C—3 D—4

d) A—1 B—2 C—3 D—4

87. Match the following and choose the correct answer

| | |
|---|---|
| A. Nallamalai group | 1. Diamonds |
| B. Mansar formation | 2. Pyrite |
| C. Bijaigarh shales | 3. Managanese |
| D. Bijawer series | 4. Lead |

a) A—2 B—3 C—1 D—4

b) A—3 B—1 C—2 D—4

c) A—4 B—3 C—2 D—1

d) A—4 B—3 C—1 D—2

88. Lignite deposits are found in association with

(1) Cuddalore sandstones of the cauvery basin

(2) Karewas of Kashmir valley

(3) Palana formation of Rajasthan

a) 1 only

b) 1 and 2 only

c) 1 and 3 only

d) All of the ahove

*89. Choose the correct stratigraphic sequence from the following

a) Suket shales
Binota shales
Needle shales
Fenestella shales
Spiti shales

b) Binota shales
Suket shales
Fenestelia shales
Needle shales
Spiti shales

c) Spiti shales
Needle shales
Fenestella shales

d) Spiti shales
Needle shales
Fenestella shales

| Binota shales | Sūket shales |
|---|---|
| Suket shales | Binota shales |

90. Which of the following conglomerates has the least age ?
   a) Banganapalli conglomerate
   b) Bangra Conglomerate
   c) Ralam Songlomerate
   d) Hemis Conglomerate

91. The physiographic features of India were shaped during
   a) Proterozic
   b) Paleozoic
   e) Mesozoic
   d) Cenozoic

92. The first upheaval of the Himalayan mountain belt took place in
   a) Lower-Middle Cretaceous
   b) U. Cretaceous-Eocene
   c) Miocene-Pliocene
   d) Pliocene-Pliestocene

93. The first stage of the Himalayan orogeny was accompanied by the emplacement of
   1. Dras volcanics
   2. Indus ophiolites
   3. Siwalik molasse
   4. Krol nappes
   a) 1 and 2 only
   b) 1 and 3 only
   c) 2 and 3 only
   d) 2, 3 and 4 only

94. In the Kashmir valley, the Neogene-Quaternary boundary is recognised on the basis of
   a) Disappearance of Hipparian
   b) Appearance of Hippopotamus
   c) Appearance of Elephas and Stegodon
   d) Appearance of Equus

95. The Neogene molassic rock of the Intermontane basins of the Himalayas are best represented by
   a) The Siwalik group
   b) The Subathu group
   c) The Karewas
   d) The Murree group

96. The Siwalik group of rocks exhibit
    a) A general coarsening upwards of sediments
    b) A general fining upwards of sediments
    c) Alternate coarsening and fining of sediments
    d) Uniformity in grain size from bottom to the top of the sequence

97. The Main Boundary fault separates
    a) Upper Vindhyans and Aravallis
    b) Siwaliks and Aravallis
    c) Siwaliks and Tertiaries
    d) Siwaliks and the Bundelkhand massifs

98. Match the following and choose the correct answer

| | |
|---|---|
| A. Upper Siwaliks | 1. Fluviatile environment |
| B. Middle Siwaliks | 2. Marshy environment |
| C. Lower Siwaliks | 3. Lacustrine environment |

    a) A—1 B—2 C—3
    b) A—3 B—2 C—1
    c) A—2 B—1 C—3
    d) A—2 B—3 C—1

*99. The Siwalik group of rocks were deposited in the Himalayan fore deep during the———— stage of Himalayan orogeney
    a) First
    b) Second
    c) Third
    d) Fourth
    e) Fifth

100. Proboscidean fossils (Mastodons, elephants and mammoths) have NOT been recorded from
    a) Kamlial formation
    b) Chinji formation
    c) Nagri formation
    d) Dhokpathan formation
    e) Tatrot formation

*101. Kyanite is found be to the most dominant heavy mineral in
    a) Kamlial formation
    b) Chinji formation
    c) Dhokpathan formation
    d) Pinjor formation

102. Which of the following formations is richest in fossils ?

a) Pinjor formation
b) Dhokpathan formation
c) Chinji formation
d) Nagri formation

103. Match the following and choose the correct answer

A. Dryopithecus — 1. Kamlial
B. Synconolophus — 2. Chinji
C. Equus sivalensis — 3. Dhokpathan
D. Elephas planiferous — 4. Tatrot
5. Pinjor

a) A—1 B—3 C—2 D—4
b) A—2 B—3 C—1 D—4
c) A—1 B—2 C—4 D—5
d) A—1 B—3 C—4 D—5

104. Match the following and choose the correct answer

| *Formation/Group* | *Characteristic fossils* |
|---|---|
| A. Subathu | 1. Nautilus danicus |
| B. Kasauli | 2. Nummulites |
| C. Niniyur | 3. Sabal major |
| D. Murree | |

a) A—2 B—3 C—1 D—3
b) A—3 B—3 C—1 D—2
c) A—2 B—3 C—1 D—3
d) A—1 B—2 C—3 D—1

*105. Remains of dinosaurs have not been recorded from
a) Bagh beds
b) Lameta beds
c) Umia beds
d) Panchet formation

*106. Which of the following is NOT correct ?
a) Murree group————Lower Miocene
b) Siwalik group————M. Miocene to L. pliestocene
c) Mahadek formation————U. Cretaceous to Eocene
d) Rajahmundry sandstone————Paleocene

107. Choose the correct sequence from the following ?
a) Blaini/Krol/Tal
b) Subathu/Kasauli/Dagshai
c) Tipam/Surma/Barail
d) Nari/Gaj/Kirthar

108. Which one the following statements about the deccan traps is correct ?

a) The lower unit is exposed in the northern parts of the Deccan

b) The lower parts show evidence of explosive volcanic activity

c) The upper parts exhibit normal polarity of flows

d) The upper parts are devoid of intertrappean beds

109. The Deccan Traps show maximum thickness in the

a) North

b) South

c) East

d) West

110. Which one of the following statements is incorrect ?

a) The Deccan Traps are younger than the Infratrappeans

b) The Intertrappeans are older than the Infratrappeans

c) The Bagh Beds are older than Deccan Traps

d) The Deccan Traps are younger than the Lameta Beds

111. The most characteristic fossil of the Intertrappean beds is

a) Cardita beaumonti

b) Rhynconella griesbachi

c) Physa princepi

d) Otoceras varaha

112. In the Deccan Traps, fossiliferous beds are characteristically absent in

a) Lower traps

b) Middle traps

c) Upper traps

d) None of the above

113. The Intertrappean beds are of

a) Volcanic origin

b) Marine origin

c) Fluvio-Lacustrine origin

d) Continental origin

114. Which of the following areas is described as the 'Field Museum in Palaeontology' ?

a) Vriddachalam (T.N.)

b) Duddukuru (A.P.)

c) Shevnoy Hills (T.N.)

d) Rapur (Punjab)

115. The SYLHET formation typically consists of
   a) Sandstones
   b) Shales
   c) Limestones
   d) Conglomerates

116. The youngest group of rocks in the Assam-Arakan region belongs to
   a) Eastern Disang group
   b) Jaintia group
   c) Duptila group
   d) Dihing group

117. Match the following and choose the correct answer

| | |
|---|---|
| A. Girujan clay | 1. Tippam group |
| B. Denwa clay | 2. Mahadeva group |
| C. Sylhet limestone | 3. Badami group |
| D. Temple Quartz Arenite | 4. Jaintia group |
| | 5. Barail group |

   a) A—1 B—2 C—5 D—3
   b) A—1 B—2 C—4 D—3
   c) A—5 B—2 C—3 D—4
   d) A—3 B—1 C—4 D—5

118. Which of the following does NOT belong to Miocene period ?
   a) Cuddalore sandstone
   b) Rajahmundry sandstone
   c) Tippam sandstone
   d) Giumal sandstone

119. The age of the Pir Panjal traps is
   a) U. Permian-L. Triassic
   b) L. Triassic-Jurassic
   c) Cretaceous-Eocene
   d) Miocene-Pliocene

120. The Mussoorie group consists of
   a) The Subathu-Dagshai-Kasauli succession
   b) The Ranikot-Laki-Kirthar succession
   c) The Blaini-Krol-Tal succession
   d) The Madh-Berwali-Lakhpat succession

121. In the Kumaon region of U.P., the youngest palaeozoic formation is described as
   a) Garbyang formation

b) Shiala formation

c) Kali formation

d) Kringkrong formation

122. Choose the correct sequence from the following

a) Martoli formation
Ralam conglomerate
Garbyang formation

b) Garbyang formation
Ralam conglomerate
Martoli formation

c) Garbyang formation
Martoli formation
Ralam conglomerate

d) Martoli formation
Garbyang formation
Ralam conglomerate

123. The Proterozoic rocks exposed in the Bastar Depression are termed as

a) Indravati group

b) Badami group

c) Chattisgarh group

d) Kolhan group

124. The Indravati group is considered to be homotaxial with

a) The Kurnool group

b) The Cuddapah supergroup

c) The Semri group

d) The Kaladgi group

125. According to the revised stratigraphy of the Cuddaph basin adopted by the G.S.I. in 1981, the 'Chitravati Group' is the new name given to

a) Papaghani group

b) Cheyair group

c) Nallamalai group

d) Kistna group

126. The revised classification of the Cuddapah basin brought changes in the William King's original classification with respect to all the groups except

a) Papaghani group

b) Cheyair group

c) Nallamalai group

d) Kistna group

127. In the Cuddapah basin evidence of volcanic activity in the form of basic lava flows is obtained from
   a) Cumbum formation
   b) Tadpatri formation
   c) Nagri formation
   d) Vaimpalli formation

128. Within the Cuddapah crecentric, sedimentation took place in two smaller graben-like basins which led to the deposition of
   a) Kurnool and Bhima groups
   b) Kurnool and Palnad groups
   c) Kurnool and Pakhal groups
   d) Penganga and Sullavai groups

129, The Vindhyan and the Kurnool basins of the late proterozoic age were separated from each other by
   a) The Satpura orogen
   b) The Eastern Ghat orogen
   c) The Aravalli orogen
   d) The Bundelkhand massifs

130. The great boundary fault lies between
   a) Siwaliks and the Murree groups of rocks
   b) Upper Vindhyans and Aravallis
   c) Upper Vindhyans and Kurnools
   d) Siwaliks and the Tertiaries

131. The great boundary fault was formed during the time period ranging between
   a) 570—300 M.Y.
   b) 900 – 570 M.Y.
   c) 1600 – 900 M.Y.
   d) 2500—1600 M.Y.

132. In the Vindhyan group of rocks, coal (vitrain) is found in association with
   a) Sirbu shales
   b) Jhiri shales
   c) Bijaigarh shales
   d) Ganugarh shales

133. Sulphurous hot springs (solfatoras) are associated with
   a) Gwalior group
   b) Bijawer group
   c) Kaladgi group

d) Kolhan group

134. The famous Panna diamond field of Madhya Pradesh is situated in the country of———group of rocks.

a) Semri

b) Kaimur

c) Rewa

d) Bhander

135. The strike of Gondwana rocks of Godavari valley is

a) N—S

b) NW—SE

c) NE—SW

d) E—W

136. In the Gondwana succession of the Godavari graben the remains of Sauropod dinosaurs were found in

a) Kota formation

b) Kamthi formation

c) Gangapur formation

d) Chikiala formation

137. The strata belonging———period appears to be absent in the Godavari graben

a) U. Carboniferous-Permian

b) Permian-Triassic

c) Triassic-L. Jurassic

d) Middle-Upper Jurassic

138. Which of the following formations does NOT contain coal ?

a) Barren measures

b) Motur formation

c) Kulti formation

d) All of the above

139. The Gondwana group of rocks were deposited during

a) U. Carboniferous-Jurassic

b) U. Carboniferous-Cretaceous

c) Permain-Jurassic

d) Permian-Eocene

140. The equivalents of the Gondwana group of rocks in Brazil is grouped under

a) The Karoo system

b) The Kuttung system

c) The Santa-Catarina system

d) The Hawkesbury system

141. HOMOXYLON is

a) A reptile belonging to the Gondwanas

b) A type of cycad belonging to the Gondwanas

c) A type of fossil wood

d) A type of floral species

142 Which of the following is the correct sequence of the U. Gondwanas of Andhra Pradesh as found near Rajahmundry ?

a) Raghavapuram shales
Tirupati sandstones
*Golapilli sandstones*
Chintalpudi sandstone

b) Tirupati sandstones
Raghavapuram shales
*Chintalpudi sandstones*
Golapilli sandstone

c) Tirupati sandstone
Raghavapuram shales
*Golapilli sandstones*
Chintalpudi sandstone

d) Chintalpudi sandstone
Raghavapuram shales
*Golapilli sandstones*
Tirupati sandstone

143. Bucklandia indica was the stem that bore the leaves of

a) Ptillophylum

b) Dicrodium

c) Gangamopteris

d) Glossopteris

*144. Which of the following groups of rocks from the extra Peninsula can be correlated with the lower Gondwanas of the Peninsula ?

a) Krol series

b) Infra-Krol series

c) Tal series

d) Daling series

*145. In the Gondwana succession evidence of cold climate is obtained from

a) Talchir formation

b) Talchir and Damuda formations

c) Talchir and Panchet formations

d) Talchir, Panchet and Mahadeva formations

146. The upper Gondwana rocks of Orissa are deseribed as
   a) Almod beds
   b) Himgir beds
   c) Durgapur beds
   d) Athgarh beds

147. The Siwaliks belong to
   a) Lesser Himalayas
   b) Greater Himalayas
   c) Outer Himalayas
   d) Central Himalayas

148. The mountain building activity towards the end of the Siwalik phase of deposition is indicated by the presence of very coarse sediments which are described as
   a) Lei conglomerate
   b) Hemis conglomerate
   c) Bagra conglomerate
   d) Po conglomerate

149. Which of the following is the correct order of succession of fossil elephants in the Upper Siwaliks of India ?
   a) Mastodon sivalensis———Stegodon ganesa———Elephas maximus———Elephas primigenius
   b) Mastodon sivalensis———Stegodon ganesa——— Elephas primigenius———Elephas maximus
   c) Stegodon ganesa———-Elephas primigenius———Elephas maximus———Mastodon sivalensis
   d) Elephas primigenius———Stegodon ganesa———Elephas maximus———Mastodon sivalensis

150. In the Eastern Himalayas of the Sikkim region, the Permian rocks are represented by
   a) Laki series
   b) Lathi formation
   c) Lachi series
   d) Lipak series

151. The 'Tso Lhamo Series' belongs to
   a) Carboniferous
   b) Permian
   c) Triassic
   d) Jurassic

152. The speckled sandstone series of the Punjab salt range belongs to

a) Cambrian

b) Carboniferous

c) Permian

d) Jurassic

153. Which of the following groups of fossils was instrumental in the biozonation of lower Triassic rocks ?

a) Cephalopods

b) Gasteropods

c) Lamellibranchs

d) Brachiopods

154. The trend of the Spiti synclinorium is similar to that of

a) The Eastern Ghats

b) The Aravallis

c) The Himalayas

d) The Satpuras

155. The rocks of the Pakhal group strike in the direction

a) N—S

b) E—W

c) NW—SE

d) NE—SW

156. In M.P., the arenaceous and argillaceous rocks of lower Proterozoic age are described as

a) Chandrapur series

b) Chandpur series

c) Chari series

d) Chail series

157 Match the following and choose the correct answer

| | |
|---|---|
| (A) Variegated series | (1) Ordovician |
| (B) Chocolate series | (2) Jurassic |
| (C) Shiala series | (3) Triassic |
| (D) Jaisalmer limestone | (4) Silurian |

a) A—1 B—2 C—4 D—3

b) A—4 B—3 C—1 D—2

c) A—1 B—3 C—4 D—2

d) A—4 B—2 C—1 D—3

158. Large blocks of Triassic limestone of varying sizes which are found lying in a confused stratigraphic order over various Mesozoic formation near Kumaon are described as

a) Exotic blocks
b) Erratics
c) Autoclastic conglomerates
d) Po series conglomerates

159. The exotic blocks of Malla Johar which comprises of white, red and pink limestones of all ages from Permian to Cretaceous are also known as
a) Chitichun facies
b) Flysch facies
c) Intermontane molasse facies
d) Euxinic facies

160. Recent studies in the Cauvery basin have established the occurrence of reefoidal limestone beds of continental origin. These beds of L. Jurassic age have been named as
a) Dalmiapuram formation
b) Pondicherry formation
c) Sivaganga beds
d) Warkali beds

*161. In the SE Rajasthan, the rocks equivalent to the Semi group are represented by
a) Nimbahera limestone and shale
b) Tirohan limestone
c) Bargawan limestone
d) Ken limestone

*162. In the Semri group of the Son Valley the basal beds constituting the oldest formations have been renamed as
a) Paisuni formation
b) Pipartola conglomerate
c) Tanawal formation
d) Arangi formation

*163. Stomatolites are recorded from the carbonate formations of
(1) Bhander group
(2) Semri group
(3) Shali group
(4) Tejam group
a) 1 and 3 only
b) 2 and 4 only
c) 2, 3 and 4 only
d) All of the above

164. The Bap and the Badhaura formations are of

a) Precambrian age
b) Cambrian age
c) Permain age
d) Jurassic age

165. In the NE Rajasthan, rocks similar to the Vindhyan rocks of the Son Valley were earlier known as the 'Trans-Aravalli Vindhans'. These have now been separated from the Vindhyans and are now described as
a) The Marwar group
b) The Chattisgarh group
c) The Chaibasa group
d) The Riaolo group

*166. Choose the set which belongs to the same age
a) Panchet——Panchmari——Kamthi——Bijori
b) Dubrajpur——Suprapanchet——Parsora——Bagra
c) Jabalpur——Chikiala——Parsora——Dubrajpur
d) Barakar——Motur——Bijori——Kamthi

*167. Which one of the following Lower Triassic Gondwana formations of India consists of vertebrate fauna which are NOT known from any other part of the Gondwanaland ?
a) Gangamopteris beds of Kashmir
b) Panchet formation of Damodar valley
c) Bijori formation of Narmada valley
d) Mangli beds of Pranhita-Godavari valley

168. In the Upper Gondwana rocks of the Godavari graben fossils are present in all the formations except
a) Yerrapalli formation
b) Bhimaram formation
c) Maleri formation
d) Kamthi formation

169. Recent studies during the last decade has led to the discovery of early Jurassic mammalian fossils from
a) Chikiala formation
b) Dharmaram formation
c) Yerrapalli formation
d) Kota formation

*170. In the Cretaceous succession of the Cauvery basin, dinosaurian bones are characteristically present in the rocks of
a) Uttattur group
b) Tiruchinapalli group
c) Ariyallur group

d) Niniyur formation

171. In India, the Potwar loess and silt deposits corresponds to
   a) I Glacial period
   b) II Glacial period
   c) III Glacial period
   d) IV Glacial period

172. The Mindel and the Riss glaciation occurred during
   a) Lower Pliestocene
   b) Middle Pliestocene
   c) Upper Pliestocene
   d) Holocene

173. The first glacial advance has been termed as
   a) Gunz
   b) Mindel
   c) Riss
   d) Wurm

174. In India the first human occupation is not earlier than the
   a) Miocene-Pliocene
   b) Lower Pliestocene
   c) Middle Pliestocene
   d) Upper Pliestocene

175. Which of the following represents the Pliestocene and recent deposits of India ?
   a) The Loess deposits of salt range
   b) The Stalagmitic cave deposits of Kurnool, A.P.
   c) The Karewas of Kashmir
   d) All of the above

## INDIAN GEOLOGY (KEY)

| | | | |
|---|---|---|---|
| 1. a | 2. c | 3. b | 4. b |
| 5. a | 6. d | 7. d | 8. b |
| 9. c | 10. a | 11. c | 12. d |
| 13. b | 14. d | 15. c | 16. b |
| 17. a | 18. d | 19. b | 20. c |
| 21. b | *22. b | 23. d | 24. b |
| 25. a | 26. d | 27. d | 28. c |
| 29. b | 30. a | *31. b | 32. b |
| 33. a | 34. c | 35. b | 36. b |
| 37. c | 38. d | 39. c | 40. d |

| | | | |
|---|---|---|---|
| 41. a | 42. c | 43. c | 44. b |
| 45. a | 46. c | 47. d | 48. c |
| 49. d | 50. b | 51. a | 52. a |
| 53. c | 54. c | 55. d | 56. c |
| 57. b | 58. a | 59. c | 60. d |
| 61. d | 62. c | 63. a | 64. b |
| 65. c | 66. d | 67. b | 68. a |
| 69. a | 70. c | 71. c | 72. a |
| 73. b | 74. b | 75. c | 76. d |
| 77. c | *78. d | 79. c | 80. a |
| 81. d | 82. b | 83. c | 84. c |
| 85. d | 86. a | 87. c | 88. d |
| *89. d | 90. d | 91. d | 92. b |
| 93. a | 94. d | 95. c | 96. a |
| 97. c | 98. a | *99. c | 100. c |
| *101. c | 102. b | 103. d | 104. a |
| *105. a | *106. d | 107. a | 108. c |
| 109. d | 110. b | 111. c | 112. b |
| 113. c | 114. a | 115. c | 116. d |
| 117. b | 118. d | 119. a | 120. c |
| 121. d | 122. b | 123. a | 124. c |
| 125. b | 126. a | 127. d | 128. b |
| 129. a | 130. b | 131. b | 132. c |
| 133. a | 134. c | 135. b | 136. a |
| 137. d | 138. d | 139. b | 140. c |
| 141. c | 142. a | 143. a | *144. b |
| *145. c | 146. d | 147. c | 148. a |
| 149. b | 150. c | 151. b | 152. b |
| 153. a | 154. c | 155. c | 156. a |
| 157. b | 158. a | 159. a | 160. c |
| 161. a | *162. d | 163. d | 164. c |
| 165. a | *166. b | 167. a | 168. b |
| 169. d | *170. c | 171. b | 172. b |
| 173. a | 174. c | 175. d | |

## EXPLANATORY NOTES

*22 (b) : Pitepani volcanies—Precambrian
Malani volcanics—Proterozic
Panjal volcanics—Permian
Dras volcanics—Paleocene

*31 (b) : The Pakhals are considered to be equivalents of Cuddapahs. The Mallampalli, Momamaram and Mulug Groups of Pakhals are correlated with Papaghani, Cheyair and Nallamai Groups of the Cuddapah Supergroup. The topmost formation of the Pakhals i.e. the Albaka sandstone is, thus, correlated with the Srisailam Quartzite.

*78 (d) : Statement (iii) is incorrect. The middle part of the carbonate succession is devoid of fossils.

*89 (d) : Binota shales—Aravallis : Suket shales—Precambrian ; Fenestella shales—L.—U. Carboniferous ; Needle shales —U. Carboniferous ; Spiti shales—Jurassic.

*99. (c) : The third Himalayan upheaval during the middle Miocene times, known as the Sirmurian orogeny, was the most pronounced of all the phases of Himalayan orogenies and led to the Siwaliks group of rocks.

*100. (c) : The lower, middle and upper Siwaliks show a definite pattern of heavy mineral composition. Staurolite, kyanite and hornblende dominate the heavy minerals in the sediments of the lower, middle and upper Siwaliks respectively.

*105. (a) : Lameta beds—Titanosaurus indicus ; Umia beds—Plesiosaurus indica ; Panchet—Epicampodon dinosaur.

*106. (d) : The Rajahmundry sandstone belongs to Miocene—Pliocene.

*144. (b) : The Infra-Krol formation consists of micro-floral remains ranging in age from carboniferous to permian. The occurrence of glossopteris flora is reported from the black shales of this formation.

*145. (c) : The first evidence of glacial action in the Gondwanas is obtained from the Talchir rocks of Orissa. The boulder bed is wide in distribution and occurs in such areas as Hazara, Simla, Garhwal, Rajasthan,M.P. and Orissa. This is suggestive of the presence of an ice sheet in India at that time. Another cold cycle was during Panchet, the evidence for which is composed in presence of undecomposed feldspar grains among the clastic sediments.

*160. (a) : The Semri sedimentation took place in three main sub-basins i.e., the Son Valley, Bundelkhand and SE Rajasthan. In the SE Rajasthan the succession includes Khardeola sandstone, Bhagwanpura limestone, Binota shales, Jiran sandstone, Nimbahera limestone and Bari shale (Nimbahera shale). In the Bundelkhand, the succession includes Pandwa Fall quartzite, Ken limestone, Palkawan shale and Tirohan limestone.

*162. (d) : The basal beds of the Semri group of the Son Valley are renamed as the Arangi formation. The standard classification for the Semri group is as follows :

Rohtas limestone

Basuhari limestone (Glauconitic beds)

Bargawan limestone

Kheinjua shale (Olive shales)

Chopan porcellanite (Porcellanite stage)

Kajrahat limestone

Arangi formation (Basal beds)

*163. (d) : Stromatolites are recorded from the Nagode limestone, Sirbu shale, U. Bhander limestone in the Bhander group : the Rohtas limestone, Kajrahat limestone, Tirbhan limestone, etc. in the Semri group. The Shali group contains well-preserved Stromatolites such as Collenia collumnaris, C. baicalica and C. symmetrica. The Deoban formation of the Tejam group of the Kumaon Himalayas comparises of a 1200 m thick succession of stromatolite-bearing c erty dolomitic limestones.

*166. (b) : The Dubrajpur formation (Rajmahal basin) ; the Supra-Panchet formation (Damodar-Son basin), the Parsora formation (Rewa basin) and the Bagra formation (Satpura basin) are of the same age e.g. Noric to Rhaetic (U. Triassic).

*167. (a) : The Gangamopteris bed fauna has a distinct European affinity. Its presence in Kashmir indicates that this part of India was in close proximity to Eurasia in early permian time allowing free migration of tetrapods from laurasia. The fauna consists of fishes and labyrinthodont amphibians.

*170. (c) : The Kallamedu Formation of Ariyalur group is characterized by the presence of the skeletal parts of a Sauropod, a Carnosaur and Stegosaur. However, the Trichinapalli group also shows evidence of a stegosaurian dinosaur—Dravidosaurus blanfordi.

# 11

# *Igneous Petrology*

1. Ijolite is a
   a) felsic alkaline rock  b) mafic alkaline rock
   c) ultramafic alkaline rock  d) volcanic rock
2. Nephelinite is the volcanic equivalent of
   a) Ijolite  b) Melteigite
   c) Nepheline syenite  d) Peridotite
3. Which of the following is an example of a potassic alkaline ultrabasic lava ?
   a) Serpentinite  b) Essexite
   c) Ijolite  d) Leucitite

*4. The change in the magma composition from ***tholeiitic to alkaline*** occurs in the oceanic islands as a result of
   a) variation in lithospheric thickness
   b) thickening of the lithosphere
   c) thinning of the lithosphere
   d) variation in the asthenospheric thickness
5. An opening rift typically consists of
   a) Alkali magmas
   b) Calc-alkalic magmas
   c) Tholeiitic magmas
   d) Bimodal magmas
6. The alkaline magmatism
   a) precedes the opening of an ocean
   b) succeeds the opening of an ocean
   c) either precedes or succeeds the ocean opening
   d) precedes and succeeds the ocean opening
7. The "Harpoon effect" is a result of
   a) reversal in the magnetic polarity of major continents
   b) reversal in the magnetic polarity of micro continents
   c) reversal in the shearing of a fault system
   d) accelerations and decelerations of plate motions
8. Match the following and choose the correct answer
   (A) calc-alkaline lamprophyre  (1) Orendite
   (B) alkaline lamprophyre  (2) Alnoiite
   (C) ultramafic lamprophyre  (3) Camptonite

(D) lamproite (4) Minette

a) A-4 B-3 C-2 D-1 b) A-3 B-4 C-2 D-1

c) A-2 B-3 C-4 D-1 d) A-1 B-3 C-2 D-4

9. The line which demarcates the charnockitic and non-charnockitic regions in Peninsular India is described as

a) Eparchaen unconformity b) Oldham's line

c) Fermor's line d) Holland's demarcation zone

10. Continental alkaline rocks are distinguished from basaltic and tholeiitic rocks by

a) their high LILE and LREE contents
b) their high LILE and low LREE contents
c) their high LILE and LREE and low HREE contents
d) their low LILE and LREE and high HREE contents

11. The ***tholeiitic*** magma series is characteristic of

a) extensional stress regimes
b) compressional stress regimes
c) both compressional and extensional stress regimes
d) none of the above

12. The Agpaitic Index is denoted by

a) $\frac{Na_2O + K_2O}{Al_2O_3}$

b) $\frac{Na_2O}{Na_2O + K_2O}$

c) $\frac{Al_2O_3}{Na_2O + K_2O}$

d) $\frac{Fe_2O_3 + FeO}{Na_2O + K_2O}$

13. Match the following and choose the correct answer.

| | Igneous Complex | | Place |
|---|---|---|---|
| A) | Mount Girnar | 1) | Russia |
| B) | Mount Erebus | 2) | Canada |
| C) | Mount Johnson | 3) | India |
| D) | Kola Peninsula | 4) | Antarctica |

a) A-3 B-2 C-1 D-4
b) A-3 B-4 C-2 D-1
c) A-4 B-3 C-1 D-2
d) A-3 B-1 C-2 D-4

14. The A-type granites are characterised by

i) high alkalis and low CaO contents
ii) high FeO/MgO
iii) enrichment in halogens
iv) depletion in HREE/LILE

a) i & iv are correct
b) i, ii & iii are correct
c) i, ii & iv are correct
d) all the above are correct

15. Match the following and choose the correct answer.

| Type of Granitoids | Mode of Derivation |
|---|---|
| A) A-type granitoids | 1) by emplacement into non-orogenic settings |
| B) M-type granitoids | 2) from melting of subducted oceanic crust |
| C) S-type granitoids | 3) from igneous or meta-igneous rocks |
| D) I-type granitoids | 4) from sources which have undergone atleast one weathering cycle |

a) A-1 B-2 C-4 D-3
b) A-2 B-4 C-1 D-3
c) A-1 B-3 C-2 D-4
d) A-4 B-3 C-1 D-2

16. The A-type granitoids are characteristic of
a) tensional environments
b) compressional environments
c) either tensional or compressional environments
d) none of the above

17. Which of the following is NOT a characteristic feature of S-type granites?
a) High $K_2O/Na_2O$
b) High $Al_2O_3/Na_2O + K_2O + CaO$
c) Low Sr87/ Sr86
d) Presence of Hornblende in mineral assemblages

18. Which of the following can be classified as an 'alkaline rock'?
a) hornblende-granite b) hornblende-syenite
c) riebeckite-granite d) biotite-granite

19. Which of the following is true of "alkaline rocks"?
1) alkaline rocks may be silica-oversaturated, saturated or unsaturated
2) they normally contain alkali-rich mafics
3) normatively, these rocks contain nepheline and/or acmite
a) 1,2 and 3 are correct
b) 1 and 3 are correct
c) 2 and 3 are correct
d) none of the above are correct

20. The 'Miaskitic' rocks are characterised by
   a) $Na_2O + K_2O > Al_2O_3$
   b) $Na_2O + K_2O < Al_2O_3$
   c) $Na_2O + K_2O = Al_2O_3$
   d) none of the above
21. According to Shand's silica saturation principle " certain minerals cannot exist in the presence of free silica. If silica were to be present, these minerals combine with it to give rise to different minerals". Considering this statement, a mineral like Albite could be formed by the combination of silica with
   a) leucite b) nepheline
   c) olivine d) biotite
22. Which of the following pairs does NOT belong to the same class as per Shand's classification?
   a) pyroxenes — amphiboles
   b) quartz — tridymite
   c) granites— syenites
   d) syenite — nepheline-syenite
23. Which of the following is an example of a non-feldspathoidal undersaturated rock?
   a) nepheline syenite b) nepheline basalt
   c) leucite basanite d) olivine gabbro
24. In the series granite-granodiorite-syenite-diorite, from left to right the proportion of
   a) Quartz and K-feldspar increases
   b) K-feldspar decreases
   c) Sodic plagioclase increases
   d) Mafic minerals decreases
25. The rocks 'syenites' are characterized by the predominance of
   a) Quartz b) Nepheline
   c) Alkali feldspar d) K-feldspar
26. Syenites in which plagioclase becomes approximately equal in amount to the alkali feldspars are called
   a) Monzonites b) Shonkinites
   c) Pulaskites d) Nordmarkites
27. In which of the following rocks the K-feldspar content is NOT equal to that of plagioclase feldspars?
   a) Adamellite b) Monzonite
   c) Toscanite d) Dacite
28. Which of the following is a glassy modification of an acidic lava?
   a) Obsidian b) Tektite
   c) Tachylite d) Limburgite

29. The composition of the rock 'gabbro' can be approximately designated as
    a) Hornblende + Andesine
    b) Olivine + Calcic plagioclase
    c) Labradorite + Orthopyroxenes
    d) Labradorite + Clinopyroxenes
30. 'Norite' is a gabbroic rock in which
    a) OPX>CPX    b) OPX<CPX
    c) OPX=CPX    d) OPX is absent
31. A rock that consists of quartz, albite and aegirine in approximately equal amounts is described as
    a) Tonalite    b) Aplite
    c) Rockallite    d) Monzonite
32. Alkali syenite which is deficient in silica but with a little nepheline is described as
    a) Pulaskite    b) Nordmarkite
    c) Shonkinite    d) Toschenite
33. Which of the following are considered to be synonymous with hydrous diabases?
    a) Diorites    b) Dolerites
    c) Dacites    d) Syenites
34. Match the following and choose the correct answer.

| | |
|---|---|
| A. Monchiquite | 1. Mafic alkaline rock |
| B. Sovite | 2. Basalt |
| C. Ijolite | 3. Carbonatite |
| D. Picrite | 4. Lamprophyre |

    a) A-4 B-1 C-3 D-2
    b) A-2 B-4 C-3 D-1
    c) A-4 B-3 C-1 D-2
    d) A-2 B-3 C-4 D-1
*35. Which of the following does not belong to the same group?
    a) Sovite    b) Rauhaugite
    c) Alvikite    d) Jacupirangite
36. The lava flows of peridotitic composition are described as
    a) Phonolites    b) Tachylites
    c) Komatiites    d) Tholeiites
37. The Komatiites have a high concentration of
    a) Mg    b) Fe
    c) Mn    d) CaO
38. Who among the following was the first person to classify rocks based on geologic occurrence?
    a) F. Zirkel (1881)
    b) H. Rossenbusch (1887)

c) A. Harker (1895)
d) J. P. Iddings (1909)

39. According to the C.I.P.W. classification of rocks, the salic: femic ratio of 3 : 5 to 5 : 7 represents
a) Persalane rocks
b) Salfemane rocks
c) Dosalane rocks
d) Dofemane rocks

40. The 'Norm' or C.I.P.W. classification is essentially a
a) Mineralogical classification
b) Chemical classification
c) Genetic classification
d) Textural classification

41. According to Shand's classification, over saturated rocks are those which contain
a) saturated minerals only
b) saturated minerals + silicon dioxide in combination
c) saturated minerals + free silica in the form of quartz
d) none of the above

42. In the Shand's classification based on the percentage of the oxides of Al, K, Na and Ca, the peraluminian rocks are indicated as
a) $Al_2O_3 > Na_2O + K_2O + CaO$
b) $Al_2O_3 < Na_2O + K_2O + CaO$
c) $K_2O + Na_2O > Al_2O_3$
d) $Al_2O_3 < K_2O + Na_2O$

43. Hornblende, epidote, biotite-pyroxene pair and hornblende-olivine pair are classified under
a) per aluminian rocks
b) meta aluminian rocks
c) sub aluminian rocks
d) per alkaline rocks

44. Select the correct order of crystallization of the following rocks
a) sub-aluminian rocks—met aluminian rocks — per alkaline rocks
b) subaluminian rocks—peralkaline rocks——metaluminian rocks
c) peralkaline rocks——metaluminian rocks——subaluminian rocks
d) metaluminian rocks———peralkaline rocks———subaluminian rocks

45. If a rock consists dominantly of such minerals as aegirine, riebeckite, arfvedsonite, etc it is described as
a) Peralkaline
b) Peraluminian
c) Subaluminian
d) Metaluminian

46. ***Spessarite*** is a lamprophyre composed of
a) Common hornblende and orthoclase
b) Common hornblende and plagioclase
c) Biotite and orthoclase
d) Barkevikite and plagioclase

47. In which of the following lamprophyres the feldspars are absent

a) Camptonite b) Vogesite
c) Malchite d) Monchiquite

48. Magmatic rocks of calcareous composition known as ***Carbonatites*** do not occur as

a) large plutons of batholithic dimensions
b) steeply dipping cone sheets
c) ring complexes
d) plugs

49. Which of the following is/are the points of differences between ***tholeiites*** and alkali olivine basalts?

i) tholeiitic basalts are unsaturated while alkali olivine basalts are saturated
ii) tholeiites are lower in $Na_2O$, $K_2O$ and $TiO_2$ as compared to alkali olivine basalts
iii) tholeiitic basalts are lower than alkali basalts in elements characteristic of granitic rocks such as Ba, La, Sr, Pb, etc.
iv) alkali basalts are more oxidized than the tholeiites

a) (i) and (iii) only
b) (i), (ii) and (iii) only
c) (ii) and (iv) only
d) (ii), (iii) and (iv) only

50. Match the following and choose the correct answer.

(A) Websterite (1) Olivine + Orthopyroxene + Clinopyroxene
(B) Wehrlite (2) Olivine + Orthopyroxene
(C) Hazburgite (3) Olivine + Clinopyroxene
(D) Lherzolite (4) Orthopyroxene + Clinopyroxene

a) A-3 B-4 C-1 D-2
b) A-4 B-3 C-2 D-1
c) A-2 B-3 C-1 D-4
d) A-1 B-2 C-3 D-4

51. In a ***Poikilitic texture***, the host crystals or oikocrysts must have

a) high concentration and should crystallize early to attain a large size
b) high concentration but the crystallization can proceed simultaneously with other constituents of low concentration
c) crystallized after all other constituents but under conditions favouring the formation of large crystals
d) crystallized before all other constituents and being insoluble with them, a large size is attained

52. Porphyritic texture is frequently found in

a) Plutonic rocks
b) Volcanic rocks

c) Hypabyssal rocks
d) Volcanic and hypabyssal rocks
e) Plutonic and hypabyssal rocks

53. When large ovoids of K-feldspar are surrounded by rims of plagioclase and are together enclosed in a medium-grained groundmass, the resulting texture is described as
a) Ophitic texture b) Rapakivi texture
c) Hiatal texture d) Miarolitic structure

54. As compared to the average granites, the ***Rapakivi granites*** are enriched in
a) K-feldspar b) Plagioclase feldspar
c) Quartz d) all of the above

55. Secondary intergrowth of two different minerals of which one exhibits a vermicular habit is described as
a) Synantectic texture b) Sagenetic texture
c) Symplektite texture d) Symplectic texture

56. The ***Myrmekite*** texture is an intergrowth between
a) orthoclase and plagioclase
b) orthoclase and quartz
c) plagioclase and quartz
d) plagioclase and vermicular quartz

57. An intergrowth texture in which plagioclase contains inclusions of potash feldspar is known as
a) Perthite b) Anti-perthite
c) Graphic texture d) Seriate texture

58. Match the following and choose the correct answer

| Texture | Rocks |
|---|---|
| A. Runic texture | 1. Dolerites |
| B. Myrmekite texture | 2. Syenites |
| C. Ophitic texture | 3. Granites |
| D. Trachytoid texture | 4. Charnockites |

a) A-3 B-4 C-2 D-1
b) A-3 B-1 C-4 D-2
c) A-3 B-4 C-1 D-2
d) A-4 B-3 C-1 D-2

59. The ***Ophitic*** texture consists of
a) laths of augite enclosed in the crystals of plagioclase
b) laths of plagioclase enclosed in the crystals of augite
c) laths of orthoclase enclosed in the crystals of plagioclase
d) an intergrowth between plagioclase and orthoclase

60. In the ***Ophitic texture***
a) plagioclase is idiomorphic and pyroxene is allotriomorphic
b) pyroxene is idiomorphic and plagioclase is allotriomorphic

c) both plagioclase and pyroxene are idiomorphic
d) both plagioclase and pyroxene are allotriomorphic

61. The texture formed by the parallel arrangements of feldspars in some syenites is described as
a) Trachytic texture
b) Trachytoid texture
c) Trachy-Ophitic texture
d) Variolitic texture

62. Which of the following exhibits a distinctive texture called ***spinifex texture***?
a) Tholeiites
b) Picrites
c) Spilites
d) Komatiites

63. In a '***Seriate***' texture
a) the phenocrysts and the matrix are approximately equal in size
b) there is a partial gradation in size from phenocrysts to the matrix
c) there is a complete range in grain size between the groundmass and the phenocrysts
d) the ground mass is made up entirely of glassy matter

64. When the contents of microlites and glass becomes more or less equal, the texture is described as
a) Vitrophyric
b) Holovitrophyric
c) Hyalopilitic
d) Holohyaline

65. When a rock is composed of mineral fragments most of which are subhedral, the texture is referred as
a) Hypidiomorphic
b) Hypautomorphic
c) Granitic
d) any of the above
e) none of the above

66. The lamprophyres characteristically exhibit
a) Allotriomorphic texture
b) Panidiomorphic texture
c) Automorphic texture
d) Hypautomorphic texture

67. A zone of reaction products surrounding a mineral and which is produced by primary magmatic reactions is most appropriately designated as a
a) Reaction rim
b) Corona
c) Kelyphitic border
d) Cognate Xenolith

68. Elongated "streaked out" patches in igneous rocks are described as
a) Schillers
b) Schlieren
c) Marbles
d) Varioles

69. Existence of several primary magmas of mafic and ultramafic composition was suggested by
a) Buffen
b) Buddington
c) Daly and Brown
d) Turner

70. Which of the following are the most active bases present in igneous magmas?
a) K and Na
b) Na and Ca
c) Na, Ca and Ba
d) K and Ca

71. Which of the following volcanic rocks contain the highest percentage of silica?
a) Basalts
b) Andesites
c) Trachytes
d) Rhyolites

72. In the Bowen's reaction series, the crystallization of minerals from the magma follows the order of increasing
i) basicity
ii) acidity
iii) solubility
iv) insolubility
a) (i) only
b) (ii) only
c) (i) and (iii) only
d) (ii) and (iv) only

73. In a continuous reaction series, there is a continuous change in
a) composition only
b) crystal structure only
c) both composition and crystal structure
d) none of the above

74. Which of the following systems was designated by Bowen as "Petrogeny's Residua system"?
a) $Mg_2SiO_4$ —$NaAlSiO_4$ system
b) $Na.AlSiO_4$——$KAlSiO_4$—$SiO_2$ system
c) $CaMgSi_2O_6$ —— $NaAlSi_3O_8$ —— $CaA_{12}Si_2O_8$ system
d) $CaMgSi_2O_6$ ——$Mg_2SiO_4$ ——$SiO_2$ system

75. Arrange the following magmas in the increasing order of the grain size of their resultant rocks
a) multicomponent — bicomponent — unicomponent
b) unicomponent— bicomponent— multicomponent
c) bicomponent— unicomponent— muiticomponent
d) multicomponent— unicomponent — bicomponent

76. In the Continuous Reaction series the Al: Si ratio shows a
a) steady decrease
b) steady increase
c) more or less constant value
d) none of the above

77 The process by which a homogenous magma breaks up into unlike fractions that ultimately form rocks of different composition is described as

a) Crystal fractionation
b) Fractional crystallization
c) Gravity separation
d) Magmatic differentiation

78. The last residues of ***magmatic crystallization*** often react with pre-crystallized minerals form new minerals that are called
a) Syntectic minerals
b) Syngenetic minerals
c) Epigenetic minerals
d) Paulopost minerals

79. The condition in which different parts of a liquid magma chamber exhibit different temperature values is described as
a) Sorel effect
b) Sorret effect
c) Anatexis
d) Incongruent melting

80. Liquid immiscibility is said to have given rise to many deposits of
a) Sulphides
b) Silicates
c) Sulphates
d) Phosphates

81. The process of crystallization in a binary magma of eutectic proportions takes place in
a) three stages
b) two stages
c) a single stage
d) none of above

82. The crystallization of a bicomponent magma of mixed crystals ceases when
a) the magma is exhausted completely
b) the eutectic temperature is attained
c) the molecular concentration of the constituents becomes zero
d) the crystal attains the composition of the melt

83. If the proportions of $Na_2O$ and $K_2O$ in an igneous rock series equals that of CaO when the $SiO_2$ exceeds 61%, the series is described as
a) Alkalic
b) Alkali-Calcic
c) Calc-Alkalic
d) Calcic

83. Incongruent melting is a process whereby a solid transforms at a definite temperature into another solid and liquid. Select the correct statement about this process.
a) the resulting solid and liquid have a higher silica content than the original solid
b) the resulting solid and liquid have a lesser silica content than the original solid
c) the resulting solid is higher is silica while the liquid is lesser in silica
d) the resulting solid is lesser in silica while the liquid is higher in silica

85. Which of the following are concordant intrusive igneous plutons found in folded terrains?
a) Laccoliths
b) Lopoliths
c) Phacoliths
d) Bysmaliths

86. Which of the following is an example of a discordant igneous intrusion?
a) Laccolith
b) Lopolith
c) Bysmalith
d) Chonolith

87. Lopolith exhibits —————— shape
    a) Plano-convex
    b) Plano-concave
    c) Concavo-.convex
    d) any of the above
    e) none of the above
88. Which of the following igneous plutons have been emplaced by the Cauldron subsidence mechanism?
    a) Bysmalith b) Batholith
    c) Chonolith d) Phacolith
89. Which of the following statements is/are true of Batholiths?
    i) the size is more than 40 sq. miles
    ii) they have steeply plunging contacts and no visible floor
    iii) they usually run across the axes of the mountains
    iv) they have a relatively homogenous composition, which is usually granitic or granodioritic
    a) (i) and (iii) only
    b) (i), (ii) and (iii) only
    c) (i), (ii) and (iv) only
    d) all of the above
90. Volcanic necks are
    a) sedimentary masses which lie towards the top of volcanic vents
    b) igneous masses which lie towards the top of volcanic vents
    c) igneous masses which seal up the vents of ancient volcanoes
    d) thin masses of igneous rocks having numerous fissures
91. Lavas in which the gas cavities are very numerous and irregular in shape are called
    a) Scoria b) Pumice
    c) Amygdales d) none of the above
92. In blocky lavas, the cavities are
    a) small and regular in shape
    b) large and regular in shape
    c) small and irregular in shape
    d) large and irregular in shape
93. ***Ignimbrites*** are
    a) Volcanic bombs b) Welded tuffs
    c) Volcanic glasses d) Volcanic breccias
94. Agglomerates are pyroclastic rocks consisting mainly of fragments with a diameter of
    a) 2-4 mm b) 4-32 mm
    c) 32 mm-2 cm d) 2 cm
95. The pillow structure is characteristically exhibited by
    a) soda-rich olivine basalts known as spilites

b) low-K tholeiitic basalts
c) alkaline olivine basalts
d) soda-rich olivine-poor basalts known as spilites

96. Embryo crystals that have not yet organized a full crystal-line structure are called
a) Crystallites b) Globulites
c) Microlites d) Margarites

97. When magmas of different composition intrude through the same channel it gives rise to
a) Multiple intrusion b) Composite intrusion
c) Forced intrusion d) Assimilation

98. The zone of igneous activity and lava formation, situated between the lithosphere and the barysphere is called
a) Asthenosphere b) Mesosphere
c) Pyrosphere d) Tectonosphere

99. Small dome-like protuberances projecting from the main body of a large igneous intrusion are described as
a) Stocks b) Bosses
c) Schillers d) Cupolas

100. A body consisting of intimately and complexly mixed igneous and metamorphic rocks is known as a
a) Harpolith b) Chonolith
c) Sphenolith d) Stromatolith

101. Fragmental volcanic (pyroclastic) material is sometimes ejected as liquid lava and solidifies during flight. Which of the following is NOT formed in this way?
a) Bombs b) Pumice
c) Tuffs d) Scoriae

102. Anorthosites are
a) Mafic b) Felsic
c) Mafelsic d) Domafic

103. The plagioclase feldspar that is characteristic of anorthosites is
a) Albite b) Oligoclase
c) Labradorite d) Bytownite

104. Monominerallic rocks are generally formed in the
a) Shallow zones of the earth
b) Intermediate zones of the earth
c) Deeper zones of the earth
d) Any of the above

105. Quartz is a diagnostic mineral in
a) Oversaturated rocks b) Saturated rocks
c) Unsaturated rocks d) Undersaturated rocks

106. Which of the following eutectic proportions (as determined by Vogt) is incorrect?
   a) Orthoclase: Quartz : : 27.5 : 72.5
   b) Orthoclase: Albite : : 42.0 : 58.0
   c) Anorthite : Olivine : : 70.0 : 30.5
   d) Enstatite: Diopside : : 55.0 : 45.0

107. An overall mineralogical and chemical gradation from the bottom to the top of an intrusion is described as
   a) Colloform banding
   b) Cryptic layering
   c) Phase layering
   d) Rhythmic layering

108. Which of the following types of layering in gabbroic rocks resemble the grading that is found in the sedimentary rocks?
   a) Cryptic layering
   b) Mineral layering
   c) Phase layering
   d) Rhythmic layering

109. In which of the following rocks quartz is present as an accessory mineral?
   a) Adamellite
   b) Granite
   c) Granodiorite
   d) Diorite

110. In the rocks consisting of plagioclase and pyroxene, the pyroxene ratio (P.R.) is calculated by the formula
   a) $P.R. = \frac{Pyx\ X\ 100}{Pyx - Plag}$
   b) $P.R. = \frac{Pyx\ X\ 100}{Pyx + Plag}$
   c) $P.R. = \frac{Pyx + Plag}{Pyx + 100}$
   d) $P.R. = \frac{Pyx + Plag}{Pyx\ X\ 100}$

111. The most abundant substance that escapes when a magma crystallizes is
   a) $CO_2$
   b) Water vapour
   c) $H_2S$
   d) $O_2 + N_2$

112. Choose the correct equivalents from the following.

| | *Volcanic* | | *Plutonic* |
|---|---|---|---|
| A. | Toscanite | I. | Nepheline syenite |
| B. | Latite | 2. | Granite |
| C. | Phonolite | 3. | Adamellite |
| D. | Liparite | 4. | Monzonite |

   a) A-4 B-3 C-2 D-1
   b) A-1 B-4 C-2 D-3
   c) A-3 B-4 C-1 D-2
   d) A-2 B-3 C-1 D-4

113. Andesites on an average are intermediate in composition between
   a) Syenites and Gabbros
   b) Basalts and Rhyolites
   c) Rhyolites and Trachytes
   d) Anorthosites and Ijolites

114. Andesites are susceptible to alteration known as
   a) Sericitization    b) Kaolinization
   c) Pyritization    d) Propylitization

115. Luxullianites are the rocks that are formed due to
   a) Martitization    b) Tourmalinization
   c) Albitization    d) Sericitization

116. ***'Greisen'*** is a pneumatolytic modification of granitic rocks. It is composed essentially of
   a) Quartz + White mica
   b) Quartz + Sericite
   c) Sericite+White mica
   d) none of the above

117. Match the following and choose the correct answer

| | |
|---|---|
| A) The Sithumpudi complex | 1. Charnockites |
| B) Shivamalai hills | 2. Anorthosites |
| C) Mor Valley | 3. Lamprophyres (Minette) |
| D) Chelima, Kurnool (A.P.) | 4. Alkaline rocks |

   a) A-2 B-1 C-4 D-3
   b) A-2 B-4 C-1 D-3
   c) A-2 B-4 C-3 D-1
   d) A-1 B-2 C-3 D-4

118. Which of the following groups of rocks are most abundant on the moon?
   a) Granites and granodiorites
   b) basalts and andesites
   c) Peridotites and picrites
   d) Anorthosites and gabbros

119. Which of the following pairs is incorrect?
   a) Banatite _______ quartz monzonite
   b) Ankaramite _______ pyroxene-rich olivine-basalt
   c) Troctolite _______ hypersthene gabbro
   d) Keratophyre _______ trachytic spilite

120. Which of the following rocks is least silicic in composition?
   a) Adamellite    b) Phonolite
   c) Kersanite    d) Alnoite

121. Which of the following is an orbicular diorite?
   a) Corsite    b) Andesite
   c) Malchite    d) Plumasite

122. Which of the following are the evidences in favour of the magmatic origin of granites?
i) perfect gradation between the country rocks and the granites
ii) chilled margins of the granite
iii) presence of xenoliths of earlier country rocks in the granites
iv) granophyric or graphic texture
v) relict bedding in granitic rocks
a) (i), (ii) and (v) only
b) (i), (ii), (iii) and (iv) only
c) (ii), (iii) and (iv) only
d) (i), (ii) and (iv) only

123. Rock forming minerals that require a high concentration of volatiles for their formation are described as
a) Pyrogenetic minerals
b) Hydrothermal minerals
c) Hydatogenetic minerals
d) Released minerals

124. The oceanic tholeiites are characterised by
a) Na/K > 10
b) Na/K < 10
c) Na/K > 3 but less than 6
d) Na/K ranges from 1 to 10

125. Among the different kinds of ***flood basalts***, the $SiO_2$ content is highest in
a) continental tholeiites
b) oceanic tholeiites
c) alkaline olivine basalts
d) spilites

126. ***Keratophyres*** are intermediate and acid lavas and minor intrusions associated with
a) continental tholeiites
b) oceanic tholeiites
c) alkaline olivine basalts
d) spilites

127. If the mineralogical composition of a rock consists of such minerals as chrysotile, antigorite, lizardite, bastite, brucite, etc, then it is most likely to be a
a) peridotite
b) serpentinite
c) keratophyre
d) komatiite

128. The most dominant mineral in Anorthosites is
a) orthoclase
b) plagioclase
c) pyroxene
d) amphibole

129. The rock ***'mangerite'*** is a
a) quartz syenite
b) pyroxene syenite
c) hypersthene granite
d) ferro-augite grano-diorite

130. Which of the following is a carbonatite with dolomitic composition?
a) rauhaugite
b) sovite
c) essexite
d) marble

131. The mafic alkaline rocks such as ijolites, jacupirangites, essexites, etc exhibit chemical characteristics which closely resemble those of the

a) continental tholeiites b) oceanic tholeiites
c) alkaline olivine basalts d) spilites

132. Which of the following rocks are most likely to be associated with carbonatites?
a) ijolites b) komatiites
c) anorthosites d) nepheline syenites

133. Which of the following is a non-feldspar bearing lamprophyre ?
a) Minette b) Vogesite
c) Kersantite d) Monchiquite

134. According to Rosenbusch's classification of lamprophyres, ***Vogesite*** is a lamprophyre that consists of
a) biotite and orthoclase b) biotite and plagioclase
c) hornblende and orthoclase d) hornblende and plagioclase

135. Match the following and choose the correct answer.

| **Lamprophyre occurrence** | **Type of Lamprophyre** |
|---|---|
| A) Chelima, Kurnool, A.P. | 1) Minette |
| B) Deccan Traps | 2) Camptonite |
| C) Mt. Girnar | 3) Monchiquite |
| D) Salem, Tamil Nadu | 4) Olivine Vogesite |

a) A-1 B-2 C-3 D-4
b) A-1 B-3 C-2 D-4
c) A-4 B-2 C-3 D-1
d) A-4 B-3 C-2 D-1

136. Most geological and geophysical evidences suggest that ***Ophiolite*** complexes represent
a) young orogenic activity
b) remains of stable Archaen plutons
c) fragments of older continental crust
d) fragments of older oceanic crust

137. In the ***Ophiolite Complexes***, a ***"fossil Moho"*** is represented by the contact between
a) pillow lavas and the sheeted dyke complex
b) low-K tholeiitic gabbros and the sheeted dyke complex
c) low-K tholeiitic gabbros and the ultramafic rocks
d) pillow lavas and the overlying graywackes

138. Which of the following rocks does NOT belong to the Nephelinite family of the alkaline rocks?
a) Nephelinite b) Melteigite
c) Ijolite d) Nepheline diorite

139. ***Latite*** is the effusive equivalent of
a) Monzonite b) Nepheline syenite
c) Gabbro d) Diorite

140. ***Alaskites*** are the granites containing
   a) quartz only
   b) quartz and alkali feldspar
   c) quartz, alkali feldspar and plagioclase feldspar
   d) none of the above
141. Which of the following syenites exhibit radioactivity?
   a) Shonkinite b) Teschenite
   c) Pullaskite d) Monzonite
142. The Rapakivi texture is essentially a
   a) Poikilitic texture b) Porphyritic texture
   c) Intergrowth texture d) Volcanic texture
143. The rock ***tephrite*** is an example of
   a) ultramafic rocks b) granitic rock
   c) dyke rock d) basalt
144. Which of the following is a mono-minerallic rock consisting almost exclusively of olivine?
   a) Hazburgite b) Dunite
   c) Wehrlite d) Picrite
145. Which of the following rocks does NOT contain olivine?
   a) Hazburgite b) Dunite
   c) Tephrite d) Picrite
146. Which of the following is NOT a volcanic rock?
   a) Liparite b) Toscanite
   c) Dacite d) Lherzolite
147. The mode of origin of granites by partial melting during high-grade regional metamorphism is described as
   a) Anatexis b) Wet granitization
   c) Stoping d) Magmatic differentiation
148. Consider the following reaction : $3Mg_2SiO_4 + SiO_2 + H_2O \leftrightarrow 2Mg_3Si_2O_5(OH)_4$
   This represents the process of
   a) Tourmalinization b) Serpentinization
   c) Martitization d) Sericitization
149. The rocks ***Serpentinites*** are most commonly associated with
   a) Keratophyres b) Spilites
   c) Peridotites d) Gabbros
150. The type of layering found in Gabbroic layered intrusions which involves sudden appearance or disappearance of a particular mineral in the rock sequence is described as
   a) Orbicular banding b) Rhythmic layering
   c) Phase layering d) Cryptic layering
151. Which of the following represents the ***Steinmann Trinity*** ?
   a) Pillow basalts, serpentinites and bedded radiolarian cherts

b) Tholeiitic basalts, serpentinites and bedded radiolarian cherts
c) Peridotites, layered gabbros and graywackes
d) Low-K tholeiites, sheeted diabase dykes and pelagic limestones

152. In terms of chemical composition the lamprophyres are closest to
a) tholeiitic basalts
b) spilites
c) alkali olivine basalts
d) anorthosites

153. Which of the following is an abundant pyroxene in the lamprophyres associated with nepheline syenites ?
a) hypersthene
b) enstatite
c) augite
d) titan-augite

154. The degree of silica saturation depends upon the ratio of
a) $SiO_2/Na_2O + K_2O$
b) $SiO_2/Na_2O$
c) $SiO_2/K_2O$
d) $Na_2O/K_2O$

155. If the ratio of $SiO_2/Na_2O$ in a magma is less than 2, the resulting rocks would be
a) Oversaturated
b) Undersaturated
c) Saturated
d) Unsaturated

156. Which of the following is a silica-unsaturated mineral?
a) tourmaline
b) hypersthene
c) sphene
d) corundum

157. Select the most suitable expression among the following.
a) the silica-oversaturated rocks have quartz in the norm
b) the silica-oversaturated rocks have quartz and hypersthene in the norm
c) the silica-oversaturated rocks have quartz, hypersthene and acmite in the norm
d) the silica-oversaturated rocks have olivine in the norm

158. The variation diagram in which the weight percentage of SiO2 is plotted on the x-axis and the concentrations of other oxides on the y-axis is described as
a) Cartesian diagram
b) Harker diagram
c) AFM diagram
d) QAP diagram

159. Rocks that are dominantly composed of quartz can be classified on the basis of
a) ACF diagrams
b) AKF diagrams
c) AFM diagrams
d) QAP diagrams

160. In a Cartesian diagram
a) the wt. Percentages of FeO and $Na_2O$ are plotted against the wt.% of $SiO_2$
b) the wt. Percentages of FeO and $K_2O$ are plotted against the wt.% of $SiO_2$
c) the wt. Percentages of FeO, $K_2O$ and $Na_2O$ are plotted against the wt.% of $SiO_2$

d) the wt. % of $Fe_2O_3$ is plotted against the wt.% of FeO

161. The ***Felsic Index*** is denoted as

a) $Na_2O + K_2O / (Na_2O + K_2O + CaO)$
b) $Na_2O + K_2O + CaO / Na_2O + K_2O$
c) $SiO_2 + Na_2O + K_2O / (Na_2O + K_2O + CaO)$
d) $SiO_2 + Na_2O + K_2O + CaO / (Na_2O + K_2O)$

162. The sum of the normative An + Mg.diopside + Forsterite + Enstatite is described as

a) the differentiation index
b) the solidification index
c) the crystallization index
d) the agpaitic index

163. The ***Differentiation Index*** is the sum of the normative compositions of

a) quartz and orthoclase
b) quartz, orthoclase and albite
c) quartz, orthoclase, albite and nepheline
d) quartz, orthoclase, albite, nepheline, leucite and kalsilite

164. The Solidification Index proposed by the Japanese petrologist H.Kuno is denoted as

a) $MgO + FeO + Fe_2O_3 + Na_2O + K_2O / MgO + FeO + Fe_2O_3$
b) $MgO + FeO + Fe_2O_3 / MgO + FeO$
c) $100\ MgO / MgO\ FeO + Fe_2O_3 + Na_2O + K_2O$
d) $MgO + FeO + Fe_2O_3 + Na_2O + K_2O / 100\ MgO$

165. Subalkaline rocks may have

a) orthopyroxene in the norm
b) orthopyroxene and quartz in the norm
c) orthopyroxene, quartz and feldspthoids in the norm
d) feldspthoids in the norm

166. If acmite appears in the norm, then the rock is most likely to be

a) alkaline
b) peralkaline
c) subalkaline
d) non-alkaline

167. ***Pitchstone*** is a

a) massive layered glass with a wáxy or resinous luster
b) pearly gray glass with perlitic fabric
c) glass with highly vesicular pumiceous fabric
d) vitrophyric glass

168. Which of the following is/are NOT characteristic of the ***S-type*** granitic rocks?

a) their initial Sr87/Sr86 ratios are >0.710
b) their normative compositions contain corumdum
c) they contain hornblende in the mode
d) they occur in catazonal migmatitic complexes

169. Tin deposits are commonly found associated with

a) A-type granitoids
b) M-type granitoids
c) S-type granitoids
d) I-type granitoids

170. Wyomingite, orendite, madupite are examples of highly potassic peralkaline rocks which are characterized by
  a) K2O/Na2O >3 b) K2O/Na2O <3
  c) K2O/Na2O >6 d) K2O/Na2O <6
171. Given below is a list of unusual rock types. Match these rocks with their dominant mineral composition and choose the correct answer.

| **Rock** | **Dominant minerals** |
|---|---|
| 1) Wyomingite | A) leucite, sanidine |
| 2) Madupite | B) diopside, leucite, phlogopite |
| 3) Orendite | C) leucite, diopside, apatite |
| 4) Katungite | D) olivine, melilite and K-rich glass |

  a) 1-D 2-C 3-B 4-A
  b) 1-C 2-B 3-A 4-D
  c) 1-B 2-D 3-A 4-C
  d) 1-A 2-D 3-C 4-B
172. Which of the following rock types has the lowest percentage of silica?
  a) Rhyolite b) Trachyte
  c) Phonolite d) Nepheline Syenite
173. In which of the following rocks, the mineral diopside does Not generally appear in the norm?
  a) granites b) syenites
  c) nepheline syenites d) diorites
174. Which of the following rocks does NOT contain nepheline in the norm?
  a) phonolite b) trachybasalt
  c) basanite d) syenite
175. The chemical formula $Na_2O.Fe_2O_3.4SiO_2$ pertains to the normative mineral——
  a) diopside b) sphene
  c) acmite d) nepheline

## IGNEOUS PETROLOGY (KEY)

| | | | | | | | | | |
|---|---|---|---|---|---|---|---|---|---|
| 1-c | 2-a | 3-d | *4-b | 5-c | 6-d | 7-c | 8-a | 9-c | 10-c |
| 11-a | 12-a | 13-b | 14-b | 15-a | 16-a | 17-d | 18-c | 19-a | 20-b |
| 21-b | 22-d | 23-d | 24-b | 25-c | 26-a | 27-d | 28-a | 29-d | 30-a |
| 31-c | 32-a | 33-b | 34-a | *35-d | 36-c | 37-a | 38-b | 39-d | 40-c |
| 41-c | 42-a | 43-b | 44-a | 45-a | 46-b | 47-d | 48-a | 49-d | 50-b |
| 51-c | 52-d | 53-b | 54-a | 55-d | 56-d | 57-b | 58-c | 59-b | 60-c |
| 61-b | 62-d | 63-c | 64-c | 65-d | 66-b | 67-b | 68-b | 69-b | 70-a |
| 71-d | 72-c | 73-a | 74-b | 75-a | 76-a | 77-d | 78-d | 79-b | 80-c |
| 81-a | 82-d | 83-d | 84-d | 85-c | 86-d | 87-c | 88-a | 89-c | 90-c |
| 91-a | 92-d | 93-b | 94-b | 95-d | 96-a | 97-b | 98-c | 99-d | 100-d |
| 101-c | 102-b | 103-c | 104-c | 105-a | 106-a | 107-b | 108-d | 109-d | 110-b |

| | | | | | | | | | |
|---|---|---|---|---|---|---|---|---|---|
| 111-b | 112-c | 113-b | 114-d | 115-c | 116-a | 117-b | 118-d | 119-c | 120-d |
| 121-a | 122-c | 123-c | *124-a | 125-a | 126-d | 127-b | 128-b | 129-b | 130-a |
| 131-c | 132-a | 133-d | 134-c | 135-a | 136-d | 137-c | *138-d | 139-a | 140-b |
| 141-d | 142-b | 143-d | 144-b | 145-c | *146-d | 147-a | 148-b | 149-c | 150-c |
| 151-a | *152-c | 153-d | 154-b | 155-b | 156-d | 157-b | 158-b | 159-b | 160-a |
| 161-a | 162-c | 163-d | 164-c | 165-b | 166-b | 167-a | 168-d | 169-c | 170-a |
| 171-b | 172-d | 173-a | 174-d | | | | | | |

## EXPLANATORY ANSWERS

*4-b : In the oceanic islands, the magmas change from tholeiitic to alkaline when the lithosphere is thickening; whereas, in continental widening rift systems, the magmas change from alkaline to tholeiitic when the lithosphere thins down.

*35-d : Jacupirangite is a "magnetite pyroxenite", which is a mafic alkaline rock; while all the others are carbonatites.

*124-a : It has been noticed by many workers that Na/K for oceanic tholeiites is greater than 10, whereas for those on the continents the ratio is generally greater than 3 but less than 6.

*138-d : Nepheline diorite belongs to the Alkali basalt family while the other three belong to the Nephelinite family as per the classification of alkaline rocks by Currie (1976).

*146-d : Lherzolite is ulramafic rock composed of olivine, orthopyroxene, clinopyroxene, garnet or sphene in decreasing order of abundance.

*152-c : Chemically, the lamprophyres are characteristically low in $SiO_2$ and high in $Na_2O$, $K_2O$, $TiO_2$, BaO, SrO, and $P_2O_5$, as are the alkali olivine basalts.

# 12

# *Sedimentary Petrology*

1. The smallest megascopic layer in a sedimentary sequence is a
   a) Bed b) Stratum
   c) Lamina d) All of the above
2. To be classified as 'laminae', the thickness of each layer should be
   a) >1 cm b) <1 cm
   c) >2cm d) <2cm
3. A bedset consists of two or more superimposed beds having identical
   a) Colour
   b) Composition
   c) Texture
   d) Internal structure
   e) Genesis
4. Which of the following statements is NOT correct?
   1) Variation in composition and texture is a characteristic feature of laminae.
   2) Internally the laminae are not megascopically layered.
   3) A lamina has a smaller (or the same) areal extent as enclosing bed.
   4) A lamina forms in a longer period of time than the enclosing bed.
   a) 1 only
   b) 1 and 4 only
   c) 1, 3 and 4 only
   d) all of the above are correct
5. The 'Flysch' sequences are characterised by
   a) Discontinuous, unequal beds, laterally variable in thickness
   b) Continuous equal or sub-equal beds, laterally uniform in thickness
   c) Continuous unequal beds, laterally uniform in thickness
   d) Discontinuous, equal or sub-equal beds, laterally variable in thickness
6. Which of the following is an example of bedding plane markings or irregularities found on the base of the bed?
   a) Load casts b) Ripple marks
   c) Rain prints d) Ball and Pillow structure
7. 'Parting Lineation' is found
   a) on the top of beds b) on the base of beds
   c) within the beds d) any of the above
8. Alternate deposition of fine and coarse-grained sediment gives rise to

a) Torrential bedding
b) Cross bedding
c) Graded bedding
d) Festoon bedding

9. Flaser bedding is
   a) a ripple bedding in which mud streaks alternate with sand and silt surfaces
   b) a ripple bedding which exhibits well-preserved sand lenses embedded within the muddy layers
   c) a discordant bedding exhibiting lateral gradation
   d) a concordant bedding exhibiting progressive fining in the upward direction
10. The most common mode of origin for cross-bedding is
    a) Migration of small and mega-ripples
    b) Deposition on the point bars of small meanders
    c) Deposition on the inclined surfaces of beaches
    d) Lee-side deposition of sand dunes
11. The bundle-wise up building of foreset laminae in a single unit is characteristically seen in
    a) Longitudinal cross-bedding
    b) Channel-fill cross-bedding
    c) Wave-ripple bedding
    d) Climbing-ripple lamination
12. Aeolian ripples are characterized by the presence of
    a) Coarser grains on the crests
    b) Finer grains on the crests
    c) Uniformly even grains on the crests and the troughs
    d) None of the above
13. The horizontal angle between the meridian and the horizontal projection of the dip line of the foreset in a cross-stratified layer is described as
    a) Scale
    b) Inclination
    c) Dip angle
    d) Azimuth
14. Ripples formed by water and wind differ in their
    a) Symmetry
    b) Scale
    c) Azimuth
    d) Ripple index
15. Ripple index (R.I.) is the wavelength of the ripple divided by its amplitude. If the R.I. > 15 it may be presumed that the ripples are formed by the action of
    a) Water
    b) Wind
    c) Waves
    d) Currents
16. Horizontal or lateral gradation of material occurs if the deposition takes place in
    a) Stagnant water
    b) Running water
    c) Underground water
    d) Shore zone
17. In flaser bedding the conditions are more favourable for the deposition

and preservation of

a) Sand
b) Mud
c) Both sand and mud
e) None of the above

18. The inter-tidal flat environments are characterised by the presence of
a) Herringbone cross-bedding
b) Flaser bedding
c) Longitudinal cross-bedding
d) Channel-fill cross-bedding

19. Graywackes typically exhibit
a) Graded bedding
b) Torrential bedding
c) Flaser bedding
d) Herringbone cross-bedding

20. The point of maximum elevation on the vertical profile through a ripple is described as
a) Summit point
b) Brink point
c) Toe point
d) Ripple height

21. In the current ripples
a) the crests are pointed but troughs are rounded
b) the crests are rounded but troughs are pointed
c) both the crests and troughs are rounded
d) both the crests and troughs are pointed

22. The 'Linguoid ripples' face
a) the direction of the current flow
b) the direction opposite to that of the current flow
c) the direction perpendicular to that of the current flow
d) none of the above

23. Textures in which the fragmental characteristics are NOT clearly visible are described as
a) Epiclastic
b) Clastic
c) Non-clastic
d) Pyroclastic

24. The degree of roundness does NOT depend upon
a) Size of the particles
b) Hardness of the particles
c) Specific gravity of the particles
d) Degree of transportation

25. Theoretically speaking sphericity and roundness of a grain are

i) Geometrically different and independent properties
ii) Gèometrically similar and mutually dependent properties
iii) Related to the size of the clastic particles
iv) Unrelated to the size of the elastic particles

a) (i) and (iii) are correct
b) (i) and (iv) are correct
c) (ii) and (iii) are correct
d) (ii) and (iv) are correct

26. In nature roundness varies between
a) 0.01 to 0.05
b) 0.05 to 0.1
c) 0.1 to 0.9
d) 1.0 to infinity

27. SORTING may be defined as
a) A measure of the spread of grain-size distribution
b) A qualitative parameter measuring the departure of a body from equidimensional character
c) A measure of the sharpness of the edges and corners of a particle
d) None of the above

28. Residual deposits are
a) Well sorted and rounded
b) Unsorted and angular
c) Ill sorted and sub-rounded
d) Assorted and well-rounded

*29. Sorting of particles is best in
a) Fluvial sediments
b) Sand dunes
c) Beach sediments
d) Loess sediments

30. Which of the following is a measure of the symmetry of the distribution of grains?
a) Sphericity
b) Roundness
c) Kurtosis
d) Skewness

31. If a grain-size distribution has excess coarse material, the sediment is said to possess
a) Positive skewness
b) Negative skewness
c) Zero skewness
d) None of the above

*32. Which of the following exhibits negative skewness?

a) Aeolian sands
b) River sands
c) Loess
d) Beach sands

33. Which of the following frequency curves shows the best sorting?

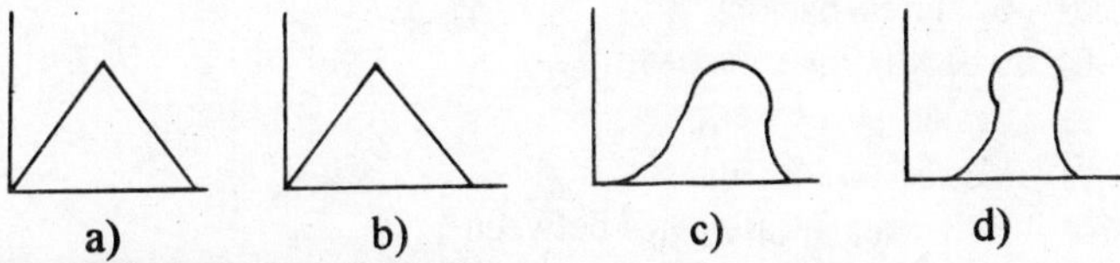


34. The Wentworth's scale is a/an
a) Arithmetic grade scale
b) Geometric grade scale
c) Logarithmic grade scale
d) Arbitrary grade scale

35. According to Wentworth's scale, particles having a diameter between 64 mm-4 mm are described as
a) Granules
b) Pebbles
c) Cobbles
d) Coarse sands

36. Cement in a detrital sedimentary rock is
a) Pre-depositional
b) Post-depositional
c) Syn-depositional
d) Any of the above
e) All of the above

37. Which of the following consists only of clasts and cement?
a) Mudstone
b) Graywacke
c) Orthoquartzite
d) Arkose

38. A conglomerate consists of detritals, matrix and cement. These are respectively
a) Authigenic, allogenic, allogenic
b) Allogenic, authigenic, allogenic
c) Allogenic, allogenic, authigenic
d) Authigenic, allogenic, authigenic

39. As compared with sandstones, shales have
a) Higher porosity and permeability
b) Lower porosity and permeability
c) Higher porosity and lower permeability
d) Lower porosity and higher permeability

40. The most compact or "tightest" packing in a sedimentary rock having minimum porosity is
   a) Cubic packing
   b) Conical packing
   c) Rhombohedral packing
   d) Spherical packing
41. Cross bedding is a ——property.
   a) Scalar
   b) Vector
   c) Either scalar or vector
   d) None of the above
42. Which of the following sedimentary properties is NOT scalar?
   a) Sphericity
   b) Roundness
   c) Sorting
   d) Striations
43. Match the following and choose the correct answer

| | |
|---|---|
| (1) Flaser bedding | (A) |
| (2) Herringbone bedding | (B) |
| (3) Climbing ripple lamination | (C) |
| (4) Convolute bedding | (D) |

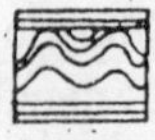

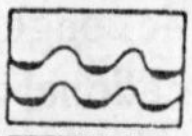

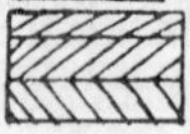

   a) 1—B 2-D 3-A 4-C
   b) 1-A 2-B 3-C 4-D
   c) 1-C 2-D 3-B 4-D
   d) 1-B 2-D 3-C 4-A
44. Match the following and choose the correct answer

| *Cause* | *Sedimentary structure* |
|---|---|
| A. Dessication | 1. Nodules |
| B. Diffusion | 2. Stylolites |
| C. Pressure solution | 3. Colour banding |
| D. Replacement | 4. Mud cracks |

   a) A-1 B-2 C-3 D-4
   b) A-4 B-1 C-2 D-3
   c) A-4 B-3 C-1 D-2
   d) A-4 B-3 C-2 D-1
45. According to the geochemical calculations and measurements of stratigraphic sections, the relative proportion of which of the following sediments is highest in the Earth's crust?
   a) Shales
   b) Limestones

c) Sandstones
d) Conglomerates

46. Arkose is derived from the disintegration of
a) Gabbro
b) Granite
c) Ultrabasic rocks
d) Marble

47. Oligomictic conglomerates are indicative of
a) Tectonically stable environments
b) Geosynclinal environments
c) Tectonically unstable environments
d) Fluctuating environments

48. Which of the following statements is NOT true about Orthoconglomerates?
a) They are formed as a result of long and multicyclic sedimentation
b) The matrix is less than 15%
c) The detritals are sub-rounded to angular
d) These are mature conglomerates

49. Which of the following is correctly matched?
a) Oligomictic conglomerates- -Heterogenous
b) Polymictic conglomerates—Homogenous
c) Intraf6rmational conglomerates—Heterogenous
d) Agglomerates — Heterogeneous

50. Which of the following are characteristic of major unconfirmities?
a) Intraformational conglomerates
b) Extraformational conglomerates
c) Polymictic conglomerates
d) Agglomerates

51. Orthoquartzites are indicative of
a) Stable environments
b) Unstable environments
c) Transitional environments
d) Any of the above
e) None of the above

52. Coarse sandstone with angular grains may be described as
a) Gannister
b) Grit
c) Arkose
d) Quartzite

53. Which of the following are characterisically associated with transitional environments?
a) Orthoconglomerates
b) Orthoquartzites

c) Arkoses
d) Graywackes

*54. Graywackes can be distinguished from Arkoses by
a) the difference in their chemical composition
b) the greater silica content
c) the percentage of matrix and cement
d) the ratio of rock fragments to feldspars

55. ONYX is a
a) Massive $CaCO_3$ deposit
b) Banded $CaCO_8$ deposit
c) Granular $CaCO_3$ deposit
d) Coralloid limestone

56. A coarse-grained biogenic limestone is termed as
a) Biomicrite
b) Biosparite
c) Oomicrite
d) Oosparite

*57. Limestones were deposited mainly in the form of micrites during
a) Precambrian
b) Cambrian
c) Ordovician
d) Tertiary

58. A limestone composed entirely of organic detritus is termed as
a) Coquina
b) Encrinite
c) Spergenite
d) Oolite

59. According to Folk's classification of limestones which of the following does NOT belong to the 'Allochems'?
a) Detritals
b) Pellets
c) Oolites
d) Micrites
e) Fossils

60. Which of the following statements about the limestones is/are incorrect?
a) they are good indicators of paleo-continental distribution
b) they are generally confined to the polar regions
c) their crystallization is favoured by high temperatures
d) all or the above
e) none of the above

61. The term 'Pelitolite' is used to describe a/an
a) Clastic sedimentary rock
b) Non-clastic sedimentary rock

c) Arenaceous rock
d) Argillaceous rock
e) Rudaceous rock

62. Which of the following are non-plastic clays?
a) China clays
b) Ball clays
c) Fire clays
d) Pottery clays

*63 The approximate current velocity of the movement of clays in shallow water is
a) 1.62 m/sec
b) 0.70 m/sec
c) 0.28 m/sec
d) 0.08 m/sec

64. Marl is a clay rock which contains
a) $CaCO_3$ only
b) $MgCO_3$ only
c) $CaCO_8$+$MgCO_3$
d) CaCO3 + MgCO3 +FeCO3

65. The process of submarine weathering by which iron is concentrated in the sediments is called
a) Palingenesis
b) Spheroidal weathering
c) Diagenesis
d) Halmyrolysis

66. In natural solutions iron is present commonly as
a) Sulphides
b) Silicates
c) Carbonates
d) Bicarbonates

67. Which of the following statements is Incorrect?
a) Bog iron ore is soft porous material accumulated on the surface of the oceans and seas
b) Iron pan is a ferruginous layer formed in tropical countries
c) $CaCO_8$ deposits due to capillary precipitation are known as cornstones
d) Iron deposited as silicate is generally of marine origin

68. Caliche is
a) Sodium nitrate
b) Sodium phosphate
c) Calcium carbonate
d) Potassium chloride

69. Which of the following is a phosphate deposit or direct organic origin?

a) Apatite
b) Collophanite
c) Guano
d) Dahlite

70. Glauconite is
a) Potassium-iron silicate
b) Na-Fe Silicate
c) K-Fe carbonate
d) Na-Fe carbonate

71. The size of the oolites commonly ranges between
a) 0 to 0.25mm
b) 0.25 to 0.50 mm
c) 0.50 to 1.0 mm
d) 1.0 to 2.0mm

72. Deformed oolites are described as
a) Pisolites
b) Spherulites
c) Spastolites
d) Stromatolites

73. The most predominant salts dissolved in sea water are
a) $CaCO_3$ and $MgCl_3$
b) $CaCO_3$ and NaCl
c) NaCl and $MgCl_3$
d) $NaNO_3$ and NaCl

74. The deposits of terrestrial salt lakes are differentiated from those of the marine salt lakes by
a) the presence of sodium carbonate
b) the absence of sodium carbonate
c) the presence of magnesium chloride
d) the absence of magnesium chloride

75. Which of the following features are the most useful in the analysis of sedimentary environments?
a) Primary structures
b) Sedimentary textures
c) Bioturbate structures
d) Colour of the sediments

76. The mineral 'gypsum' is indicative of
a) Normal saline conditions
b) Hypersaline conditions
c) Hyposaline conditions
d) Any of the above
e) All of the above

77. Which of the following sedimentary environments is characterized by

the abundance of evaporites?

a) Tidal flats
b) Intracratonic
c) Sabkha
d) Reef complex

78. Which of the following does NOT belong to the Shore zone environment?

a) Delta
b) Lagoon
c) Estuary
d) Tidal flats
e) None of the above

79. High-sinuosity multi-channel flows are typical of

a) Straight river channels
b) Meandering river channels
c) Braided rivers channels
d) Anastomosing river channels

80. Which of the following is a flood basin deposit?

a) Point bar deposits
b) Lag deposits
c) Crevasse splay deposit
d) Marsh deposits

81. In a lake basin, sandstones and conglomerates are deposited

a) at the centre
b) towards the margins
c) throughout the concentric basin
d) none of the above

82. The 'Paludal environment' is typical of

a) Swamps
b) Marshes
c) Tidal flats
d) Lagoons

83. The main part of the tidal flat deposit is formed in the

a) Foreshore
b) Backshore
c) Shore face
d) Swashzone

84. Which of the following types of bedding is characterized by reversed foreset laminae in one coset?

a) Tidal bedding
b) Flaser bedding
c) Herringbone bedding
d) Convolute bedding

*85. Which one of the following sedimentary structures is referred as a ***'Signature of intertidal deposits'***?
a) Herringbone structure
b) Stromatolites
c) Flaser bedding
d) Mudcracks

86. Red shales are the characteristic feature of
a) Intertidal zone
b) Supratidal zone
c) Subtidal zone
d) Swash zone

87. Which of the following has the coarsest grains ?
a) Green mud
b) Blue mud
c) Black mud
d) Red mud

88. Red beds were the elastic depositional products of ——— period
a) Permo-Carboniferous
b) Permo-Triassic
c) Triassic-Jurassic
d) Jurassic-Cretaceous

89. The Rann of Kutch is an example of
a) non-marine evaporite basin
b) marine evaporite basin
c) ephemeral evaporite lake
d) Playa

90. The evaporite that is typical of non-marine basins is
a) Gypsum
b) Halite
c) Trona
d) Calcite

91. Evaporites are generally deposited by cyclic repetition of sediments. The potash salts are deposited during
a) Normal marine stage
b) Penesaline stage
c) Saline stage
d) Supersaline stage

92. Which of the following depicts the correct advancing cycle (Sabkha cycle) of evaporites?
a) Calcite-gypsum-halite-potash
b) Calcite-halite-gypsum-potash
c) Gypsum-anhydrite-calcite-potash
d) Potash-halite-gypsum-calcite

93. The chief characteristic of euxinic environment is
   a) its oxidizing conditions
   b) its reducing conditions
   c) presence of large volumes of stagnant water
   d) presence of large volumes of stagnant water and black carbonaceous pyritic shales
94. When black shales occur in eugeosynclines, they constitute
   a) Euxinic environment
   b) Fondo environment
   c) Parallic environment
   d) Limnic environment
95. An ideal single turbidite sequence (the Bouma sequence) is made up of
   a) Three units  b) Four units
   c) Five units  d) Six units
96. Which of the following denotes the Bouma sequence?
   a) Massive grading—Parallel lamination—Rippled lamination——Parallel lamination——Interturbidite sequence
   b) Massive grading —— Rippled lamination —Parallel lamination —— Inter-turbidite sequence- —Parallel lamination
   c) Parallel lamination —Massive grading ——Rippled lamination——Interturbidite sequence
   d) Parallel lamination ——Massive grading ——Parallel lamination ——Rippled lamination ——Interturbidite sequence
97. Petrographically the turbidite sequences may occur in
   a) Graywackes alone
   b) Proto-quartzites alone
   c) Graywackes and proto-quartzites
   d) Graywackes, proto-quartzites and layered gabbros
98. Bodies of sand in which the ratio of the width to thickness varies between 50: 1 and 1000 : 1 are called
   a) Blankets  b) Shoestrings
   c) Prisms  d) Fans
99. A sedimentary facies may be described in terms of
   a) Lithologic characteristics only
   b) Palaeontologic characteristics only
   c) Both lithologic and palaeontologic characteristics
   d) None of the above
100. 'Syneresis' is a diagenetic process which involves
   a) elimination of free oxygen from the sediments by bacterial action
   b) reduction in the bulk volume of sediments
   c) de-watering of sediments under pressure
   d) deposition of minerals in the cavities and pore spaces
101. The diagnostic features of the mineral quartz that is derived from the

volcanic rocks are

i) absence of inclusions
ii) unstrained nature
iii) wavy extinction
iv) equidimensional habit

a) (i),(ii) and (iii) only
b) (i), (ii) and (iv) only
c) (i), (iii) and (iv) only
d) All of the above

102. The characteristic feldspar of the acid volcanic rocks is
a) Sanidine
b) Orthoclase
c) Microcline
d) Perthite

103. Minerals that dissolve in water are formed during
a) the early stages of magmatic crystallization
b) the later stages of magmatic crystallization
c) the differentiation of a magma
d) regeneration of a magma

104. Zircon and quartz usually occur as
a) Unaltered minerals
b) Altered minerals
c) Insoluble minerals
d) Soluble minerals

105. When calcium carbonate is inorganically precipitated from seawater it assumes the form of
a) Calcite
b) Aragonite
c) Dolomite
d) Ankerite

106. Sedimentary sulphides, coal and petroleum are designated as
a) Reduzates
b) Resistates
c) Hydrolysates
d) Oxidates

107. 'Chamosite' is a/an
a) Iron silicate
b) K-Fe silicate
c) Hydrous iron silicate
d) Hydrous potassium-iron silicate

108. Which of the following is NOT a borate?
a) Tincal

b) Borax
c) Ulexite
d) Kieserite

109. Muds that contain abundant organic matter are known as
a) Sapropelites
b) Stromatolites
c) Pelitolites
d) None of the above

110. Sapropelites are generally deposited in
a) Oxidizing environments
b) Reducing environments
c) Partly oxidizing and partly reducing environments
d) None of the above

111. If the rain falls obliquely, the rain print formed will have a ridge that is
a) high on the leeward side
b) high on the windward side
c) equally high on both the sides
d) none of the above

112. Cavities lined with easily separable secretions are termed as
a) Druses
b) Geodes
c) Dendrites
d) Agates

113. 'Psammitic' rocks are the metamorphic equivalents of
a) Rudaceous rocks
b) Arenaceous rocks
c) Argillaceous rocks
d) Non-clastic rocks

114. Which one of the following is an important constituent of the marine clay ironstones?
a) Bog iron ore
b) Siderite
c) Chamosite
d) Glauconite

115. Bog Iron Ore is
a) Iron carbonate
b) Iron bicarbonate
c) Iron oxide
d) Iron hydroxide

116. Bog Iron Ore is of
a) Organic origin
b) Chemical origin
c) Biochemical origin

d) Biological origin

117. 'Jet' is a/an

a) Ejection of water with considerable force from underground aquifers
b) Ejection of dissolved substances along with hot water from a thermal spring
c) Compact coaly substance
d) Crinoidal limestone deposit

118. Which of the following types of coals are dominantly composed of algal and fungal matter?

a) Lignite
b) Anthracite
c) Torbanite
d) Cannel coal

119. Which of the following types of coal consists of very thin alternating bright and dull laminae?

a) Fusain
b) Vitrain
c) Durain
d) Clarain

120. 'Clarite' is an association of

a) Vitrinites only
b) Vitrinite and exinite
c) Exinite and inertinite
d) Vitrinite and inertinite

121. The environment between the highest and the lowest levels of spring tides is described as

a) Littoral environment
b) Neritic environment
c) Continental shelf
d) Bathyal environment

122. A lake environment that is affected by marine conditions is described as

a) Lagoonal
b) Paralic
c) Limnic
d) Paludal

123. A group of cross-bedding exhibiting a uniform direction of current flow is termed as

a) Bedset
b) Coset
c) Composite bedset
d) Parallel lamination

124. Which of the following sandstones is slightly flexible in thin slabs?
   a) Grit
   b) Gannister
   c) Itacolumite
   d) Ignimbrite
125. The term 'Pinch out' is used when
   a) a lithosome thins progressively to extinction
   b) some bodies of sediment disappear and are lost in laterally adjacent masses
   c) a lithosome is terminated by replacement of their lithologic characteristics by another lithosome
   d) two different sediment types grade from one to the other
126. Which of the following directional structures are best suited for the analysis of a turbidite basin?
   a) Cross Bedding
   b) Ripple lamination
   c) Flute marks
   d) Scours
   e) Slump structures
127. The dispersal patterns peculiar to the Deltaic environments are
   a) unimodal
   b) bimodal
   c) polymodal
   d) random
128. Which of the following are NOT erosional current structures?
   a) Channel marks
   b) Rill marks
   c) Scour marks
   d) Swash marks
129. Tongue-shaped ripples that resemble mudflows and point up-current are called
   a) Linguoid ripples
   b) Lunate ripples
   c) Rhomboid ripples
   d) Undulatory ripples
130. Small, sub-spherical, concretionary algal stromatolite forms are called
   a) Oncolites
   b) Weedia crusts
   c) Collenia
   d) Crypotozoans

*131. In most of the sedimentary rocks
   a) K>Na
   b) K<Na

c) K=Na
d) None of the above

132. Which of the following is richest in polymetallic nodules?
a) The Indian ocean
b) The Pacific ocean
c) The Atlantic ocean
d) The Arctic ocean

133. Sheet-like reef structures composed of the hard parts of benthonic organisms are called
a) Bioherms
b) Biostromes
c) Bioceonose
d) Thanatoceonose

134. Modern hermatypic hexacorals are found in the warm surface waters extending between the latitudes
a) 15°N to 15°S
b) 30°N to 30°S
c) 45o N to 450 S
d) 600 N to 600 S

135. The f-scale (Phi-scale) which was devised by W.C. Krumbein in 1938 to classify sediments according to their grades is a/an
a) Arithmetic scale
b) Geometric scale
c) Logarithmic scale
d) Arbitrary scale

136. Shale is made up of
a) clasts only
b) clasts + matrix
c) clasts + cement
d) clasts + matrix + cement

137. In the direction of transport the grain size
a) increases
b) decreases
c) increases or decreases depending upon the duration of transport
d) remains constant

138. In the Krumbein's f-scale (Phi-scale), pebbles range in value from
a) -1 to zero
b) -1 to -5
c) -5 to -8
d) -8 to 1

139. The 'Frequency Curve Scale' of size is a
a) logarithmic scale
b) geometric scale
c) arithmetic scale
d) random scale

140. According to the Folk and Ward formula, the mean grain size is denoted as

a) $M = \frac{P25 + P50 + P75}{3}$

b) $M = \frac{j16 + j50 + j84}{3}$

c) $M = \frac{j25 + j50 + j75}{3}$

d) none of the above

141. A sediment having excess coarse material exhibits

a) negative skewness
b) positive skewness
c) no skewness
d) either of the above

142. Near-symmetrical grains exhibit skewness of

a) -0.10 to -0.30
b) +0.10 to -0.10
c) +0.30 to +0.10
d) >+0.30

143. The term 'Katagenesis' is used to describe

a) the conversion of sediments into a sedimentary rock
b) the accumulation of sediments over a period of time
c) the alteration of a sedimentary rock before the onset of metamorphism or weathering
d) the formation of the initial sedimentary material

144. Diagenesis is a

a) syn-depositional change
b) post-depositional change
c) pre-depositional change
d) none of the above

145. According to the "Mineral Stability Series" of Goldich, the first mineral to be decomposed is

a) Olivine
b) Pyroxene
c) Amphibole
d) Biotite

146. High maturity of sediments is indicated by

a) Well rounding and good sorting
b) Well rounding , good sorting and complete removal of unstable minerals
c) Well rounding , good sorting and retention of unstable minerals
d) Well rounding , good sorting, complete removal of unstable minerals and optical continuity of stable minerals

147. The presence of polycrystalline quartz with mineral inclusions and exhibiting strain effects is indicative of a——— source for the sediments.

a) Plutonic
b) Volcanic
c) Metamorphic
d) Sedimentary

148. Free-rolling or mobile Stromatolites are described as

a) Oncolites
b) Collenia
c) Weedia Crusts
d) Columnaria

149. The 'Reef' and 'Back Reef' environments may be classified under

a) Clastic, Shore zone sedimentary environments

b) Non-clastic, Shore zone sedimentary environments
c) Inland basin sedimentary environments
d) Marine sedimentary environments

150. Which of the following is NOT a characteristic of the Alluvial environment?
a) The deposited material is heterogenous, ranging from boulders to clay
b) The sorting of grains range from well sorted to ill-sorted
c) Structures such as current bedding, ripple marks and mudcracks are characteristic
d) In the fluvial cycle, coarsening upwards of the particles in a sequence takes place

151. The Meandering river channels are typically
a) Unimodal with low sinuosity
b) Unimodal with high sinuosity
c) Bimodal with low sinuosity
d) Bimodal with high sinuosity

*152. In the fining upward sequences of a meandering river channel, the sedimentary structures found on the top of the sequence are
a) Small-scale cross-bedding
b) Large-scale cross-bedding
c) Parallel lamination
d) Ripple marks

153. 'Levee' and 'Crevasse Splay' deposits are
a) Channel deposits
b) Channel-fill deposits
c) River Bank deposits
d) Flood basin deposits

154. Laterally extensive sheets of coarse braided alluvium generally only a few hundred feet thick and blanketing continental shields is described as
a) Molasse facies
b) Flysch facies
c) Nubian facies
d) Alluvial facies

*155. Which of the following deltas consist of fine sediments?
a) Nile
b) Ganges
c) Indus
d) Mississipi

156. The Tidal bedding is composed of
a) a layered bedding of very-fine grained clay deposited over coarse sand zones
b) a ripple bedding with mud streaks in troughs and crests
c) irregular cross-bedding characterized by reversed foreset laminae in one coset
d) organo-sedimentary structures formed by the activity of blue-green algae

157. The Stromatolites that are typical of inter-tidal flats are
a) Columnaria
b) Conophyta
c) Symmetrica
d) Collenia

158. Which of the following are characteristic features of the Supratidal Zone?
    1) Mudcracks
    2) Red Shales
    3) Salt pseudomorphs
    a) 1 only    b) 1 and 2 only
    c) 2 and 3 only    d) all of the above

159. Biogenic oozes, red clays and poly-metallic nodules are the typical sediments found in
    a) the Littoral zone    b) the Neritic zone
    c) the Bathyal zone    d) the Abyssal zone

160. The "Bar Theory" is used to explain the origin of
    a) Stromatolites    b) Turbidites
    c) Evaporites    d) Poly-metallic nodules

161. Which of the following symmetries is not exhibited by sedimentary rocks?
    a) Isotropic    b) Axial
    c) Orthorhombic    d) Monoclinic

162. As one moves from older to younger beds, the number of mineral species
    a) increases
    b) decreases
    c) either increases or decreases
    d) neither increases or decreases

## SEDIMENTRY PETROLOGY (KEY)

| | | | | | | | | | |
|---|---|---|---|---|---|---|---|---|---|
| 1-c | 2-b | 3-e | 4-b | 5-b | 6-a | 7-c | 8-a | 9-a | 10-a |
| 11-c | 12-a | 13-d | 14-d | 15-b | 16-b | 17-a | 18-c | 19-c | 20-a |
| 21-c | 22-a | 23-c | 24-c | 25-b | 26-c | 27-a | 28-b | *29-c | 30-d |
| 31-b | *32-d | 33-d | 34-b | 35-b | 36-b | 37-c | 38-c | 39-c | 40-c |
| 41-b | 42-d | 43-d | 44-d | 45-a | 46-b | 47-a | 48-c | 49-d | 50-b |
| 51-a | 52-b | 53-c | *54-c | 55-b | 56-b | 57-a | 58-c | 59-d | 60-b |
| 61-d | 62-a | 63-d | 64-c | 65-d | 66-d | 67-a | 68-a | 69-c | 70-a |
| 71-c | 72-c | 73-c | 74-a | 75-a | 76-b | 77-c | 78-e | 79-d | 80-d |
| 81-b | 82-a | 83-a | 84-c | *85-b | 86-b | 87-a | 88-b | 89-b | 90-c |
| 91-d | 92-a | 93-d | 94-b | 95-c | 96-a | 97-d | 98-b | 99-c | 100-c |
| 101-c | 102-a | 103-a | 104-a | 105-b | 106-a | 107-c | 108-d | 109-a | 110-c |
| 111-a | 112-b | 113-b | 114-c | 115-d | 116-c | 117-c | 118-c | 119-d | 120-b |
| 121-a | 122-b | 123-b | 124-c | 125-a | 126-c | 127-a | 128-d | 129-b | 130-a |
| *131-a | 132-b | 133-b | 134-b | 135-a | 136-a | 137-b | 138-b | 139-a | 140-b |
| 141-a | 142-b | 143-c | 144-b | 145-a | 146-d | 147-c | 148-a | 149-b | 150-d |
| 151-b | *152-c | 153-c | 154-c | *155-d | 156-a | 157-a | 158-d | 159-d | 160-c |
| 161-d | 162-a | | | | | | | | |

## EXPLANATORY ANSWERS

*29- (c): According to the folk and Ward formula

$$ø = \frac{ø\,84 - ø\,16}{4} + \frac{ø\,95 - ø\,5}{6.6}$$

(where ø is the standard deviation. Sediments whose standard deviation values are

<3. 5 ø are the best sorted. e.g. Texas Beach sand -0.25 ø to -0.35 ø.

*32- (d): Beach sands have a negative skew as fine components are removed by the persistent wave-action. River sands, on the other hand, are often positively skewed, since much silt and clay is not removed by the currents.

*54- (c): Graywackes consist of 15% of the matrix components and the chemical cement is absent. In Arkoses the matrix is absent or is <15% but chemical cement is present. The silica % is <75 in both graywackes and arkoses.

*57- (a): As there was least biological activity and abundance of chemical precipitation during the Precambrian, extensive lime muds were deposited e.g. limestone deposition in the Cuddapah basin.

*85- (b): Stromatolites are formed by the activity of the blue-green algae in the tidal zone. They are so reliable and diagnostic criteria for the recognition of intertidal environments that they are referred to a "signatures of intertidal deposits".

*131 (a) Sodium and potassium are held largely in feldspars. These are not stable in the weathering environment and consequently passes into solution. As plagioclase has relatively lower stability than the potash feldspars, Na passes into solution more rapidly than K. The effect of chemical weathering is to increase the K/Na ratio in the resistrate fraction. Simultaneously, K enters into newly formed clay minerals while Na tends to remain in solution, almost indefinitely. Thus, most sediments are rich in K relative to Na.

*152 (c): The topmost beds in a fining-upwards sequence are composed of finer sands, silts and mudstones which characteristically exhibit parallel lamination.

*155-(d): Fine sediments are deposited in a bird's foot digitate deltas like that of the Mississipi; whereas coarse sediments are deposited in arcuate deltas like those of the Nile, Ganges and the Indus.

****

# 13

# *Metamorphic Petrology*

1. Which of the following is/are characteristic of the process of metamorphism?
   a) under normal conditions the bulk chemistry of the rock remains unchanged
   b) there is no large-scale liquid at any given time
   c) there is no order of crystallization
   d) all of the above
2. The main type of chemical reaction involved in the process of metamorphism is
   a) Solid → Solid + Vapour
   b) Solid + Liquid → Solid + Gas
   c) Liquid + Liquid → Solid
   d) Liquid + Gas → Solid
3. The thermal or contact metamorphism is characterized by
   a) High temperature, low pressure, low strain and variable fluid pressure
   b) High temperature, high pressure, low strain and variable fluid pressure
   c) High temperature. low pressure, variable strain and variable fluid pressure
   d) High temperature, high pressure, high strain and high fluid pressure
4. Which of the following rocks have NOT resulted due to contact metamorphism 7
   a) Spotted slates b) Hornfelses
   c) Skarns d) Augen gneisses
5. Hornfelses generally possess
   a) Porphyroblastic fabric
   b) Granoblastic fabric
   c) A combination of porphyroblastic and granoblastic fabric
   d) Foliated fabric
6. Thermal metamorphism of dolomitic limestones with small siliceous impurity leads to the formation of
   a) Forsterite marble b) Brucite marble

c) Serpentine marble d) Tremolite marble

7. Metamorphism involving substantial addition or removal of materials is generally termed as
   a) Contact metamorphism b) Autometamorphism
   c) Metasomatism e) Pneumatolysis
8. The metamorphism involving the combined effect of uniform pressure and heat is described as
   a) Plutonic metamorphism
   b) Dynamothermal metamorphism
   c) Cataclastic metamorphism
   d) Contact metamorphism
9. The development of tectonites is widespread in
   a) Dynamothermal metamorphism
   b) Plutonic metamorphism
   c) Cataclastic metamorphism
   d) Pyrometamorphism
10. Which of the following is matched correctly ?
    a) Pyrometamorphism——low temperatures
    b) Contact metamorphism——high temperatures
    c) Kinetic metamorphism ——directed pressure predominant
    d) Pneumatolysis ——changes due to solution action
11. Choose the correct statement from the following
    a) Uniform pressure acts on both liquids and solids
    b) Directed pressure causes a change in volume
    c) Uniform pressure leads to a change of surface
    a) All of the above are correct
12. When uniform or hydrostatic pressure acts on a body, its
    a) Volume decreases
    b) Specific gravity decreases
    c) Volume decreases but specific gravity increases
    d) Volume increases but specific gravity decreases
13. Which of the following effects takes place when a body is subjected to directed pressure ?
    i) Dilation
    ii) Distortion
    iii) Formation of uni-dimensional grains
    iv) Growth of the body in a direction perpendicular to the direction of stress
    v) Solubility of the minerals is decreased
    a) (i), (iii), (iv) and (v) only
    b) (i), (iii) and (iv) only
    c) (ii), (iii), (iv) and (v) only
    b) (;ii), (iii) and (iv) only

14. Oriented mineral grains and foliated appearance of the rocks is a characteristic feature of the rocks belonging to
    a) Epizone b) Mesozone
    c) Katazone d) Epizone and Katazone
15. Which of the following rocks are completely unfoliated?
    a) Slates b) Schists
    c) Phyllites d) Hornfelses
16. Minerals whose fields of stability on a P-T diagram are extended on the introduction of a non-hydrostatic stress are described as
    a) Stress minerals b) Anti-Stress minerals
    c) Elastic minerals d) Plastic minerals
17. Stress minerals are characterized by
    (i) High packing index of crystal lattice
    (ii) Low molar volume
    (iii) High density
    (iv) Equidimensional habit
    a) (i) and (iii) only b) (i) and (ii) only
    c) (i). (ii) and (iii) only d) (ii), (iii) and (iv) only
18. Which of the following is a stress mineral?
    a) Andalusite b) Sillimanite
    c) Kyanite d) Cordierite
19. Stress minerals are generally absent in
    a) Epizone b) Mesozone
    c) Katazone d) None of the above
20. Minerals that develop their crystal form in spite of the resistance of a solid medium are described as
    a) Xenoblastic b) Idioblastic
    c) Crystalloblastic d) Granoblastic
21. Choose the mineral sequence that is arranged in the correct idioblastic order
    a) Orthoclase—— Plagioclase- —Mica— Garnet—Sphene
    b) Calcite- -Quartz—Pyroxene— Garnet—Epidote
    c) Orthoclase—Calcite—Mica—Quartz— Garnet
    d) Sphene—Garnet- - Epidote— Calcite—Quartz
22. In the metamorphic rocks the most commonly occurring potash feldspar is microcline and NOT orthoclase.because
    a) Orthoclase cannot withstand high temperatures and pressure
    b) Orthoclase transforms into its high temperature form 'Sanidine'
    c) Metamorphic conditions generally favour the crystallization of the form with lower symmetry
    d) Metamorphic conditions generally favour the crystallization of the ordered form

23. Plagioclase minerals are unstable at
    a) High temperatures
    b) High pressures
    c) Both high temperatures and high pressures
    d) None of the above
24. The minerals Glaucophane and Jadeite are indicative of
    a) High pressures and low temperatures
    b) Low pressures and high temperatures
    c) High pressure and high temperature
    d) Low pressures and low temperatures
25. Which of the following minerals has the largest stability field ?
    a) Andalusite   b) Sillimanite
    c) Kyanite   d) Cordeirite
26. The most commonly occurring silicates in the metamorphic rocks are
    a) Ionosilicates—Phyllosilicates—Nesosilicates
    b) Ionosilicates — Phyllosilicates — Tectosilicates
    c) Nesosilicates — Phyllosilicates —Tectosilicates
    d) Nesosilicates — Sorosilicates —Cyclosilicates
27. The principal carrier of sodium in Eclogites is
    a) Albite   b) Paragonite
    c) Glaucophane   d) Jadeite
28. The 'Zone concept' in metamorphism was proposed by
    a) Grubermann and Becke   b) Barrow
    c) Tilley   d) Eskola
29. The type of metamorphism dominant in the Epizone is
    a) Cataclastic metamorphism
    b) Thermal metamorphism
    c) Plutonic metamorphism
    d) Dynamothermal metamorphism
30. Barrow established a series of zones of progressive regional metamorphism in the Dalriadan Schist Field of the Scottish Highlands. The lowest grade of metamorphism is marked by the mineral
    a) Chlorite   b) Biotite
    c) Garnet   d) Kyanite
31. The 'Buchan Zones' described from the Northeast of Scotland consists of the following sequence in the order of increasing metamorphic grade
    a) Chlorite zone -Biotite zone -Garnet zone -Staurolite zone-Kyanite zone-Sillimanite zone
    b) Chlorite zone -Biotite zone -Staurolite zone -Garnet zone-Kyanite zone -Sillimanite zone
    c) Chlorite zone - Biotite zone -Cordeirite zone -Andallusite zone-Sillimanite zone

d) Chlorite zone- Cordeirite zone -Biotite zone- Andallusite zone- Sillimanite zone

32. Match the following and choose the correct answer.

| | |
|---|---|
| A) Ultrametamorphism | 1) Indurating and burning effects produced by lavas on the rocks in their contact |
| B) Autometamorphism | 2) Mechanical breakdown of rocks due to predominant effects of stress |
| C) Optalic Metamorphism | 3) Complete melting of rocks with the formation of new magmas under the influence of hot, ascending gases |
| D) Kinetic Metamorphism | 4) Alteration of a magmatic rock by volatiles and solution given off in the cooling process |

a) A-3 B-4 C-1 D-2
b) A-3 B-1 C-4 D-2
c) A-4 B-2 C-1 D-3
d) A-4 B-3 C-1 D-2

33. In the Dynamothermal Metamorphism
a) recrystallization is brought about by heat
b) deformation of rocks is brought about by directed pressure
c) formation of new structures takes place
d) all of the above statements are correct

34. The "Snowball Garnets" are the products of
a) Contact metamorphism
b) Cataclastic metamorphism
c) Dynamothermal metamorphism
d) Plutonic metamorphism

35. Flaser rocks are the chief products of
a) Thermal metamorphism
b) Dynamothermal metamorphism
c) Plutonic metamorphism
d) Cataclastic metamorphism

36. Choose the correct sequence of deformation and metamorphism
a) Shale à Slate → Schist → Phyllite → Gneiss
b) Shale à Slate → Phyllite → Gneiss → Schist
c) Shale à Slate → Phyllite → Schist → Gneiss
d) Slate à Shale → Phyllite → Schist → Gneiss

37. In the transition : Shalé à Gneiss, the grain size of the rocks
   a) Increases
   b) Decreases
   c) Remains constant
   d) Increases or decreases depending upon the conditions of metamorphism
38. 'Greisening' is a process of
   a) Pneumatolytic alteration b) Magmatic alteration
   c) Metasomatic alteration d) Metamorphism alteration
39. Fenite is a rock of syenitic composition formed due to
   a) Alkali metasomatism b) Lime metasomatism
   c) Load metamorphism d) Retrograde metamorphism
40. Adinoles are compact fine-grained, slaty rocks consisting mainly of
   a) Muscovite +Quartz
   b) Albite +Quartz
   c) Microcline +Albite
   d) Orthoclase +Muscovite +Clay
41. Adinoles are the products of
   a) Cataclastic metamorphism b) Dislocation metamorphism
   c) Metasomatism d) Diapthoresis
42. Match the following and choose the correct answer.

| | |
|---|---|
| A) Greisen | 1) Quartz+Muscovite+Topaz |
| B) Pinite | 2) White Mica+Chlorite |
| C) Saussurite | 3) Albite+Epidote |
| D) Cymatolite | 4) Albite + Muscovite |

   a) A-1 B-2 C-4 D-3
   b) A-1 B-2 C-3 D-4
   c) A-2 B-1 C-3 D-4
   d) A-2 B-1 C-4 D-3
43. A 'Migmatite' is a mixed rock in which
   a) an older metamorphic rock is mixed with younger granitic material
   b) an older granitic rock is mixed with newly metamorphosed sediments
   c) a primary gneiss mixes with metamorphosed gneissic sediments
   d) fragments of igneous, sedimentary and metamorphic rocks are complexly mixed together
44. Migmatites are a result of
   a) Retrograde metamorphism b) Ultrametamorphism
   c) Metasomatism d) Palingenesis
45. Migmatites in which traces or patches of thc original fabric elements are completely destroyed are termed as
   a) Agmatites b) Nebulites
   c) [illegible]rsites d) Anatexites

46. Match the following and choose the correct answer

| | | | |
|---|---|---|---|
| A) | Hartschiefer | 1) | A brucite-bearing limestone |
| B) | Lit-par-lit Gneiss | 2) | Finely banded mylonitized rock |
| C) | Kakirite | 3) | A crush breccia in which the fragments are enclosed in rock powder |
| D) | Predazzite | 4) | A migmatite in which the two phases are separated into distinct parallel layers |

a) A-4 B-2 C-3 D-1
b) A-2 B-4 C-l D-3
c) A-4 B-2 C-l D-3
d) A-2 B-4 C-3 D-1

47. The structure characterized by spotted appearance due to incipient crystallization of minerals under contact metamorphic conditions is termed as
a) Maculose structure
b) Myrmekitic structure
c) Flaser structure
d) Cataclastic structure

48. The typical texture exhibited by augen gneisses in which lenticular grains are enclosed in granulated material is termed as
a) Maculose structure
b) Mortar structure
c) Mosaic structure
d) Helicitic structure

49. Contorted trails of inclusions in poikiloblastic crystals in the regionally metamorphosed rocks gives rise to a
a) Helicitic structure
b) Decussate texture
c) Saccharoidal texture
d) Nematoblastic texture

50. Maculose structure is typical of
a) Argillaceous rocks
b) Arenaceous rocks
a) Rudaceous rocks
d) Calcareous rocks

51. If a foliated structure is dominated by prismatic minerals it is described as
a) Nematoblastic
a) Lepidoblastic
b) Idioblastic
c) Xenoblastic

52. The 'Muth Quartzites' of the Devonian age characteristically exhibit
a) Hornfelsic texture
b) Cataclastic texture
c) Granulose texture
d) Schistose texture

53. Metamorphic facies are defined by
a) The conditions of temperature and pressure
b) A single dominant rock type
c) Peculiar textures and structures of the rock types
d) Critical mineral assemblages

54. Which of the following metamorphic facies is characterized by maximum temperatures and minimum pressures?
   a) Blueschist facies  b) Greenschist facies
   c) Sanidinite facies  d) Hornblende hornfels facies
55. The metamorphic facies that represents a transition between diagenesis and regional metamorphism is
   a) Zeolite facies  b) Greenschist facies
   c) Hornfels facies  d) Sandinite facies
56. The most voluminous and widespread of the metamorphic rocks belongs to
   a) Hornfels facies  b) Greenschist facies
   c) Amphibolite facies  d) Granulite facies
57. Which of the following rocks are formed under abnormally low geothermal gradients?
   a) Blue schists  b) Green schists
   c) Eclogites  d) Amphibolites
58. The Garnet-Sillimanite Schists of the Eastern Ghat Province, better known as the 'Khondalites', are the characteristic rocks of
   a) Amphibolite facies  b) Granulite facies
   c) Eclogite facies  d) Greenschist facies
59. The mineral assemblage Larnite-Merwinite-Spurrite is typical of
   a) Sanidinite facies  b) Zeolite facies
   c) Glaucophane schist facies  d) Granulite facies
60. Match the following and choose the correct answer

| *Metamorphic facies* | *Diagnostic mineral* |
|---|---|
| A) Zeolite facies | 1) Lawsonite |
| B) Sanidinite facies | 2) Chlorite |
| C) Blueschist facies | 3) Laumontite |
| D) Greenschist facies | 4) Sanidine |

   a) A-3 B-2 C-1 D-4
   b) A-1 B-2 C-3 D-4
   c) A-3 B-4 C-1 D-2
   d) A-1 B-4 C-3 D-2
61. The amphibole that is characteristic of high pressure low temperature environments is
   a) Hornblende  b) Tremolite
   c) Actinolite  d) Glaucophane
62. Hydroxyl bearing minerals are essentially absent in the rocks belonging to
   a) Hornfels facies  b) Granulite facies
   c) Eclogite facies  d) Zeolite facies
63. Pick out the least silicic mineral assemblage from the following.
   a) Quartz—orthoclase—plagioclase—cordierite

b) Calcite—wollastonite—diopside
c) Calcite --diopside—forsterite
d) Calcite—forsterite—periclase

64. Which of the following seta of minerals serve as good 'geological barometers' by giving a fairly precise indication of the pressure and the depth of crystallization?
a) Enstatite-orthoferrosilite-hypersthene
b) Forsterite-olivine-fayalite
c) Andalusite-sillimanite-kyanite
d) Diamond-graphite

65. In the environment of the blueschist facies the eclogites occur as
a) Layers and lenses
b) Streaks and bands
c) Inclusions sharply bounded to the enclosing rock
d) Isolated tectonic blocks along shear zones

66. Omphacite, the diagnostic clinopyroxene of Eclogites, is characterized by
a) High Na2O/$K_2O$ and very low A12O3
b) High $Na_2O/Al_2O_3$ nd very low $K_20$
a) High $K_2O$/A12O3 and very low $Na_2O$
b) Low $Na_2O/K_2O$ and high $Al2O_3/Fe_2O_3$

67. In terms of chemical composition the Eclogites are closely similar to
a) Continental tholeiites
b) Oceanic tholeiites
c) Spilites
d) Granodiorites

68. With a drop in temperature and increase in water pressure the rocks belonging to the granulite facies suffer a retrogressive metamorphism and a transition is effected to amphibolite fades. The correct retrogressive sequence is
a) Pyroxene-granulite facies— Hornblende-granulite facies- - Amphibolite facies
b) Hornblende-granulite facies — Pyroxene-granulite facies — Amphibolite facies
c) Granulite facies — Eclogite facies — Amphibolite facies
d) Granulite facies — Blueschist facies — Amphibolite facies

69. Which of the following factors favour an igneous origin for the rocks 'Charnockites'?
1) Presence of mesoperthitic feldspar in the rocks
2) Prevalence of the rare 'Albite- Ala' twins
3) Occurrence of pyrrhotite as an accessory mineral
4) Apophyses and veins of Charnockites into the surrounding rocks
a) l, 2 and 3only
b) l, 2 and 4only
c) 2, 3 and 4 only
d) All of the above

70. 'Enderbites' are a type of Charnockites of
   a) Acidic composition
   b) Basic composition
   c) Ultrabasic composition
   d) Any of the above
71. Gneissic rocks that are metamorphosed sediments are described as
   a) Orthogneisses
   b) Paragneisses
   c) Augen gneisses
   d) Phyllonites
72. Which of the following criteria can be used to establish the sedimentary Origin of metamorphic rocks?
   i) Excess of carbon in the norm
   ii) A very high percentage in silica
   iii) $Na_2O$ in excess of $K_20$

   a) (i) and (ii) only
   b) (i) and (iii) only
   c) (ii) and (iii) only
   d) All of the above
73. In the expression for the Mineralogical Phase Rule P=C+F -2, the conditions for metamorphism are ideal when
   a) F=0
   b) F=1
   c) F>2
   a) Any of the above
74. 'Buchite' is a
   a) Partially fused hornfelsic rock occurring as xenoliths
   b) Fine-grained, flinty-looking rock formed by extreme dislocation metamorphism
   c) Rock formed by mechanical disintegration of graywackes
   d) Fine-grained granulitic rock
75. Metamorphic textures of argillaceous rocks are described as
   a) Blasto-psephitic
   b) Blasto-psammitic
   c) Blasto-pelitic
   d) Blasto-porphyritic
76. Relicts of original minerals and textures found in metamorphic rocks are described as
   a) Palimpsest textures
   b) Nematoblastic textures
   c) Lepidoblastic textures
   d) Decussate textures

77. When idioblastic crystals are embedded in fine-grained groundmass the resulting texture is termed as
a) Blastoporphyritic b) Blastopoikilitic
c) Porphyroblastic d) Granoblastic
78. Any reversal of metamorphism which produces rocks of a lower metamorphic grade from one of higher grade is termed as
a) Diapthoresis b) Anatexis
c) Palingenesis d) Consanguinity
79. The retrogressive metamorphism of cordierite yields
a) Sericite b) Pinite
c) Saussurite d) Luxullianite
80. Which of the following types of folds are commonly associated with migmatites?
a) Reclined folds b) Supratenuous folds
c) Ptygmatic folds d) Piercing folds
81. In the triangular ACF diagrams used to designate, the mineralogical and chemical composition of metamorphic facies, the alphabet 'A' represents
a) $A12O_3$
b) $A12O_3 + Fe_2O3 - Na_2O + K_20$
c) $Na_2O + K_20$
d) $A1_20_3 + Na_2O - Fe2O_3 + CaO$
82. The process of complete mixing and melting of rocks with the production of neomagma is described as
a) Anatexis h) Palingenesis
c) Ultrametamorphism d) Autometamorphism
83. Palingenesis is shown to start at a temperature of about
a) 500°C b) 800°C
c) 1000°C d) 1500°C
84. Which of the following is NOT an effect of alkali metasomatism ?
a) Formation of myrmeckite
b) Formation of greisen
c) Alteration of granite to fenite
d) Replacement of hornblende by biotite in the granitization of amphibolite
85. Tourmalinization is the combined effect of
a) Water and fluorine
b) Water, boron and fluorine
c) Water, carbondioxide and chlorine
d) Water, carbondioxide and hydrogen
86. When quartzofeldspathic rocks are subjected to dynamothermal metamorphism a distinct mineralogical change takes place where in the alkali feldspar breaks down into white mica and quartz. This change is described as

a) Feldspathisation b) Silicification
c) Kaolinization d) Sericitization

87. Which of the following metamorphic rocks are essentially composed of hornblende and plagioclase?
a) Amphibolite b) Hornblendite
b) Hornfels c) Blueschist

88. When shales are subjected to thermal metamorphism the products formed are?
a) Hornfelses b) Novaculite
c) Phyllites d) Schists

89. Contact metamorphism of a dolomite with alumina and silica as impurities leads to the formation of
a) Tremolite marble b) Forsterite marble
c) Brucite marble d) Forsterite and spinel

90. 'Leptynite' is a
a) Fine-grained acidic meta-igneous rock with a granulose structure
b) Coarse grained acidic meta-igneous rock with a granulose structure
c) Ferruginous meta-sedimentary rock
d) Migmatised granulite

91. Choose the correct sequence of deformation of hard brittle rocks due to the effects of cataclastic metamorphism
a) Mortar structure → Augen structure → Flaser rock → Cataclasite → Mylonite
b) Mortar structure→ Augen structure → Cataclasite → Flaser rock → Mylonite
c) Mortar structure → Cataclasite → Augen structure → Flaser rock → Mylonite
d) Flaser rock → Mortar structure → Cataclasite → Augen structure → Mylonite

92. Which of the following rocks do NOT exhibit granulose structure ?
a) Adinole b) Marble
c) Skarn d) Khondalite

93. The mineralogy of which of the following contact metamorphic facies is similar to that of the high grade amphibolite facies?
a) Albite-epidote-hornfels facies
b) Hornblende-hornfels facies
c) Pyroxene hornfels facies
d) Sanidinite facies

94. The amphibolite facies covers the entire span of medium to high grade zones in many classic areas of progressive regional metamorphism. The subfacies that indicates the highest grade is
a) Staurolite-Almandine

b) Kyanite- Almandine-Muscovite
c) Sillimanite-Almandine-Muscovite
d) Sillimanite-Almandine- Orthoclase

95. The most common accessory mineral in Eclogites is
a) Rutile b) Zoisite
c) Ilmenite d) Sphene

96. In the paired metamorphic belts associated with subduction complexes, the HT-LP land-ward member is constituted by
a) Blueschists b) Amphibolites
c) Eclogites d) Granulites

97. The conversion : Eclogites → Amphibolites is an example of
a) Progressive regional metamorphism
b) Retrogressive metamorphism
c) Autometamorphism
d) Load metamorphism

98. The acid division of Charnockites does not include
a) Enderbites b) Alaskites
c) Birkremites d) Norites

99. Which of the following is formed by the metasomatic exchange of alkalies?
a) Myrmekite b) Fenite
c) Adinole d) Sericite

100. 'Agmatites' may be defined as
a) Migmatitic rocks which show traces or patches of the original fabric elements
b) Migmatitic rocks in which traces of original fabric elements have been completely destroyed
c) Migmatitic rocks in which the introduced material appears to form a network of veins throughout the host rock
d) None of the above

101. Which of the following is NOT matched correctly.
a) Mylonitization —— deep shear zones
b) Contact metamorphism —— magmatic intrusions
c) Dynamothermal metamorphism — orogenies and igneous intrusions
d) None of the above
e) All of the above

102. Thermal metamorphism leads to
a) change in mineralogical composition only
b) change in fabric only
c) change in mineralogical composition and fabric
d) none of the above

103. Which of the following involves flow of large quantities of fluid through an 'open' rock body along significant temperature or composition gradients?
a) Metamorphic differentiation b) Metasomatism
c) Diapthoresis d) Kinetic metamorphism

104. Which of the following is widespread around carbonatite intrusions?
a) Sericitization b) Tourmalinisation
c) Fenitization d) Mylonitization

105. Segregation of Mg,Fe,Ca,Al, Si and O from the homogenous matrix of a shale to form a large garnet grain is an example of
a) Mylonitization
b) Metasomatism
c) Metamorphic differentiation
d) Diapthoresis

106. The Dynamothermal metamorphism is a ———— metamorphism.
a) Pre-tectonic b) Post-tectonic
c) Syntectonic d) None of the above

107. The high-pressure polymorphs of silica, coesite and stishovite, may be associated with
a) Burial metamorphism b) Regional metamorphism
c) Sub-sea floor metamorphism d) Impact metamorphism

108. Which of the following has the most well-developed crystal outlines?
a) Sphene b) Garnet
c) Epidote d) Calcite

109. Consider the following reaction:

$$\underset{\text{K-feldspar}}{KalSi_3O_8} + \underset{\text{Corundum}}{Al_2O_3} + H_2O \rightarrow \underset{\text{muscovite}}{KAl_2AlSi_3O_{10}(OH)_2}$$

This is a possible illustration of :
a) Thermal metamorphism
b) Dynamothermal metamorphism
c) Retrograde metamorphism
d) Metasomatism

110. The assemblage : quartz + laumontite + chlorite is diagnostic of
a) Zeolite facies b) Prehnite-pumpellyite facies
c) Blueschist facies d) Granulite facies

111. Any penetrative set of more or less parallel lines is described as
a) Foliation b) Lineation
c) Schistosity d) Gneissosity

112. Migmatite is an example of
a) strongly foliated rock b) weakly foliated rock
c) non- foliated rock a) none of the above

113. The chief mineral constituents in ***mylonites*** are
a) Quartz and feldspar

b) Quartz , feldspar and mica
c) Quartz , feldspar, mica and hornblende
d) Quartz , feldspar, mica, hornblende and pyroxene

114. Which is the most widespread rock type in regionally metamorphosed Precambrian shield areas.
a) Gneiss b) Schist
c) Granulite d) Amphibolite

115. When the gneiss is derived from a magmatic source it is described as
a) Orthogneiss b) Paragneiss
c) Augengneiss d) Schist

116. In the granulite facies assemblages the feldspars are characterised by
a) antiperthitic plagioclases and mesoperthitic K-feldspars
b) perthitic plagioclases and mesoperthitic K-feldspars
c) antiperthitic plagioclases and perthitic K-feldspars
d) none of the above

117. Dedolomitisation, the destruction of dolomite to form calcite, takes place as a result of
a) Thermal metamorphism
b) Dynamothermal metamorphism
c) Retrograde metamorphism
d) Kinetic metamorphism

118. Thermal metamorphism of a calcareous rock with an aluminous impurity gives rise to
a) recrystallised limestone b) wollastonite
c) anorthite d) periclase

119. Match the following and choose the correct answer.

| List 1 | List 2 |
|---|---|
| A) Serpentine marble | 1) Pencatite |
| B) Brucite marble | 2) Ophicalcite |
| C) Charnockite | 3) Skarn |
| D) Lime silicate rock | 4) Hypersthene granite |

a) A-2 B-1 C-4 D-3
b) A-1 B-2 C-3 D-4
c) A-2 B-1 C-3 D-4
d) A-4 B-1 C-2 D-3

120. Thermal metamorphism of dolomitic limestones with intermediate amount of silica gives rise to
a) Brucite marble b) Forsterite marble
c) Serpentine marble d) Tremolite marble

121. When the feldspars in a granite subjected to cataclastic metamorphism are crushed to give rise to a pulverized form it is described as
a) Mortar structure b) Augen structure
c) Cataclasite d) Mylonite

122. When there is a change of form due to recrystallization, the deformation is described as
a) blastic
b) plastic
c) elastic
d) clastic
123. The pyroxene which is characteristic of the rocks belonging to the ***Charnockite series*** is
a) diopside
b) hypersthene
c) enstatite
d) omphacite
124. The *schillerisation* exhibited by the orthopyroxene in the granulite facies due to the presence in it of the exsolution lamellae of clinopyroxene is indicative of the presence of
a) high temperature, high pressure and rapid cooling
b) high temperature, low pressure and slow cooling
c) high temperature, high pressure and slow cooling
d) none of the above
125. Which of the following is a cyclosilicate occurring as a contact mineral in slates and hornfelses?
a) Andalusite
b) Sillimanite
c) Cordierite
d) Tourmaline
126. *Soapstones* are the metasomatic products of
a) Peridotites
b) Anorthosites
c) Spilites
d) Tholeiites
127. The *Staurolite Zone* established by Barrow is restricted to rocks having high content of
a) copper
b) magnesium
c) iron
d) manganese
128. Which of the following facies was NOT included in the original classification of facies by Eskola?
a) Sanidinite facies
b) Hornfels facies
c) Greenschist facies
d) Blueschist facies
129. Contact metamorphism develops in the contact aureoles of plutonic rocks. From the actual intrusive contact to the country rocks, the temperature
a) decreases
b) increases
c) first increases than decreases
d) first decreases than increases
130. Among the facies of Contact metamorphism, the one which represents the highest temperatures and lowest pressures is
a) Sanidinite facies
b) Pyroxene hornfels facies
c) Hornblende hornfels facies
d) Albite-epidote hornfels facies

131. In which of the following facies of Contact metamorphism the mineral assemblages is similar to that of the greenschist facies?
a) Sanidinite facies
b) Pyroxene hornfels facies
c) Hornblende hornfels facies
d) Albite-epidote hornfels facies

132. The development of the pyroxene-hornfels facies is favoured by
a) high pressures and high temperatures
b) low pressures and low temperatures
c) high pressures and low temperatures
d) low pressures and high temperatures

133. The presence of anhydrous minerals like chlorite, actinolite, epidote and albite are characteristic of
a) Greenschist facies b) Hornblende hornfels facies
c) Sanidinite facies d) Blueschist facies

134. The following reaction marks the transition from the pyroxene-hornfels facies to the sanidinite facies.
$CaMgSi_2O_6 + Mg_2SiO_4 + 2CaCO_3 \leftrightarrow 3X + 2CO_2$
The mineral X can be identified as
a) Diopside b) Monticellite
c) Rankinite d) Sanidine

135. The transition between the zeolite facies and the zone of diagenesis is marked by the mineral
a) Laumontite b) Heulandite
c) Prehnite d) Pumpellyite

136. The mineral ***pumpellyite***, which is characteristic of the Prehnite-Pumpellyite facies is a
a) pyrope garnet b) diopside-jadeite
c) Mg-epidote d) Fe-epidote

137. The rocks of the Greenschist facies develop under
a) normal geothermal gradient
b) normal to steep geothermal gradient
c) abnormally low geothermal gradient
d) abnormally high geothermal gradient

138. Which of the following minerals is stable at low pressures?
a) wollastonite b) grossularite
c) vesuvianite d) none of the above

139. The Fe+2/Mg ratio in ortho and clinopyroxenes can be used as a geothermometer. This is given as

$$\frac{(Fe+2/Mg)\ OPX}{(Fe+2/Mg)\ CPX} = KD = \text{Distribution coefficient.}$$

The distribution coefficient for igneous rocks is
a) 1.0 b) 1.4

c) 1.6 d) 1.8

140. Eclogites occur as streaks and bands in the environment of

a) greenschist facies b) granulite facies

c) dunites d) migmatite gneisses

141. The diagnostic mineral in the blueschist facies is

a) lawsonite b) laumontite

c) glaucophane d) jadeite

142. In the ultrabasic division of the Charnockite suite of rocks, the dominant pyroxene is

a) hypersthene b) enstatite

c) bronzite d) diopside

143. The composition of "Sericite" is

a) white mica + chlorite b) albite + epidote

c) albite + muscovite d) quartz + muscovite + topaz

144. Which of the following is the correct progressive sequence of the Barrow's zones which were established in the Dalradian rocks of SW Scottish Highlands?

a) Chlorite zone → Biotite zone → Garnet zone → Staurolite zone → Kyanite zone → Sillimanite zone

b) Chlorite zone → Garnet zone → Biotite zone → Staurolite zone → Kyanite zone → Sillimanite zone

c) Sillimanite zone → Kyanite zone → Staurolite zone → Garnet zone → Biotite zone → Chlorite zone

d) Sillimanite zone → Staurolite zone → Kyanite zone → Garnet zone → Biotite zone → Chlorite zone

145. Autometasomatic replacement of such minerals as feldspar, augite and olivine, by carbonates in igneous rocks is known as

a) Lime metasomatism

b) Carbondioxide metasomatism

c) Alkali metasomatism

d) None of the above

146. The conversion of hornblende-labradorite assemblage of an amphibolite into a greenschist assemblage of albite-epidote-chlorite-calcite illustrates the process of

a) Normal metamorphism

b) Retrograde metamorphism

c) Metasomatism

d) Deuteric alteration

147. Fixation of alkalis by silicates of iron and magnesium, illustrated by the replacement of hornblende by biotite in the granitization of amphibolite, is a result of

a) Lime metasomatism

b) Carbondioxide metasomatism

c) Alkali metasomatism
d) Retrograde metamorphism

148. Which of the following is the most common Garnet of gneisses and schists?
a) Almandine
b) Pyrope
c) Grossularite
d) Andradite

149. Which of the following statements is true?
a) Plagioclase is stable at low temperatures and pressures
b) Plagioclase is stable at high temperatures and pressures
c) Plagioclase is stable at high temperatures but is unstable at high pressures
d) Plagioclase is stable at low temperatures but is unstable at low pressures

150. With the increase in the grade of metamorphism, the anorthite content in plagioclases
a) increases
b) decreases
c) either increases or decreases
d) there is no relationship between the anorthite content of plagioclases and the grade of metamorphism

## METAMORPHIC PERTOLOGY (KEY)

| | | | | | | | | | |
|---|---|---|---|---|---|---|---|---|---|
| 1-d | 2- a | 3- a | 4-d | 5-c | 6- a | 7-c | 8- a | 9-a | 10- c |
| 11-a | 12- c | 13-d | 14- b | 15-d | 16- a | 17-c | 18- c | 19-c | 20-b |
| 21-a | 22- d | 23-b | 24- a | 25-b | 26- a | 27-d | 28-a | 29-a | 30-b |
| 31-c | 32-a | 33-d | 34- c | 35-d | 36-c | 37-a | 38-c | 39-a | 40- b |
| 41-c | 42-b | 43-a | 44-b | 45-d | 46-d | 47-a | 48- b | 49-a | 50-a |
| 51-a | 52-b | 53-d | 54-c | 55-a | 56-b | 57-a | 58- b | 59-a | 60-c |
| 61-d | 62-b | 63-d | 64-c | 65-d | 66-b | 67-b | 68-a | 69-d | 70-a |
| 71-b | 72-d | 73-c | 74-a | 75-c | 76-a | 77-c | 78-a | 79-b | 80- c |
| 81-b | 82-b | 83-d | 84-b | 85-b | 86- d | 87-a | 88-a | 89-d | 90-b |
| 91-b | 92-a | 93-b | 94-b | 95-a | 96-b | 97-b | 98-d | 99-a | 100-c |
| 101-d | 102-c | 103-b | 104-c | 105-c | 106-c | 107-d | 108-a | 109-c | 110-a |
| 111-b | 112-b | 113-a | 114-a | 115-a | 116-a | 117-a | 118-c | 119-a | 120-d |
| 121-c | 122-a | 123-b | 124-c | 125-c | 126-a | 127-c | 128-d | 129-a | 130-a |
| 131-d | 132-d | 133-a | 134-b | 135-b | 136-c | 137-a | 138-d | 139-b | 140-c |
| 141-a | 142-c | 143-a | 144-a | 145-b | 146-b | 147-c | 148-a | 149-c | 150-a |

# 14

# *Economic Geology*

1. A body of oil and gas or both occurring in a separate reservoir and under a single pressure system is described as
   a) pool
   b) field
   c) province
   d) region
2. Which of the following represents "solid petroleum"?
   a) Tar
   b) Asphalt
   c) Wax
   d) None of the above
   e) All of the above
3. In situ deposits of petroleum formed by heavier residues as a result of the evaporation of lighter fractions are known as
   a) Inspisatted deposits
   b) Bitumen exudates
   c) Tar Sands
   d) Vein bitumins
4. *Ozokerite* is an example of
   a) Oil shale
   b) Tar sand
   c) Mineral wax
   d) Coal
5. The correct sequence of encountering oil in a drill well is
   a) gas-oil-water
   b) water-gas-oil
   c) gas-water-oil
   d) oil-water-gas
6. The number of sedimentary basins in India is
   a) 15
   b) 22
   c) 27
   d) 33
7. Match the following and choose the correct answer.

| Basin | Type |
|---|---|
| A. Cambay Basin | 1. Onshore and offshore |
| B. Bombay High | 2. Onshore |
| C. Krishna-Godavari Basin | 3. Offshore |
| D. Bengal Basin | |

   a) A-2 B-3 C-3 D-1
   b) A-2 B-3 C-1 D-1
   c) A-3 B-3 C-2 D-1
   d) A-2 B-1 C-1 D-2

8. Match the following and choose the correct answer.

| *Basin* | *Structure* |
|---|---|
| A. Bombay High | 1. Doubly plunging anticline |
| B. Cambay Basin | 2. Elongated structural dome |
| C. Assam Shelf | 3. Folded asymmetrical anticlines |
| D. Tripura-Cachar Basin | 4. Anticline |

a) A-3 B-4 C-2 D-1
b) A-2 B-3 C-4 D-1
c) A-1 B-3 C-2 D-4
d) A-1 B-2 C-3 D-4

9. The Digboi oil field is associated with the *Tippam Sandstones*, which are of
a) Eocene age b) Miocene age
c) Oligocene age d) Pliocene age

10. In India oil was struck for the first time in limestones at
a) Ankleshwar b) Kaikalur
c) Ravva d) Bombay High

11. The *Gandhar field* in Gujarat is famous for the occurrence of
a) oil only b) gas only
c) both oil and gas d) none of the above

12. The reservoir rocks at the *Bombay High* are predominantly
a) biomicrites b) biosparites
c) oosparites d) oomicrites

13. According to the Oil and Natural Gas Corporation (ONGC), the "low risk-high return" sedimentary basins of India are classified under
a) Category I b) Category II
c) Category III d) Category IV

14. Which of the following is NOT matched correctly?
a) Krishna-Godavari Basin ……Kaikalur
b) Cauvery Basin ….. Narimanam
c) Bombay High….. Heera & Panna
d) Cambay Basin …. Razole

15. *Cambay Shale*, the oil-producing horizon of the Gandhar onland field, is of ——— age.
a) Eocene b) Miocene
c) Oligocene d) Pleistocene

16. The theory that primarily owes the origin of petroleum to reactions between alkali metals and carbondioxide at high temperatures was postulated by
a) Sebatiers and Sederen (1902) b) Berthelot (1917)
c) Byasson (1871) d) None of the above

17. The *Sausar & Sakoli Series* is associated with the deposition of
    a) diamonds  b) lead and copper
    c) manganese  d) iron
18. The *Karewa formation* in the Kashmir valley is renowned for its —— deposits.
    a) Gypsum  b) Barytes
    c) Lignite  d) Ochre
19. In Khetri and Singhbhum, copper mainly occurs as
    a) Cuprite  b) Chalcopyrite
    c) Chalcocite  d) Covellite
20. Kalahandi in the state of Orissa is famous for its deposits of
    a) Bauxite  b) Gypsum
    c) Galena  d) Diamonds
21. In the Khetri Copper Belt the main ore minerals are
    a) Chalcopyrite-pyrite-pyrrhotite
    b) Chalcopyrite-pyrite-bornite
    c) Chalcopyrite-cuprite-sphalerite
    d) Sphalerite-chalcocite-malachite
22. The bornite-chalcopyrite deposits of Bhilwara region in Rajasthan are of—— origin.
    a) Epithermal  b) Mesothermal
    c) Hypothermal  d) Telethermal
23. The main ore mineral in the Mailaram Copper Belt of the Khamman district in Andhra Pradesh is
    a) Chalcopyrite  b) Cuprite
    c) Bornite  d) Chalcocite
24. Drilling of major copper deposits encounter various zones. The Zone of Oxidation consists of
    a) a crust of limonite
    b) soluble salts of copper removed in solution
    c) minerals like malachite, azurite, chrysocolla
    d) minerals like covellite, bornite, tetrahedrite and chalcocite
25. The Nallamalai Group of rocks of the Cuddapah Supergroup is characterized by mineralisation of
    a) Copper  b) Lead
    c) Copper and lead  d) Copper, lead and tin
26. The gold-quartz mineralization of the KGF took place during
    a) Dharwar orogeny
    b) Dharwar geosynclinal period
    c) Eastern Ghat orogeny
    d) Cuddapah geosynclinal period

27. The Indian gold deposits are of —origin

a) Epithermal b) Mesothermal

c) Hypothermal d) Xenothermal

28. The Champion Lode of the KGF is composed of

a) Au-Quartz veins

b) Au-Quartz sulphide veins

c) Massive replacement deposits

d) Fissure veins

29. Given below are some or the well-known gold occurrences of the world. Match these with their proper mode of their occurrence

| *Gold occurrences* | *Kinds of deposits* |
|---|---|
| A. Witwatersand (S. Africa) | 1. Magmatic deposit |
| B. Kalgoorlie (Australia) | 2. Saddle Reefs |
| C. Bendigo (Australia) | 3. Fissure veins |
| D. Homestake (S. Dakota) | 4. Replacement lode |

a) A-1 B-2 C-3 D-4

b) A-4 B-3 C-2 D-1

c) A-1 B-3 C-2 D-4

d) A-4 B-1 C-2 D-3

30. The paragenetic sequence of gold formation in the Kolar Golf Field (KGF) is

a) Sulphide formation→recrystallization of silicates→ intrusion of quartz→crystallization of gold

b) Sulphide formation→intrusion of quartz→recrystallization of silicates→crystallization of gold

c) Sulphide formation→recrystallization of silicates→crystallization of gold→intrusion or quartz

d) Intrusion of quartz→recrystallization of silicates→ sulphide formation→crystallization of gold

31. Which of the following minerals are associated with the gold deposits of Kolar?

(i) Pyrite (ii) Pyrrhotite (iii) Arsenopyrite (iv) Galena

a) (i) only

b) (i) and (ii) only

c) (i), (ii) and (iii) only

d) all of the above

32. Match the following and choose the correct answer

| | |
|---|---|
| A. Chennabhavi mines (A.P.) | 1. Copper |
| B. Ghatsila | 2. Lead |
| C. Bawdin mines (Burma) | 3. Uranium |
| D. Jaduguda | 4. Gold |

a) A-4B-1 C-2 D-3
b) A-4 B-2 C-1D-3
c) A-1 B-4 C-2 D-3
d) A-1 B-3 C-4 D-2

33. A typical bauxite deposit has
   a) a laterite capping and lithomarge base
   b) a lithomarge capping and laterite base
   c) a limonite capping and laterite base
   d) None of the above

34. Bauxite occurs as
   (i) Blanket deposits (ii) Interstratified deposits
   (iii) Pocket deposits (iv) Transported deposits
   a) (i) and (ii) only
   b) (i) and (iii) only
   c) (i), (ii) and (iii) only
   d) all of the above

35. Bauxite of detrital origin occur at
   a) Katni (M.P.) b) Kolaba (Maharashtra)
   c) Kutch (Gujarat) d) Kalahandi (Orissa)

36. In the Karst regions bauxite deposits occur as
   a) Blanket deposits b) Interstratified deposits
   c) Pocket deposits d) Transported deposits

37. The East Coast bauxite extends between
   a) the Cauvery and the Godavari basins
   b) the Cauvery and the Mahanadi basins
   c) the Godavari and the Mahanadi basins
   d) the Godavari and the Damodar basins

38. Most or the domestic bauxite deposits occurring along the sea coasts are composed of
   a) bohemite
   b) diaspore
   c) gibbsite
   d) a mixture of diaspore and gibbsite

39. The average thickness of the East Coast bauxite is
   a) 2m b) 4.5m
   c) 5.5m d) 8.5m

40. Banded manganese ores are generally
   a) Epigenetic
   b) Syngenetic
   c) both epigenetic and syngenetic
   d) Paragenetic

41. The manganese deposits of Andhra Pradesh occur in association with
   a) Gondites b) Kodurites

c) Charnockites d) Lateroid concentrations

42. The manganese deposits of Kalahandi and Koraput regions are associated with
 a) Archaen gondites b) Precambrian iron ore series
 c) Dharwarian meta sediments d) Vindhyan sandstones

43. The hardest among the following manganese minerals is
 a) Braunite b) Cryptomelane
 c) Psilomelane d) Pyrolusite

44. The mineral 'Hausmonite' is chemically represented as
 a) $Mn_4O_6$ b) $Mn3O_4$
 c) $Mn_2O_4$ d) $Mn_2O_3$

45. Kudremukh in Karnataka is famous for
 a) Copper deposits b) Hematite deposits
 c) Magnetite deposits d) Ilmenite deposits

46. The purest form of iron is
 a) Native iron b) Pig iron
 c) Wrought iron d) Steel

47. The Bailadila iron ores are associated with
 a) Banded hematite quartzites b) Banded hematite jaspers
 c) Hematite schists d) Archaen gneisses

48. Iron is deposited in lakes mainly due to
 a) Magmatic concentration b) Mechanical concentration
 c) Residual concentration d) None of the above

49. Malachite has the formula
 a) $CuCO_3 2Cu(OH)_2$ b) $CuCO_3 Cu(OH)_2$
 c) $2 CuCO_3 . Cu(OH)2$ d) $2 CuCO_3 .2Cu(OH)_2$

50. 'Atacamite' is an ore mineral of
 a) Copper
 b) Lead
 c) Zinc
 d) Tungsten
 e) Tin

51. In the Agnigundala mineralized belt of Andhra Pradesh, the mineralization of Pb-Cu-Zn is reported from
 a) Vaimpalli limestones b) Nagri quartzites
 c) Tadpatri shales d) Cumbum phyllites

52. Which one of the following is NOT a lead mineral ?
 a) Galena b) Anglesite
 c) Sphalerite d) Cerrusite

53. The Indian lead deposits chiefly occur as
 a) Fissure fillings b) Breccia deposits
 c) Replacement lode deposits d) Sedimentary deposits

54. In the Zawar region of Rajasthan the country rocks for lead and zinc mineralization are the
   a) Quartzites  b) Phyllites
   c) Carbonates  d) Schists

55. In the Zawar region, the principal ore minerals of lead are
   a) galena and sphalerite  b) galena and pyrrhotite
   c) galena and arsenopyrite  d) pyrrhotite and sphalerite

56. Which of the following is a carbonate ore mineral of lead?
   a) Galena  b) Cerrusite
   c) Anglesite  d) Sphalerite

57. Which of the following does NOT contain lead?
   a) Chrome yellow  b) Chrome Red
   c) Lithophone  d) Pewter

58. Type Metal, Solder and Pewter are the alloys of lead with other minerals. Which of the following depicts the correct order of increasing lead content in these alloys?
   a) Type metal >Solder>Pewter
   b) Type metal > Pewter>Solder
   c) Solder >Pewter > Type metal
   d) Pewter >Type metal >Solder

59. Which of the following is NOT an ore mineral of zinc?
   a) Franklinite
   b) Smithsonite
   c) Hemimorphite
   d) Calamine
   e) None of the above

60. Match the following and choose the correct answer.

| *Place* | *Deposits* |
|---|---|
| A. Goramahisani (Orissa) | 1. Old workings of copper |
| B. Kalva-Gani (Andhra Pradesh) | 2. Iron |
| C. Pitchli (Bihar) | 3. Chrysotile asbestos |
| D. Khemaru (Rajasthan) | 4. Pitchblende |

   a) A-1 B-2 C-3 D-4
   b) A-4 B-2 C-3 D-1
   c) A-2 B-1 C-3 D-4
   d) A-2 B-1 C-4 D-3

61. Solder is an alloy of
   a) Lead and zinc  b) Lead and tin
   c) Lead, zinc and tin  d) Lead, zinc and antimony

62. Brass consists of
   a) Cu-80% Zn-20%  b) Cu-20% Zn-80%
   c) Cu-80% Zn-10% Sn-10%  d) Cu-70% Zn-20% Sn-10%.

63. 'Electrum' is a/an
   a) Natural alloy of gold and sliver
   b) Natural alloy of mercury with non-metallic minerals
   c) Alloy of mercury and gold
   d) None of the above

64. The Chemical composition of chromite is
   a) $Cr_2O_3$ b) $Cr_2O_4$
   c) $2FeO.Cr_2O_3$ d) $FeCr_2O_4$

65. The ratio of FeO:Cr2O3 in chromite is
   a) 84:16 b) 65:35
   c) 40:60 d) 68:32

66. The principal mineral constituent of drilling mud is
   a) Bauxite b) Barytes
   c) Beryl d) Corundum

67. Chemically barytes is composed of
   a) BaO-65.7% SO2-34.3% b) BaO-55% $SO_2$-45%
   c) BaO-12% SO2-28% d) BaO-80% SO2-20%

68. Lithophone is a white pigment composed of
   a) $PbSO_4 + ZnSO_4$ b) $ZnSO_4 + BaSO_4$
   c) $ZnS + BaSO_4$ d) $PbS + BaSO_4$

69. Match the following and choose the correct answer

| | | | |
|---|---|---|---|
| A. | Mangampeta (A P.) | 1. | Cement |
| B. | Lapsa Baru (Bihar) | 2. | Bauxite |
| C. | Churk (U.P.) | 3. | Barytes |
| D. | Salem Agency Tracts (A.P.) | 4. | Kyanite |

   a) A-3 B-1 C-4 D-2
   b) A-2 B-4 C-1 D-3
   c) A-2 B-1 C—4 D-3
   d) A-3 B-4 C-1 D-2

70. In India, Ruby mica occurs in
   a) Bihar b) Andhra Pradesh
   c) Rajasthan d) Tamil Nadu

71. The main source of commercial ruby mica in the Bihar Schist belt is
   a) Pegmatites occurring within granites
   b) Pegmatites found within schists
   c) Pegmatites occurring in shear zones
   d) Pegmatites occurring within metamorphic rocks

72. Which of the following is a hydrous mica?
   a) Muscovite b) Illite
   c) Montmorillonite d) Kaolinite

73. The Gypsum deposits of Tiruchirapalli (TN.) are of——— age.
   a) Cretaceous b) Eocene
   c) Miocene d) Pliocene

74. 'Plaster of Paris' is represented as

a) $CaSO_4.H_2O$ b) $2CaSO_4.2 H_2O$

c) $2CaSO_4.2H_2O$ d) $CaSO_4. 1/2H_2O$

75. Gypsum changes to 'Plaster of Paris' by losing——of its water molecules

a) 1/4 b) 1/2

c) 2/3 d) 3/4

76. Thorium is extracted from Monazite, which is a complex

a) Sulphate b) Silicate

c) Phosphate d) Carbonate

77. 'Smarskite' is an ore mineral or

a) Uranium b) Thorium

c) Cobalt d) Osmium

78. Uranium deposits of Jaduguda are of

a) Magmatic origin b) Sedimentary origin

c) Metamorphic origin d) Hydrothermal origin

79. Which of the following is matched correctly?

a) Pitchblende b) Uraninite

c) Carnotite d) Torbarnite

80. In the Singhbhum Shear Zone (Copper Belt Thrust) there is a definite sequence of mineralization of iron, copper and uranium. The correct Sequence of mineralization is

a) $Fe \rightarrow Cu \rightarrow U$ b) $Cu \rightarrow U \rightarrow Fe$

c) $U \rightarrow Fe \rightarrow Cu$ d) $Cu \rightarrow Fe \rightarrow U$

81. Match the following and choose the correct answer

A. Chalk Hills (T.N.) 1. Uranium

B. Araku Valley (A.P.) 2. Coal

C. Ranjit Valley (Sikkim) 3. Bauxite

D. Bhuj Formation (Gujarat) 4. Magnesite

a) A-4 B-3 C-1 D-2

b) A-4 B-2 C-1 D-3

c) A-4 B-1C-2 D-3

d) A-3 B-2 C-1D-4

82. The Gollapilli beds of Andhra Pradesh are famous for

a) Layered bands of magnesite-bearing rocks

b) Sandstone type of uranium deposits

c) Sedimentary type or uranium deposits

d) Placer deposits of gold

83. The cement made from magnesite is called

a) Puzzolona cement b) Portland cement

c) Sorrel cement d) Dolomitic cement

84. Nickel ores are found in

a) Singhbhum (Bihar) b) Guntur (A.P.)

c) Kheonjhar (Orissa) d) Khetri (Rajasthan)

85. Which of the following is an ore of tungsten?
   a) Scheelite  b) Tungstenite
   c) Wolframite  d) Hubnerite

86. Which or the following is associated with the discovery of tungsten by the GSI in the early 90s?
   a) Agucha (Rajasthan)  b) Thar desert (Rajasthan)
   c) Sironi (Rajasthan)  d) Tuen Sang (Nagaland)

87. Uranium is mostly obtained from
   a) Acid igneous rocks  b) Basic igneous rocks
   c) Sedimentary rocks  d) Metamorphic rocks

88. Which of the following is the most persistent of elements?
   a) Silver  b) Cobalt
   c) Gold  d) Nickel

89. Which of the following elements is used as a pathfinder for gold?
   a) Arsenic  b) Platinum
   c) Antimony  d) Mercury

90. Gold deposits range in age from
   a) Precambrian-Lower Tertiary
   b) Cambrian-Lower Tertiary
   c) Precambrian-Jurassic
   d) Precambrian -Recent

91. Which of the following is NOT a silver mineral?
   a) Polybasite  b) Pyrargyrite
   c) Proustite  d) None of the above

92. In India, silver is found in small quantities along with the deposits of
   a) Gold  b) Lead
   c) Zinc  d) Gold, lead and zinc

93. Which of the following is generally NOT used in the smelting of aluminum ores?
   a) Bohemite  b) Gibbsite
   c) Cryolite  d) Diaspore

94. Which of the following asbestos minerals is NOT abundant in India?
   a) Chrysotile  b) Amosite
   C) Actinolite  d) Crocidolite

95. Corundum occurs in association with
   a) Norites  b) Peridotites
   c) Nepheline syenites  d) Syenites

96. Ochres are generally composed of
   a) Laterite and pyrite
   b) Kaolin and manganese oxides
   c) Hematite and manganese oxides
   a) Magnetite pyrite and limonite

97. The Sehota deposit in Khetri (Rajasthan) is famous for
a) Copper
b) Lead-zinc
c) Realgar
d) Pyrrhotite

98. Orpiment is chemically represented as
a) AsS
b) $As_2S_3$
c) Fe.As.S
d) $Sb_2S_3$

99. Graphite of workable quantities can be extracted from
a) Khondalites
b) Charnockites
c) Gondites
d) Kodurites

100. Magmatic magnetic deposits usually occur in association with
a) Anorthosites
b) Dunites
c) Syenites
d) Peridotites

101. At a temperature of 1020°C sphalerite inverts to
a) Anglesite
b) Pentlandite
c) Wurtzite
d) Pyrrhotite

102. Which one of the following is used as a sound insulator?
a) Muscovite
b) Vermiculite
c) Steatite
d) Biotite

103. The main raw material for ceramic industry is
a) Silica
b) Clay
c) Gypsum
d) Terracotta

104. "Bleaching clays" are
a) Fire clays
b) Bentonites
c) China clays
d) Fuller's earth

105. Coal seams are often found to be associated with
a) China clays
b) Fire clays
c) Pottery clays
d) Bentonites

106. China clay deposits of economic significance are associated with
a) Talchir formation
b) Karharbari formation
c) Barakar formation
d) Ranigunj formation

107. Which of the following is a Tertiary coal field?
a) Korba (M.P.)
b) Jharia (Bihar)
c) Godavari (A.P.)
d) Palana (Rajasthan)

108. The semi-coking coal product found in the Godavari valley is called
a) Petzite
b) Kolsite
c) Cannel coal
d) Torbanite

109. Miocene coal fields are found in
a) Antartica
b) Russia
c) China
d) India

110. In India, the main coal-forming epoch was
a) Carboniferous
b) Permian
c) Tertiary
d) Cretaceous

111. In which of the following places coal is associated with Numulitic

limestones ?

a) Rajasthan
b) Spiti Valley
c) Jammu & Kashmir
d) West Bengal

112. 'Black lignite' is the name given to

a) Cannel coal
b) Anthracite
c) Bituminous coal
d) Sub-bituminous coal

113. The world's largest producer of lignite is

a) Nevyeli
b) Germany
c) Russia
d) Scandinavia

114. The age of the Neyvelli Lignite deposits is

a) Cretaceous
b) Eocene
c) Miocene
d) Oligocene
e) Pleistocene

115. Apart from Neyveli, lignite is also found in

a) Palana (Rajasthan}
b) Kothagudem (A.P.)
c) Rant Valley (Sikkim)
d) Lapsa Baru, (Bihar)

116. Petroleum is found in rocks ranging in age from

a) LowerTertiary to Recent
b) Paleocene to Pleistocene
c) Cretaceous to Miocene
d) Precambrian to Pleistocene

117. The most common reservoir rocks for petroleum are

a) Limestones
b) Sandstones
c) Shale
d) Conglomerates

118. Which of the following statements about Bombay High is correct?

i) It is a regional uplift
ii) Oil has been struck in limestones for the first time in India
iii) Oil is found in formations younger than Miocene for the first time
iv) It contributes over 2/3 rd of the needs of the Country

a) (i) and (ii) only
b) (i) and (iii) only
c) (i), (ii) and (iii) only
d) All of the above

119. The specific gravity of petroleum is

a) 1.0
b) 1.7
c) 0.7
d) 0.07

120. The largest onland oil and gas field in India is

a) Bombay High
b) Daman
c) Gandhar
d) Ankleshwar

121. In the Krishna-Godavari basin oil is accumulated in

a) Stratigraphic traps
b) Structural traps
c) Combination traps
d) None of the above

122. Which one of the following is NOT an oil refinery?

a) Barauni (Bihar)
b) Haldia (West Bengal)
c) Noonannati (Assam)
d) None of the above

123. The most dominant constituent of the underwater polymetallic nodules is

a) Ni
b) Co
c) Fe
d) Mn
e) Mo

124. Malanjkhand is famous for

a) the first open-cast mine in India
b) the biggest open-cast mine in India
c) the first sponge iron plant in India
d) Hot springs and geysers

125. India's first sponge iron plant is located at

a) Kothagudem (A.P.)
b) Kurnool (A.P.)
c) Salem (T.N.)
d) Bolangir (Orissa)

126. The parent rocks from which the asbestos deposits originate are generally very rich in

a) Olivine
b) Pyroxenes
c) Feldspars
d) Silica

127. Which of the following is an asbestos belonging to the serpentine group?

a) Crocidolite
b) Chrysotile
c) Amosite
d) Actinolite

128. The topmost producer of asbestos in the country is

a) Bihar
b) Madhya Pradesh
c) Orissa
d) Rajasthan
e) Andhra Pradesh

129. Uranium deposits are found in

a) Epicontinental environments
b) Marine environments
c) Coastal environments
d) all of the above
e) none of the above

130. Which of the following is NOT a carbonate ore mineral?

a) Bornite
b) Cerrusite
c) Smithsonite
d) Witherite

131. Which of the following is NOT a siliceous abrasive?

a) Quartzite
b) Novaculite
c) Tripoli
d) Rouge

132. Which of the following is NOT associated with the occurrence of manganese?

a) Jhunjhuna
b) Midnapur
c) Banswara
d) Mayurbhanj

133. Koduritic manganese ores are characteristically developed in

a) Balaghat-Nagpur belt
b) Singhbhum-Keonjhar belt
c) Eastern Ghats
d) Mysore-South Bombay

134. Which of the following constituent is present in Kodurites but not in Gondites?

a) Quartz
b) Potash feldspar
c) Manganese garnet
d) Manganese pyroxene

135. In the Zawar area of Rajasthan, the mineralization took place in the rocks belonging to

a) Aravalli system
b) Delhi group
c) Post-Delhi group
d) Paleocene rocks

136. In India, gold deposits are associated with rocks of

a) Archean age
b) Proterozoic age
c) Cambrian age
d) Mesozoic age

137. Match the following and choose the correct answer

A. Hausmaunite — 1. Zn
B. Hemimorphite — 2. Mn
C. Tetrahedrite — 3. Cu
D. Anglesite — 4. Ag
5. Pb

a) A-2 B-1 C-3D-4
b) A-1 B-2 C-3 D-5
c) A-2 B-lC-4 D-3
d) A-2 B-1 C-3 D-5

138. Which of the following affords an example or low-lying bauxite deposits?

a) East-Coast bauxite
b) Kolaba and Kolhapur (Maharashtra)
c) Shahdol-Bilaspur-Durg (M.P.)
d) Katni (M.P.)

139. The bauxite deposits of Madhya Pradesh and Bihar are composed of

a) monohydrates only
b) trihydrates only
c) a mixture or monohydrates and trihydrates
d) None of the above

140. A large part of the East-Coast Bauxite is formed due to

a) Sub-aerial weathering and decomposition of Deccan traps
b) Weathering of Khondalites
c) Weathering of Charnockites
d) Decomposition of Vindhyan sandstones and quartzites

141. The East-Coast Bauxite differs from the other bauxite occurrences in the country in one or more of the following respects

1) Greater average thickness of deposits
2) Large size of individual deposits
3) Occurrence at much higher elevations
4) Composition is entirely gibbsitic

a) 1 and 3 are correct
b) 1 and 4 are correct
c) 1, 2 and 3 are correct
d) 1, 2 and 4 are correct

142. The East-Coast Bauxite covers the states of

a) Andhra Pradesh and Tamil Nadu
b) Andhra Pradesh and Orissa
c) Andhra Pradesh, Orissa and Tamil Nadu
d) Andhra Pradesh, Orissa and Kerala

143. The Chromite deposits of Andhra Pradesh are associated with

a) Quartzites
b) Enstatite-Peridotites
c) Serpentinized ultrabasic charnockites
d) Khondalites

144. The coal seams formed by in situ origin are characterized by

a) the presence of coal beds
b) the presence of coal beds of wide aerial extent
c) the presence of coal beds of uniform thickness and wide aerial extent
d) the presence of partings in the coal seams

145. The 'Clinton type' of ironstones are

a) Hematite-chamosite-siderite rocks
b) Limonite- chamosite-siderite rocks
c) Magnetite- chamosite-siderite rocks
d) Banded magnetite-hematite quartzites

146. 'Carbanados' are

a) massive deposits of Cannel coal
b) thick successions or graphite
c) jet black variety of industrial diamond
d) abrasives made of siliceous carbide

147. Which of the following ore minerals is the chief source of thorium in the world?

a) Thorianite
b) Allanite
c) Pitchblende
d) Monazite

148. Which of following is used in making the "gypsum plate" for petrological microscopes?

a) Selenite
b) Satinspar

c) Alabastar d) Anhydrite

149. Which of the following is used as a "barium meal" while taking x-rays for detecting tumors and growths in the stomach and intestines?
a) barium sulphide b) barium chlorite
c) arsenic sulphide d) lead dioxide

150. The first cement factory was built in India in 1904 at
a) Ranchi b) Chennai
c) Hazaribagh d) Ahmedabad

151. Who among the following is considered as the 'Father of Economic Geology'?
a) Georgius Agricola b) Descartes
c) Nicolaus Steno d) Von Oppel

152. All ore deposits that are directly crystallized from magma are termed as
a) Magmatic segregation deposits
b) Contact metamorphic deposits
c) Pegmatites
d) Pyroxomatic deposits

153. Syngenetic deposits are crystallized
a) before the host rocks
b) after the host rocks
c) simultaneously with host rocks
d) any of the above

154. Which of the following is an endogenetic deposit?
a) Oxidation and Supergene sulphide enrichment
b) Residual concentration deposits
c) Sedimentation deposits
d) Fumarolic deposits

155. Sublimation is a process in which compounds are formed by
a) Mixing of mineralizing solutions
b) Hydrothermal alteration at high pressures and temperatures
c) Volatilization and redeposition from vapour at LP-LT
d) All of the above

156. The process of formation of deposits from channels at or near the surface is described as
a) Epigenetic b) Syngenetic
c) Paragenetic d) none of the above

157. The deposits formed at the end of magmatic differentiation are termed as
a) Orthotectic deposits b) Orthomagmatic deposits
c) Metasomatic deposits d) Hydrothermal deposits

158. If gangue minerals such as chalcedony and adularia are found in association with the hydrothermal ores, then the ores are of
a) Hypothermal type b) Mesothermal type

c) Epithermal type d) Xenothermal type

159. The optimum temperature range for the formation of hypothermal deposits is

a) 300-600°C b) 300-450°C
c) 200-300°C d) 50-200°C

160. The disseminated or "porphyry" copper deposits are of

a) Hypothermal origin b) Mesothermal origin
c) Epithermal origin d) Telethermal origin

161. Which of the following deposits are most abundant amongst young rocks?

a) Epithermal b) Mesothermal
c) Hypothermal d) Xenothermal

162. Bonanzas of native gold and electrum are generally deposited under

a) Epithermal conditions b) Mesothermal conditions
c) Hypothermal conditions d) Telethermal conditions

163. The telethermal deposits are generally

a) Syngenetic b) Diagenetic
c) Epigenetic d) Any of the above

164. The mineralogy of telethermal deposits is similar to that of

a) Xenothermal deposits b) Epithermal deposits
c) Mesothermal deposits d) Hypothermal deposits

165. The 'Xenothermal' deposits are formed under the conditions of

a) HP-HT b) HP-LT
c) LP-HT d) LP-LT

166. Closely spaced parallel fractures which are filled with mineral matter are described as

a) Composite veins b) Linked veins
c) Branched veins d) Sheeted veins

167. A vertical dyke showing transverse veins is known as

a) Stockwork b) Saddle reef
c) Ladder vein d) Vug

168. Veins that are essentially parallel to one another but dipping in different directions are referred to as

a) Parallel veins b) Linked veins
c) Cognate veins d) Radial veins

169. Deposits that result from the mineralization in the openings along anticlines are called

a) Shear zone deposits b) Stockworks
c) Fahlbands d) Saddle reefs

170. Gash veins are

a) horizontal or gently inclined cave deposits
b) large fracture zones composed of several parallel veins
c) mineralized portions of shear dikes

d) mineralized vertical solution joints in carbonate country works

171. Zoning around a nucleus by rhythmic precipitation of mineral matter results in a/an
   a) Colloform texture
   b) Intergrowth texture
   c) Spherulitic texture
   d) Mimetic texture

172. Pseudomorphs are very common in—deposits.
   a) Replacement
   b) Cavity filling
   c) Fissure filling
   d) Metasomatic

173. The 'Sideronitic texture' consists of crystals that exhibit
   a) perfect cleavage planes
   b) fractures with broken corners
   c) a number of well-developed faces
   d) all of the above

174. Sideronitic texture is typically associated with
   a) Hydrothermal deposits
   b) Contact metasomatic deposits
   c) Late magmatic deposits
   d) Early magmatic deposits

175. Placer deposits are formed as a result of
   a) Residual liquid segregation
   b) Residual concentration
   c) Mechanical concentration
   d) All of the above

176. A placer deposit formed at the site of the destruction of the primary source rock is called
   a) Aeolian placer
   b) Alluvial placer
   c) Elluvial placer
   d) Colluvial placer

177. Placer gold deposits are mostly
   a) Alluvial
   b) Elluvial
   c) Pluvial
   d) Colluvial

178. Important placer deposits of monazite and ilmenite occur along
   a) The East Coast of India
   b) The West Coast of India
   c) The East and the West coasts of India
   d) None of the above

179. Hydrothermal replacement process controls
   a) Oxidation
   b) Metasomatism
   c) Supergene enrichment
   d) All of the above

180. If an ore body undergoing oxidation and supergene enrichment lies in limestone country rocks, then
   a) Enrichment is complete
   b) Enrichment is partial
   c) Enrichment does not take place
   d) any of the above

181. If the colour of the gossan is black, it indicates the presence of
a) Iron b) Manganese
c) Copper d) Graphite

182. Filter pressing movements are usually seen in
a) Immiscible liquid injected deposits
b) Residual liquid injected deposits
c) Immiscible liquid seggregated deposits
d) Residual liquid seggregated deposits

183. The best example or magmatic seggregated ores is
a) Bauxite deposits b) Iron deposits
c) Chromium deposits d) Phosphate deposits

184. According to the Schrumann's series, which or the following has least solubility?
a) Hg b) Ag
c) Mn d) Cu

185. Pegmatites are
a) massive mineral deposits of sedimentary origin
b) huge outcrops of fine-grained volcano-plutonic rocks
c) very coarse-grained igneous or metamorphic rocks
d) bonanzas of gold and silver

186. In a complex Pegmatite, the zone that is richest in economic minerals is
a) Border zone b) Wall zone
c) Intermediate zone d) Core zone

187. Which of the following deposits are commonly associated with fumaroles?
a) Sulphides b) Phosphates
c) Carbonates d) Oxides

188. A rich streak of ore, especially of gold and silver is termed as
a) Protore b) Vug
c) Bonanza d) Ore shoot

189. The mineral that responds to a change in the environment by the process of molecular exchange is
a) Chalocite b) Covellite
c) Cassiterite d) Diamond

190. Tactite is a/an
a) Complex carbonate rock b) Complex silicate rock
c) Iron meteorite d) Stony meteorite

191. The surface expression of a major copper deposit underneath is given by a crust of
a) Limonite b) Malachite
c) Covellite d) Chalcocite

192. Itabinite, jaspilite and specularite are types of
a) Banded iron formations b) Iron stones

c) Precious metals
d) Precious stones

193. The great bulk of Banded iron formations (BIFs) was laid down during the time interval of
a) 3500-2600 ma.
b) 2600-1800 ma.
c) 1800-950 ma.
d) 950-570 ma.

194. Which of the following types of BIFs are economically most important?
a) Oxide facies
b) Carbonate facies
c) Silicate facies
d) Sulphide facies

195. The 'Algoma type' of BIFs was the most characteristic of
a) Orogenic Mountain belts
b) Proterozoic geosynclines
c) Archaen greenstone belts
d) Upper Jurassic lava flows

196. According to Lindgren's classification of ore deposits, the pegmatite deposits can be described as
a) deposits formed by chemical processes in a magma by differentiation at moderate to high pressures and high temperatures
b) deposits formed by chemical processes in bodies of rocks without introduction of foreign substances
c) deposits of huge size formed by mechanical processes
d) direct magmatic deposits formed at low temperatures and low-moderate pressures

197. Who introduced the term 'Xenothermal' to Lindgren's classification of ore deposits?
a) Graton (1933)
b) Buddington (1935)
c) Bateman (1942)
d) Stanton (1972)

198. Wàter that is trapped in sediments at the time of their deposition is known as
a) Meteoric water
b) Juvenile water
c) Connate Water
d) Mine water

199. Which of the following does not represent exogenetic mineral deposits?
a) Residual deposits
b) Sedimentary deposits
c) Oxidised and secondary enriched deposits
d) Fumerolic deposits

200. The 'Pyrosomatic deposits' are formed by
a) metamorphic processes
b) residual alteration
c) contact metasomatism
d) pneumatolytic alteration

201. Mineral deposits formed in which of the following suites are difficult to be recognised by temperature-pressure critèria owing to mixing of HT and LT mineral suites
a) Hypothermal
b) Mesothermal

c) Epithermal
d) Telethermal
e) Xenothermal

202. Match the following and choose the correct answer.

| Type of deposit | Example |
|---|---|
| A. Magmatic deposits | 1. Sulphur deposits of Chile |
| B. Fumerolic deposits | 2. Uranium-vanadium deposits of Colarado, USA |
| C. Residual deposits | 3. Kyanite deposits of Singhbhum |
| D. Telethermal deposits | 4. Diamond deposits of Majhagawan, M.P. |

a) A-4 B-1 C-3 D-2
b) A-2 B-1 C-3 D-4
c) A-1 B-4 C-3 D-2
d) A-4 B-1 C-2 D-3

203. The Early Magmatic Deposits can be formed by
I. Simple crystallization without concentration.
II. Segregation of early formed crystals
III. Injection of the minerals formed elsewhere by differentiation
a) I is correct
b) I & II are correct
c) II & III are correct
d) I, II & III are correct

204. The diamond deposits of Wajrakarur in the Anantapur district of Andhra Pradesh are an example of
a) Early magmatic dissemination deposits
b) Early magmatic segregation deposits
c) Early magmatic injection deposits
d) Residual liquid segregation deposits

205. A "Cockade ore" is formed when
a) the breccia fragments are surrounded by crusts
b) the ore is arranged in successive layers
c) fissure fillings are composed of alternate bands of quartz and altered country rocks
d) small crystals are developed haphazardly within cavities

206. Mineralized portions of shear zones that are formed as a result of cavity filling or metasomatic replacement are described as
a) Lodes
b) Stockworks
c) Fahlbands
d) Saddle Reefs

207. A thin layer of clayey material that commonly lines the fissure vein deposits is described as
a) Gouge or Selvage
b) Vug or druse

c) En echelon lense, d) Gallery

208. The term 'Leptothermal' was introduced by L.C.Graton to describe the hydrothermal deposits which grade from
   a) Hypothermal→Mesothermal b) Mesothermal→Epithermal
   c) Epithermal→Telethermal d) Telethermal→Xenothermal

209. Colloform textures are typically exhibited by
   a) Hypothermal deposits b) Mesothermal deposits
   c) Epithermal deposits d) Telethermal deposits

210. An ore mineralogy comprising of minerals like polybasite, stepnanite, pearcite, pyrargyrite, proustite, petzite, slyvite, calavarite, etc is most likely to be formed under
   a) Hypothermal conditions b) Mesothermal conditions
   c) Epithermal conditions d) Telethermal conditions

211. Which of the following are the deepest of all the hydrothermal deposits?
   a) Hypothermal deposits b) Mesothermal deposits
   c) Epithermal deposits d) Telethermal deposits

212. The chemical equation: $ZnS + CuSO_4 \rightarrow CuS + ZnSO_4$ is an illustration of
   a) Replacement b) Cavity Filling
   c) Deuteric alteration d) Pneumatolytic alteration

213. Which of the following is true of a "massive" deposit?
   I. It results due to the entire replacement of the host or country rock.
   II. These deposits end abruptly against the host or country rock.
   III. They generally preserve the structure and texture of the original rock.
   IV. They vary in size and are extremely irregular in form.
   a) only I is correct b) only I & II are correct
   c) only I, II & III are correct d) all of the above are correct

214. Which of the following rocks are most suitable for the formation of replacement deposits?
   a) Limestones b) Sandstones
   c) Quartzites d) Granites

215. Match the following and choose the correct answer.

| **Mineral deposit** | **Associated Rocks** |
|---|---|
| A. Platinum | 1. Anorthosites |
| B. Chromium | 2. Norites and Peridotites |
| C. Titanomagnetite | 3. Gabbros |
| D. Diamonds | 4. Syenites |
| | 5. Kimberlites |

   a) A-1 B-2 C-3 D-5 b) A-2 B-1 C-4 D-5
   c) A-4 B-2 C-3 D-5 d) A-2 B-1 C-3 D-5

216. The "Porphyry" Copper deposits of USA are a result of deposition at
   a) divergent plate boundaries b) convergent plate boundaries
   c) cratonic rift systems d) cratonic basins

217. Accumulation of ore minerals like pyrrhotite, chalcopyrite and pentlandite at the bottom of the magma chamber indicate that the mineral deposits have originated by
  a) Early Magmatic Dissemination
  b) Late Magmatic Residual Liquid Segregation
  c) Late Magmatic Residual Liquid Injection
  d) Immiscible Liquid Segregation

218. The Sideronitic texture is generally associated with deposits formed by
  a) Early Magmatic Dissemination
  b) Late Magmatic Residual Liquid Segregation
  c) Late Magmatic Residual Liquid Injection
  d) Immiscible Liquid Segregation

219. The Banded Magnetite Quartzite deposits of Salem district in Tamil Nadu have resulted from
  a) Hydrothermal processes
  b) Magmatic processes
  c) Contact Metasomatism
  d) Residual Concentration
  e) Sedimentation

220. Gossans are good indicators of the form and size of the underlying ore minerals. Which of the following is NOT matched correctly?

| | **Colour of Gossans** | | **Possible Presence of** |
|---|---|---|---|
| a) | Yellow or brown or red | ……. .._ | Iron |
| b) | Green or blue | ……… | Copper |
| c) | Waxy green | ……… | Silver |
| d) | Dark green | ……… | Manganese |

221. The correct sequence of mineralization of uranium at Jaduguda in Singhjhum district of Bihar is
  a) Apatite→Magnetite→Copper→Uranium
  b) Magnetite→Apatite→Copper→Uranium
  c) Uranium→CopperàMagnetite→Apatite
  d) Apatite→Magnetite→Uranium→Copper

222. Which of the following is NOT an example of sandstone type of Uranium deposit?
  a) Mahadek Formation, Meghalaya
  b) Mottur Beds, Madhya Pradesh
  c) Bhuj Formation, Gujarat
  d) Badam Pahar, Orissa

223. Which of the following is neither a paint nor a pigment?
  a) Lithophone  b) White Lead
  c) Chrome Red  d) Plaster of Paris

224. 'Spiegeleisen' is an alloy of
  a) Magnesium  b) Manganese

c) Arsenic d) Thorium

225. Choose the incorrect pair from the following.
   a) Azurite — copper
   b) Anglesite — lead
   c) Goethite — iron
   d) Rhodonite — manganese
   e) Saxonite — chromium

226. In which of the following places in Assam, oil has been discovered in 1998.
   a) Mohanbari b) North Balijan
   c) South Kathaloni d) All of the above

227. During 1997-98 Oil has been discovered in Deohal, Bhogpara, Shalmari and Jeraipathar structures in the Lakadong- Therria region in Assam. These structures are of
   a) Eocene age b) Oligocene age
   c) Miocene age d) Pliocene age

228. In the recent years Oil India Limited (OIL) is redirecting its exploratory efforts towards prospects in
   a) Deeper Paleocene and lower Eocene
   b) Lower Eocene and Upper Miocene
   c) Lower Eocene to Upper Oligocene
   d) Pleistocene to Recent

229. The actual gold extracted in India during 1997-98 was
   a) 446.29 kg b) 514.1 kg
   c) 688.24 kg d) 937.92 kg

230. The per capita consumption of copper metal in India is about
   a) 0.21 kg b) 1.0 kg
   c) 5.6 kg d) 13.14 kg

231. Which country has the highest per capita consumption of copper (13.14 kg) in the world
   a) United States b) France
   c) Japan d) Germany

232. Copper represents ——of the total ore resources in India.
   a) 58.7% b) 42.9%
   c) 33.3% d) 25%

233. Which of the following is the most widely consumed metal in the world?
   a) Iron b) Aluminum
   c) Zinc d) Copper

234. In which of the following countries, the consumption of zinc is highest in the world?
   a) Japan b) Korea
   c) China d) USA

235. In India the consumption of zinc is highest in
   a) Die-castings b) Alloys
   c) Galvanising d) Batteries

236. A major multi-metal world-class deposit of zinc, lead and associated metals had been discovered in
   a) Chicholi in Durg district of Madhya Pradesh
   b) Rampura-Agucha belt in Bhilwara district of Rajasthan
   c) Jhani and Bareonata in Kangra district of Himachal Pradesh
   d) Histau in Hazaribagh district of Bihar

237. The Oil India Limited (OIL) was started to develop the oil fields in
   a) Nahorkatia b) Moran
   c) Ankleshwar d) Bombay High

238. The Anandapur-Bhukia block in the Banswara district of Rajasthan is associated with the recent discovery of
   a) Coal Bed Methane b) Gold ore
   c) Diamond pipes d) Clay deposits

239. The Anumpalle kimberlite is located in
   a) Andhra Pradesh b) Rajasthan
   c) Bihar d) Karnataka

240. Match the following and choose the correct answer.

| | Place/Block | | Mineral Deposit |
|---|---|---|---|
| a) | South Sindeswar district, Rajasthan Ridge Block, Rajsamand | 1. | Diamonds |
| b) | B. Mundayur block, Attapady valley, Kerala | 2. | Pt+Pd+Au |
| C. | Boula-Nuasahi prospect, Sukinda belt, Orissa | 3. | Gold ores |
| D. | Dibbasanipalle kimberlite, Anantpur district, Andhra Pradesh | 4. | Pb+Zn |

   a) A-1 B-2 C-3 D-4
   b) A-4 B-2 C-3 D-1
   c) A-4 B-3 C-2 D-1
   d) A-1 B-3 C-2 D-4

241. Oil was first struck in India in
   a) 1866 b) 1882
   c) 1889 d) .1893

242. With a view to accelerate exploration efforts for oil and gas, GOI has formulated an accelerated exploration programme (APEX). The main components of the APEX programme are:

i) Deep water exploration
ii) Frontier area exploration
iii) National seismic survey programme
iv) Acquisition of foreign acreages/equity oil
a) i & ii are correct
b) ii, iii & iv are correct
c) i, iii & iv are correct
d) All of the above are correct

243. The giant oil field of 'Ankleshwar' was discovered by the ONGC in
a) 1956 b) 1958
c) 1960 d) 1963

## ECONOMIC GEOLOGY (KEY)

| | | | | | | | | | |
|---|---|---|---|---|---|---|---|---|---|
| 1-a | 2-e | 3-a | 4-c | 5-a | 6-c | 7-b | 8-d | 9-b | 10-d |
| 11-c | 12-a | 13-b | 14-d | 15-a | 16-b | 17-c | 18-c | 19-b | 20-a |
| 21-a | 22-b | 23-a | 24-c | 25-c | 26-a | 27-c | 28-a | 29-c | 30-d |
| 31-d | 32-a | 33-a | 34-d | 35-a | 36-c | 37-c | 38-c | 39-d | 40-b |
| 41-b | 42-b | 43-a | 44-b | 45-c | 46-c | 47-a | 48-c | 49-b | 50-a |
| 51-d | 52-c | 53-a | 54-a | 55-a | 56-b | 57-c | 58-c | 59-e | 60-d |
| 61-c | 62-a | 63-a | 64-d | 65-d | 66-b | 67-a | 68-c | 69-d | 70-a |
| 71-d | 72-b | 73-c | 74-d | 75-d | 76-c | 77-b | 78-d | 79-b | 80-a |
| 81-c | 82-b | 83-c | 84-c | 85-c | 86-a | 87-a | 88-c | 89-a | 90-a |
| 91-d | 92-d | 93-c | 94-d | 95-c | 96-c | 97-C | 98-B | 99-a | 100-c |
| 101-c | 102-b | 103-b | 104-b | 105-b | 106-c | 107-d | 108-b | 109-a | 110-b |
| 111-c | 112-d | 113-b | 114-c | 115-a | 116-d | 117-b | 118-d | 119-c | 120-c |
| 121-c | 122-d | 123-d | 124-b | 125-a | 126-a | 127-b | 128-d | 129-d | 130-a |
| 131-d | 132-a | 133-c | 134-b | 135-a | 136-a | 137-d | 138-d | 139-c | 140-b |
| 141-d | 142-b | 143-c | 144-c | 145-a | 146-c | 147-d | 148-a | 149-a | 150-c |
| 151-a | 152-a | 153-c | 154-d | 155-c | 156-a | 157-d | 158-c | 159-a | 160-b |
| 161-a | 162-a | 163-d | 164-b | 165-c | 166-d | 167-c | 168-c | 169-d | 170-d |
| 171-a | 172-a | 173-b | 174-d | 175-c | 176-c | 177-a | 178-b | 179-c | 180-c |
| 181-b | 182-b | 183-c | 184-a | 185-c | 186-c | 187-a | 188-c | 189-a | 190-b |
| 191-a | 192-a | 193-b | 194-a | 195-c | 196-a | 197-c | 198-c | 199-d | 200-c |
| 201-e | 202-a | 203-d | 204-a | 205-a | 206-c | 207-a | 208-b | 209-c | 210-c |
| 211-a | 212-a | 213-d | 214-a | 215-d | 216-b | 217-d | 218-c | 219-e | 220-d |
| 221-d | 222-d | 223-d | 224-b | 225-e | 226-d | 227-a | 228-a | 229-b | 230-a |
| 231-d | 232-a | 233-a | 234-d | 235-c | 236-b | 237-a | 238-b | 239-a | 240-c |
| 241-c | 242-d | 243-c | | | | | | | |

# 15

# Applied Geology

## A. EXPLORATION METHODS

1. The outcrops of manganese deposits exhibit ———colour.
   a) Black
   b) Brown
   c) Maroon
   d) Waxy green
2. If an outcrop of a mineral deposit shows a prominent pinkish black colour it indicates the presence of —— deposits underneath.
   a) Nickel
   b) Uranium
   c) Arsenic
   d) Copper
3. A surface outcrop of limonite indicates the presence underneath of
   a) Copper deposits only
   b) Copper and some sulphide deposits
   c) Copper and some oxide deposits
   d) Oxides, Carbonates and Sulphides
4. The ochreous orange colour of limonite indicates the presence of
   a) Galena
   b) Molybdenite
   c) Sphalerite
   d) Bornite
5. The shapes of sulphide voids in certain leached outcrops are good indicators of the characteristic predecessor mineral. Match the shape of the voids with their correct indicative mineral deposit.

| **Shape of Voids** | **Indicative Mineral** |
|---|---|
| A. Spear-shape | 1. Rhodochrosite |
| B. Blades | 2. Molybdenite |
| C. Plates | 3. Arsenopyrite |
| D. Rhombs | 4. Enargite |

   a) A-4 B-3 C-2 D-1
   b) A-3 B-4 C-2 D-1
   c) A-4 B-3 C-1 D-2
   d) A-3 B-4 C-1 D-2
6. As one moves from the Tropical zones towards the Polar regions, the concentration of limestones
   a) increases

b) decreases
c) first increases then decreases
d) first decreases then increases

7. Chloritic alteration of wall rocks is NOT a feature of
a) Clastic sedimentary rocks
b) Non-clastic sedimentary rocks
c) Acid igneous rocks
d) Metamorphic rocks

8. Extensive dolomitization is a characteristic feature of
a) molybdenum deposits b) copper deposits
c) gold-silver deposits d) lead-zinc deposits

9. Titanium-magnetite deposits are often found associated with
a) acid plutonic rocks b) acid volcanic rocks
c) basic plutonic rocks d) basic volcanic rocks

10. The Structure of Bombay High is
a) Doubly plunging anticline b) Doubly plunging syncline
c) Non-Plunging anticline d) Non-plunging syncline

11. The Bauxite deposits of Orissa are associated with
a) positive topographic forms
b) negative topographic forms
c) structural elevations
d) tectonic deeps

12. Certain minerals show their oxidation products in the form of distinct 'blooms' that can be spotted from air if in sufficiently high concentrations. Pink 'blooms' are exhibited by
a) cobalt b) vanadium
c) uranium d) lead

13. Which of the following minerals are most likely to occur together?
a) Va-Cu-Ni b) Ni-Co-Au
c) Pb-Zn-Cu d) Va-U-Fe

14. The Graphite deposits of Kalahandi, the Garnet deposits of Rajasthan and the Kyanite deposits of Singhbhum are associated with
a) Igneous rocks b) Sedimentary rocks
c) Metamorphic rocks d) All of the above

15. 'Trenching' involves
a) linear excavations
b) linear excavations in which one dimension is longer than the other
c) drilling wedge-shaped bore holes
d) digging even-sized pits

16. 'Drilling' is employed for
1) determining the stratigraphic or rock sequence

2) locating and evaluating substances of economic value
3) extracting economically valuable substances
4) obtaining data for correlation
a) 2 and 3 are correct
b) 1, 2 and 3 are correct
c) 2, 3 and 4 are correct
d) all of the above are correct

17. Diamond drilling can be used to bore holes in
a) horizontal direction only
b) vertical direction only
c) horizontal and vertical directions
d) all directions

18. The most useful type of drilling for penetrating hard or abrasive ground is
a) Diamond drilling b) Rotary drilling
c) Percussion drilling d) Churn drilling

19. Which of the following rotary drilling cannot assess?
a) Type of the rock b) Mineralogy of a rock
c) Texture of a rock d) All of the above

20. Which of the following is a common method used for sampling of placer deposits?
a) Churn drilling b) Auger drilling
c) Jet drilling d) Rotary drilling

21. Churn drilling or cable-tool drilling is capable of drilling holes in
a) horizontal directions only b) vertical directions only
c) inclined directions only d) all of the above

22. Caliper logging is employed for determining
a) physical properties of rock formations
b) stratigraphic sequence of rock formations
c) variations in bore-hole diameters
d) porosity of reservoir fluids

23. The 'conductivity' or rock formations can be measured by means of
a) Electric logging b) Induction logging
c) Sonic logging d) Radiation logging

24. A Natural Gamma-ray log exhibits a "high response" from
a) shales b) limestones
c) dolomites d) coal

25. The Self-potential (SP) and Induced potential (IP) surveys to locate ore bodies are done by
a) Seismic Methods b) Magnetic Methods
c) Radiometric Methods d) Electromagnetic Methods

26. The average gravitational force of the earth is
a) 98cm/s2 b) 980cm/s2

c) 9800 cm/s2 d) 980 cm/s

27. The value of Universal Gravitational Constant (G) in the S.I. units is
   a) $6.67 \times 10^{-8}$ b) $6.67 \times 10^{-9}$
   c) $6.67 \times 10^{-10}$ d) $6.67 \times 10^{-12}$
28. The difference between the observed and the predicted gravity values after applying all the corrections is described as
   a) Free-air anomaly b) Free-air correction
   c) Bouguer anomaly d) Bouguer correction
29. Which of the following is an example of "unstable" gravimeter?
   a) Hartley gravimeter b) Ising gravimeter
   c) Worden gravimeter d) Wright gravimeter
30. The gravity prospecting methods are useful in the exploration of oil and gas to determine
   a) the thickness of sedimentary columns
   b) the nature of trap rocks
   c) the potential of oilfields
   d) none of the above
31. Which of the following is commonly used as a "drilling mud" in the exploration of oil and gas?
   a) Heavyspar b) Satinspar
   c) Flourspar d) Jasper
32. The total magnetic field strength is weakest at
   a) 0° latitude b) 30°N -30°S latitudes
   c) 90° latitude d) 60°N-60° S latitudes
33. The unit used to express the Magnetic field strength in the M.K.S. system is called
   a) Oersted b) Tesla
   c) Weber d) Maxwell
34. Magnetic materials having extremely high magnetic susceptibility are described as
   a) Ferromagnetic b) Paramagnetic
   c) Diamagnetic d) Non-magnetic
35. Pick out the sequence of minerals that are arranged according to their increased magnetic susceptibility values.
   a) Quartz—Anhydrite—Chromite—Magnetite
   b) Quartz—Chromite—Anhydrite —Magnetite
   c) Magnetite—Quartz—Chromite—Anhydrite
   d) Anhydrite —Quartz —Chromite—Magnetite
36. The magnetism that is produced during the formation of iron-bearing minerals at low temperatures is called
   a) Thermo-remnant magnetism
   b) Depositional remnant magnetism
   c) Viscous remnant magnetism

d) Chemical remnant magnetism

37. Detailed survey of oil and gas is done by
a) Seismic reflection
b) Seismic refraction
c) Geomagnetic Methods
d) Electromagnetic Methods

38. 'Hodograph' is
a) an instrument used to record the time taken by a seismic wave to travel from the point of explosion to a seismograph
b) a curve plotted on graphs which depicts the time taken by a seismic wave to travel from the point of explosion to a seismograph
c) an instrument which records the intensity of earthquake shocks
d) a graph showing the variation of seismic velocities with depth

39. The 'hodograph' of a reflected wave is
a) hyperbolic
b) parabolic
c) rectilinear
d) curvilinear

40. Which of the following geophysical methods of exploration is of particular significance in defining oilfield traps such as anticlines and salt domes?
a) Seismic refraction
b) Geomagnetic Method
c) Electromagnetic Method
d) Gravity method

41. The presence of a well-defined "structural high" near Lunej in the Cambay basin has been brought to light as a result of
a) Resistivity surveys
b) Bouguer Correction
c) Seismic Reflection Survey
d) Studying the thermo-remnant magnetism

42. The P-wave velocities are highest in
a) Air
b) Water
c) Sand
d) Granite

43. In terms of electric conductivity, granites can be described as
a) Good conductors
b) Bad conductors
c) Moderate conductors
d) Semi-conductors

44. Which of the following has the least electrical resistivity?
a) Diamond
b) Sulphur
c) Clay
d) Anthracite

45. In the Resistivity Sounding method
a) the position of electrodes is fixed while that of the central point is changed
b) the position of electrodes is changed with respect to a fixed central point
c) the position of all electrodes is changed simultaneously
d) none of the above

46. Electric Profiling Method is used to locate
a) Horizontal discontinuities

b) Vertical discontinuities
c) Both Horizontal and Vertical discontinuities
d) None of the above

47. The 'Schlumberger Configuration' is used in
a) Electric profiling
b) Magnetic profiling
c) Geomagnetic profiling
d) Seismic profiling

48. The Self-Potential or Spontaneous Polarization (SP) Method is most suitable for the exploration of
a) sulphides
b) sulphates
c) oxides
d) carbonates

49. The boundaries of the Barakar rocks in the Kamthi Coal field of Maharashtra was delineated using
a) Gravity surveys
b) Resistivity surveys
c) Seismic Reflection surveys
d) Geomagnetic surveys

50. Dispersion of elements is favoured by
I. Low melting and boiling points
II. Odd atomic numbers and atomic weights
III. Very low atomic volumes
IV. Strong polarization of ions in the lattice
a) I and III only
b) I, II and III only
c) I, II and IV only
d) All of the above

51. Elements which are found in the dispersed state generally have
a) very large ionic radii
b) very small ionic radii
c) very large or very small ionic radii
d) intermediate ionic radii

52. The primary geochemical environment is characterized by
a) high temperatures and low pressures
b) high temperatures and pressures
c) high temperatures and pressures and restricted circulation of fluids
d) high temperatures and pressures, restricted circulation of fluids and low free-oxygen content

53. Which of the following pairs of elements do not commonly form an association?
a) K-Rb
b) Ca-Sr
c) Al-Si
d) Zr-Hf

54. In the mafic igneous rocks, the most common geochemical association is
a) Si-K-Na
b) Fe-Mg-Ti
c) Ag-Au-As
d) Zn-Pb-Ba-F-Sr

55. The common indicator elements in the Skarn deposits are

a) Cu-Mo
b) Pt-Cr-Ni
c) Mo-Zn-Cu
d) Au-Ag

56. In the uranium deposits of the sandstone type, the pathfinder is

a) Se
b) Cu
c) As
d) Zn

57. Syngenetic dispersion patterns are NOT formed due to

a) Igneous activity
b) Granitisation
c) Sedimentation
d) Hydrothermal solutions

58. Geochemical anomalies are termed as 'Significant' when

a) they are related to the ore deposits
b) they are unrelated to the ore deposits
c) confirm ore deposits
d) none of the above

59. 'Significant anomalies' are characterized by

a) very high concentration of indicator elements
b) very high concentration of path finders
c) very low concentration of indicator elements
d) very low concentration of path finders

60. In stream-sediment sampling, the geochemical anomalies are noticed at

a) upstream of the deposit
b) downstream of the deposit
c) both upstream and downstream of the deposit
d) none of the above

61. Mercury vapour sampling is used as a guide to

a) Metallic ore deposits
b) Non-metallic ore deposits
c) Sulphide ore deposits
d) Carbonate ore deposits

62. Match the following and choose the correct answer

| Vapour | Mineral deposits |
|---|---|
| A. Mercury | 1. Mercury deposits |
| B. Carbondioxide | 2. Uranium deposits |
| C. Radon | 3. Skarn and Greisen |
| D. Fluorine | |

a) A-1B-3C-2D-3
b) A-1B-2C-4D-3
c) A-1B-4C-2D-3
d) A-2B-3C-4D-1

63. Geochemical soil anomalies are mostly due to

a) Hydrothermal accumulation
b) Heavy Mineral accumulation

c) Residual accumulation
d) Selective concentration

64. Match the following and choose the correct answer.

| | Sample source | | Cause of Anomalies |
|---|---|---|---|
| A. | Stream sediments | 1. | Selective concentration |
| B. | Vegetation | 2. | Heavy mineral accumulation |
| C. | Marine sediments | 3. | Primary dispersion |
| D. | Sea water | 4. | Secondary dispersion |

a) A-1B-2C-3D-4
b) A-4B-3C-2D-1
c) A-2B-3C-4D-1
d) A-2B-1C-4D-3

65. Geobotanical prospecting involves
a) visual observation of plants as a guide to buried ore
b) chemical analysis of plant organs as a guide to buried ore
c) study of the distribution patterns of various plant species
d) study of the growth rates of plants in various mineralized zones

66. The plants which indicate the presence of groundwater are described as
a) Hydrophytes b) Halophytes
c) Phreatophytes d) Xerophytes

67. Halophytes are the plants that are indicative of
a) Saline deposits b) Fresh water deposits
c) Hydrocarbons d) Non-metallic ores

68. Match the following and choose the correct answer.

| | Plant Indicator | | Deposits |
|---|---|---|---|
| A. | *Alfalfa* | 1. | Groundwater |
| B. | *Salicornea* | 2. | Borate-rich saline deposits |
| C. | *Calamine violet* | 3. | Zinc |
| D. | *Astragalus* | 4. | Selenium |

a) A-2B-1C-4 D-3
b) A-4B-2C-3 D-1
c) A-4B-2C-1 D-3
d) A-1B-2C-3 D-4

69. "*Ocimum homblai*" is a universal plant indicator for
a) Copper b) Zinc
c) Iron d) Uranium

70. Match the following and choose the correct answer.

| | Plant indicator | | Element |
|---|---|---|---|
| A. | *Gypsophilia patrini* | 1. | Lead |
| B. | *Eriogonum inflatum* | 2. | Boron |
| C. | *Salsola nitraria* | 3. | Gypsum |

D. *Erianthus gigantus* 4. Copper

a) A-4B-3C-2 D-1
b) A-3B-4C-1 D-2
c) A-1B-3C-2 D-4
d) A-2B-1C-3 D-4

71. Microfossils are characterised by
a) restricted geographical but wide-spread geological distribution
b) restricted geological but wide-spread geographical distribution
c) restricted geological and geographical distribution
d) wide-spread geological and geographical distribution

72. '*Thenatocoenose*' is
a) the typical microfossil assemblage of Upper Cretaceous period
b) the typical microfossil assemblage of Lower Eocene period
c) the total fossil assemblage in a bed
d) the total fossil assemblage in a formation

73. Microfossils are useful in
I. Intra-basin correlation
II. Inter-continental correlation
III. Micro-zonation
IV. Reconstruction of Paleo-environments
a) I, III & IV are correct
b) II, III & IV are correct
c) I, II & III are correct
d) All of the above are correct

74. Which of the following depicts the correct sequence of occurrence of microfossils as one moves from the Littoral zone to the Abyssal Zone?
a) Diatoms — Ostracods — Foraminifers —Radiolarians
b) Diatoms — Foraminifers — Ostracods —Radiolarians
c) Foraminifers — Diatoms— Ostracods — Radiolarians
d) Radiolarians— Foraminifers — Diatoms— Ostracods

75. Indication about the Paleo-temperatures can be obtained by
a) studying the dextral/sinistral coiling ratios of some microfossils
b) drilling bore holes in microfossil assemblages of different ages
c) calculating the abundance of fossil spores from different horizons
d) utilising the depth sensitivity of some foraminifers

76. Which one of the following microfossils is typical of deep waters?
*a) Elphidium*
*b) Globigerina*
*c) Ammonia*
*d) Nodosaria*

77. The systematic micro-paleontological studies carried out by the ONGC in the Jaisalmer area have revealed that a marine transgression started in the area in Paleocene and continued till upper Middle Eocene. The

beginning and end of this marine transgression is represented respectively by the microfossils

a) *Globorotalia* and *Globotruncana*
b) *Globorotalia* and *Globogerapsis*
c) *Globigerina* and *Nummulites*
d) *Nummulites* and *Nodosaria*

78. Which of the following instruments is NOT used in geochemical prospecting?
a) Atomic Absorption Spectrophotometer
b) Emission Spectrograph
c) Flame Photometer
d) Optical Microscope

79. Gold prospecting in both geophysical and geochemical methods is most effectively done by
a) Electromagnetic surveys
b) Remote Sensing
c) Seismic Profiling
d) Soil analysis

80. Which sampling method is resorted to when the exposure of outcrops is scarce?
a) Bulk Sampling
b) Drilling
c) Trenching
d) Well-logging

81. A good stratigraphic guide for the prospecting of Pb-Zn ores in India is the presence of
a) Impure calcareous rocks
b) Ultramafic rocks
c) Ultra-metamorphosed rocks
d) Sheared and migmatised zones

82. Air-borne electromagnetic method is best suited for the prospecting of
a) Sulphide ores
b) Ferro-magnesian ores
c) Native metals
d) Base metals

83. "DOVEMAP" is one of the activities undertaken by the Geological Survey of India (GSI). This relates to
a) Extensive seismotectonic studies and consultancy services
b) Geochemical mapping for mineral targeting in concealed terrains
c) Development of Village Economy Through Mineral Appraisal Programme
d) Geoenvironmental studies of Sedimentary basins

84. The Geological Mapping undertaken by GSI is on the scale of
a) 1:50,000
b) 1:65,000
c) 1:40,000
d) 1:25,000

85. So far, the Geological Survey of India has completed ——— of the systematic geological mapping of the country.
a) 92.4%
b) 95.6%
c) 96.8%
d) 97.94%

86. Exploratory drilling for base metals and gold has been carried out by the Geological Survey of India in Karimanahalli, Mata Dungri, Rudiyo-Khokhro and Rengatur sectors. Match these sectors to the states to which they belong.

| Sector | State |
|---|---|
| 1. Karimanahalli | A. Maharashtra |
| 2. Mata Dungri | B. Bihar |
| 3. Rudiyo-Khokhro | C. Rajasthan |
| 4. Rengatur | D. Karnataka |

a) 1-D 2-C 3-B 4-A
b) 1-A 2-C 3-B 4-D
c) 1-A 2-B 3-C 4-D
d) 1-D 2-B 3-C 4-A

87. Multi-sensor airborne surveys were carried out by GSI over parts of Kotri lineament in the states of Madhya Pradesh and Orissa to locate
a) kimberlite pipe zones
b) gold-bearing lodes
c) multi-mineral deposits of Pb, Zn and Cu
d) hydrocarbon-bearing limestones

88. The Puga Geothermal field is located in
a) Kinnaur district, Himachal Pradesh
b) Salbardi area, Madhya Pradesh
c) Chamoli district, U.P.
d) Ladakh district, J & K

89. During 1997-98 dinosaurian nests of megalalithids type and probable multiple tract sites were discovered by the GSI in Maharashtra from the sediments of
a) Triassic age
b) Jurassic age
c) Cretaceous age
d) Permo-Carboniferous age

90. GSI was established in
a) 1831
b) 1851
c) 1947
d) 1952

## (B) MINERAL PROCESSING, MINING GEOLOGY AND MINERAL ECONOMICS

1. Screening is an ore beneficiation process which takes advantage of the differences in the
a) Size of the particles
b) Cleavage or fractures of the particles
c) Specific gravity of the particles
d) Hardness of the particles
e) Volatility and solubility of the particles

2. *Comminution* is the process of

a) Size reduction
b) Concentration of valuables
C) Liberation of valuable minerals from the gangue
d) Sorting of the particles

3. Sorting of grains is generally carried out by instruments called
a) Jaw crushers
b) Cone crushers
c) Classifiers
d) Vibrating screens

4. Hydraulic classifiers sorts out materials according to
i) Size specification
ii) Specific gravity
iii) Share of the grains
a) (i) only
b) (i) and (ii) only
c) i) and (iii) only
d) All of the above

5. Gravity separation is effective when the difference in the specific gravity of the valuable and the gangue is
a) 1
b) <1
c) >1
d) None of the above

6. In the sink-float process or the heavy media separation
a) the waste separates as float while the valuable as sink products
b) the valuable separates as float while the waste separates as sink products
c) the density of the medium is greater than that of the gangue and the valuables
d) None of the above

7. *Jigging* is a type of gravity separation process that is applicable to the particles of the size range
a) -34 to -8 mesh
b) -8 to +28 mesh
c) -28 to +200 mesh
d) None of the above

8. In a Wilfley Table used for the gravity Separation of the ores, the heavier particles settle down
a) in the direction of the motion of the table
b) in the direction of the Gravitational force
c) in the direction of the Resultant force
d) None of the above

9. Which of the following is the most successful method for chrome ore beneficiation?
a) Spiraling
b) Jigging
c) Heavy-media separation
d) Tabling

10. Arrange the minerals : Franklinite - Magnetite - Ilmenite - Hematite-Pyrite in the increasing order of their magnetic susceptibilities

a) Ilmenite<Franklinite<Pyrite<Hematite<Magnetite
b) Franklinite<Pyrite<Ilmenite<Hematite<Magnetite
c) Pyrite<Hematite<Ilmenite<Franklinite<Magnetite
d) Pyrite<Hematite<Franklinite<Ilmenite<Magnetite

11. Which of the following is the most feebly magnetic among the following?
a) Quartz b) Rutile
c) Magnesite d) Galena

12. Permanent magnets are used in
a) Low-intensity magnetic separators
b) High-intensity magnetic separators
c) Wet magnetic separators
d) None of the above

13. Ilmenite is separated from chromite by
a) Gravity separation b) Magnetic separation
c) Electrostatic separation d) Froth flotation process

14. Which of the following floatable minerals possess hydrophobic surfaces?
j) Coal
ii) Graphite
iii) Sulphur
v) Diamond
a) (i), (ii) and (iv) only
b) (i) (ii) and (iii) only
c) (ii), (iii) and (iv) only
d) All of the above

15. The object of the froth flotation process is to make certain ore minerals
a) Hydrophobic
b) Hydrophilic
c) Either hydrophobic or hydrophilic
a) None of the above

16. Briquetting, sintering and pelletising are the methods of
a) Amalgamation b) Agglomeration
c) Comminution d) Froth flotation

17. Which of the following is a process of agglomerating the fine particles by the application of pressure?
a) Briquetting b) Sintering
c) Pelletising d) Nodulising

18. In the open-pit mining, the ratio of the waste rock mined to ore is known as the
a) Working ratio b) Stripping ratio
c) Recovery ratio d) None of the above

19. A horizontal entry into an ore body is called

a) Adit b) Shaft
c) Bench d) Pit

20. Which of the following are more suitable for underground mining of ore bodies situated at great depths?
a) Horizontal shafts b) Inclined shafts
c) Vertical shafts d) None of the above

21. Which of the following is an underground mining method using artificially supported openin$_g$s?
a) Open stoping b) Shrinkage opening
c) Room and pillar mining d) Longwall mining

22. Breast stoping is employed for the working of
a) moderately thick horizontal or low-dipping ore bodies
b) very thick horizontal or low-dipping ore bodies
c) steeply dipping ore bodies with strong ore and strong wall
d) stratified or bedded deposits

23. Minerals in which a country has total inadequacy and depend upon foreign sources for its needs are described as
a) Strategic minerals b) Critical minerals
c) Essential minerals d) Expendable minerals

24. Which of the following minerals are strategic for India?
a) Tin, zinc, copper, chromium and graphite
b) Tin, sulphur, nickel, cobalt and silver
c) Petroleum, lead, zinc, copper and vanadium
d) Uranium, thorium, ilmenite, zircon and tungsten

25. Strategic and critical minerals that are mined only by the state and its agencies are called
a) A-Schedule minerals b) B-Schedule minerals
c) C-schedule minerals d) None of the above

26. Which of the following types of jigs are employed to treat coals below 1.2cm in size?
a) Harz jig b) Baum jig
c) Feldspar jig d) All of the above

27. 'Collection' is the process of
a) Sorting of minerals according *to* their grain size
b) Sorting of minerals according to their specific gravity
c) Changing the surface tension of a solvent
d) Producing a water-repellent coating on the mineral particles

28. 'Xanthates' and 'aerofloats' are the examples of
a) Collectors b) Frothers
c) Modifiers d) Dispersants

29. 'Amines' are
a) Cationic collectors b) Anionic collectors
c) Oily collectors d) Frothers

30. Chemically the frothing agents are
    a) Polar inorganic compounds
    b) Non-polar inorganic compounds
    c) Polar organic compounds
    d) Non-polar organic compounds
31. Nodulising is the process of agglomeration involving
    a) Application of pressure
    b) Incipient fusion of particles
    a) Rolling of particles to form small pellets
    b) Partial fusion of particles at low temperatures
32. The term 'Liquidation grade' is applied to
    a) a grade of an ore which is infested with gangue impurities
    b) a grade of an ore whose price is forced
    c) a grade of an unworkable ore deposit
    d) none of the above
33. According to the Hoofer's classification of ore reserves, the term 'Possible Reserve' implies
    a) a geologically identified ore body
    b) a geologically presumed ore body
    c) a tentatively calculated reserve
    d) a suspected reserve
34. The ore reserves that are computed from evidence provided by pits, trenches and drill holes are described as
    a) Measured reserves  b) Indicated reserves
    c) Inferred reserves  d) None of the above
35. The feasibility report of a mine is based on
    a) Measured reserves  b) Indicated reserves
    c) Inferred reserves  d) None of the above
36. According the classification of ore reserves adopted by the Geological Survey of India, the C2 category denotes
    a) Developed reserves  b) Proved reserves
    c) Probable reserves  d) Possible reserves
37. Which of the following statements is Incorrect?
    a) India produces 89 minerals out of which 4 are fuel minerals, 11 are metallic, 52 are non-metallic and 22 are minor minerals.
    b) India ranks 3rd in production of coal & lignite and barytes
    c) India is the world's largest producer of mica blocks and mica splitting.
    d) India has reached a self-sufficiency level of 55% in crude oil production.
38. At the 1996-97 level of production, the life span of the crude oil reserves in India is estimated as
    a) 10 years  b) 15 years

c) 20 years d) 30 years

39. The GSI has to its credit geological mapping, covering an area of approximately
 a) 3.146 million sq.km b) 4.456 million sq.km
 c) 7.289 million sq.km d) 13.862 million sq.km
40. In Antarctica, India has completed geological cartography and thematic mapping of an area of
 a) 10,000 sq. km b) 15,000 sq.km
 c) 20,000 sq.km d) 25,000 sq.km
41. India's recoverable reserves of copper ore are of the order of
 a) 4300 million tonnes b) 7200 million tonnes
 c) 650 million tonnes d) 3.52 million tonnes
42. Arrange the following in the increasing order of ore abundance in India.
 a) Iron ore>Manganese ore>Magnesite>Chromite
 b) Magnesite >Iron ore>Chromite >Manganese ore
 c) Magnesite >Chromite >Iron ore >Manganese ore
 d) Manganese ore>Chromite>Magnesite>Iron ore
43. The Department of Mines is NOT responsible for
 a) coal
 b) petroleum & natural gas
 c) atomic minerals
 d) All of the above
 e) None of the above
44. The Old Bisanatham Mine is engaged in the mining of
 a) Gold b) Silver
 c) Copper d) Uranium
45. Which of the following is renowned for the mining of barytes in India?
 a) Hutti Gold Mines Limited
 b) Rajasthan State Mines and Minerals Limited
 c) Andhra Pradesh Mining Development Corporation
 d) National Mineral Development Corporation
46. The Yappamana Mine is famous for mining of
 a) Diamonds b) Copper
 c) Silver d) Gold
47. The Old Bisanatham and Chigargunta gold mines are located in
 a) Karnataka b) Andhra Pradesh
 c) Uttar Pradesh d) Bihar
48. The Indian Bureau of Mines (IBM)is the principal government agency responsible for
 a) the survey and exploration of all minerals, other than natural gas, petroleum, and atomic minerals
 b) the assessment of geological and regional mineral resources of the country

c) compiling exploration data and mineral maps and for providing access to the latest information is respect of mineral resources in the country
d) detailed exploration of various minerals/ores by drilling, and exploratory mining

49. The main constraint in Lead production in the country is
a) inadequacy of lead ore reserves
b) lack of proper machinery for drilling
c) inadequate technology
d) low tenor of the lead ores

50. The highest production of zinc in the country is from the smelter
a) Debari (Rajasthan)
b) Vizag (Andhra Pradesh)
c) Chanderiya (Rajasthan)
d) Tundoo (Bihar)

51. The role of Mineral Exploration Corporation Limited (MECL) can be best described as
a) Survey and Exploration
b) Regulation and Conservation
c) Mining and Processing
d) Export and Import

52. The Mines Act was promulgated in
a) 1950
b) 1952
c) 1957
d) 1964

53. The ' Minor minerals' come under the purview of
a) The Mineral Concession Rules, 1960
b) The Mineral Conservation and Development Rules, 1988
c) The Union Government
d) The State Governments

54. According to the First Schedule of the Mines and Minerals (regulation and development) Act, 1957, which of the following are NOT specified as "Atomic Minerals"?
a) Beryl and other beryllium-bearing minerals
b) Lithium-bearing minerals.
c) Minerals of the "rare earths" group containing Uranium and Thorium
d) Niobium-bearing minerals
e) None of the above

## (C) HYDROGEOLOGY

1. Water that is formed at the time of consolidation of magma is termed
a) Connate water
b) Vadose water
c) Meteoric water
d) Juvenile water

2. The underground water that occurs within the zone of aeration is termed
a) Plutonic water
b) Meteoric water

c) Vadose water d) Connate water

3. The water that is entrapped in sedimentary rocks during their formation is called
   a) Underground water
   b) Surface-runoff
   c) Vadose water
   d) Connate water
   e) Meteoric water
4. Within the lithosphere, water is found in
   a) Liquid state only
   b) Solid state only
   c) Vaporous state only
   d) Liquid and vaporous states only
   e) Liquid, solid and vaporous states
5. A 'Pluviometer' is an instrument that is used to measure
   a) Amount of precipitation b) Permeability of rocks
   c) Porosity of rocks d) Safe Yield of UGW
6. Hail is a type of precipitation that consists of
   a) Ice crystals of delicate, feathery structure
   b) Hard pellets of ice
   c) Droplets of water deposited on objects near the ground
   d) Minute crystals of ice deposited on a cold surface
7. A dense mass of water drops on smoke or dust particles in the lower atmospheric layers constitute
   a) Fog b) Mist
   c) Frost d) Blizzard
8. In a 'mist' the visibility is
   a) <1km b) >1km
   c) >1km but <2km d) >2km
9. A majority of the aquifers consists or
   a) Sands and gravels
   b) Sandstones
   c) Limestones
   d) Porous breccias and conglomerates
10. In a 'free' or phreatic aquifer
    i) The water table serves as the upper surface or the zone of saturation
    ii) The water table varies in form and slope
    iii) Changes in the water table correspond to changes in the volume of water
    iv) Artesian conditions does not exist
    a) (i), (ii) and (iii) only
    b) (i), (ii) and (iv) only

c) (i), (iii) and (iv) only
d) All of the above

11. An impermeable formation that neither contains nor transmits water is called
    a) Aquifer b) Aquiclude
    c) Aquifuge d) Aquitard
12. An impermeable formation that may contain water but is incapable of transmitting it in significant quantities is described as
    a) Confined aquifer b) Aquiclude
    c) Aquitard d) Aquifuge
13. Which of the following formations are encountered in an artesian basin from its top to the bottom?
    a) Aquiclude - Aquifer - Aquitard
    b) Aquiclude - Aquitard - Aquifer
    a) Aquitard - Aquifer- Aquiclude
    b) Aquifer- Aquitard- Aquiclude
14. Which of the following serves as an Aquitard ?
    a) Loam b) Sandy loam
    c) Clay d) Solid granite
15. In Karst region the underground water flow is
    a) Laminar b) Parallel
    c) Turbulent d) None of the above
16. The Linear law of Infiltration was given by
    a) H. Darcy b) White
    c) Charles Linnaeus d) Weber
17. The 'Hydrolic conductivity' of a rock is expressive of its
    a) Permeability b) Porosity
    c) Water retentivity d) Yield potential of UGW
18. Permeability is measured in units called
    a) Stokes b) Poise
    c) Darcy d) Weber
19. In general, fine-grained rocks have
    a) High porosity
    b) High permeability
    c) High porosity and permeability
    d) High porosity and low permeability
20. The porosity of a rock depends upon
    i) Shape of the 'grains
    ii) Size of the grains
    iii) Packing of the grains
    iv) Grading of the particles
    a) (i) and (ii) only
    b) (i), (ii) and (iii) only

a) (ii), (ii) and (iv) only
b) All or the above

21. Which of the following materials has the highest porosity?
a) Clay b) Slit
c) Gravel d) Sandstone

22. An influent stream is one which
a) flows into a parent stream
b) flows parallel to a consequent stream
c) recharges the ground water
d) receives discharges from the groundwater

23. The 'Safe yield' of a groundwater basin is
a) The total quantity of water that can be withdrawn from it annually
b) The total quantity of water obtained from a basin less the water lost on account of evapo-transpiration
c) The amount of water that can be withdrawn from it annually without producing an undesired result
d) None of the above

24. An 'Overdraft' results when
i) Safe yield exceeds the mean annual supply of water to a basin
ii) Ground water occurs very near to the surface
iii) Pumping is done in a coastal aquifer
iv) Poor quality water is drawn into a pumping aquifer
a) (i), (ii) and (iii) only
b) (i), (iii) and (iv) only
c) (I), (ii) and (iv) only
d) (ii), (iii) and (iv) only

25. The release of water over the ground surface in order to increase the quantity of underground water is called
a) Water spreading b) Sewage recharge
c) Induced recharge d) Shaft recharge

26. Deeply confined aquifers can be recharged by means of
a) Water spreading b) Inverted wells
c) Pits and shafts d) Induced recharge

27. The usable reserves of ground water in India are estimated at
a) 20 mhm b) 27 mhm
c) 35 mhm d) 44 mhm

28. Perched water tables are commonly seen at
a) Jaisalmer b) Bokaro
c) Jabalpur d) Kolaba

29. Which of the following formations consists of several prolific aquifers?
a) Golapilli sandstones b) Cuddalore sandstones
c) Vaimpalli limestones d) Nimar limestone

30. Which of the following formations is/are a good source of ground water 7
    a) Bhuj sandstones (Jurassic)
    b) Lathi sandstones (Jurassic)
    c) Umia sandstones
    d) All of the above
    e) None of the above
31. In the crystalline rock area of the Peninsula, the minimum depth at which the water table is encountered is
    a) 0.66 m
    b) 1.22 m
    c) 3.75 m
    d) 6.1 m

## (D) ENGINEERING GEOLOGY

1. The 'Brazilian test' is used to determine
   a) Tensile strength
   b) Compressive strength
   c) Shear strength
   d) Modulus of elasticity
2. Which of the following has the highest compressive strength?
   a) Granite
   b) Dacite
   c) Quartzite
   d) Marble
3. Which of the following material has the least stiffness?
   a) Limestone
   b) Sandstone
   c) Quartz
   d) Basalt
4. If is the specific gravity of a rock piece is § and the specific gravity of its powder is v the porosity coefficient is given as
   a) $P = \frac{(\S - v)\,100}{\S}$
   b) $P = \frac{(\S - v)\,100}{v}$
   c) $P = \frac{(v - \S)\,100}{v}$
   d) $P = \frac{(v - \S)\,100}{\S}$
5. In the case of diamond drilling, the core that is recovered in sticks longer than 10cm is expressed as a percentage of the rock quality designation (RQD). High quality materials have RQD values greater than
   a) 0
   b) 25
   c) 50
   d) 75
6. 'Grouting' may be described as
   a) The injection of suitable material into the earth's crust to seal any open fissures

b) A process of determining the coefficient or water saturation of a material
c) A process of lining of tunnels to support the pressure exerted by the material in which the tunnel is excavated
d) None of the above

7. The type of dam preferred where the river section is wide and the foundation is unsound is
a) Gravity dam b) Embankment dam
c) Arch dam d) Multiple arch dam

8. Excavation is very easy in the rocks belonging to the
a) Archaen period b) Precambrain period
c) Paleozoic period d) Holocene period

9. Match the following and choose the correct answer

| | |
|---|---|
| A. Taj Mahal | 1. Granite |
| B. Sanchi Stupa | 2. Gondwana sandstone |
| C. Jagannath Temple (Puri) | 3. Vindhyan sandstone |
| D. Golconda Fort | 4. Jaisalmer marble |
| | 5. Makrana marble |

a) A-5 B-3 C-2 D-1 b) A-5 B-2 C-3 D-1
c) A-4 B-3 C-2 D-1 d) A-5 B-4 C-2 D-3

10. The lime:clay ratio in cement in approximately
a) 1:4 b) 4:1
c) 2:3 d) 3:2

11. The Portland cement was discovered by
a) Aspdin b) Frankland
c) Stokes d) Heisenberg

12. Which of the following is NOT a fire-resisting material?
a) Chrysotile
b) Diatomite
c) Vermiculite
d) Powdered mica
e) None or the above

13. In the manufacture of coloured glasses selenium is used to impart ———colour to the glass
a) Green b) Blue
c) Red d) Pink

14. The Alluvial soils are deficient in
a) Nitrogen
b) Humus
c) phosphorus
d) None of the above
e) All of the above

15. The soil that is most characteristic of the Archaen basement of the peninsular India is
   a) Alluvial soil  b) Lateritic soil
   c) Red soil  d) Regur

## APPLIED GEOLOGY (KEY)

### (A) EXPLORATION METHODS

| | | | | | | | | | |
|---|---|---|---|---|---|---|---|---|---|
| 1-a | 2-d | 3-b | 4-a | 5-b | 6-b | 7-b | 8-d | 9-c | 10-a |
| 11-b | 12-a | 13-c | 14-c | 15-b | 16-d | 17-d | 18-c | 19-c | 20-a |
| 21-b | 22-c | 23-b | 24-a | 25-d | 26-b | 27-c | 28-c | 29-a | 30-a |
| 31-a | 32-a | 33-b | 34-a | 35-d | 36-d | 37-a | 38-b | 39-b | 40-a |
| 41-c | 42-d | 43-c | 44-d | 45-b | 46-b | 47-a | 48-a | 49-b | 50-c |
| 51-c | 52-d | 53-c | 54-b | 55-c | 56-a | 57-d | 58-a | 59-a | 60-b |
| 61-c | 62-c | 63-c | 64-d | 65-a | 66-c | 67-a | 68-d | 69-a | 70-a |
| 71-b | 72-c | 73-d | 74-a | 75-a | 76-d | 77-b | 78- d | 79-d | 80-c |
| 81-a | 82-d | 83-c | 84-a | 85-d | 86-a | 87-a | 88-d | 89-d | 90-b |

### (B) MINERAL PROCESSING, MINING GEOLOGY & MINERAL ECONOMICS

| | | | | | | | | | |
|---|---|---|---|---|---|---|---|---|---|
| 1-a | 2-a | 3-c | 4-d | 5-c | 6-a | 7-b | 8-c | 9-d | 10-c |
| 11-d | 12-a | 13-b | 14-d | 15-a | 16-b | 17-a | 18-b | 19-a | 20-c |
| 21-d | 22-a | 23-a | 24-b | 25-a | 26-c | 27-d | 28-a | 29-a | 30-c |
| 31-b | 32-b | 33-b | 34-b | 35-b | 36-d | 37-d | 38-b | 39-a | 40-a |
| 41-d | 42-b | 43-d | 44-a | 45-c | 46-d | 47-b | 48-c | 49-a | 50-c |
| 51-a | 52-b | 53-d | 54-e | | | | | | |

### (C) HYDROGEOLOGY

| | | | | | | | | | |
|---|---|---|---|---|---|---|---|---|---|
| 1-d | 2-c | 3-d | 4-c | 5-a | 6- b | 7-a | 8-c | 9-a | 10-d |
| 11-c | 12-b | 13-a | 14-c | 15-C | 16-a | 17-a | 18- C | 19-d | 20-d |
| 21-a | 22-c | 23-c | 24-b | 25-a | 26- b | 27-b | 28-a | 29-b | 30-d |
| 31-b | | | | | | | | | |

### (D) ENGINEERING GEOLOGY

| | | | | | | | | | |
|---|---|---|---|---|---|---|---|---|---|
| 1-a | 2-c | 3-c | 4-c | 5-d | 6-a | 7-d | 8-c | 9-a | 10-b |
| 11-a | 12-c | 13-d | 14-e | 15-c | | | | | |

# 16

# *Questions Based on Assertion and Reasoning*

Directions to answer the questions in this section

1 Mark 'a' if both the Assertion and Reason are correct and if Reason explains Assertion

2. Mark 'b' if both the Assertion and the Reason are right but the Reason does not explain Assertion.

3. Mark 'c' if Assertion is Right and Reason is wrong
4. Mark 'd' if both Assertion and Reason are wrong
5. Mark 'e' if Assertion if false but Reason is true.

Assertion=A Reason=R

1. A : Bouguer anomalies over oceans have high positive values

   R : Oceans have mass deficiency and high subsurface density.

2. A : The linear magnetic anomalies of the oceans are caused by the normal and reversed magnetizations in the basaltic layer-2 of the oceanic crust.

   R : These anomalies cannot be caused by the sediment layers, which are essentially non magnetic.

3. A : The free-air gravity anomalies are negative in the shield areas

   R : Low-velocity zones are absent or poorly developed in these areas.

4. A : Mantle material is the major contributor of magnetic anomalies.

   R : Temperatures in the mantle are always maintained below the Curie point of natural magnetic materials.

5. A : Andes mountains are the most extensive modern continental volcanic mountain chains in the world.

   R : They are formed by the collision of oceanic and continental lithospheric plates.

6. A : Several worst natural disasters are caused by the volcanic activity occurring in the island-arc systems.

   R : It is due to the fissure type volcanism which causes the erupted magma to spread over large areas.

7. A : According to Pratt's hypothesis of Isostasy, compensation takes place by adjustment of crustal base.

   R : The thickness of the crust varies by the top and the bottom.

8. A : The Ganga Basin represents a large-scale depression on the Vindhyan syeneclise.
   R : It is a northward continuation of the Vindhyan sediments
9. A : Cratonic basins provide an ideal setting for the accumulation of oil and gas.
   R : High geothermal gradients and increasing pressure due to burial of sediments facilitates the conversion of organic matter into oil and gas.
10. A : Gypsum has lower density than Anhydrite
   R : Hydrated minerals tend to expand many times the original volume of the anhydrous minerals.
11. A : Presence of sea caves, arches and stacks indicate submergence of the coastlines.
   R : These coastlines are produced by the subsidence of land or rise in sea levels.
12. A : Drumlins show features opposite to those of the Roches Mountonees.
   R : The stross side is steep and the lee side is roughened.
13. A : Cinnabar has more reflectivity then diamond.
   R : Reflectivity depends on the Refractive index. As cinnabar has higher R.I. than diamond, it exhibits more reflecting power (luster) than diamond.
14. A : Calcite shows red, pink or yellow fluorescence.
   R : These colours are a result of the substitution of rare-earth elements for a small amount of calcium.
15. A : The transformation $\alpha$ Qtz $\rightleftharpoons$ $\beta$ Qtz is reversible.
   R : Change is only due to the alteration of the crystal symmetry without causing the breaking up of $SiO_4$ tetrahedral links.
6. A : **Microcline has higher crystal symmetry than Sanidine.**
   R : Microcline has poor ordering of ions in the crystal structure.
7. A : In a hand specimen, quartz can be easily distinguished from orthoclase.
   R : Orthoclase is biaxial negative and has lower refractive index than quartz.
8. A : L-waves are responsible for most of the destructive earthquakes.
   R : They are the surface waves and are developed in the immediate vicinity of the epicentre.
9. A : Glacial valleys have transverse U-shaped profiles.
   R : Glaciers cut their sides at an equal or faster rate than their bases.
). A : Subsequent valleys are those streams which have shifted

from an original consequent position to belts oi erosible rock.

R : Subsequent valleys represent structurally adjusted stream courses.

21. A : Minerals with —OH components such as hornblende and biotite are characteristic of plutonic igneous rock.

R : Hydroxyl-bearing minerals tend to break-down in a volcanic environment.

22. A : Continents have drifted in the past.

R : The Glossopteris and Gangamopteris flora have been found on all the Southern continents and India.

23. A : Beach sand can be distinguished from river sands by the symmetry of their distribution.

R : Beach sands are symmetrically distributed while river sands show marked asymmetry.

24. A : Sodalite and leucite do not commonly occur in association with quartz.

R : These are incompatible with free silica under magmatic consitions.

25. A : Correlation based on paleontologic criteria is based on the Law of Irreversibility of biological evolution.

R : Any given species is never repeated again in geologic time.

26. A : In the Cuddapah Supergroup each division begins with coarse sediments and terminates in the finer sediments.

R : The deposition took place in several cycles in an auto-geosyncline.

27. A : The Upper Vindhyan rocks are of aeolian origin and were deposited under dry climatic conditions.

R : Texture is fine and the colour of the sediments is red.

28. A : Even large lithospheric plates are able to move without undergoing internal deformation.

R : Length to thickness ratio is such that neither compressional nor tensional stress could be transmitted from one end to another end of the plate.

29. A : Base of the Cambrian or the Pre-cambrian-Cambrian boundary cannot be specifically located in India.

R : The beds formed during the transitional period are conformable.

30. A : In many parts of the world the Permian succession is characterised by the presence of red beds.

R : There was a world-wide regression at the close of the Permian.

31. A : Fracture cleavage can be used to determine the direction of shear of beds.

R : Acute angle between bedding and fracture cleavage points in the directions of relative shear.

32. A : Strike-slip faults show horizontal movement of blocks on either side.

R : Strike-slip faults have no hanging wall or footwall.

33. A : Pyroclasts erupted from continental side of Andesite Line are more voluminous.

R : Volcanoes on continental side of Andesite line are supplied with high pressure gas.

34. A : Rift valleys of MOR are regions of highest heat flow on the earth.

R : The are located on the ascending limbs of the convection cells.

35. A : Perlitic cracks are partially concentric cracks formed due to shrinkage during cooling.

R : Perlitic cracks are typical of felsites and form due to devitrification.

## XVI. REASONING AND ASSERTION (KEY)

| | | | |
|---|---|---|---|
| 1. c | 2. a | 3. b | 4. d |
| 5. b | 6. a | 7. e | 8. b |
| 9. a | 10. a | 11. b | 12. b |
| 13. a | 14. c | 15. a | 16. d |
| 17. e | 18. a | 19. a | 20. a |
| 21. a | 22. b | 23. c | 24. a |
| 25. a | 26. a | 27. b | 28. a |
| 29. a | 30. a | 31. a | 32. b |
| 33. a | 34. a | 35. b. | |

# SECTION—III

## PAST PAPERS OF COMPETITIVE EXAMINATIONS

**Note :** *The papers given in this section are based on the memory recollections of the candidates who have actually appeared in these examinations. These papers have been cross-checked thoroughly and edited by the author. While utmost care has been taken to maintain the original format of these papers, the author holds no responsibility for any discrepancy in the same.*

# 1. Examination 2000 (Unsolved)

## GEOLOGY

1. In a sulphide ore containing chlorite, covellite, chlorapatite, chalcopyrite and chalcocite, the gangue minerals include :
   (a) Chlorite and chalcocite
   (b) Covellite and chalcocite
   (c) Chalcocite and chlorapatite
2. Match List I (Properties) with List II (Ore minerals) and select the correct answer using the codes given below the lists:

| List I | List II |
|---|---|
| A. ZnS; Colour-light to dark brown ; H = 4; Streak = light-brown Sp. gr = 4 | 1. Stibnite |
| B. FeAsS; colour silver or tin white, H = 5.2; Streak = black; Sp. gr = 4.6 | 2. Pyrrhotite |
| C. $Sb_2S_3$; Colour-gray; H = 2; Streak = black; will mark paper; Sp. gr = 4.6 | 3. Galena |
| D. $Fe_{1-x}s$; Colour- brownish bronze; H=4, Streak = black; Sp. gr = 4.6 | 4. Arsenopyrite |
| | 5. Sphalerite |

Codes:

| | A | B | C | D |
|---|---|---|---|---|
| (a) | 4 | 2 | 1 | 3 |
| (b) | 5 | 4 | 1 | 2 |
| (c) | 5 | 4 | 2 | 1 |
| (d) | 1 | 2 | 4 | 3 |

3. Which one of the following features does not represent the criterion of replacement?

   (a) Doubly terminated crystals

(b) Ore abutting bedding

(c) Comb structure, vug with latest crystals

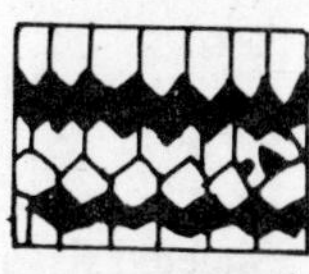

(d) Preserved folded structure

4. Consider the following statements:
   Localisation of mineral deposits is controlled by:
   1. Availability of structures and suitable tectonic setting.
   2. Availability of suitable host rocks.
   3. Metamorphic differentiation.
   4. Metamorphic facies.
   5. Appropriate stratigraphic trap.
   6. Attainment of optimum physico-chemical parameters.

   Which of the above statements are correct?
   (a) 1, 2, 3 and 4
   (b) 1, 2, 5 and 6
   (c) 1, 2, 3, 4 and 6
   (d) 3, 4, 5 and 6
5. Match List I with List II (figures) and select the correct answer using the codes given below the lists:

**List I** **List II**

A. Parallel fissure system

1 

B. Conjugated fissure system

2 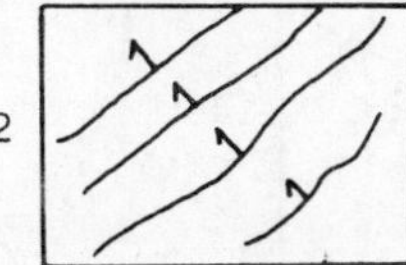

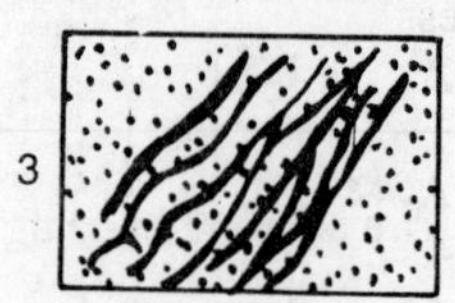


C. Chambered fissure vein

D. Sheeted fissure veins

Codes:

| | A | B | C | D |
|---|---|---|---|---|
| (a) | 4 | 2 | 1 | 3 |
| (b) | 2 | 4 | 3 | 1 |
| (c) | 2 | 4 | 1 | 3 |
| (d) | 4 | 2 | 3 | 1 |

6. Match List I (Mineral deposits) with the locations marked 1, 2, 3 and 4 in the map and select the correct answer using the codes given below the map:

**List I**

A. Lead-Zinc
B. Lignite
C. Petroleum+ gas
D. Iron-Ore

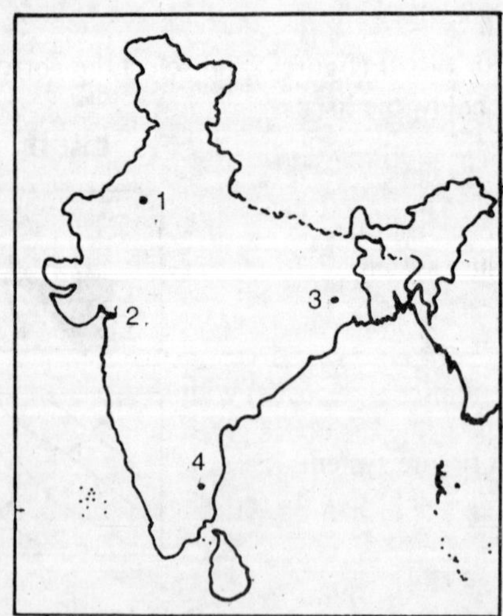

Codes:

| | A | B | C | D |
|---|---|---|---|---|
| (a) | 1 | 4 | 2 | 3 |
| (b) | 4 | 1 | 2 | 3 |
| (c) | 1 | 4 | 3 | 2 |
| (d) | 4 | 1 | 3 | 2 |

7. The Zawar Pb-Zn-Ore deposit of Rajasthan occurs within the:
   (a) Delhi Group (b) Aravalli Supergroup
   (c) Marwar Supergroup (d) Banded Gneissic Complex
8. The host rock of the auriferous quartz veins in the Kolar gold field is:
   (a) Champion Gnesis (b) Charnockite
   (c) Hornblende Schist (d) Chamundi Granite
9. Which one of the following structures has acted as the trap for petroleum and gas in Digboi Oilfield?
   (a) Asymmetric anticline (b) Fault zone
   (c) Salt dome (d) Nonconformity
10. Which one of the following rock formations contains the largest manganese ore deposits of india?
    (a) Sausar (b) Sakoli
    (c) Dharwar (d) Aravalli
11. Match List I (deposits) with List II (locations) and select the correct answer using the codes given below the lists:

| List I | List II |
|---|---|
| A. Bauxite | 1. Byrapur |
| B. Monazite | 2. Agucha |
| C. Zinc | 3. East coast |
| D. Chromite | 4. Malabar coast |

Codes:

| | A | B | C | D |
|---|---|---|---|---|
| (a) | 4 | 3 | 2 | 1 |
| (b) | 3 | 4 | 1 | 2 |
| (c) | 3 | 4 | 2 | 1 |
| (d) | 4 | 3 | 1 | 2 |

12. Match List I (Age of coal deposits) with List II (Coalfields) and select the correct answer using the codes given below the lists:

| List I | List II |
|---|---|
| A. Permian | 1. Kota stage, Maharashtra |
| B. Jurassic | 2. Raniganj, Jharia and Bokaro |

| | | | |
|---|---|---|---|
| C. | Lower Ecocene | 3. | Khasi, Garo and Mikir Hills, N-E region |
| D. | Oligocene Milocene | 4. | Jaipur, Nazira and Makum, N-E region |

Codes:

| | A | B | C | D |
|---|---|---|---|---|
| (a) | 3 | 2 | 1 | 4 |
| (b) | 2 | 1 | 3 | 4 |
| (c) | 1 | 2 | 4 | 3 |
| (d) | 3 | 4 | 1 | 2 |

**Directions**: The following 16 (sixteen) items consist of two statements, one labelled the 'Assertion A'and the other labelled the 'Reason R'. You are to examine these two statements carefully and decide if the Assertion A and the Reason R are individually true and if so, whether the Reason is a correct explanation of the Assertion. Select your answer to these items using the codes given below and mark your answer sheet accordingly.

Codes:

(a) Both A and R are true and R is the correct explanation of A.
(b) Both A and R are true but R is not a correct explanation of A.
(c) A is true but R is false.
(d) A is false but R is true.

13. Assertion (A): APW curves are the loci of the past magnetic poles of the Earth.
Reason (R) : APW curves have been constructed by measuring the change in position of the magnetic north pole during the last 100 million years.

14. Assertion (A): The oldest dated rocks in the modern oceans are not more than 2000 million years old.
Reason (R) : Plate Tectonics did not operate before 200 million years.

15. Assertion (A) : The age of the Earth is around 4600 million years.
Reason (R) : The radiometric age of the oldest crustal rocks is 4600 million years.

16. Assertion (A) : Global seismograph records reveal a ring-like shadow zone on the Earth's surface, free from P and S waves.
Reason (R) : The earth has a core which acts like a spherical lens by refracting the waves.

17. Assertion (A): Reef-building corals flourished well within the Arctic Circle during Lower Palaeozoic times.

Reason (R) : Palaezoic reef-building corals thrived in cold waters.

18. Assertion (A) : In the Cuddapah Supergroup each divison begins with coarse sediments and terminates in the finer sediments.

Reason (R) : The deposition took place in several cycles in an autogeosyncline.

19. Assertion (A): The base of the Cambrian or the Precambrian boundary cannot be specifically located in India.

Reason (R) : The beds formed during the transitional period are conformable.

20. Assertion (A): The Upper Vindhyan rocks are of aeolian origin and were deposited under dry climatic condition.

Reason (R) : In the Vindhyan rocks, texture is fine and colour is red.

21. Assertion (A): Prismatic cleavage angle in pyroxenes is nearly 90° and in amphiboles it is nearly 60°-120°.

Reason (R) : Presence of single chains of $SiO_4$ tetrahedra in pyroxenes and of double chains of the same in amphiboles directly control the cleavage angle.

22. Assertion (A): Ultramafic rocks occur in much the same tectonic setting as basalts and these as well as anorthosites are sometimes monomineralic.

Reason (R) : This feature, together with the virtual absence of aphanitic volcanic equivalents, suggests that ultramafic rocks and anorthosites in general are crystal differentiates of basaltic magmas.

23. Assertion (A): Several granites and granodiorites have developed through regional refusion of ancient geneissic masses.

Reason (R): The metasomatic and additive processes often result in reconstitution of rocks involving melting under appropriate P-T conditions with the production of new bodies of magma capable of injection.

24. Assertion (A): Griesening is a process of metasomatic alteration due to action of superheated steam and fluorine.

Reason (R): The alkali felspars of quartzo-felspathic rocks break down with the formation of white mica and quartz.

25. Assertion (A): Rocks dredged from the sea floor, especially in the vicinity of topographic scarps and transform faults, include greenstones, serpentinites and rare amphibolites.

Reason (R): Geochemical characteristics, particularly isotopic composition of such rocks attest to hydration of magmatic rocks through interaction with sea-water.

26. Assertion (A): In oxidation and supergene enrichment process, pyrite is finally converted into limonite forming gossan.
    Reason (R): Water with dissolved and entangled oxygen is the most powerful oxidizing agent which reacts with the minerals to yield sulphuric acid.
27. Assertion(A): Polymetallic veins are common in hydrothermal deposits.
    Reason (R): Fissures and cracks within rocks are good receptacles for ore fluid migration and localization.
28. Assertion(A) : Undisturbed coal formations occur along rifted river basins in the Lower Gondwana sequence of India.
    Reason (R): The Gondwana sequence of rocks were deposited in block faulted basins which also generated resurgent tectonic setting.
29. The upper 50-250 km thick layer within the upper mantle is called as:
    (a) Lithosphere (b) Tectosphere
    (c) Low velocity zone (d) Mesosphere
30. The relative abundance of oxides in the earth's crust is such that:
    (a) $Al_2O_3 > SiO_2 > Fe_2O_3 + FeO > CaO$
    (b) $Al_2O_3 > SiO_2 > CaO > Fe_2O_3 + FeO$
    (c) $SiO_2 > Al_2O_3 > Fe_2O_3 + FeO > CaO$
    (d) $SiO_2 > Al_2O_3 > CaO > Fe_2O_3 + FeO$
31. The lowest point on the continents is in the:
    (a) Baikal Lake (b) Dead Sea
    (c) Black Sea (d) Caspian Sea
32. A radioactive isotope A decays to isotope B with a half-life of 1000 years. If initially 500 atoms of A were present, how many atoms of A and B will be respectively present after 2000 years?
    (a) 250 and 250 (b) Nil and 500
    (c) 375 and 125 (d) 125 and 375
33. Match List I and List II and select the correct answer using the codes given below the lists:

| List I | List II |
|---|---|
| A. Mohorovicic discontinuity | 1. Crust-mattle boundary |
| B. Conrad discontinuity | 2. Mantle-core boundary |
| C. Gutenberg discontinuity | 3. Granodiorite-basalt boundary |
| D. Lehman discontinuity | 4. Inner core-outer core boundary |

Codes:

| | A | B | C | D |
|---|---|---|---|---|
| (a) | 2 | 4 | 3 | 1 |
| (b) | 3 | 4 | 2 | 1 |

(c) A B C D
1 3 2 4
(d) A B C D
2 1 3 4

34. Consider the following statements:
1. Orthogeosynclines include eugeosynclines and miogeosynclines.
2. Miogeosynclines are rich in volcanic rocks.
3. Eugeosynclines are devoid of volcanic rocks.
Which of the above statement(s) is/are correct?
(a) 1 and 3 (b) 1, 2 and 3
(c) 1 alone (d) 3 alone

35. Consider the following statements:
1. Majority of active volcanoes are closely related to seismic belts.
2. Hawaiian volcanoes are on a plate boundary.
3. Volcanoes are present only along subduction zones.
4. There is extensive volcanism along the Alpine-Himalayan mountain belt.
Which of the above statement(s) is/are correct?
(a) 1 alone (b) 1 and 2
(c) 1 and 3 (d) 1, 3 and 4

36. Match List I and List II and select the correct answer using the codes given below the lists:

| List I | List II |
|---|---|
| A. Travertine | 1. Wind |
| B. Kame | 2. River |
| C. Yardang | 3. Ground water |
| D. Inselberg | 4. Glacier |
| | 5. Wave |

Codes:
(a) A B C D
3 4 1 2
(b) A B C D
2 3 4 1
(c) A B C D
5 1 2 3
(d) A B C D
3 1 5 2

37. Match List I and List II and select the correct answer using the codes given below the lists:

| List I | List II |
|---|---|
| A. Alluvium | 1. Gravity |
| B. Colluvium | 2. Glacier |
| C. Till | 3. River |
| D. Eluvium | 4. Weathering |

Codes:

| | A | B | C | D |
|---|---|---|---|---|
| (a) | 1 | 2 | 3 | 4 |
| (b) | 3 | 1 | 2 | 4 |
| (c) | 3 | 4 | 2 | 1 |
| (d) | 4 | 3 | 1 | 2 |

38. Consider the following types of residual hills:
   1. Monadnocks
   2. Hums
   3. Mogotes

   Those, which remain on a karst plain after other features have been destroyed by erosion, will include:

   (a) 1, 2 and 3 (b) 1 and 2
   (c) 2 and 3 (d) 1 and 3

39. In the case of constant volume deformation, a prolate strain ellipsoid represents a strain state where:

   (a) $\lambda_1 = \lambda_2 > \lambda_3$ (b) $\lambda_1 > \lambda_2 > \lambda_3$
   (c) $\lambda_1 > \lambda_2 = 1 > \lambda_3$ (d) $\lambda_1 > \lambda_2 = \lambda_3$

40. Consider the following statements:

   The effect of increasing confining pressure on the rheology of rocks is to:
   1. increase the ductility of the rocks.
   2. increase the magnitude of permanent deformation.
   3. increase the ultimate strength of the rocks.
   4. decrease the ultimate strength of the rocks.
   5. increase the rupture strength of the rocks.
   6. decrease the rupture strength of the rocks.
   7. bring no change in the rupture strength of the rocks.

   Which of the above statements are correct?

   (a) 1 and 2 (b) 1, 2, 3 and 5
   (c) 1, 2, 4 and 6 (d) 1, 2, 7 and 8

41. The cleavage-bedding relation, shown in the figure given below, indicates:

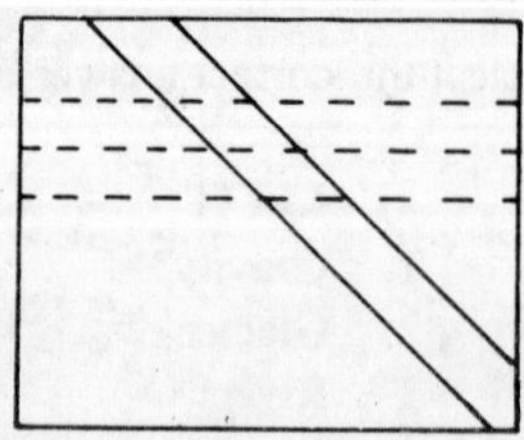

- - - - Axl, Pl, Cleavage
——— Bedding

(a) a synformal closure towards right.
(b) the upper limit of a neutral fold closing to the right.
(c) an antiformal closure to the right.
(d) the lower limb of a neutral fold closing to the right.

42. Which one of the following is correctly represented in the diagram given below?

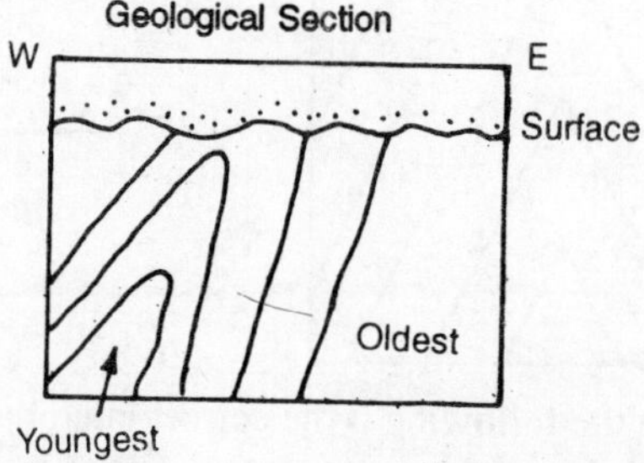


(a) Antiformal anticline
(b) Antiformal syncline
(c) Synformal syncline
(d) Synformal anticline

43. Match List I (Types of fold) with List II (Diagrams showing the stereographics plots of the fold axes and axial-plane poles) and select the correct answer using the codes given below the lists:

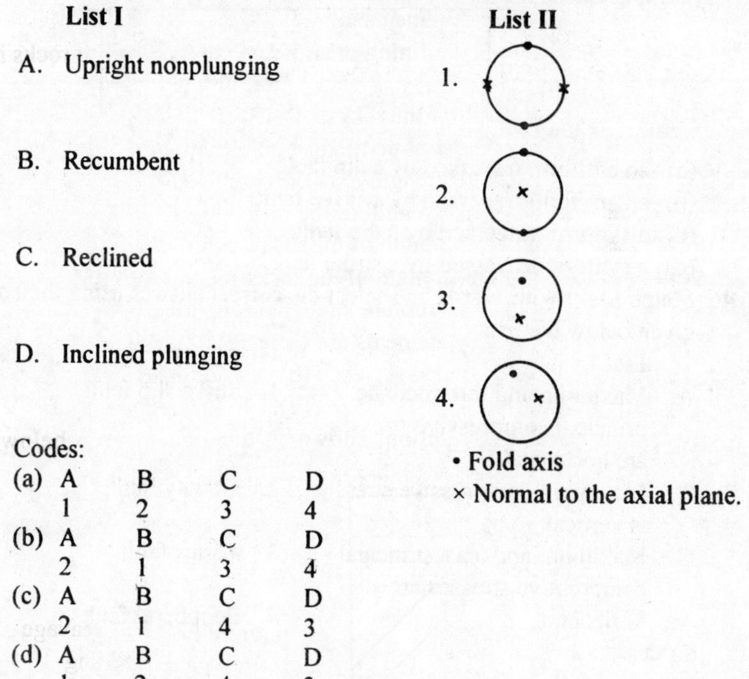

**List I**

A. Upright nonplunging

B. Recumbent

C. Reclined

D. Inclined plunging

**List II**

1.

2.

3.

4.

• Fold axis
× Normal to the axial plane.

Codes:

| | A | B | C | D |
|---|---|---|---|---|
| (a) | 1 | 2 | 3 | 4 |
| (b) | 2 | 1 | 3 | 4 |
| (c) | 2 | 1 | 4 | 3 |
| (d) | 1 | 2 | 4 | 3 |

44. The diagram, given below, represents the structure contours of a single horizon.

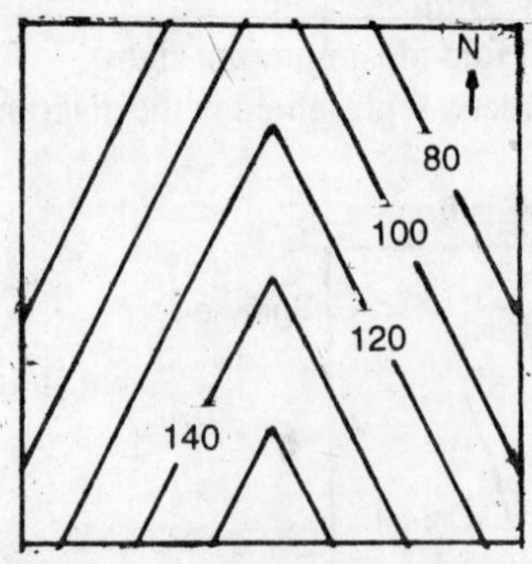


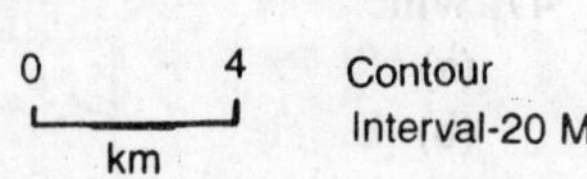


Which one of the following is the correct interpretation of the same.

(a) Antiform plunging north (b) Antiform plunging south

(c) Synform plunging north (d) Synform plunging south

45. The feature depicted in the map given below is:

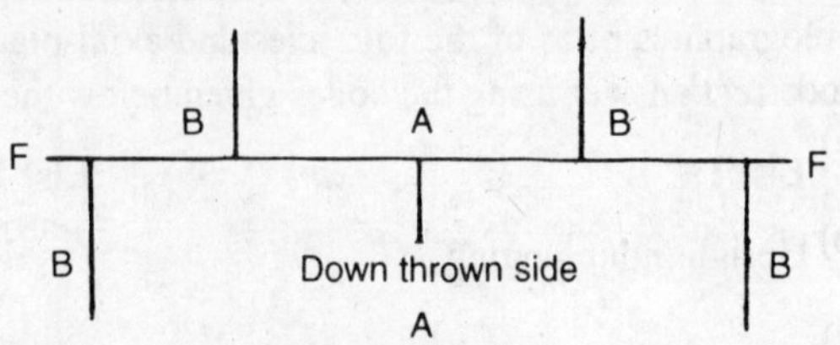


(a) an antiform traversed by a dip fault.

(b) an antiform traversed by a strike fault.

(c) a synform traversed by a dip fault.

(d) a synform traversed by a strike fault.

46. Match List I with List II and select the correct answer using the codes given below the lists:

| List I | List II |
|---|---|
| A. Maximum and intermediate principal compressive stresses are horizontal | 1. Strike slip fault |
| B. Maximum compressive stress is vertical | 2. Gravity fault |
| C. Maximum and least principal compressive stresses are horizontal | 3. Thrust fault |
| | 4. Peripheral fault |

Codes:

| | A | B | C |
|---|---|---|---|
| (a) | 1 | 3 | 4 |

(b) A B C
3 1 2

(c) A B C
3 2 1

(d) A B C
4 3 1

47. Which one of the following pairs is *not* correctly matched?

(a) Rhyolite - Pillow structure

(b) Basalt - Columnar joint

(c) Komatite - Spinifex texture

(d) Trachyte - Flow structure

48. Match List I with List II and select the correct answer using the codes given below the lists:

| List I | List II |
|---|---|
| A. Metamorphism | 1. Conglomerate |
| B. Magmatism | 2. Marble |
| C. Unconformity | 3. Pegmatite |
| D. Volcanism | 4. Pumice |
| | 5. Shale |

Codes:

(a) A B C D
1 2 5 4

(b) A B C D
2 3 4 1

(c) A B C D
1 2 4 5

(d) A B C D
2 3 1 4

49. Consider the following statements:

1. Competence contrast of layers affects the shape of boudins.
2. Smooth convex sides of mullions are directed to the layers with lower viscosity.
3. Rodding of minerals is common in ductile shear zones.

Which of the above statements are correct?

(a) 1, 2 and 3 (b) 1 and 2

(c) 2 and 3 (d) 1 and 3

50. In which one of the following rocks, is crenulation cleavage most commonly developed?

(a) Granulite (b) Gneiss

(c) Quartzite (d) Phyllite

51. Consider the following structures:

1. Wave ripple marks
2. Current ripple marks
3. Torrential cross bedding

4. Load cast

Those, which are useful in determining the top of beds, would include:

(a) 1 and 2 (b) 1 and 3

(c) 1, 2 and 4 (d) 1, 3 and 4

52. What is the correct sequence of the following geological events on the basis of the given map (the oldest being at the bottom) ?

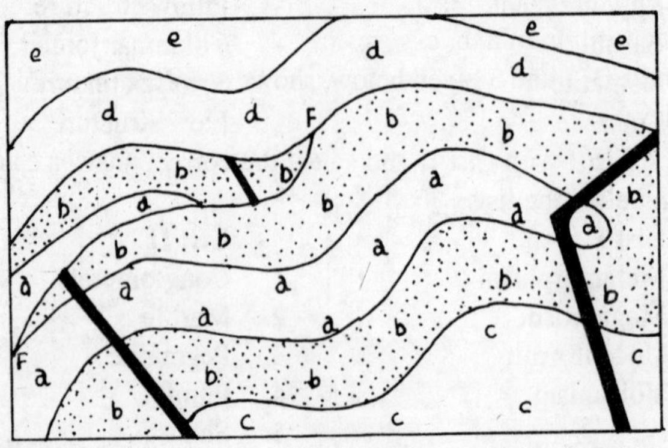


FF-Fault

(a) Folding
Deposition of d, e
Unconformity
Faulting
Dyke intrusion
Folding
Deposition of a, b, c

(b) Folding
Deposition of d, e
Unconformity
Folding
Faulting
Dyke intrusion
Deposition of a, b, c

(c) Folding
Deposition of d, e
Unconformity
Faulting
Dyke intrusion
Folding
Dyke intrusion
Deposition of a, b, c

(d) Folding
Deposition of d, e
Unconformity
Folding
Dyke intrusion
Faulting
Dyke intrusion
Deposition of a, b, c

53. Diagrams 1, 2 and 3 given below, shows outcrops of a coal seam across a valley.

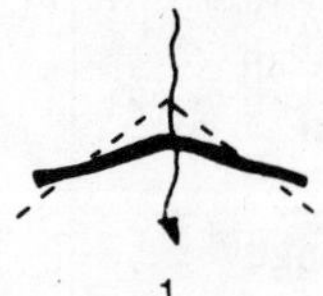

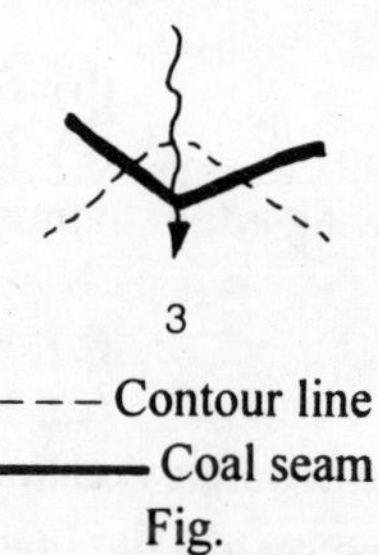


--- Contour line
—— Coal seam

Fig.

Which one of the following correctly represents the directions of the dip?

(a) 1-Northerly dip, 2-Northerly dip, 3-Southerly dip
(b) 1-Northerly dip, 2-Southerly dip, 3-Northerly dip
(c) 1-Southerly dip, 2-Northerly dip, 3-Northerly dip
(d) 1-Southerly dip, 2-Southerly dip, 3-Northerly dip

54. A major stratigraphic gap is noted in peninsular India between:

(a) Late Proterozoic and Upper Carboniferous.
(b) Upper Devonian and Lower Permian.
(c) Upper Permian and Middle Triassic.
(d) Lower Cretaceous and Upper Cretaceous.

55. A sedimentary body that is likely to be diachronous is:

(a) an ash fall deposit
(b) a fluvial deposit
(c) a storm deposit
(d) a turbidite

56. Spectacular rise of global sea level occurred during the:
(a) Triassic (b) Late Mesozoic
(c) Late Tertiary (d) Middle Riphean

57. Throughout the world stratigraphy, the boundary between the Cretaceous and the Teritiary is well known for:
(a) Glacial episodes (b) Tectonic activity
(c) Mass extinction (d) Coal formation

58. The three main divisions of the Jurassic system—Black Jura, Brown Jura and White Jura, correspond respectively to:
(a) Malm, Lias and Dogger (b) Lias, Dogger and Malm
(c) Malm, Dogger and Lias (d) Lias, Malm and Dogger

59. In which one of the following periods did fishes flourish?
(a) Cambrian (b) Ordovician
(c) Devonian (d) Permian

60. The type succession of Muth Quartzite is in:
(a) Kashmir (b) Salt Range
(c) Spiti (d) Kinnaur.

61. Which one of the following formations represents the euxinic facies that is regarded as a marker horizon in the Phanerozoic Stratigraphy of the Himalayans?
(a) Sulcacutus Beds (b) Spiti Shales
(c) Products Limestone (d) Otoceras Beds

62. Which one of the following is the stratigraphic equivalent of the Siwalik Group?
(a) Dwarka Beds (b) Bagh Beds
(c) Bugti Beds (d) Karaikal Beds

63. The Nagari stage in the Siwaliks is characterized by the:
(a) absence of Proboscidea and the first appearance of Primates.
(b) absence of Proboscidea.
(c) absence of Primates.
(d) appearance of Proboscidea and Equus.

64. Which one of the following is the correct ascending order of succession?
(a) Talchir, Karharbari, Barren Measure, Barakar, Raniganj.
(b) Talchir, Barkar, Barren Measure, Karharbari, Raniganj.
(c) Talchir, Karharbari, Barren Measure, Raniganj, Barakar.
(d) Talchir, Karharbari, Barakar, Barren Measure, Raniganj.

65. Which one of the following pairs is correctly matched?
(a) Glossopteris Flora - Raniganj Formation
(b) Physa princepi - Early Carboniferous
(c) Lignite deposits - Intertrappean Beds
(d) Syringothyris cuspidata - Bhander Limestone

66. The correct sequence of the Upper Siwaliks is:
(a) Tatrot, Pinjore

(b) Pinjore, Boulder Conglomerate, Tatrot
(c) Tatrot, Boulder Conglomerate, Pinjore
(d) Tatrot, Pinjore, Boulder Conglomerate

67. Which one of the following factors contributes most to make a fossil population an index fossil?
(a) Evolutionary gap (b) Iterative evolution
(c) Rapid evolution (d) Convergent evolution

68. Which one of the following genera is an example of 'Living Fossil'?
(a) Calceda (b) Eurydesma
(c) Bellerophon (d) Nautilus

69. Match List I with List II and select the correct answer using the codes given below the lists:

| | List I | | List II |
|---|---|---|---|
| A. | A mark left on bed surface by a shell fragment dragged by current | 1. | Trace fossil |
| B. | A mark left on bed surface by a twig of a plant swung by wind | 2. | Tool mark |
| C. | Cross-bedding | 3. | Fossilized bedform |
| D. | Ripples | 4. | Current structure |

Codes:

| | A | B | C | D |
|---|---|---|---|---|
| (a) | 2 | 1 | 3 | 4 |
| (b) | 3 | 2 | 4 | 1 |
| (c) | 3 | 2 | 1 | 4 |
| (d) | 2 | 1 | 4 | 3 |

70. The preservation of delicate primary features in a fossil despite extensive mineral replacement is best described as:
(a) Orthogenesia (b) Carbonisation
(c) Petrifaction (d) Cast

71. Match List I (Genera) with List II (Division) and select the correct answer using the codes given below the lists:

| | List I | | List II |
|---|---|---|---|
| A. | Globotruncana | 1. | Brachiopoda |
| B. | Calymene | 2. | Foraminifera |
| C. | Terebratula | 3. | Echinoidea |
| D. | Stygmatophygus | 4. | Trilobita |

Codes:

| | A | B | C | D |
|---|---|---|---|---|
| (a) | 2 | 3 | 4 | 1 |

(b) A B C D
2 4 1 3

(c) A B C D
4 1 2 3

(d) A B C D
1 3 2 4

72. Which one of the following living habits is indicated by a deep pallial sinus in a lamellibranch?
(a) Planktonic (b) Nektonic
(c) Benthic vagrant (d) Benthic sessile

73. In all the regular echinoids the:
(a) anus lies outside the apical disc.
(b) tests posses bilateral symmetry.
(c) mouth lies at the centre of the oral side and the anus is located at the centre of apical system.
(d) tests are not spheroidal in shape.

74. Which one of the following geological ages can be attributed to a formation that contains leaves of angiosperms and some cheek teeth of Mesohippus?
(a) Pliocene (b) Late Miocene
(c) Early Miocene (d) Oligocene

75. The correct sequence of evolution of horses is:
(a) Miohippus, Hyracotherium, Hipparian
(b) Hipparian, Hyracotherium, Miohippus
(c) Hyracotherium, Miohippus, Hipparian
(d) Miohippus, Hipparian, Hyracotherium

76. The complete history of evolution of which one of the following organisms can be reconstructed from the vertebrate remains in the Himalayan Neogene-Quarternary sequence?
(a) Hominids (b) Perissodactyles
(c) Proboscideans (d) Giraffoides

77. Following are some representatives of the Gondwana flora:
1. Ptilophyllum 4. Williamsonia
2. Palmoxylon 5. Nilssonia
3. Alethopteris 6. Nilssonia
Which of the above assemblages have a post-Permian affinity?
(a) 1, 2 and 5 (b) 2, 4 and 6
(c) 2, 3 and 4 (d) 4, 5 and 6

78. Which one of the following is primarily phosphatic?
(a) Ostracoda (b) Radiolaria
(c) Conodonta (d) Nannoplankton

79. Consider the following statements:
Micropalaeontology has become exceedingly important since the

inception of microscope and the advent of oil exploration because:

1. all the sediments contains microfossils.
2. a small sample of rocks may contains hundreds of complete and identifiable microfossils.
3. microfossils give the age of the formation of rock strata.
4. all the microfossils are facies independent.

Which of the above statements are correct?

(a) 1, 2, 3 and 4 (b) 2, 3 and 4
(c) 2 and 3 (d) 3 and 4

80. Which one of the following foraminifera did *not* undergo heterophasic life-cycle that became common among foraminifera since the Late Palezoic?

(a) Planktonic foraminifera
(b) Large benethic foraminifera
(c) Foraminifera with agglutinated tests
(d) Fresh-water foraminifera

81.

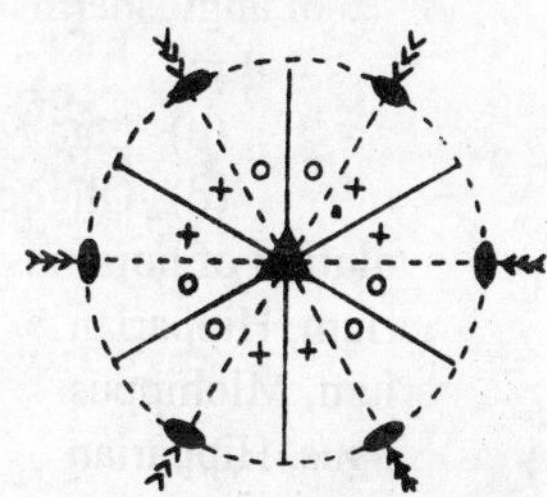

Which crystal class is represented by the above stereogram?

(a) Hexagonal-normal (b) Hexagonal-scalenohedral
(c) Hexagonal-rhombohedral (d) Hexagonal-trapezohedral

82. Consider the following parameters:

1. Axial angle 2. Axial length
3. Axial ratio 4. Axial intercept

Which of these can be directly and accurately measured by X-ray diffraction?

(a) 1 and 2 (b) 2 and 3
(c) 3 and 4 (d) 1 and 4

83. The figure given below shows a twin crystal having twin axis ⊥ (001) and composition plane (001). The type of twin shown is:

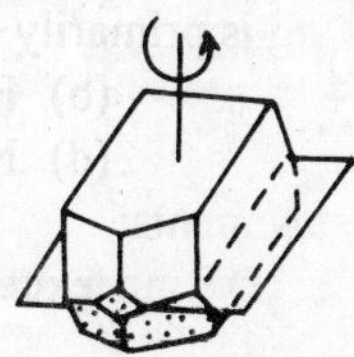

(a) Carlsbad (penetration) (b) Pericline
(c) Manebach (d) Baveno

84. In the drawing of barite crystal, given below, the Miller Indices of the face marked 'd' would be

d

(a) (001) (b) (011)
(c) (101) (d) (110)

85. Match List I ( Minerals) with List II (Composition) and select the correct answer using the codes given below the lists:

| List I | List II |
|---|---|
| A. Anhydrite | 1. NaCl |
| B. Gypsum | 2. KCl |
| C. Halite | 3. $CaSO_4$ |
| D. Sylvite | 4. $CaSO_4$ $2H_2O$ |

Codes:

| | A | B | C | D |
|---|---|---|---|---|
| (a) | 4 | 3 | 2 | 1 |
| (b) | 3 | 4 | 1 | 2 |
| (c) | 3 | 4 | 2 | 1 |
| (d) | 4 | 3 | 1 | 2 |

86. The different forms of quartz are given below:
1. High Quartz 2. Low Quartz
3. Tridymite 4. Crystobalite

The correct order in which these appear with decreasing temperature is:
(a) 1, 2, 3, 4 (b) 4, 3, 2, 1
(c) 4, 3, 1, 2 (d) 2, 3, 1, 4

87. Match List I with List II and select the correct answer using the codes given below the lists:

| List I | List II |
|---|---|
| A. Calcite | 1. $BaCO_3$ |
| B. Rhodocrosite | 2. $CaCO_3$ |
| C. Witherite | 3. $MnCO_3$ |
| D. Cerrusite | 4. $PbCO_3$ |

Codes:

| | A | B | C | D |
|---|---|---|---|---|
| (a) | 1 | 2 | 3 | 4 |
| (b) | 2 | 3 | 4 | 1 |

| (c) A | B | C | D | (d) A | B | C | D |
|---|---|---|---|---|---|---|---|
| 2 | 3 | 1 | 4 | 3 | 4 | 1 | 2 |

88. The given figure shows anortnite content in different zones.

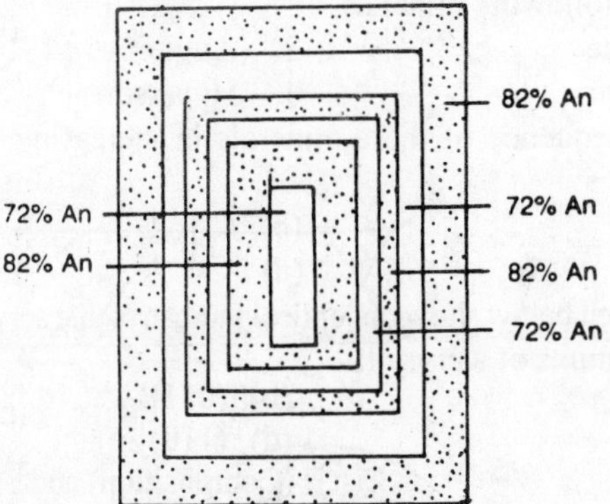


It also shows a plagioclase grain with

(a) reverse zoning (b) oscillatory zoning

(c) normal zoning (d) complex zoning

89. The clinopyroxenes are considered as members of the four component system-Diopside-Hedenbergite-Enstatite-Ferrosilite. Subcalcic augite belongs to which one of the following fields shown in the figure given below?

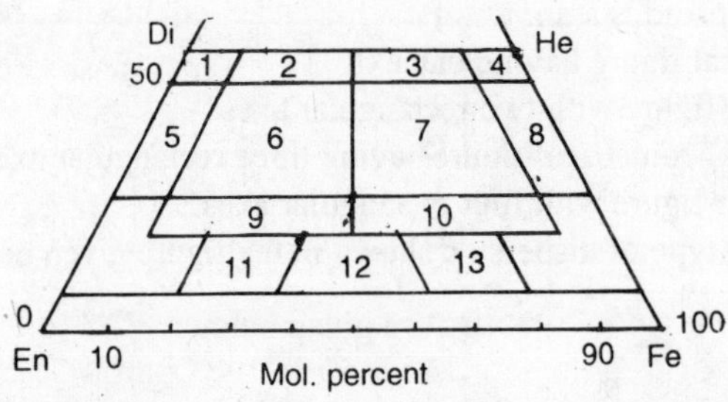


(a) 6 (b) 9 (c) 7 (d) 10

90. The figure given below is a cross-section of a mineral showing a typical lozenge-shaped marking formed due to the impurities enclosed during formation.

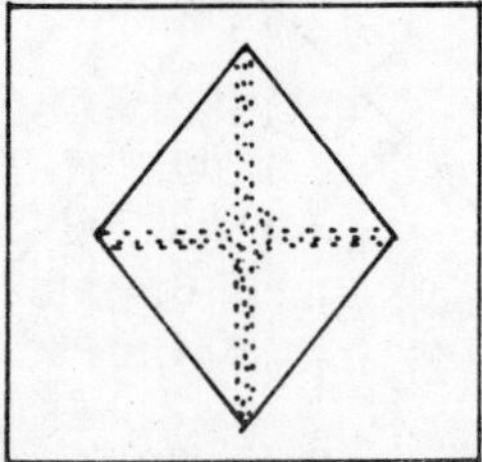

Which one of the following minerals exhibits such a feature?

(a) Fibrolite (b) Staurolite

(c) Andalsite (d) Melilite

91. Consider the following minerals:

1. Orthoclase 2. Augite

3. Muscovite 4. Hypersthene

The correct sequence of these minerals in increasing order of their birefringence is:

(a) 4, 1, 3, 2 (b) 4, 1, 2, 3

(c) 1, 4, 3, 2 (d) 1, 4, 2, 3

92. The figure given below shows an optical indicatrix that serves to illustrate the optical features of a crystal:

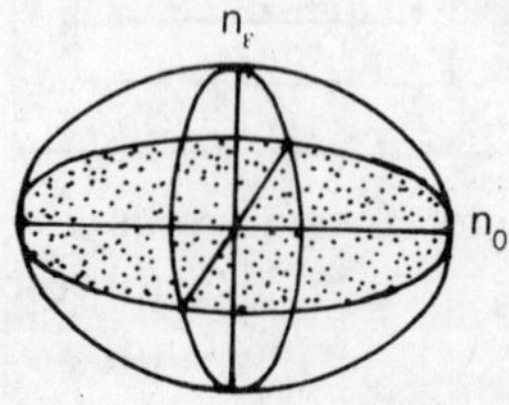


Which one of the following minerals shows such optical behaviour?

(a) Calcite (b) Quartz

(c) Olivine (d) Gypsum

93. Optical ellipsoid is a/an:

(a) spherical figure having one axis.

(b) curved figure with two rectangular axes.

(c) curved geometrical figure having three rectangular axes.

(d) circular figure with four rectangular axes.

94. What is the type of dispersion shown in the figure given below?

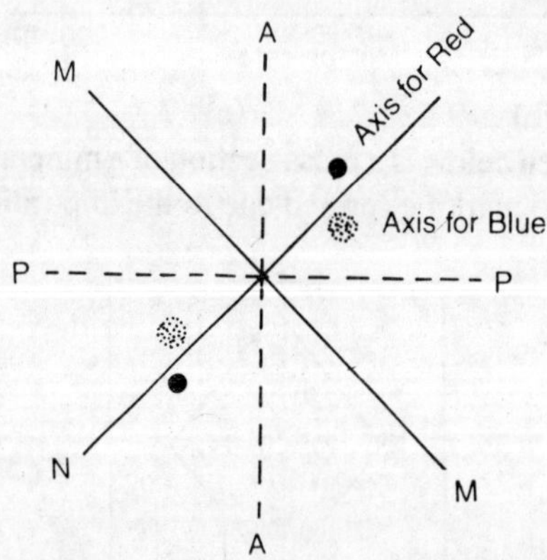

(a) Rhombic (b) Horizontal
(c) Crossed (d) Inclined

95. Lavas of basic composition with a smooth rounded, undulating surface, sometimes with a ropy appearance, are called:
(a) a flows (b) block lava
(c) pillow lava (d) pahoehoe

96. Match List I with List II and select the correct answer using the codes given below the lists:

| | List I | List II |
|---|---|---|
| A. | Intergrowth of orthoclase and quartz | 1. Rapakivi |
| B. | Intergrowth of quartz and garnet | 2. Symplectite |
| C. | Intergrowth of vermicular quartz and oligoclase | 3. Myrmakite |
| D. | Orthoclase with a mantle of sodic plagioclase | 4. Graphic growth |

Codes:

| | A | B | C | D |
|---|---|---|---|---|
| (a) | 2 | 4 | 3 | 1 |
| (b) | 4 | 2 | 3 | 1 |
| (c) | 2 | 3 | 1 | 4 |
| (d) | 3 | 2 | 1 | 4 |

97. Which one of the following pairs is not correctly matched?
(a) Tholeiite basalt-quartz dolerite association : Non-orogenic continental
(b) Spilite-Keratophyre association : Miogeosyncline
(c) Basalt-andesite-rhyolite association : Orogenic belts
(d) Leucite basalt-potassic trachybasalt association : Continental rift

98. Match List I with List II and select the correct answer using the codes given below the lists:

| | List I | | List II |
|---|---|---|---|
| A. | Websterite | 1. | Olivine + Orthopyroxene + Clinopyroxene |
| B. | Wehrlite | 2. | Olivine + Orthopyroxene |
| C. | Harzburgite | 3. | Olivine + Clinopyroxene |
| D. | Lherzolite | 4. | Orthopyroxene + Clinopyroxene |

Codes:

| | A | B | C | D |
|---|---|---|---|---|
| (a) | 3 | 4 | 1 | 2 |
| (b) | 4 | 3 | 2 | 1 |

(c) A B C D
2 3 1 4

(d) A B C D
3 2 4 1

99. Consider the following statements regarding "eutectic" crystallisation in a binary system:
1. It is an isobarically invariant point in a T-X space.
2. It is the minimum temperature at which crystal-melt equilibrium is possible.
3. More than two solid phases can crystalize simultaneously at a constant proportion.
4. Eutectic ratio at which the solid phases crystalize, is not dependent on pressure.
5. The system becomes univarient subsequent to crystallization of the melt.

Which of the above statements are correct?
(a) 1, 2 and 3 (b) 1, 3 and 4
(c) 2, 4 and 5 (d) 1, 2 and 5

100. Consider the following statements:
The upward migration of igneous melt is fasicilated by:
1. density difference between magma and surrounding rocks.
2. degree of crystallinity of the melt.
3. dominance of fractional crystallization over equibrium crystallization.
4. pre-existing deep fracture systems in the surrounding rocks.
5. temperature difference between the melt and its surrounding rocks.

Which of the above statements are correct?
(a) 1, 2, 3 and 4 (b) 1, 2, 4, and 5
(c) 2, 3 and 4 (d) 2, 3 and 5

101. Consider the following statements:
1. Basalt can originate from lower crustal rocks.
2. Basalt is formed by quasi-eutectic melting.
3. Basalt forms through partial melting of upper mantle ultramafic rocks.
4. Basalt cannot form by melting of rocks other than eclogite and peridotite.
5. Partial melting of wet mantle produces alkali basalt.

Which of the above statements are correct?
(a) 1, 2 and 4 (b) 1, 3 and 5
(c) 2, 3 and 4 (d) 2, 4 and 5

102. Consider the following statements:
Kimberlites
1. are products of continental intra-plate magmatism.

2. occur in oceanic environments or young fold belts.
3. do not contain crystals of fragmented crustal and mantle xenoliths.
4. show distinctive inequigranular textures due to the presence of macrocrysts in a fine-grained matrix.
5. are volatile-rich potassic ultrabasic igneous rocks with high concentration of incompatible elements.

Which of the above statements are correct?

(a) 2, 3 and 5 (b) 2, 4 and 5
(c) 1, 3 and 5 (d) 1, 4 and 5

103. The correct sequence in which loose sediments are converted into a sedimentary rock is:

(a) Cementation, authigenesis, pressure solution, compaction
(b) Authigenesis, pressure solution, compaction, cementation
(c) Compaction, pressure solution, authigenesis, cementation
(d) Compaction, pressure solution, cementation, authigenesis.

104. Match List I with List II and select the correct answer using the codes given below the lists:

| | List I | | List II |
|---|---|---|---|
| A. | Well-rounded sandgrains in a sandstone | 1. | Transportation by river |
| B. | Facetted and striated pebbles in a conglomerate | 2. | Sediments derived from a cold provenance |
| C. | Assorted sands in a sedimentary basin | 3. | Transportation of grains by wind |
| D. | Undecomposed laths of feldspars in a sandstone | 4. | Transportation by glacier |

Codes:

| | A | B | C | D |
|---|---|---|---|---|
| (a) | 4 | 3 | 2 | 1 |
| (b) | 3 | 4 | 2 | 1 |
| (c) | 3 | 4 | 1 | 2 |
| (d) | 4 | 3 | 1 | 2 |

105. Match List I (Rock name, type of cement) with List II (Textural maturity) and select the correct answer using the codes given below the lists:

| | List I | | List II |
|---|---|---|---|
| A. | Medium-grained, quartz-cemented, quartz arenite | 1. | Mature |
| B. | Fine-grained clay -cemented arkose | 2. | Submature |

C. Pebbly, calcite-cemented lithic conglomerate — 3. Immature
D. Coarse-grained calcite-cemented sub-arkose — 4. Supermature

Codes:

| | A | B | C | D |
|---|---|---|---|---|
| (a) | 4 | 3 | 2 | 1 |
| (b) | 4 | 3 | 1 | 2 |
| (c) | 3 | 4 | 1 | 2 |
| (d) | 3 | 4 | 2 | 1 |

106. Frequency curves for four sand samples A, B, C and D are shown below:

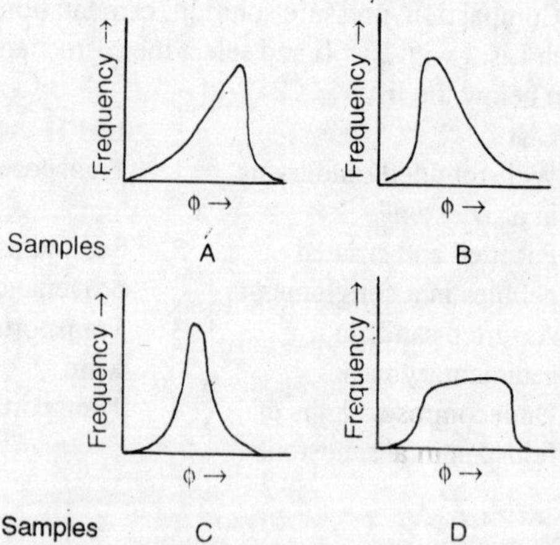


The prominent features of these curves can be described as:
(a) A-Coarse skewed, B-Fine skewed, C-Leptokurtic, D-Platykurtic
(b) A-Fine skewed, B-Coarse skewed, C-Leptokurtic, D-Platykurtic
(c) A-Platykurtic, B-Leptokurtic C-Coarse skewed, D-Fine skewed
(d) A-Leptokurtic, B-Platykurtic, C-Fine skewed, D-Coarse skewed

107. A grain-supported limestone with intergranular spaces filled with micrite is termed as:
(a) mudstone (b) packstone
(c) wackestone (d) grainstone

108. Match List I (feature) with List II (deposit) and select the correct answer using the code given below the lists:

| | List I | | List II |
|---|---|---|---|
| A. | Grain rounding not a function of sediment transport | 1. | Glacial lake |
| B. | Thinly laminated mudstone with outsize clasts | 2. | Submarine debris flow |
| C. | Repetitive occurrence of desiccation cracks in mud dominated sequence | 3. | Stable platform |
| D. | Orthoquartzite-carbonate association | 4. | Tidal flat |

Codes:

| | A | B | C | D |
|---|---|---|---|---|
| (a) | 1 | 2 | 3 | 4 |
| (b) | 2 | 1 | 4 | 3 |
| (c) | 1 | 2 | 4 | 3 |
| (d) | 3 | 4 | 2 | 1 |

109. For the determination of palaeocurrent direction from alluvial gravel, the most useful scalar parameter would be the:
   (a) decreasing mean pebble size.
   (b) increasing pebble roundness.
   (c) decreasing maximum pebble size.
   (d) increasing pebble sphericity.
110. Interbedded sandstone and shale is typical of shelf deposits dominated by:
   (a) tide
   (b) storm
   (c) wave
   (d) periodic tectonic disturbance
111. 'Syneresis' is a diagenetic process which involves:
   (a) elimination of free oxygen from the sediments of bacterial action.
   (b) deposition of minerals in the cavities and pore spaces.
   (c) reduction in the bulk volume of sediments.
   (d) dewatering of sediments under pressure.
112. Match list I (Characteristics) with List II (Name) and select the correct answer using the codes given below the lists:

| | List I | | List II |
|---|---|---|---|
| A. | Processes of physical and chemical changes within a sediment after its deposition, but prior to the onset of either metamorphism or weathering | 1. | Authigenesis |
| B. | Modification of mineral textures during diagenesis | 2. | Halmyrolysis |

C. Precipitation of mineral water during diagenesis — 3. Neomorphism

D. The earliest changes at the sediment surface often through inherent disquilibrium between sedimentary particles and the depositional medium. — 4. Diagenesis

Codes:

(a) A B C D
4 3 1 2

(b) A B C D
3 2 4 1

(c) A B C D
4 1 3 2

(d) A B C D
1 2 3 4

113. Match List I with List II and select the correct answer using the codes given below the lists:

| List I | List II |
|---|---|
| A. A sandstone containing extrabasinal carbonate | 1. Calcirudite |
| B. Field term to differentiate limestone composed of sand-sized carbonate grains from a sandstone composed of silicate grains | 2. Calcilutite |
| C. Limestone equivalent of mud rock | 3. Calcarenite |
| D. Limestone equivalent of conglomerate | 4. Calclithite |

Codes:

(a) A B C D
1 2 3 4

(b) A B C D
2 1 3 4

(c) A B C D
3 4 1 2

(d) A B C D
4 3 2 1

114. Match List I with List II and select the correct answer using the codes given below the lists:

| List I | List II |
|---|---|
| A. Ultra-metamorphism | 1. Induration and burning effects produced by lavas on the rocks at the contact. |

| | | | |
|---|---|---|---|
| B. | Auto-metamorphism. | 2. | Mechanical breakdown of rocks due to predominant effects of stress. |
| C. | Optalic metamorphism | 3. | Complete melting of rocks with the formation of new magmas under the influence of heat and ascending gases. |
| D. | Dynamic metamorphism | 4. | Alteration of magmatic rocks by volatiles and solution given off in the cooling process. |

Codes:

| | A | B | C | D |
|---|---|---|---|---|
| (a) | 3 | 4 | 1 | 2 |
| (b) | 1 | 2 | 4 | 3 |
| (c) | 4 | 2 | 1 | 3 |
| (d) | 1 | 3 | 2 | 4 |

115. Match List I (Mineral assemblages) with List II (Facies) and select the correct answer using the codes given below the lists.

| | List I | | List II |
|---|---|---|---|
| A. | Albite + Epidote + Chlorite + Pyrophyllite in peltic rocks. | 1. | Eclogite |
| B. | Hornblends + Plagioclase in mafic rocks. | 2. | Greenschist |
| C. | Augite + Orthopyroxene + Plagioclase + Mg-Fe garnet | 3. | Amphibolite |
| D. | Jadeite-rich Clinopyroxene+ Pyrope-rich Garnet in mafic rocks, feldspar free | 4. | Granulite |

Codes:

| | A | B | C | D |
|---|---|---|---|---|
| (a) | 2 | 3 | 1 | 4 |
| (b) | 3 | 2 | 4 | 1 |
| (c) | 2 | 3 | 4 | 1 |
| (d) | 3 | 2 | 1 | 4 |

116. Consider the following statements:

Metamorphic differentiation is the process whereby a rock:

1. develops contrasted mineral assemblages in adjacent domains from an initial homogenous mineralogy.
2. undergoes metasomatic alteration through influx of hydrothermal solutions along planar structures in the rock.
3. often develops gneissic banding.
4. undergoes recrystallization in response to building up of a chemical potential gradient during dynamothermal metamorphism.

Which of the above statements are correct?

(a) 1, 2 and 3 (b) 1, 3 and 4
(c) 2, 3 and 5 (d) 3, 4 and 5

117. In high-grade rocks of granulite facies where muscovite is unstable, sillimanite coexists with:

(a) Na-feldspar (b) Biotite
(c) K-feldspar (d) Phlogopite

118. Consider the following characteristics.

1. Felsic veins show no dilation.
2. Contacts are gradational.
3. Granitic material occurs in isolated patches.
4. Orientation of sedimentary relics is compatible with surrounding rocks.
5. Abundance of perthitic feldspars.
6. Feldspars present, are of the subsolvus type.

Which of the above characteristics are diagnostic of migmatites?

(a) 1, 3, 4 and 5 (b) 1, 2, 5 and 6
(c) 1, 2, 4 and 6 (d) 2, 3, 4 and 6

119. Which one of the following pairs is *not* correctly matched?

(a) The longest dimension of tabular crystals aligned sub-parallel to schistosity. -Lepidoblastic
(b) Equigranular non-schistose hornfels showing orientation of minerals at random. -Xenoblastic
(c) Preservation of altered psammitic and pelitic sediments in garnetiferous biotite-hornfels. -Palimpsest
(d) Porphyroblasts of chiastolite in a fine-grained groundmass exhibiting strain-slip cleavage. -Maculose

120. A granoblastic aggregate of quartz and muscovite (or lepidolite) with topaz, tourmaline, fluorite, rutile, cassiterite and wolframite, formed by metasomatic hydrothermal alteration of granitic rocks, is called:

(a) skarn (b) granofels
(c) greisen (d) leptynite.

# 2. Examination 1999 (Unsolved)

## GEOLOGY

1. Match figures A, B, C, and D with List I and select the correct answer using the codes given below the List:

| Figure | List I |
|---|---|
| A | 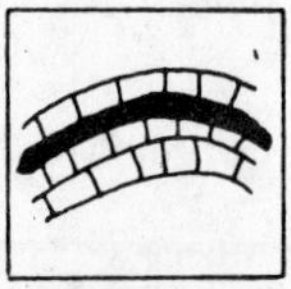 1. Bysmalith |
| B |  2. Lopolith |
| C |  3. Phacolith |
| D |  4. Laccolith |

Codes:

| | A | B | C | D |
|---|---|---|---|---|
| (a) | 3 | 4 | 1 | 2 |
| (b) | 4 | 3 | 2 | 1 |
| (c) | 3 | 4 | 2 | 1 |
| (d) | 4 | 3 | 1 | 2 |

2. Match List I with List II and select the correct answer using the codes given below the list:

| List I | List II |
|---|---|
| A. In lavas glass occupies minute interspaces between microlites of feldspar in random orientation. | 1. Intersertal |
| B. In lavas angular interspaces between feldspars are occupied by olivine, pyroxene or iron oxides in random orientation. | 2. Pilotaxitic |

C. The interstices in lavas are filled with glass, cryptocrystalline material or non-granular deuteric and secondary minerals. — 3. Intergranular

D. The crowded microlites of feldspars in lavas are dispersed in a sub-parallel manner as a result of few and their interstices are occupied by micro or cryptocrystalline material. — 4. Hyalopilitic

Codes:

| | A | B | C | D |
|---|---|---|---|---|
| (a) | 4 | 3 | 1 | 2 |
| (B) | 3 | 4 | 2 | 1 |
| (C) | 4 | 3 | 2 | 1 |
| (D) | 3 | 4 | 1 | 2 |

3. The lithologic association of altered ultramafic rocks with massive gabbro and dolerite dykes as a parallel sheeted complex, overlain by pillow lavas of spilite and altered basalt, greywacke, and chert is referred to as the:
   (a) ophiolite suite  (b) alpine-type suite
   (c) layered complex  (d) greenstone belt
4. Match list I with list II and select the correct answer using the codes given below the lists:

| List I | List II |
|---|---|
| A. Majhgawan | 1. Carbonalic |
| B. Mt. Girnar | 2. Kimberlite |
| C. Ambadongar | 3. Ultramafites |
| D. Sukinda | 4. Granophyre |

Codes:

| | A | B | C | D |
|---|---|---|---|---|
| (a) | 2 | 4 | 1 | 3 |
| (B) | 1 | 3 | 2 | 4 |
| (C) | 3 | 1 | 2 | 4 |
| (D) | 3 | 1 | 4 | 2 |

5. Consider the following statements:
   1. A basic igneous rock in which olivine shows reaction relation with Ca-poor pyroxene is a tholeitic basalt.
   2. A magnesian volcanic rock with high concentration of cumulus olivine is a komatite.
   3. A rock with more than 85% albitic plagioclase is an anorthosite.

4. Lamprophyres are basic to ultrabasic rocks containing phenocrysts of hydrous matic minerals.

Which of the above statements are correct?

(a) 1, 2 and 3 (b) 1 and 4
(c) 2, 3 and 4 (d) 2 and 3

6. Consider the following statements:
1. Anorthosite does not have an exact volcanic counterpart.
2. Leucite is virtually unknown in plutonic rocks.
3. Pigeonite is common in basic igneous rocks.
4. Biotite is common in volcanic rocks.

Which of the above statements are correct?

(a) 1, 2 and 3 (b) 1, 3 and 4
(c) 2, 3 and 4 (d) 1, 2 and 4

7. Consider the following statements regarding crystallisation of a binary melt of two pure end-members under isobaric condition:
1. It can have one peritectic and one eutectic.
2. It can have two eutectics.
3. The position of the eutectic point(s) in a T-X space is a function of the ratio of the two components in the melt.
4. The position of the eutectic point(s) in a T-X space is a function of $P_{Total}$ and $P_{H_2O}$.

Which of the above statements are correct?

(a) 1, 2 and 3 (b) 1, 2 and 4
(c) 1, 3 and 4 (d) 2, 3 and 4

8. Consider the following statements:

A variety of magmatic differentiation trends may be caused by:
1. initial magma composition.
2. depth of magma formation.
3. depth of magmatic differentiation.
4. $CO_2$ - content of the magma.
5. oxygen contents of the magma.

Which of the above statements are correct?

(a) 2, 3 and 4 (b) 2, 4 and 5
(c) 1, 2 and 3 (d) 1, 3 and 5

9. Match List I with List II and select the correct answer using the codes given below the Lists:

| | List I | List II |
|---|---|---|
| A. | Mol proportion of $Al_2O_3$ >mol proportion of ( $Na_2O + K_2O + CaO$): excess alumina forms muscovite, biotite, corundum | 1. Subaluminous |
| B. | Mol proportion of $Al_2O_3$ > ($Na_2O_2 + K_2O$) but less than that of alkalies + lime: | 2. Peralkaline |

hornblende, epidote

C. Mol proportion of $Al_2O_3 < (NA_2O + K_2)$: reibeckite, aegirite — 3. Metaluminous

D. Little or no excess $Al_2O_3$ over the requisite amount to crystallise feldspars and feldspathoids: oviline, orthopyroxene, diopside — 4. Peraluminous

Codes:

| | A | B | C | D |
|---|---|---|---|---|
| (a) | 4 | 3 | 2 | 1 |
| (b) | 3 | 4 | 2 | 1 |
| (c) | 2 | 3 | 1 | 4 |
| (d) | 3 | 2 | 4 | 1 |

10. Match figures A, B, C and D with List I and select the correct answer using the codes given below the List:

**Figure** — **List I**

A. — 1. Reef knoll

B. 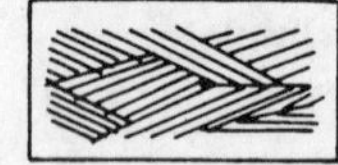 — 2. Cyclic bedding

C.  — 3. Dune bedding

D. 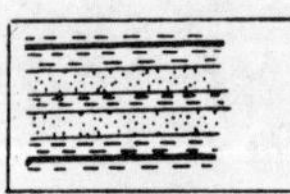 — 4. Current bedding

Codes:

| | A | B | C | D |
|---|---|---|---|---|
| (a) | 4 | 3 | 2 | 1 |
| (b) | 3 | 4 | 1 | 2 |
| (c) | 4 | 3 | 1 | 2 |

(d) A B C D
3 4 2 1

11. Match List I with List II and select the correct answer using the codes given below the List:

| | List I | | List II |
|---|---|---|---|
| A. | Vectorial sedimentary structure. | 1. | Convolute bedding |
| B. | Non-vectorial sedimentary structure. | 2. | Flaser bedding |
| C. | Post-depositional sedimentary structure. | 3. | Hummocky cross stratification. |
| D. | Structures from storm deposition | 4. | Asymmetric ripples. |

Codes:

| | A | B | C | D | | A | B | C | D |
|---|---|---|---|---|---|---|---|---|---|
| (a) | 2 | 3 | 1 | 4 | (b) | 2 | 4 | 1 | 3 |
| (c) | 4 | 1 | 2 | 3 | (d) | 4 | 2 | 3 | 1 |

12. The most reliable imprint of aeolian deposition is the:
(a) translatent ripple lamination.
(b) large-scale cross-bedding.
(c) hierarchical arrangement of bounding surfaces of cross-beds.
(d) high degree of sorting and roundness of sands.

13. Match List I with List II and select the correct answer using the codes given below the Lists:

| | List I | | List II |
|---|---|---|---|
| A. | Crystals tend to be uniform in size and may be simply or irregularly sutured | 1. | Gneisses and schists |
| B. | Somewhat elongate crystals displaying subparallel alignment; highly sutured contacts common | 2. | Metamorphic quartzites |
| C. | Grains equicrystalline but may show variation in other characters. | 3. | Plutonic igneous rocks |
| D. | Grains flattening and irregular; suturing observed with undulose extinction of the quartz crystals | 4. | Quartzites |

Codes:

| | A | B | C | D | | A | B | C | D |
|---|---|---|---|---|---|---|---|---|---|
| (a) | 1 | 3 | 4 | 2 | (b) | 3 | 1 | 4 | 2 |
| (c) | 3 | 1 | 2 | 4 | (d) | 1 | 3 | 2 | 4 |

14. Which one of the following lithologic association is *unlikely* to occur in a sedimentary sequence?
    (a) Quartzarenite-limestone.
    (b) Greywacke-radiolarian chert-shale.
    (c) Oolitic limestone-radiolarian chert red clay.
    (d) Arkose-shale-coal-basic volcanics.
15. Which one of the following pairs is not correctly matched?
    (a) Precipitation of mineral matter during diagenesis : Authigenesis
    (b) Total removal of mineral without its substitution by another : Solution
    (c) Textural modification with composition change during lithification : Recrystalization
    (d) Modification of mineral textures during diagenesis : Neomorphism
16. Dominance of lithologically homogenous, well-sorted, matrix-free and well-rounded sediments with high-angle wedge-planar cross stratification represents:
    (a) littoral deposits (b) piedmont deposits
    (c) aeolian deposits (d) valley flat deposits
17. Consider the following features:
    1. Poorly sorted, hard, impure sandstone with 15-75% matrix.
    2. Presence of sedimentary structures formed by turbidity currents.
    3. Arenite containing 15% detrital matrix.
    4. Formation in areas of maximum uplift.
    5. Abundant fine-grained micaceous and chloritic minerals.

    The features which characterise graywackes include:
    (a) 1, 3 and 4 (b) 1, 2 and 5
    (c) 2, 3 and 5 (d) 2, 4 and 5
18. Consider the following characteristics:
    1. Shallow-water origin.
    2. Organisms that thrive on accession of soluble silica to the seawater.
    3. Protection by the growth of coral reefs over them.
    4. Association with abyssal red clay.
    5. Association with submarine pillow lavas.

    The characteristics which define radiolarian cherts include:
    (a) 1, 3 and 4 (b) 1, 2 and 5
    (c) 2, 3 and 5 (d) 2, 4 and 5
19. Match List I with List II and select the correct answer using the codes given below the lists:

| List I | List II |
|---|---|
| A. Phyllite | 1. Contact metamorphism |
| B. Glaucophane Schist | 2. High pressure metamorphism |

C. Pyroxenehornfels — 3. Granulite facies

D. Enderbite — 4. Greenschist facies

Codes:

(a) A B C D — 4 2 1 3
(b) A B C D — 3 4 2 1
(c) A B C D — 4 2 3 1
(d) A B C D — 3 4 1 2

20. Which one of the following statements is NOT correct?
When present in a metamorphic environment, circulating free water:
(a) helps in diffusion of ions.
(b) acts as a catalytic agent.
(c) helps in nucleation of new mineral(s).
(d) helps in release of ions.

21. Match List I and List II and select the correct answer using the codes given below the Lists:

| List I | List II |
|---|---|
| A. Calcareous rocks: tremolite muscovite, glaucophane | 1. Amphibolite facies |
| B. Pelites; chlorite, muscovite albite, quartz | 2. Epidoteamphibolite facies, |
| C. Mafic igneous rock; hornblende, epidote, albite, almandine, garnet, quartz | 3. Blueschist facies |
| D. Pelite; garnet, biotite, muscovite, sillimanite, quartz | 4. Greenschist facies |

Codes:

(a) A B C D — 4 3 2 1
(b) A B C D — 4 3 1 2
(c) A B C D — 2 1 4 3
(d) A B C D — 3 4 2 1

22. Consider the following characters:
1. A composite rock with a banded structure defined by leucocratic bands alternating with dark coloured bands.
2. A composite rock of closely associated heterogeneous rock masses.
3. Evidence of cataclasis and strain recovery during recrystallisation.
4. Petrographic evidence of melting reaction of precursor mineral(s) at higher metamorphic grade and *in situ* melt-segregation.
5. Metamorphic differentiation.

The set of characters essential to designate a rock-mass as migmatite would include:
(a) 1, 3 and 5
(b) 1 and 4

(c) 2 and 5 (d) 2, 3 and 4

23. The process of at least partial mobilization, involving addition and removal of diffusing elements such that a rock becomes mechanically mobile, is called:

(a) cataclasis (b) anatexis
(c) ultrametamorphism (d) rheomorphism

24. Match List I with List II and select the correct answer using the codes given below the Lists:

| List I | List II |
|---|---|
| A. Recrystallised mossaic of equidimensional anhedral grains; inequant grains, if present, are randomly oriented. | 1. Poikioloblastic |
| B. Abundance of platy minerals grains—micas, talc, chlorite, graphite with strong planar orientation. | 2. Nematoblastic |
| C. Abundance of acicular or columnar grains displaying linear preferred dimensional orientation. | 3. Lepidoblastic |
| D. Metacrysts containing numerous small randomly oriented inclusion of other minerals. | 4. Granoblastic |

25. Match List I with List II and select the correct answer using the codes given below the Lists:

| List I | List II |
|---|---|
| A. Banded metasedimentary rocks containing manganiferous semipelites with spessartite or rhodonite | 1. Khondalite |
| B. Metamorphic rocks composed of almandine-pyrope, omphacite kyanite, rutile showing garnet porphyroblasts and crude alignment of pyroxene prisms | 2. Charnockite |
| C. Pelites with the assemblage quartz-sillimanite-garnet-perthite-graphite showing high metamorphic grade | 3. Gondite |
| D. Metamorphic rocks having granitic composition with even-grained granular textures and containing hypersthene | 4. Eclogite |

Codes:

| | A | B | C | D |
|---|---|---|---|---|
| (a) | 4 | 3 | 1 | 2 |
| (b) | 3 | 4 | 2 | 1 |
| (c) | 3 | 4 | 1 | 2 |
| (d) | 4 | 3 | 2 | 1 |

26. Match List I with List II and select the correct answer using the codes given below the lists:

| List I | List II |
|---|---|
| A. The most dominant rock of the Indian Peninsula | 1. Amphibolite |
| B. An important rock in Wajra-karur kimberlites | 2. Eclogite |
| C. A significant rock in the Eastern Ghats granulites | 3. Granite gneiss |
| D. A prominent rock in the Delhi Supergroup | 4. Leptinite |

Codes:

| | A | B | C | D |
|---|---|---|---|---|
| (a) | 1 | 2 | 3 | 4 |
| (b) | 1 | 2 | 4 | 3 |
| (c) | 3 | 2 | 4 | 1 |
| (d) | 3 | 2 | 1 | 4 |

27. Consider the following statements:
   1. Nickel is extracted from teconite.
   2. Thorium is extracted from monazite.
   3. Titanium is extracted from rutile.
   4. Lithium is extracted from rhodonite.

   Which of the above statements are correct?

   (a) 1, 2 and 4 (b) 1 and 3
   (c) 2 and 3 (d) 3 and 4

28. Which one of the following pairs is correctly matched?

   (a) Zincite : Residual
   (b) Asbestos : Sedimentary
   (c) Chromite : Magmatic
   (d) Gypsum : Hydrothermal

29. Consider the following deposits:
   1. Mesothermal
   2. Epithermal
   3. Hypothermal

   The correct sequence in the ascending order of temperatures at which these are formed, is:

   (a) 1, 3, 2 (b) 2, 1, 3
   (c) 3, 1, 2 (d) 2, 3, 1

30. A gossan having high Co : Ni ratio is derived from:
    (a) gabbro (b) magnetite
    (c) pyritiferous shale (d) pyroxenite
31. Which one of the following is epigenetic?
    (a) Ores of hydrothermal derivation.
    (b) Ores of low metal concentration.
    (c) Ores formed at the same time as the host rock.
    (d) Ores formed later than the host rock.
32. Which one of the following statements is correct?
    (a) The gold placers occurring along the Subarnarekha river are of commercial importance.
    (b) India has the largest manganese ore reserves in the world.
    (c) The blue-quartz veins are selectively auriferous in Kolar.
    (d) The manganese ores of India generally have very low iron content.
33. The correct sequence in which the metal content in the following groups varies progressively from low to high, is:
    (a) Pyrolusite, Manganite, Psilomelane
    (b) Realgar, Arsenopyrite, Orpiment
    (c) Chalcopyrite, Bornite, Chalcocite
    (d) Cerussite, Boyrnonite, Galena
34. Consider the following deposits:
    1. Titaniferous magnetite deposit of Mayurbhanj.
    2. Siderite deposit of Kutch.
    3. Banded hematite quartzite of Rajhara.
    4. Nodular iron-stones of Bihar.

    The correct sequence according to their geological ages is:
    (a) 1, 2, 4, 3 (b) 2, 3, 4, 1
    (c) 3, 4, 1, 2 (d) 1, 3, 4, 2
35. Which one of the following pairs is correctly matched?
    (a) Nagaur gypsum : Upper Vindhyan
    (b) Trichinapally gypsum : Cretaceous
    (c) Jamnagar gypsum : Miocene
    (d) Bikaner gypsum : Triassic
36. Gold mineralization in Kolar occurs in veins and fissures confined to:
    (a) schistose amphibolites and banded iron formation.
    (b) metamorphosed basic volcanics.
    (c) metamorphosed basic volcanics and sulphidic banded iron formation.
    (d) orthoquartzites.
37. Which one of the following mineral assemblages represents the group of acid refractories?
    (a) Magnesite, fire clay, Bauxite, Graphite.
    (b) Chromite, Asbestos, Graphite, Zircon.
    (c) Fire clay, Quartz, Kyanite, Sillimanite.

38. The mineral assemblage expected in common acid pegmatite is:
   (a) Topaz, biotite, Calcioplagioclase, Tourmaline
   (b) Quartz, Magnetite, Ilmenite, K-feldspar
   (c) Cassiterite, Magnetite, Quartz, Wolframite
   (d) Quartz, Muscovite, K-feldspar, Beryl
39. Which one of the following indicates the optimum by-product availability during coal carbonisation?
   (a) Moisture (b) Volatile matter
   (c) Fixed carbon (d) Ash

**Directions:** The following 11(eleven) items consist of two statements. One labelled the section 'A' and the other labelled the Reason R: .You are to examine these two statements carefully and decide if the Assertion A and the Reason R are individually true and if so, whether the Reason is a correct explanation of the Assertion. Select your answers to these items using the codes given below and mark your answer sheet accordingly.

Codes:
(a) Both A and R are true and R is the correct explanation of A.
(b) Both A and R are true but R is NOT a correct explanation of A.
(c) A is true but R is false.
(d) A is false but R is true.

40. Assertion (A): $^{87}Rb/^{87}Sr$ method of dating is unsuitable for very young rocks.
   Reason (R) : $^{87}$ Rb has a comparatively short half-life.
41. Assertion (A) : Only a few tributaries of the ganga are antecedent.
   Reason (R) : Most of the tributaries of the Ganga originate from glaciers on the southern side of the Himalayan crest line, while only a few originate in Tibet.
42. Assertion(A): Salt diapirs result from the intrusion of liquid halite into surrounding sediments.
   Reason (R) : The intrusion results form the difference in density between halite and the overlying sediments.
43. Assertion (A): Although crystals possess a regular orderly internal structure, some of their properties vary continuously with direction within the crystals.
   Reason(R) : Different crystal planes and directions within them have differences in atomic environments.
44. Assertion (A): For monodinic crystals, the positive end of 'a-axis' projects upon the lower hemisphere.
   Reason (R) : It forms an obtuse angle with c-axis.
45. Assertion (A): A melt of basaltic composition may produce a basalt-andesite-rhyolite association, but a melt of granitic composition will yield a rhyolite or granite.

Reason (R) : If the crystals are completely or partially prevented from reacting with the melt and get segregated, fractional crystallization may take place and magmatic differentiation may occur.

46. Assertion (A): The relative abundance of quartz as compared to feldspar in a sedimentary rock is a measure of mineralogical maturity.

Reason (R) : Mineralogical maturity is the extent to which sands approach the ultimate end-product by the formative processes that operate upon it prior to lithification.

47. Assertion (A): From similar mineral assemblages the world over. It can be assumed that they have formed over a range of P-T conditions. In such cases P or T can vary independently and the same mineral assemblage persist—a case of divariant equilibrium.

Reason(R) : There are two degree of freedom for a metamorphic system .When P=C as per phase rule.

48. Assertion (A): Mid-Oceanic Ridge Basalt (MORB) may show metamorphism from zeolite to amphibolite facies.

Reason (R) : The MORB suffers hydrothermal metamorphism due to injection of hydrothermal fluid from the mantle below.

49. Assertion (A): According to Indian manufacturers, the domestic manganese dioxide ore is NOT suitable for battery cells in which high current flow is required.

Reason (R) : Domestic ore contains insoluble metallic impurities.

50. Assertion (A): The alumina content of blast furnace limestone must always be less than 2%.

Reason (R) : During production of pig iron, the blast furnace slag must contain 15%-22% $Al_2O_3$.

51. Which one of the following groups of elements constitute more than 80% by weight of the Earth's crust?
   (a) Silicon, Iron and Aluminium
   (b) Iron, Silicon and Oxygen
   (c) Oxygen, Silicon and Aluminium
   (d) Aluminium, Iron and Oxygen.

52. Consider the following events:
   1. Continent-Continent collision.
   2. Formation of an intracontinental mountain belt.
   3. Closing of an ocean basin.

   The correct sequence of these events during a Wilson cycle is:
   (a) 1, 2, 3 (b) 1, 3, 2
   (c) 3, 1, 2 (d) 2, 3, 1

53. Seismic velocity of P-waves is the highest in:
(a) Granite (b) Sandstone
(c) Basalt (d) Clay

54. Which one of the following is the characteristic of the Mohorovicic discontinuity?
(a) Sudden decrease in the velocity of P- and S-waves.
(b) Sudden increase in the velocity of P- and S-waves.
(c) Shear waves do not pass through.
(d) Increase in the velocity of P-wave.

55. A mobile belt characterised by very thick sediments and intermittent volcanic activity is:
(a) a miogeosyncline (b) a parageosyncline
(c) an eugeosyncline (d) an exogeosyncline

56. Consider the following four orogenies:
1. Hereynian 2. Caledonian
3. Himalayan 4. Alpine
The correct chronological order of these orogenies is:
(a) 2, 1, 4, 3 (b) 1, 3, 2, 4
(c) 1, 2, 3, 4 (d) 3, 2, 1, 4

57. Under tropical climate the rate of weathering of:
(a) Granite > Basalt> Diorite (b) Basalt > Granite > Diorite
(c) Diorite > Basalt > Granite (d) Basalt > Diorite > Granite

58. In deserts, insolation is most pronounced when there is:
(a) high temperature and low moisture content.
(b) sharp diurnal fluctuation in temperature and moisture content.
(c) sharp diurnal fluctuation in temperature.
(d) sharp fluctuation in moisture content.

60. Match List I and List II and select the correct answer using the codes given below the lists:

| List I | List II |
|---|---|
| A. Rills | 1. Tidal current |
| B. Off-shore bar | 2. Refracting waves |
| C. Inlet | 3. Backwash |
| D. Spit | 4. On-shore sediment transport |

Codes:

| | A | B | C | D |
|---|---|---|---|---|
| (a) | 4 | 3 | 1 | 2 |
| (b) | 3 | 4 | 2 | 1 |
| (c) | 4 | 3 | 2 | 1 |
| (d) | 3 | 4 | 1 | 2 |

61. Match List I and List II and select the correct answer using the codes given below the lists:

| List I | List II |
|---|---|
| A. Warm and arid | 1. Tillite |
| B. Warm and humid | 2. Quartzarenite |
| C. Semi-arid | 3. Dolomite |
| D. Glacial | 4. Concretion |

Codes:

| | A | B | C | D | | A | B | C | D |
|---|---|---|---|---|---|---|---|---|---|
| (a) | 3 | 2 | 1 | 4 | (b) | 3 | 2 | 4 | 1 |
| (c) | 2 | 3 | 1 | 4 | (c) | 2 | 3 | 4 | 1 |

62. The drainage pattern that develops when the bed rocks have uniform resistance to erosion and exercise on control over valley growth, is called:
(a) dendritic (b) trellis
(c) rectangular (d) deranged

63. Under which one of the following climates does laterite form?
(a) Tropical rainy (b) Arctic
(c) Temperate humid (d) Arid and semi-arid

64. The orientation of stress ellipsoid at the time of formation of conjugate fault is given below:

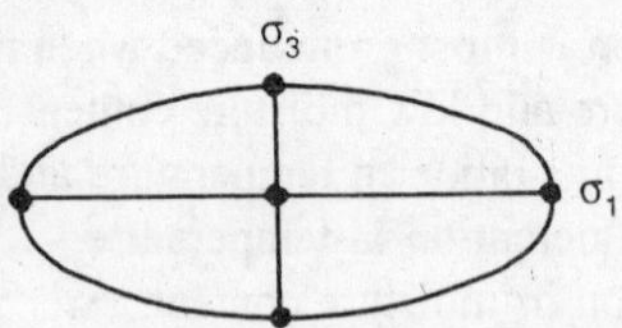


Which one of the following types of fault was formed?
(a) Normal (b) Reverse
(c) Strike-slip (d) Gravity

65. The vertical section of two limbs of a fold plunging 50° towards north is shown in the given figure:

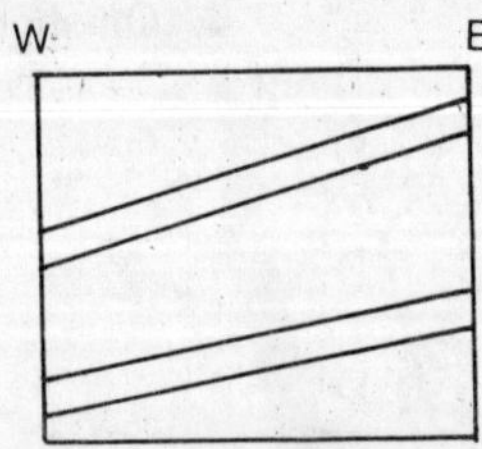


The fold is:
(a) an antiform (b) a synform
(c) a reclined fold (d) an anticline

66. Conjugate vertical strike-slip faults strike N 60° W and N 60° E respectively. Which one of the following gives the orientation of the strain axes at the time of their formation?
(a) X=N-S, Y= vertical , Z=E-W
(b) X=vertical , Y=N-S, Z=F-W
(c) X=N-S, Y=E-W, Z=vertical
(d) X=vertical, Y=E-W, Z=N-S

67. The given map shows a fault crossing a coal seam without any offset:

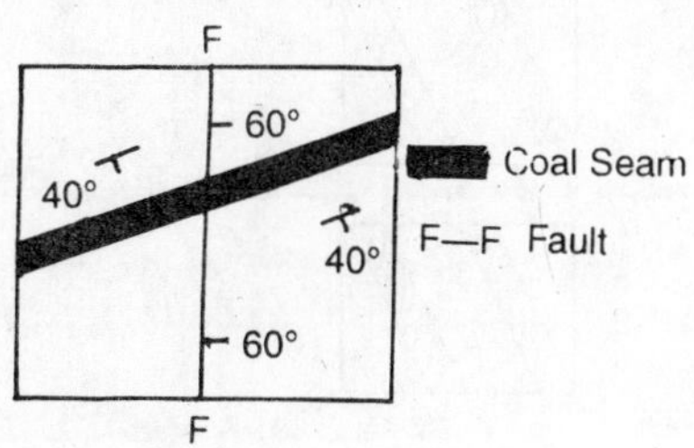


The best explanation is that:
(a) the coal seam is younger than the fault.
(b) there is no strike-slip component.
(c) net slip is parallel to the trace of the seam on the fault.
(d) such a situation is not possible.

68. Match List I with List II and select the correct answer using the codes given below the Lists:

| List I | List II |
|---|---|
| A. Puckering | 1. Quartz veins |
| B. Rodding | 2. Shear plane |
| C. Striation | 3. Amphibolite |
| D. Boudinage | 4. Phyllite |

Codes:

| | A | B | C | D | | A | B | C | D |
|---|---|---|---|---|---|---|---|---|---|
| (a) | 4 | 1 | 3 | 2 | (b) | 1 | 3 | 2 | 4 |
| (c) | 4 | 1 | 2 | 3 | (d) | 1 | 3 | 4 | 2 |

69. A limb of a fold traversed by axial plane foliation is shown in the block diagram:

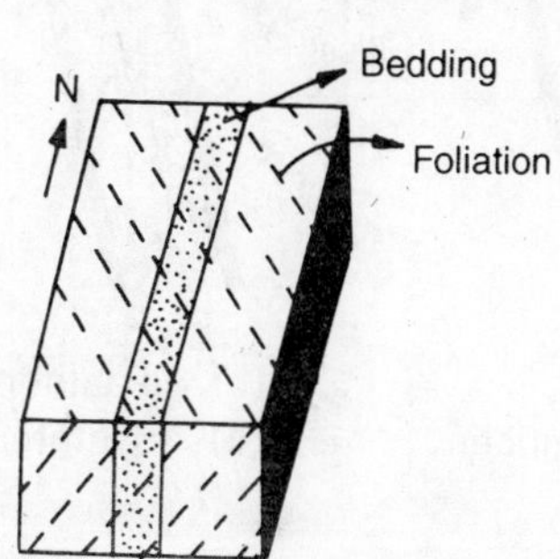

Which one of the following is the correct interpretation?
(a) Syncline to the east, fold plunges south.
(b) Syncline to the east, fold plunges north.
(c) Anticline to the east, fold plunges south.
(d) Anticline to the west, fold plunges north.

70. Cross bedding in sandstone shown below in vertical sections indicates that:

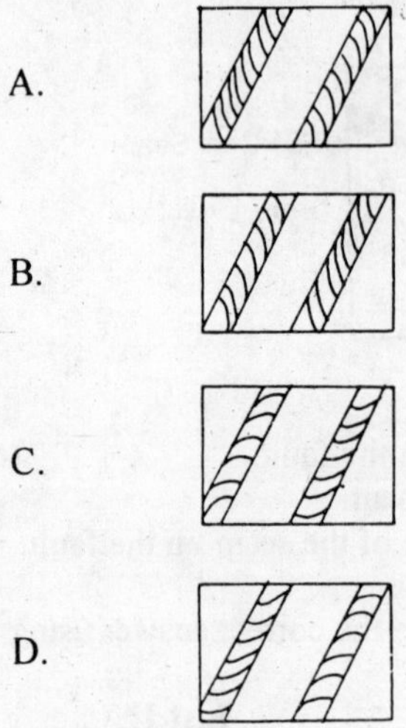


(a) A and C are synclines and B and D are anticlines.
(b) A and B are anticlines and C and D are synclines.
(c) A and D are synclines and B and C are anticlines.
(d) A and C are anticlines and B and D are syncline.

71. Minor fold pattern in the limbs of a fold in a vertical section is shown below:

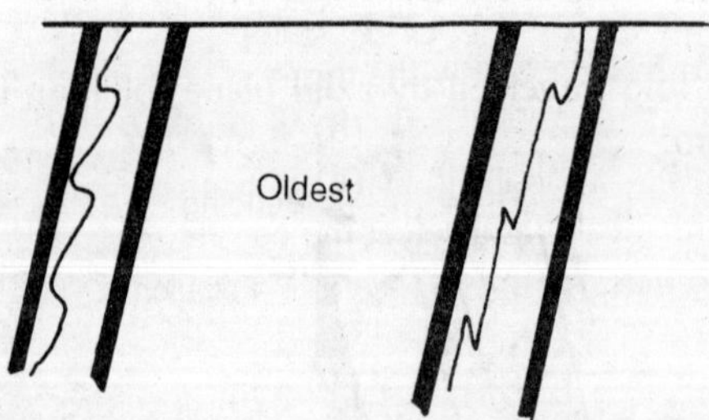


The fold is:
(a) an antiform
(b) an antiformal syncline
(c) a synformal anticline
(d) a synform

72. The correct chronology of geological events in the area shown in the given map (the oldest being at the bottom) is:

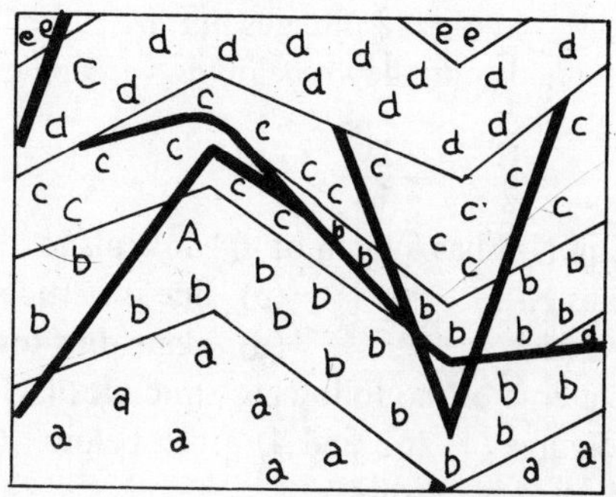

(a) Intrusion of dyke c
Folding
Deposition of d,e
Intrusion of dyke b
Intrusion of dyke a
Faulting
Deposition of a, b, c

(b) Intrusion of dyke c
Folding
Intrusion of dyke b
Intrusion of dyke a
Faulting
Deposition of a, b, c, d, e

(c) Intrusion of dyke c
Folding
Deposition of d, e
Intrusion of dyke b
Faulting
Intrusion of dyke a
Deposition of a, b, c

(d) Intrusion of dyke c
Folding
Intrusion of dyke a
Deposition of d, e
Intrusion of dyke b
Faulting
Deposition of a, b, c

73. The feature formed when the rocks above a thrust are completely surrounded by the rocks below the thrust is:

(a) an outlier (b) a window
(c) a klippe (d) an inlier

74. Match List I with List II and select the correct answer using the codes given below the lists:

| List I | List II |
|---|---|
| A. N 60°W | 1. 270° |
| B. S 10° E | 2. 220° |
| C. S 40°W | 3. 170° |
| D. West | 4. 300° |

Codes :

(a) A B C D
4 3 2 1

(b) A B C D
4 3 1 2

(c) A B C D
3 4 1 2

(d) A B C D
3 4 2 1

75. If a line trending N-S has a pitch of 0° on a plane, the:

(a) plane strikes N-S (b) line is vertical
(c) plane is vertical (d) plane dips towards north

76. The outcrop patterns of the following structures (A, B, C and D) are shown in the figures (1, 2, 3 and 4) given below : (the directions of plunge are indicated by arrows)

A. Antiform with culmination
B. Antiform with depression
C. Synform with culmination
D. Synform with depression

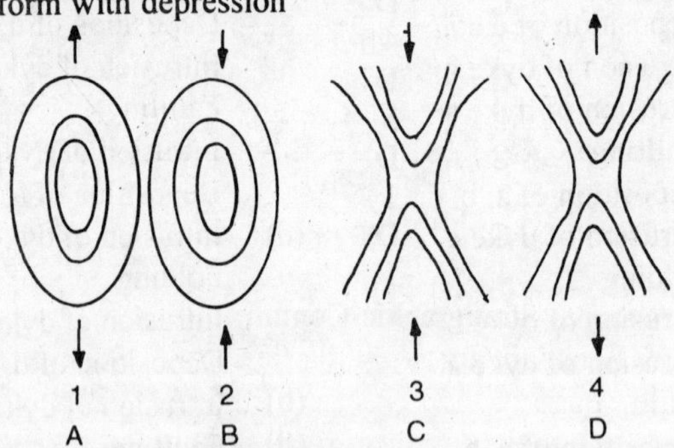


Match the structures with the figures and select the correct answer using the codes given below:

Codes:

(a) A B C D
1 2 4 3

(b) A B C D
4 3 1 2

(c) A B C D
1 2 3 4

(d) A B C D
1 3 4 2

77. A contact exposed in a stream channel is shown below:

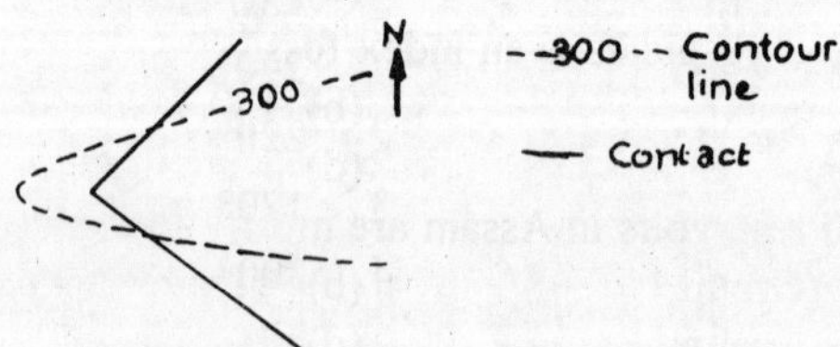

Now consider the following statements:

1. The contact dips towards east.
2. The contact dips towards west.
3. The slope is towards east.
4. The slope is towards west.

Which of the above statements are correct?

(a) 1 and 3 (b) 2 and 3
(c) 1 and 4 (d) 2 and 4

78. Match List I with List II and select the correct answer using the codes given below the list:

| List I | List II |
|---|---|
| A. Series | 1. Era |
| B. Stage | 2. Period |
| C. System | 3. Epoch |
| D. Erathem | 4. Eon |
| | 5. Age |

Codes:

| | A | B | C | D |
|---|---|---|---|---|
| (a) | 3 | 5 | 2 | 4 |
| (b) | 4 | 2 | 1 | 5 |
| (c) | 3 | 5 | 2 | 1 |
| (d) | 2 | 3 | 4 | 1 |

79. The International Stratigraphic Commision has placed the Archaean-Proterozoic boundary at:

(a) 1600 m.y. ago (b) 2000 m.y. ago
(c) 2200 m.y. ago (d) 2500 m.y. ago

80. Which one of the following is characteristic of the Ordovician?

(a) Trinucleus, Ogygia, Ampyx
(b) Illaenus, Phacops, Calymene
(c) Dalmanites, Phillipsia, Redlichia
(d) Ogygiopsis, Microdiscus, Agnostus

81. Which one of the following fossils indicates a Middle Devonian age?

(a) Stylina (b) Zaphrentis
(c) Calceola (d) Fevosites.

82. Syringothyris cuspidata is an index fossil found in the:

(a) Zewan beds (b) Lipak series
(c) Po series (d) Kuti shale

83. Gas and oil reservoirs in Assam are mostly found in the:

(a) Dihing Group (b) Tipam Group
(c) Disang Group (d) Jaintia Group

84. Match List I with List II and select the correct answer using the codes given below the lists:

| List I | List II |
|---|---|
| A. Stromatiolites | 1. Raniganj formation |
| B. Glossopteris flora | 2. Niniyur formation |
| C. Trigonia ventricosa | 3. Kaladgi group |
| D. Nautilus danicus | 4. Umia formation |

Codes:

| | A | B | C | D |
|---|---|---|---|---|
| (a) | 1 | 2 | 3 | 4 |
| (b) | 3 | 1 | 4 | 2 |
| (c) | 1 | 2 | 4 | 3 |
| (d) | 3 | 1 | 2 | 4 |

85. The base of Bagh beds in the Narmada Valley is represented by:
(a) Oyster bed
(b) Nodular limestone
(c) Deola marls
(d) Nimar sandstone

86. Which one of the following is the equivalent of Kurnools in Bastar Basin?
(a) Bhima Group
(b) Indravati Group
(c) Raialo Group
(d) Muth Quartzite.

87. Besides gold, iron, copper, chromium, and lead, the Dharwar Supergroup, contains important deposits of:
(a) Mercury
(b) Gypsum
(c) Tungsten
(d) Manganese

88. To which one of the following does the Bababudan Group belong?
(a) Aravalli
(b) Dharwar
(c) Sargur
(d) Kurnool

89. Which one of the following groups consists of alternate beds of red shales and sandstones?
(a) Semri
(b) Kaimur
(c) Rewa
(d) Bhander

90. Evidences of arid climate are found in:
    (a) Talchir Group
    (b) Barakar Group
    (c) Mahadeva Group
    (d) Rajamahal Group
91. Upper Siwalik sediments are:
    (a) fluviatile
    (b) marine
    (c) evaporite
    (d) volcanic
92. The specimen of a species first reported is called:
    (a) topotype
    (b) paratype
    (c) holotype
    (d) syntype
93. The eye has a tubercle with one or two facets in:
    (a) encrinurus
    (b) harpes
    (c) phacops
    (d) encrinuroides
94. Consider the following genera:
    1. Littorina
    2. Thecosmilia
    3. Olenellus
    4. Dictyonema

    The correct chronological order of these genera is:
    (a) 4, 3, 2, 1
    (b) 3, 4, 1, 2
    (c) 3, 4, 2, 1
    (d) 4, 3, 1, 2
95. Brachial skeleton consisting of two short curved processes is found in:
    (a) Rhynchonella
    (b) Terebratula
    (c) Stringocephalus
    (d) Magellina
96. Zaphrentis is distinguished by the:
    (a) presence of septa.
    (b) absence of septa.
    (c) presence of well-marked tabulae.
    (d) presence of well-marked cardinal fossula.
97. Which one of the following pairs is NOT correctly matched?
    (a) Arca : Taxodont
    (b) Hippurites : Isodont

(c) Trigonia : Schizodont
(d) Ostrea : Dysodont

98. In Exogyra, the unbones are:
   (a) curved towards the anterior margin
   (b) straight
   (c) direct posteriorly
   (d) spiral

99. Which one of the following triobites has a large pygidium, enlarged cephalon and only two free segments?
   (a) Olenus
   (b) Paradoxides
   (c) Trinucleus
   (d) Agnostus

100. Mid-Palaeozoic ammonoids have:
   (a) Septal necks pointing forward
   (b) Ventral siphuncles
   (c) Dorsal siphuncles
   (d) Septal necks pointing backward

101. In which one of the following shells are the whorls coiled symmetrically in one plane so that each flank of the coil is the mirror image of the other?
   (a) Helix
   (b) Euomphalous
   (c) Trochus
   (d) Bellerophon

102. Which one of the following sets is characteristics of the Raniganj Group?
   (a) Gangamopteris whittiana, glossopteris browniana, phyllotheca indica.
   (b) Glossopteris ampla, Gangamopteris cyclopteroids, Schizoneura wardi.
   (c) Gangamopteris Kashmirensis, Schizoneura wardi, phyllotheca griesbachi.
   (d) Glossopteris indica, Sphenopteris polymorpha, Gangamopteris Kashmirensis.

103. Which one of the following is a regular echinoid?
   (a) Cidaris
   (b) Holaster
   (c) Hermiaster
   (d) Micraster

104. The position of stipes in the most highly evolved species of Didymograptus is:
   (a) scandent

(b) pendent
(c) declined
(d) horizontal

105. Consider the following statements:
The demarcation of the pliocene/pleistocene boundary in the Siwalik is based on the:
1. extinction of Hipparion.
2. appearance of Equus.
3. simultaneous occurrence of Equus, Elphas, Bos and Cervides.
4. marked decline in the population of Hippopotamus.
Which of the above statements are correct?
(a) 1, 2 and 4
(b) 2, 3 and 4
(c) 1, 3 and 4
(d) 1, 2 and 3

106. Which one of the following genera is recorded from the Middle Siwalik formation?
(a) Dinotherium
(b) Elephas
(c) Tetrabelodon
(d) Mastodon

107. In foraminifera, a megalospheric form denotes a:
(a) large test with a large proloculum.
(b) small test with a small proloculum.
(c) large test with a small proloculum.
(d) small test with a large proloculum.

108. A crystal has one tetragonal axis of symmetry, 4 horizontal axes of binary symmetry, one horizontal and 4 vertical planes of symmetry and a centre of symmetry. Which one of the following classes of symmetry does it belong to?
(a) Normal class of monoclinic system.
(b) Tetrahedral class of isometric system.
(c) Normal class of tetragonal system.
(d) Normal class of orthorhombic system.

109. The given diagram is a stereogram of:

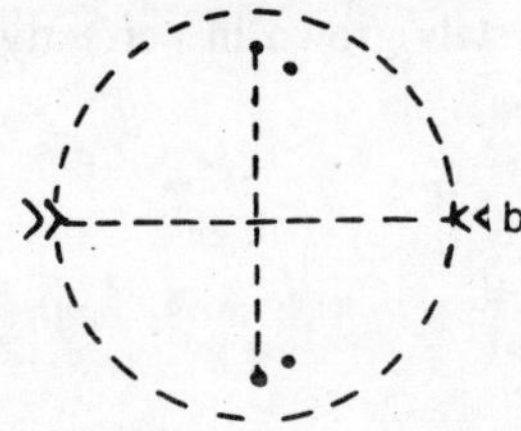

(a) Orthorhombic system with symmetry 2.
(b) Monoclinic system with symmetry 2.
(c) Orthorhombic system with symmetry m.
(d) Monoclinic system with symmetry m.

110. Match List I with List II and select the correct answer using the codes given below the lists:

| List I | List II |
|---|---|
| A. Gypsum | 1. Tarton twin |
| B. Pyrite | 2. Swallow tail twin |
| C. Staurolite | 3. Greek cross |
| D. Microcline | 4. Iron cross |

Codes:

| | A | B | C | D |
|---|---|---|---|---|
| (a) | 2 | 3 | 4 | 1 |
| (b) | 2 | 4 | 3 | 1 |
| (c) | 1 | 3 | 2 | 4 |
| (d) | 3 | 2 | 4 | 1 |

111. Match List I (Crystal form) with List II (Miller's symbol) and select the correct answer using the codes given below the lists:

| List I | List II |
|---|---|
| A. Tetrahexahedron | 1. (221) |
| B. Trisoctahedron | 2. (321) |
| C. Trapezohedron | 3. (210) |
| D. Diploid | 4. (211) |

Codes:

| | A | B | C | D |
|---|---|---|---|---|
| (a) | 1 | 2 | 3 | 4 |
| (b) | 3 | 1 | 4 | 2 |
| (c) | 3 | 1 | 2 | 4 |
| (d) | 1 | 3 | 4 | 2 |

112. Numerous small crystals grown in perfectly circular outline is an example of:
(a) barnal crystal
(b) rosette crystal
(c) cockscomb crystal
(d) reticulated crystal

113. Which one of the following general formulae represents the chemical composition of the mica group of minerals?
(a) $X_{2-3}Y_5Z_8O_{22}$ (OH)$_2$
(b) $X_2Y_{2-4}Z_8O_{20}$(OH)$_4$
(c) $X_2Y_{4-6}Z_8O_{20}$(OH, F)$_4$
(d) $X_{2-3}$ $Y_5Z_8O_{22}$
114. Which one of the following minerals possessing $Si_4O_{11}$ group, is an inosilicate?
(a) Garnet
(b) Beryl
(c) Muscovite
(d) Arfvedsonite
115. When the scheme of pleochroism is such that X= dark green, Y=light green, Z=yellow, the mineral is:
(a) Aegirine
(b) Spodumene
(c) Diopside
(d) Enstatite
116. Extinction occurs when the vibration direction of light from the polarizer coincides exactly with one of the vibration directions of the crystal. In this situation, light passes through the crystal as either O-ray or E-ray to be completely eliminated by the analyzer. Now, if the mineral is rotated, the grain will:
(a) remain extinct throughout the 360° rotation.
(b) brighten up, the maximum brightness being at 180° and next extinction being at 360°.
(c) brighten up, the maximum brightnesss being at 90° and next extinction being at 180°.
(d) brighten up, the maximum brightness being at 45° and the next extinction being at 90°.
117. The extinction angle in a biaxial mineral is defined as the angle between
(a) the length of the mineral and one vibration direction.
(b) one crystallographic direction and the slowest vibration direction.
(c) any cleavage direction and one vibration direction.
(d) one crystallogrpahic direction and the fastest vibration direction.
118. The main difference between pleochroism and play of colours in minerals is that pleochroism depends upon:
(a) the selective absorption of transmitted light whereas play of colours is interference of reflected light.
(b) the interference of transmitted light whereas play of colours is selective absorption.
(c) the interference of reflected light whereas the play of colours is

the selective absorption of transmitted light.

(d) the selective absorption of incident light whereas the play of colours is the interference of transmitted light.

119. In an optically positive orthorhombic crystal where the a, b and c crystallogrpahic axes coincide respectively with the Z, Y and X vibrations, the sections that will have to be selected to obtain an acute optic axis figure will be a section cut normal to the:
(a) a-axis
(b) b-axis
(c) c-axis
(d) a-b plane

120. Which one of the following statements is NOT CORRECT?
(a) If the dispersion is rhombic,the interference figures are symmetrical to optic plane and the plane normal thereto.
(b) If the dispersion is inclined, the interference figure is symmetrical to the centre.
(c) If the dispersion is horizontal, the interference figures are symmetrical to the plane normal to optic plane.
(d) If the dispersion is inclined, the interference figures are symmetrical to optic plane.

# 3. Civil Services Examination, 1998
## Questions And Answers

1. The most important objection to the Nebular hypothesis as propounded by Laplace is that it.
   (a) fails to explain the present-day spacing among the planetary bodies.
   (b) violates the present-day distribution of momentum in the chief members of the solar system.
   (c) does not attempt to adduce an origin for the parent Nebula.
   (d) fails to explain the contraction process of Earth.
2. Which one of the following has been regarded as the most important evidence of continental drift since the theory was advanced?
   (a) Palaeomagnetic evidence.
   (b) Similarity in palaeozoic sequence in all part of Gondwanal and.
   (c) Faunal and floral similarities in different land masses.
   (d) Similarities in the coastal outlines of West Africa and Eastern South America.
3. Zircon is the most widely used mineral for uranium-lead dating because it has:
   (a) high resistance to post-informational thermal events.
   (b) a long half life.
   (c) a very high uranium content.
   (d) a very high thorium content.
4. The deep-focus earthquakes are restricted practically to the:
   (a) Himalayan belt.
   (b) Mediterranean-Alpine belt.
   (c) Mid-oceanic ridge in the Atlantic.
   (d) Deep-sea trenches of the Pacific.
5. Vesuvian type of volcanic eruption is particularly violent because:
   (a) Lava is not very mobile.
   (b) Lava is highly charged with gases.
   (c) of growth of an obstructive dome near the crater.
   (d) of high viscosity and delayed explosion.
6. Which one of the following sets of conditions is most favourable to a luxuriant growth of corals?
   (a) Shallow, warm and turbid water.
   (b) Shallow, warm and clear water.
   (c) Deep, cold and turbid water.
   (d) Deep, cold and clear water.
7. Which one of the following best explains the phanerozoic orogenies?
   (a) Geosynclinal theory (b) Plate tectonics
   (c) Contraction theory (d) Expansion hypothesis

8. The fields labelled I, II, III, and IV in the given figure refer respectively to:
   (a) transverse, parabolic longitudinal and Barchan types of dunes.
   (b) transverse, parabolic Barchan and longitudinal types of dunes.
   (c) parabolic, transverse, Barchan and longitudinal types of dunes.
   (d) parabolic, transverse, longitudinal and Barchan types of dunes.

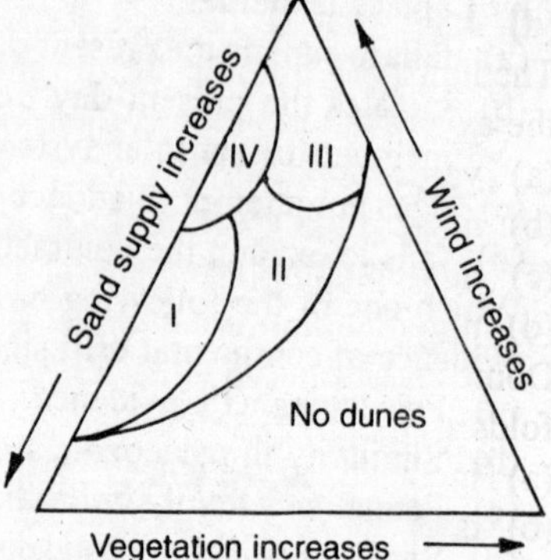


9. Darcy is the unit of:
   (a) permeability
   (b) transmissivity
   (c) porosity
   (d) hydraulic conductivity
10. According to the mineral stability series in weathering, rocks composed of weathering residues commonly contain:
   (a) olivine and anorthite
   (b) quartz and muscovite
   (c) augite and anorthite
   (d) olivine and albite
11. In an area, sandstone ridges having northerly escarpments and southerly dip slope trend east-west for five kilometers. Then they abruptly turn north-south at the western end developing easterly escarpments and westerly dip slope. Which one of the following structures does this indicate?
   (a) Antiform plunging southwest
   (b) Antiform plunging northeast
   (c) Synform plunging southwest
   (d) Synform plunging northeast
12. A well-developed soil profile in humid climate will have which one of the following sequences from top to bottom?
   (a) Humus horizon—eluvial horizon—illuvial horizon
   (b) Illuvial horizon—eluvial horizon—humus horizon
   (c) Eluvial horizon—illuvial horizon—humus horizon
   (d) Humus horizon—illuvial horizon—eluvial horizon
13. Match List I (Feature) with List II (environment) and select the correct answer using the codes given below the lists:

| List 1 | List II |
|---|---|
| A. Scoria mound | 1. Arid |
| B. Guyot | 2. Karst |
| C. Hums | 3. Volcanic |
| D. Bolson | 4. Marine |

Codes:

| | A | B | C | D |
|---|---|---|---|---|
| (a) | 3 | 4 | 2 | 1 |
| (b) | 3 | 4 | 1 | 2 |
| (c) | 4 | 3 | 1 | 2 |
| (d) | 4 | 3 | 2 | 1 |

14. The limbs of a fold dip 40° towards N 110° and N 290° respectively. If the axial plane is vertical, then the fold is:
    (a) vertical
    (b) non-plunging symmetrical
    (c) non-plunging symmetrical antiform
    (d) non-plunging symmetrical synform
15. Dome and basin type interference pattern result when earlier and later folds are:
    (a) inclined and trend perpendicular to each other.
    (b) coaxial.
    (c) upright and trend perpendicular to each other.
    (d) isoclinal and trend perpendicular to each other.
16. A vertical fault strikes N 40° W. Horizontal slickensides on the fault plane trend N 320°. Then the fault will be:
    (a) a strike fault (b) an oblique fault
    (c) a dip-slip fault (d) a strike-skip fault
17. The principal axes of stress are:
    (a) Normal to surface on which no shear stress acts.
    (b) Directions to application of forces.
    (c) Directions of maximum compression.
    (d) Directions of maximum tension.
18. Which of the following pairs of fold classes and nature of dip isogons in Ramsay's classification are correctly matched?
    1. Class 1............. Converge downward
    2. Class 2............. Parallel to one another
    3. Class 3............ Diverge downward

    Select the correct answer using the codes given below:
    Codes:
    (a) 1, 2 and 3 (b) 1 and 2
    (c) 2 and 3 (d) 1 and 3
19. Congruous boundinage structures are seen in the limbs of a non-plunging flexural fold. The fold strikes N 100°. Boudin lines will be:
    (a) horizontal, trending N10°E (b) vertical
    (c) horizontal, trending S 80°E (d) inclined, trending N 10°W
20. If N-S striking vertical beds have developed axial plane foliation dipping 60° towards E, then relatively younger beds can be expected towards the:

(a) North (b) East
(c) South (d) West

21. The intrusive body shown in the map is:
(a) sill
(b) lacclith
(c) lopolith
(d) phacolith

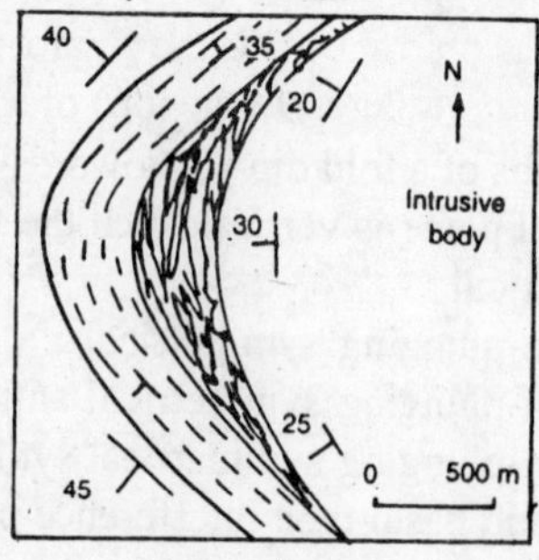


22. In a railway cutting, conformable beds exhibit mud cracks as shown in the figure. Which of the following statements are true regarding the section shown?
1. A is the oldest bed.
2. Beds are overturned
3. Beds do not agree with the law of the order of superposition of the beds.
Select the correct answer using the codes given below:
(a) 1, 2 and 3 (b) 1 and 2
(c) 2 and 3 (d) 1 and 3

A B C D E F

23. In a stereographic projection, the axial plane of a recumbent fold is represented by:
(a) a great circle (b) a small circle
(c) the primitive circle (d) the diameter of the projection

24. Consider the following statements regarding the map shown in the figure:

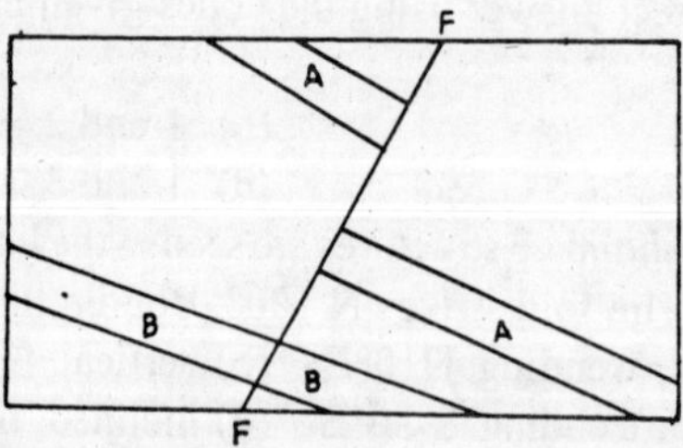


1. Bed A is inclined
2. Bed B is vertical
3. FF is an oblique-slip fault

Of these statements:

(a) 1, 2 and 3 are correct (b) 1 and 2 are correct

(c) 2 and 3 are correct (d) 1 and 3 are correct

25. As per the map shown in the figure, the correct sequence (oldest at the bottom and youngest at the top) of the given events is:

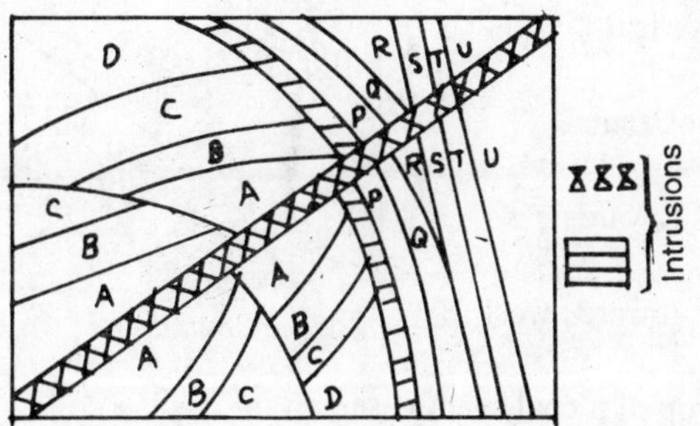


(a) Deposition of S,T,U
unconfirmity
Intrusion
Faulting
Folding
Intrusion
Deposition of P,Q,R
Unconformity
Faulting
Deposition of A,B,C,D

(b) Intrusion
Deposition of S,T,U
Unconformity
Faulting
Folding
Intrusion
Deposition of P,Q,R
Unconformity
Faulting
Deposition of A, B, C, D

(c) Intrusion
Deposition of S,T,U
Unconformity
Folding
Faulting
Intrusion
Deposition of P,Q,R
Unconformity
Faulting
Deposition of A,B,C,D

(d) Intrusion
Faulting
Folding
Intrusion
Deposition of P,Q,R,S,T,U
Unconformity
Faulting
Deposition of A,B,C,D

26. Consider the following statements:

The plunge of a line on a plane is equal to its pitch when the:

1. pitch is 0°
2. pitch is 90°
3. plane is vertical

4. pitch is 45°

Of these statements:

(a) 1 alone is correct (b) 1 and 3 are correct

(c) 2 and 3 are correct (d) 4 alone is correct

27. In the map shown in the figure, the contact between beds A and B is exposed on an oval-shaped hill. The contact:

(a) is vertical

(b) is horizontal

(c) dips towards east

(d) dips towards west

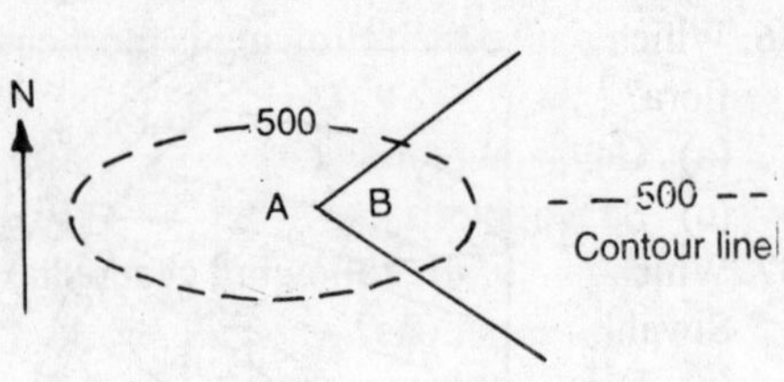


28. The outcrop of a coal seam is shown in the figure. The surface has no relief. Which one of the following is the correct interpretation of the figure?

(a) Faulted antiform plunging east.

(b) Faulted antiform plunging west.

(c) Faulted synform plunging east.

(d) Faulted synform plunging west.

29. Trills pattern is developed in an area which is characterised by:

(a) rocks affected by parallel faults.

(b) horizontal sedimentary rocks.

(c) rocks having domal structure.

(d) rocks with joints intersecting at acute angle.

30. In echinoids, the pore pair in the plates of the ambulacral areas may be surrounded by an elevated rim known as:

(a) Peripodium (b) Loven's plane

(c) Branchiae (d) Pedicellariae

31. In brachiopods, the hinge line which is much less than the width of the shell and is strongly curved is referred to as:

(a) Megathyrid (b) Spiriferid

(c) Submegathyrid (d) Terebratulid

32. When the beaks are directed towards one another in pelecypod shells, it is referred to as:

(a) Orthogyre (b) Prosogyre

(c) Opisthogyre (d) Prosocline

33. The geological range of bivalves (Lamellibranches) is:

(a) Ordovician to Recent (b) Devonian to Cretaceous

(c) Triassic to Cretaceous (d) Jurassic to Recent

34. Which one of the following best defines high and low tide limits in the marine environment?
(a) Neritic (b) Neritopelagic
(c) Bathyal (d) Littoral

35. The fauna of upper Siwalik deposits is dominated by:
(a) Reptiles (b) Birds
(c) Mammals (d) Invertebrates

36. Which one of the following is most characteristic of Lower Gondwana flora?
(a) Gangamopteris (b) Dictyozamites
(c) Sphenopteris (d) Ptilophyllum

37. Which one of the following genera has been reported from the Middle Siwalik formations?
(a) Bramapithecus (b) Semnopithecus
(c) Sugrivapithecus (d) Sivapithecus

38. The type of dentition in Bivalvia in which the teeth radiate downward from the umbo is referred to as:
(a) teleodont (b) dysodont
(c) schizodont (d) actinodont

39. The branchidium came into existence during the:
(a) Ordovician and Silurian (b) Carboniferous
(c) Permian (d) Cambrian

40. Which of the following characters within bivalves seem to have adaptive significance?
1. Shell structure and composition.
2. Structure of hinge and dentition.
3. Ornamentation.
4. Shape and general morphology.
Select the correct answer using the codes given below:
Codes
(a) 1, 2, 3 and 4 (b) 2, 3 and 4
(c) 3 and 4 (d) 1 and 2

41. Which one of the following groups of organisms is preserved by the process of carbonisation?
(a) Ammonoids (b) Corals
(c) Graptolites (d) Trilobites

42. Consider the following statements:
The first appearance of a marine fossil taxon in a succession may represent:
1. an evolutionary event.
2. a migration event from another biogeographic province.
3. A change in environment.
4. A marine regression event.

Of these statements:

(a) 1, 2 and 4 are correct (b) 1, 3 and 4 are correct
(c) 1, 2 and 3 are correct (d) 2, 3 and 4 are correct

43. The unifying concept of stratigraphy is the:
(a) geological time-scale (b) magnetostratigraphy
(c) correlation (d) biostratigraphy

44. Ideally a "group" in stratigraphic nomenclature consists of two or more successive and contiguous:
(a) periods (b) zones
(c) systems (d) formations

45. The oldest isotopic age of rocks from the Earth so far reported is:
(a) 4.2 billion years (b) 2 billion years
(c) 600 million years (d) 3.8 billion years

46. The age of Bagh Beds of Narmada Valley is:
(a) Early Cretaceous (b) Early Tertiary
(c) Late Cretaceous (d) Late Jurassic- Early Cretaceous

47. The Nummulitic Limestone in India is of the:
(a) Eocene-Oligocene (b) Palaeocene-Eocene
(c) Miocene-Pilocene (d) Pliocene-Pleistocene

48. Which one of the following correctly represents the stratigraphic sequence of the Cuddappah Supergroup?
(a) Papaghani, Nallamalai, Cheyair, Kistna
(b) Nallamalai, Cheyair, Papaghani, Kistna
(c) Papaghani, Cheyair, Nallamalai, Kistna
(d) Nallamalai, Papaghani, Cheyair, Kistna

49. A subdivision of the Cenozoic that encompasses the Palaeocene, Eocene and Oligocene Epochs is termed as:
(a) Tertiary (b) Palaeogene
(c) Neogene (d) Quaternary

50. The Precambrian rocks of India can be best correlated on the basis of their:
(a) lithostratigraphy (b) mineral deposits
(c) isotopic ages (d) palaeogeographical studies

51. The basal part of the Sargur Group is represented by:
(a) conglomerate, orthoquartzite and mafic lavas.
(b) phyllite, quartzite and greywacke.
(c) mafic and ultramafic flows and anorthosite.
(d) granite and gneisses, conglomerate and quartzite.

52. The stratigraphic classification of Jurassic rocks of Kutch has been broadly made on the basis of fossils. Which one of the following has been used for this purpose?
(a) Pelecypods (b) Ammonoids

(c) Echinoids (d) Mammals

53. The Semri Group is:

(a) calcareous and argillaceous (b) arenaceous and calcareous
(c) mainly arenaceous (d) mainly argillaceous

54. The Peninsular Indian Gondwana sequence has an age span of:

(a) Premian to Lower Jurassic
(b) Late Carboniferous to Late Cretaceous
(c) Carboniferous to Triassic
(d) Early Permian to Early Cretaceous

55. Consider the following statements:

The Talchir Group represents an assemblage of:

(1) fluvial and Lacustrine deposits
(2) glacial, glacio-marine and fluvio-glacial deposits
(3) turbidite and lacustrine deposits

Of these statements:

(a) 1 and 2 are correct (b) 1 and 3 are correct
(c) 2 and 3 are correct (d) 1, 2 and 3 are correct

56. The principle of superposition in stratigraphy denotes:

(a) space significance (b) time significance
(c) lateral continuity of strata (d) regional correlation

57. The eutectic intergrowth of orthoclase and quartz produces:

(a) perthitic granite (b) myrmekitic granite
(c) graphic granite (d) granophyre

58. Match List I (Silica saturation in magma) with List II (crystalline material) and select the correct answer using the codes given below the lists:

| List I | List II |
|---|---|
| A. $Sio_2/Na_2O<6$ | 1. Ab |
| B. $Sio_2/Na_2O=6$ | 2. Ab+ Ne |
| C. $Sio_2/Na_2O>6$ | 3. Ab not present |
| D. $Sio_2/Na_2O<2$ | 4. Ab + qz |

Codes

| | A | B | C | D |
|---|---|---|---|---|
| (a) | 2 | 1 | 4 | 3 |
| (b) | 2 | 1 | 3 | 4 |
| (c) | 1 | 2 | 3 | 4 |
| (d) | 1 | 2 | 4 | 3 |

59. Peralkaline igneous rocks are those in which:

(a) $Al_2O_3$ is less than $Na_2O$.
(b) $Al_2O_3$ exceeds ($CaO + Na_2O + K_2O$).
(c) $Al_2O_3$ is almost equal to ($Na_2O + K_2O$).
(d) ($NA_2O$+ $K_2O$) is greater than $Al_2O_3$.

60. Match List I with List II and select the correct answer using the codes given below the List:

| List I | List II |
|---|---|
| A. Kimberlite | 1. A type of mica-peridotite |
| B. Komatiite | 2. A close associate of ultrapotassic rock |
| C. Tractolite | 3. An Archaean rock |
| D. Carbonatite | 4. An olivine-plagioclase rock |

Codes:

| | A | B | C | D |
|---|---|---|---|---|
| (a) | 1 | 4 | 3 | 2 |
| (b) | 2 | 4 | 3 | 1 |
| (c) | 1 | 3 | 4 | 2 |
| (d) | 2 | 1 | 4 | 3 |

61. Which of the following stages are involved in the process of crystallisation of binary melt under isobaric conditions?
1. Cooling of magma prior to crystallisation.
2. Crystallisation of one constituent with falling temperature.
3. Supercooling of magma with crystallisation of one solid phase.
4. Simultaneous crystallisation of two solid phase with further lowering of temperature.
5. Simultaneous crystallisation of two solid phases at constant temperature.

Select the correct answer using the codes given below:

Codes

(a) 2, 3 and 4 (b) 1, 3 and 4
(c) 1, 2 and 5 (d) 1, 3 and 5

62. Which of the following silicate systems exhibit isobaric peritectic relation?

1. Leucite-quartz 2. Nepheline-quartz
3. Forsterite-quartz 4. Anorthite-quartz

Select the correct answer using the codes given below:

Codes

(a) 1, 2 and 3 (d) 2, 3 and 4
(c) 1 and 3 (d) 2 and 4

63. Consider the following statements:

Bowmen's Reaction Principle:
1. helps in predicting the order of crystallisation of minerals from any crystallising magma.
2. helps in predicting the nature of assimilation of country rocks by an anhydrous basic magma.
3. is based on calculations of free energy of formation of minerals from a crystallising magma.
4. is based on experimental results on simple silicate systems and field observations of co-magmatic rock associations.

Of these statements:

(a) 1, 2 and 3 are correct (b) 2, 3 and 4 are correct

(c) 1 and 3 are correct (d) 2 and 4 are correct

64. Match List I (Process) with List II (Name of the Process) and select the correct answer using the codes given below the Lists:

| List I | List II |
|---|---|
| A. Melting of preexisting rock. | 1. Assimilation |
| B. Separation of magma into two or more fractions which ultimately consolidate as different rocks. | 2. Segregation |
| C. Physical seperation of particular mineral or group of minerals during different stages of magmatic cooling. | 3. Differentiation |
| D. Intruded rock material is partially digested by an invading magma. | 4. Palingenesis |

Codes:

| | A | B | C | D |
|---|---|---|---|---|
| (a) | 2 | 4 | 3 | 1 |
| (b) | 4 | 2 | 3 | 1 |
| (c) | 4 | 2 | 1 | 3 |
| (d) | 2 | 4 | 1 | 3 |

65. The characteristics features of a glacial deposit is the presence of:

(a) unaltered feldspar in a sandstone with poor sorting.

(b) angular debris consisting of single rock type, often with striation marks

(c) mixed angular and rounded fragments of various rocks of all size, many with criss-cross striations.

(d) a mixture of clay and coarse material with graded bedding and sole marks.

66. Primary sedimentary structures are formed at the time of deposition or shortly thereafter. The character of these structures depends mainly on which of the following factors?

1. Current velocity
2. Water depth
3. Composition of source rocks
4. Grain size
5. Structures of the provenance
6. Sedimentation rate

Select the correct answer using the codes given below:

Codes:

(a) 2, 3, 4 and 5 (b) 1,2, 4 and 6

(c) 1, 3 5 and 6 (d) 3, 4, 5 and 6

67. Which one of the following sections can clearly distinguish between planar and trough cross-stratification?

(a) Section on plan

(b) Section transverse.to current direction

(c) Section longitudinal to current direction
(d) None of the above

68. That stromatolite is not a product of purely physical process is best established by its:
(a) virtual restriction to carbonate rocks.
(b) preserved organic skeletal structure.
(c) convex-upward laminations, locally exceedings angle of repose.
(d) widespread development in Precambrian sequences.

69. Match List I (Source rock types) with List II (Heavy mineral suites) and select the correct answer using the codes given below the lists:

| List I | List II |
|---|---|
| (a) Rocks of reworked sediments | 1. Augite, reddish brown hornblend olivine |
| (b) High-grade metamorphic rock | 2. Brownish-green hornblend, monazite, tourmaline (small pink euhedra) |
| (c) Acid igneous rock | 3. Rounded rutile, rounded zircon, rounded tourmaline |
| (d) Mafic igneous rock | 4. Kyanite, sillimanite, staurolite |

Codes:

| | A | B | C | D |
|---|---|---|---|---|
| (a) | 4 | 3 | 1 | 2 |
| (b) | 3 | 4 | 2 | 1 |
| (c) | 4 | 3 | 2 | 1 |
| (d) | 3 | 4 | 1 | 2 |

70. A residual weathering product composed of varying amounts of aluminium hydroxides, boehmite, diaspore and gibbsite is known as:
(a) diopside (b) bauxite
(c) ganister (d) laterite

71. Shelf storm sands are distinguished from turbidites by:
(a) interlayering of sand and mud
(b) wave ripples and wave ripple lamination
(c) erosive base with diffused tops
(d) sole marks

72. Beach sequences are unlikely to show:
(a) low-angle seaward dipping cross-stratification
(b) heavy mineral streaks
(c) profuse presence of mud-draped ripples
(d) vertical burrow

73. Which of the following are the consequences of diagenetic changes in mud-rocks in the depth range of 2000 to 3700 m?
1. Decrease in smectite content
2. Increase in illite content
3. Decrease in calcite content

4. Increase in fine quartz/chert content

Select the correct answer using the codes given below:

Codes:

(a) 1, 2, 3 and 4 (b) 2 and 3

(c) 2, 3 and 4 (d) 1, 3 and 4

74. Consider the following statements:

Greywackes differ from arkose in that:

1. FeO >$Fe_2O_3$ 2. $Fe_2O_3$> $F_3O$

3. MgO>CaO 4. KwO> $Na_2O$

5. $Na_2O$>$K_2O$

6. Greywacke has synorogenic character

Of these statements

(a) 1, 3, 5 and 6 are correct (b) 1, 2, 4 and 5 are correct

(c) 2, 3, 4 and 5 are correct (d) 3, 4, 5 and 6 are correct

75. Under low grade metamorphism, a plagioclase may be pseudomorphed by aggregates of microgranular mixture of albite, zoisite (or epidote) along with variable amounts of prehnite, sericite, chlorite and actinolite. This alteration is called:

(a) chloritisation (b) sericitisation

(c) fenitisation (d) saussuritisation

76. Which one of the following is the correct arrangement of rocks as per increasing grade of metamorphism?

(a) Slate-phyllite-granulite-amphibolite

(b) Phyllite-slate-amphibolite-granulite

(c) Slate-phyllite-amphibolite-granulite

(d) Phyllite-slate-eclogite-granulite.

77. The following mineral assemblages are present in the metamorphites of a terrain:

1. Quartz-Albite-Muscovite-Chlorite
2. Quartz-Phengite-Talc-Kyanite
3. Quartz-Albite-Phengite-chlorite
4. Quartz-Albite-Muscovite-Kyanite-Almandine

The correct sequence of these mineral assemblages in increasing order of pressure of regional metamorphism is:

(a) 1, 2, 4, 3 (b) 3, 1, 4, 2

(c) 3, 1, 2, 4 (d) 1, 3, 4, 2

78. Match List I with List II and select the correct answer using the codes given below the lists:

| List I | List II |
|---|---|
| A. Gneissic metapelite | 1. Alkali metasomatism |
| B. Spillite | 2. Metamorphic differentiation |

C. Albitite | 3. Ocean-floor metamorphism
D. Pyroxene hornfels | 4. Contact metamorphism

Codes:

| | A | B | C | D |
|---|---|---|---|---|
| (a) | 3 | 2 | 1 | 4 |
| (b) | 2 | 3 | 1 | 4 |
| (c) | 2 | 3 | 4 | 1 |
| (d) | 3 | 2 | 4 | 1 |

79. Which of the following pairs are correctly matched?
 1. A large crystal that has grown in a metamorphic rock .......... Porphyroblast
 2. A relict crystal in a metamorphosed rock .......... Xenoblast
 3. A large conspicous elastic fragment in a poorly sorted sedimentary rock .......... Phenoclast
 4. A large recrystallised mineral that contains numerous small inclusions .......... Poikiloblast

Select the correct answer using the codes given below:

Codes:

(a) 1, 2 and 4 (b) 1, 3 and 4
(c) 1, 2 and 3 (d) 2, 3 and 4

80. Which one of the following textural relationships can be considered as evidence for prograde metamorphism?
(a) Garnet rimmed by biotite
(b Orthopyroxene relicts in hornblende
(c) Relicts of hornblende in orthopyroxene
(d) Quartz-plagioclase symplectite

81. As per the figure shown, which one of the following is the correct sequence of the order of the crystalloblastic series?
(a) Quartz (Q) -Cordierite (Cor)-Muscovite (Mu)-Garnet (Ga)
(b) Garnet -Muscovite- Quartz-Cordierite
(c) Muscovite- Garnet-Cordierite-Quartz
(d) Garnet -Quartz- Cordierite-Muscovite

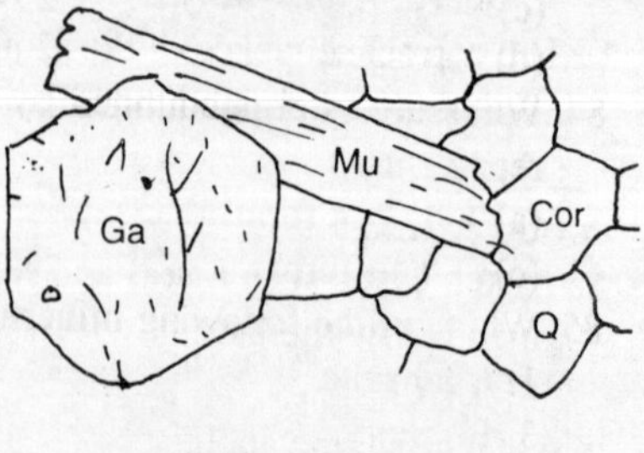

82. Match List I (character) with List II (rock) and select the correct answer using the codes given below the lists:

| | List I | List II |
|---|---|---|
| A. | Even grained rock containing Quartz, feldspar, pyroxene and garnet as a granoblastic texture | 1. Eclogite |
| B. | Coarse grained thermally metamorphosed rock, with equidimensional grains, in random orientation | 2. Phyllonite |
| C. | Medium grained rock mainly composed of omphacite and a pyrope-rich garnet | 3. Granulite |
| D. | Fine grained streaky or banded rock with silky surface, produced by granulation or flowage of coarser rocks. | 4. Granofels |

Codes:

| | A | B | C | D |
|---|---|---|---|---|
| (a) | 3 | 4 | 2 | 1 |
| (b) | 4 | 3 | 2 | 1 |
| (c) | 3 | 4 | 1 | 2 |
| (d) | 4 | 3 | 1 | 2 |

83. Match List I with List II and select the correct answer using the codes given below the lists:

| List I (Mineral) | List II (System of Crystallization) |
|---|---|
| A. Rhodonite | 1. Tetragonal |
| B. Cassiterite | 2. Orthorhombic |
| C. Chalcocite | 3. Isometric |
| D. Gold | 4. Triclinic |

Codes:

| | A | B | C | D |
|---|---|---|---|---|
| (a) | 1 | 4 | 2 | 3 |
| (b) | 4 | 1 | 2 | 3 |
| (c) | 4 | 1 | 3 | 2 |
| (d) | 1 | 4 | 3 | 2 |

84. When a crystal is bounded by three pairs of unlike faces, each pair represents:

(a) an open form
(b) a closed form
(c) a unit form
(d) a fundamental form

85. Which of the following minerals may show repeated twinning?

1. Chabazite
2. Fluorite
3. Rutile
4. Staurolite

Select the correct answer using the codes given below:

Codes

(a) 1 and 4 (b) 3 and 4
(c) 1 and 2 (d) 1 and 3

86. Match List I with List II and select the correct answer using the codes given below the lists:

| List I (Form) | List II (Miller's symbol) |
|---|---|
| A. Hexoctahedron | 1. (210) |
| B. Octahedron | 2. (111) |
| C. Tetrahexahedron | 3. (321) |
| D. Trapezohedron | 4. (211) |

Codes:

| | A | B | C | D |
|---|---|---|---|---|
| (a) | 2 | 3 | 1 | 4 |
| (b) | 2 | 3 | 4 | 1 |
| (c) | 3 | 2 | 4 | 1 |
| (d) | 3 | 2 | 1 | 4 |

87. The Silicate structure shown in the figure is:

(a) Nesosilicate
(b) Sorosilicate
(c) Inosilicate
(d) Tectosilicate

Silicon
o Oxygen

88. Match List I with List II and select the correct answer using the codes given below the lists:

| List I (Mineral) | List II (Chemical composition) |
|---|---|
| A. Leucite | 1. $NaAlSiO_4$ |
| B. Nepheline | 2. $NaAlSi_3O_8$ |
| C. Albite | 3. $KAlSiO_8$ |
| D. Orthoclase | 4. $KalSi_2O_6$ |

Codes:

| | A | B | C | D |
|---|---|---|---|---|
| (a) | 1 | 4 | 3 | 2 |
| (b) | 1 | 4 | 2 | 3 |
| (c) | 4 | 1 | 2 | 3 |
| (d) | 4 | 1 | 3 | 2 |

89. Which of the following statements are correct?

1. Protoenstatite is the stable $MgSiO_3$ polymorph between 985°C and 1385°C.
2. Pure protoenstatite melts incongruently to forsterite and liquid at 1557°C.
3. Protoenstatite solid solution $En_{76}Di_{24}$ melts incongruently to forsterite

and a more siliceous liquid at 1386°C.

4. High temperature protoenstatite has monoclinic symmetry.

Select the correct answer using the codes given below:

Codes:

(a) 2, 3 and 4 (b) 1, 3 and 4

(c) 1, 2 and 4 (d) 1, 2 and 3

90. The general formula of calcium-rich amphiboles can be expressed as $X_2Y_5Z_8O_{22}(OH)_2$, where X stands for:

(a) Ca, Na, K (b) Ca, Na, K, Al

(c) Ca, Na, K, $Fe^{+2}$ (d) Ca, $Fe^{+2}$, $Fe^{+3}$, Na

91. In the Compositional triangle:

MgO-CaO-$SiO_2$ shown in the figure, the position of forsterite is at the point labelled:

(a) 1

(b) 2

(c) 3

(d) 4

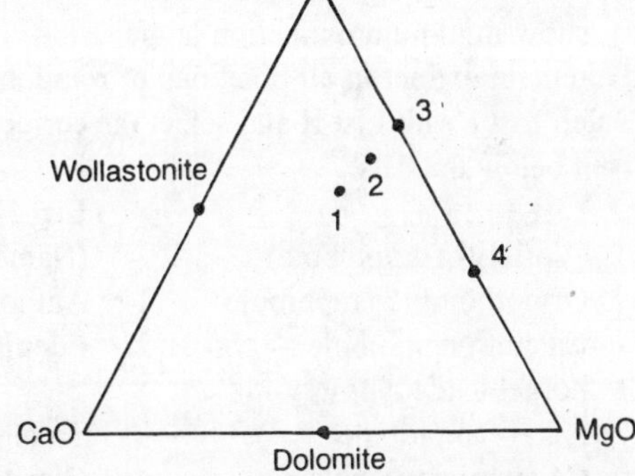


92. Biotite differs from phlogopite, in having:

(a) Mg in the Y site (b) Fe in the Y site

(c) Li in the Y site (d) Al in the Y site

93. A petrological microscope is shown in the figure. The parts ocular, Bertrand lens. Objective and Polarizer have respectively been labelled in the figure as:

(a) A, B, D and C

(b) A, B, C and D

(C) B, A, C and D

(D) B, A, D and C

94. The figure shows the optic axial plane that coincides with the single crystallographic plane of symmetry, parallel to (010) in:

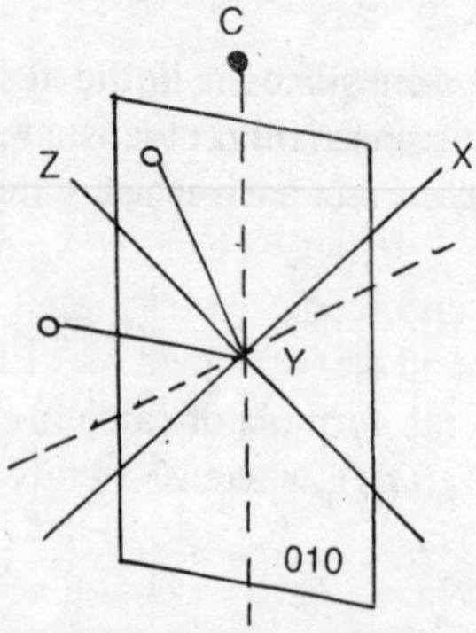

(a) clinopyroxene

(b) hornblende

(c) hypersthene

(d) bronzite

95. Some monoclinic crystals possess a single symmetry plane parallel to (010). Hence sections which are cut normal to this plane would:
(a) show straight or symmetrical extinction
(b) show maximum extinction angle
(c) show minimum extinction angle
(d) remain extinct in all positions of rotation

96. Match List I with List II and select the correct answer using the codes given below the lists:

| | List I (Variety of amphibole) | | List II (Name of mineral) |
|---|---|---|---|
| A. | Orthorhombic amphibole | 1. | Anthophyllite |
| B. | Calcium amphibole — horneblende type | 2. | Edenite |
| C. | Alkali amphibole | 3. | Tremolite |
| D. | Calcium amphibole — Non-hornblende type | 4. | Glaucophane |

Codes:

| | A | B | C | D |
|---|---|---|---|---|
| (a) | 2 | 1 | 3 | 4 |
| (b) | 2 | 1 | 4 | 3 |
| (c) | 1 | 2 | 3 | 4 |
| (d) | 1 | 2 | 4 | 3 |

97. Ore minerals include:
(a) spessartine and sphene
(b) spessartine and sheelite
(c) sheelite and siderite
(d) sphene and siderite

98. Porphyry copper deposits are rare in the Precambrian terrains because of:
(a) lack of suitable intrusives
(b) the low level of erosion of the surface
(c) lack of plate subduction
(d) the high degree of deformation and metamorphism

99. The four zones shown in the figure and labelled as M, L, C and N represent respectively:
(a) Leached zone; Enriched zone; Oxidized ore; Water table
(b) Enriched zone; Leached zone; Oxidized ore; Water table
(c) Enriched zone; Water table ; Oxidized ore; Leached zone
(d) Water table; Oxidized ore; Enriched zone; Leached zone

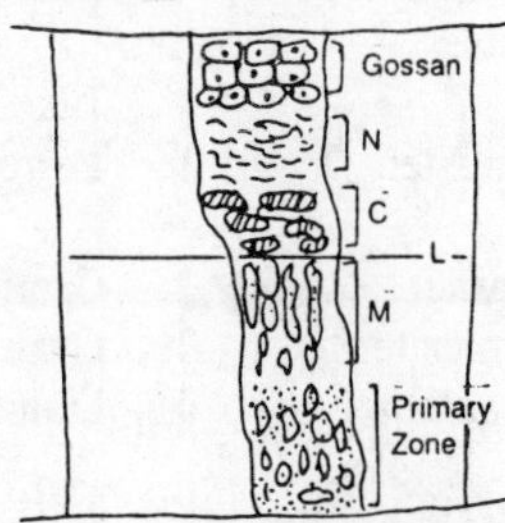


100. Which of the following minerals is expected in a placer deposit?
(a) Bornite (b) Fluorspar
(c) Columbite (d) Smithsonite

101. Kimberlite pipes are mostly found in:
(a) spreading sites (b) subduction zones
(c) transform faults (d) intraplate regions

102. Gondite is best defined as a:
(a) silicate rock associated with manganese ores
(b) Mn-oxide rock containing quartz
(c) Mn-siliate rock dominated by spessartine-rich garnet and quartz and containing rhodonite with minor Mn-oxide.
(d) Mn-silicate-carbonate rock with minor Mn-oxides

103. Match List I with List II and select the correct answer using the codes given below the lists:

| List I (Mineralisation) | List II (Deposit) |
|---|---|
| A. Sulphide mineralisation in garnet-chlorite schist in the Himalaya | 1. Rangpo |
| B. Galena-chalcopyrite bodies in mica-schists and phyllites in the western part of Singbhum shear zone | 2. Dariba-Rajpura |
| C. Lead-zinc sulphide bodies within graphitic schists along the contact between schists and tremolite marble in the Aravalli Supergroup | 3. Sandur |
| D. Copper sulphide and sphalerite in schist belt in shield area | 4. Sargipalli |

Codes:

| | A | B | C | D |
|---|---|---|---|---|
| (a) | 4 | 1 | 3 | 2 |
| (b) | 4 | 1 | 2 | 3 |
| (c) | 1 | 4 | 2 | 3 |
| (d) | 1 | 4 | 3 | 2 |

104. Match List I with List II and select the correct answer using the codes given below the lists:

| List I | List II |
|---|---|
| A. Cassiterite; tungsten minerals | 1. Peridotite |
| B. Sillimanite; garnet | 2. Granite-granodiorite |
| C. Chromite; garnierite | 3. Granulite |
| D. Chalcopyrite; galena | 4. Dolomite |

Codes

| | A | B | C | D |
|---|---|---|---|---|
| (a) | 2 | 3 | 1 | 4 |
| (b) | 3 | 2 | 4 | 1 |
| (c) | 3 | 2 | 1 | 4 |
| (d) | 2 | 3 | 4 | 1 |

105. Which of the following pairs are correctly matched?

1. Ferro-alloy minerals..................Sukinda
2. Atomic minerals...................Bassein
3. Non-ferrous minerals...........Chitradurga
4. Aluminium silicates................ Sonapahar

Select the correct answer using the codes given below:

Codes:

(a) 1, 2 and 4 (b) 1, 3 and 4
(c) 1, 2 and 3 (d) 2, 3 and 4

106. The mineralisation at Zawar as sheeted zones, veins, stringers and disseminations, forming lenticular bodies arranged in an enechelon pattern, restricted solely within the dolomitie horizon. The form of the ore bodies may therefore be described as:

(a) stratiform (b) bedded
(c) strata-bound (d) tabular

107. Formation of bauxite is facilitated by which one of the following groups of parameters?

(a) Steeply dipping surface; presence of sulphuric acid; and temperature below 10°.
(b) Gently undulating surface; source rock high in Al-silicates, low in Fe and free quartz; and availability of organic acids.
(c) Steeply dipping surface; absence of acids; and source rock low in alumina and high in quartz.
(d) Steeply dipping surface; ultrabasic source rock; colder climate; and absence of acids

108. The mobility of gold in the zone of oxidation depends on the concentration of:
(a) $SO_4^{2-}$ ion (b) $H^+$ and $Ci^-$ ions
(c) $Mn^{2+}$ ion (d) $K^+$ and $Na^+$ ions

109. Ash in coal is a measure of:
(a) fixed carbon (b) mineral matter
(c) mineral matter × 1.1 (d) mineral matter : 1.1

**Directions**: The following 11 (eleven) items consist of two statements, one labelled the 'Assertion A' and the other labelled the 'Reason R'. You are to examine these two statements carefully and decide if the Assertion A and the Reason R are individually true and if so, whether the reason is a correct explanation of the assertion. Select your answer to these items using the codes given below and mark your answer sheet accordingly.

Codes:
(a) Both A and R are true and R is the correct explanation of A
(b) Both A and R are true but R is not a correct explanation of A
(c) A is true but R is false
(d) A is false but R is true

110. *Assertion (A)*: The petroliferous basins in the Arabian Sea, off the coast of Bombay are known as "Bombay High".
*Reason (R)*: "Bombay High" is no named because it is a region of high magnetic and gravity anomalies.

111. *Assertion (A)*: A dip fault can be a wrench fault.
*Reason (R)*: Along a dip fault, the net-slip can be equal to strike-slip, dip-slip or oblique-slip.

112. *Assertion (A)*: Percolating groundwater generally destroys the remains of past organisms. It is also capable of making them preservable as fossils.
*Reason (R)*: Precipitation of minerals inside the shell cavity and filling of silica in the woody tissue enable them to withstand crushing or disintegration.

113. *Assertion (A)*: Marine organisms were generally fossilised rather than the terrestrial organisms.
*Reason (R)* : Land plants and animals were less in number than those that lived in the sea.

114. *Assertion (A)*: Chronostratigraphy is concerned with the age of strata and their time relationship.
*Reason (R)*: Formations, member and bed are recognised as units of chronostratigraphy.

115. *Assertion (A)*: The reaction between olivine and an equilibrating liquid results in the enrichment of the liquid in silica and quartz may appear as a released mineral.

*Reason (R)*: The partial failure of reaction between olivine and liquid gives rise to the final product, a mixture of olivine, pyroxene and quartz, where olivine does not coexist with quartz.

116. *Assertion (A)*: Failure to attain complete reaction of early formed crystal is responsible for zoned plagioclase.
*Reason (R)*: If equilibrium between one early phase, a new stable phase and the melt is not maintained, then the earlier crystals are encased by a mantle of the newer stable phase.

117. *Assertion (A)*: Among the products of dedolomitisation during diagenesis, irregular mosaic (pseudomorphs) of calcite after dolomite crystals have been reported.
*Reason (R)*: During dedolomitisation magnesium ions in dolomite tend to react more readily than the calcium ions with impurities such as silica or clay minerals.

118. *Assertion (A)*: Andalusite-bearing metapelites are restricted to thermally metamorphosed terrains.
*Reason (R)*: Andalusite is favoured in metapelite only in metamorphic environment when T/P ratio is high.

119. *Assertion (A)*: There is increasing recognition of multiepisodes of reworkings particularly in the Proterozoic metamorphic belts.
*Reason (R)*: Relics of Archaean granulite facies rocks occur in metamorphic belts of younger ages.

120. *Assertion (A)*: The angle between any two faces of a crystal is termed as interfacial angle.
*Reason (R)*: The interfacial angles of corresponding faces of different crystals of a mineral are constant provided they have identical chemical composition and measurements are made at the same temperature.

## ANSWER

| | | | | | | | |
|---|---|---|---|---|---|---|---|
| 1. (b) | 2. (a) | 3. (a) | 4. (d) | 5. (d) | 6. (b) | 7. (b) | 8. (d) |
| 9. (a) | 10. (b) | 11. (a) | 12. (a) | 13. (a) | 14. (a) | 15. (c) | 16. (d) |
| 17. (a) | 18. (a) | 19. (a) | 20. (d) | 21. (d) | 22. (b) | 23. (d) | 24. (a) |
| 25. (b) | 26. (b) | 27. (c) | 28. (c) | 29. (a) | 30. (a) | 31. (d) | 32. (b) |
| 33. (a) | 34. (d) | 35. (c) | 36. (a) | 37. (c) | 38. (c) | 39. (d) | 40. (c) |
| 41. (c) | 42. (c) | 43. (a) | 44. (d) | 45. (a) | 46. (c) | 47. (b) | 48. (c) |
| 49. (b) | 50. (c) | 51. (c) | 52. (b) | 53. (a) | 54. (b) | 55. (a) | 56. (b) |
| 57. (c) | 58. (a) | 59. (d) | 60. (c) | 61. (b) | 62. (a) | 63. (d) | 64. (b) |
| 65. (a) | 66. (b) | 67. (c) | 68. (a) | 69. (b) | 70. (b) | 71. (a) | 72. (a) |
| 73. (a) | 74. (a) | 75. (d) | 76. (c) | 77. (c) | 78. (b) | 79. (b) | 80. (c) |
| 81. (b) | 82. (c) | 83. (b) | 84. (a) | 85. (b) | 86. (d) | 87. (c) | 88. (c) |
| 89. (d) | 90. (d) | 91. (c) | 92. (a) | 93. (a) | 94 . (a) | 95. (a) | 96. (d) |
| 97. (c) | 98. (c) | 99. (c) | 100. (c) | 101. (d) | 102. (c) | 103. (c) | 104. (a) |
| 105. (b) | 106. (c) | 107. (b) | 108. (b) | 109. (a) | 110. (a) | 111. (d) | 112. (b) |
| 113. (c) | 114. (c) | 115. (d) | 116. (c) | 117. (a) | 118. (a) | 119. (a) | 120. (b) |

# 4. Civil Services (Preliminary)
## EXAMINATION 1997

1. Match List I with List II and select the correct answer using the codes given below the lists:

| | List I *(Essential symmetry)* | | List II *(Name of crystal system)* |
|---|---|---|---|
| A. | 4 axes of trigonal symmetry | 1. | Trigonal Division of Hexagonal System |
| B. | 1 vertical axis of 4-fold symmetry | 2. | Orthorhombic System |
| C. | 1 vertical axis of binary symmetry | 3. | Tetragonal System |
| D. | 1 axis of trigonal symmetry | 4. | Isometric System |

Codes:

| | A | B | C | D |
|---|---|---|---|---|
| (a) | 3 | 4 | 2 | 1 |
| (b) | 4 | 3 | 1 | 2 |
| (c) | 4 | 3 | 2 | 1 |
| (d) | 3 | 4 | 1 | 2 |

2. The given figure is a stereogram showing the symmetry elements of :

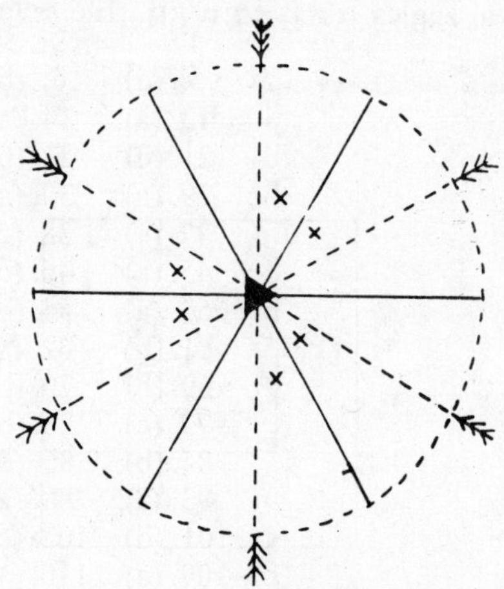

(a) Calcite (b) tourmaline

(c) alpha-quartz (d) phenacite

3. The axial ratio of an orthorhombic crystal is:

$$a : b : c = 1.5 : 1 : 075$$

The Miller index of crystal face which cuts a, b, and c axes at 0.6 mm, 0.8 mm and 0.6 mm respectively will be:

(a) (1 2 1)

(b) (2 1 1)

(c) (2 1 2)

(d) (1 2 2)

4. Match List I (Form) with List II (Miller index) and select the correct answer using the codes given below the lists:

| | List I | | List II |
|---|---|---|---|
| A. | Trapezohedron | 1. | (3 2 1) |
| B. | Trisoctahedron | 2. | (2 1 0) |
| C. | Tetrahexahedron | 3. | (2 1 1) |
| D. | Hexoctahedron | 4. | (2 2 1) |

Codes:

| | A | B | C | D |
|---|---|---|---|---|
| (a) | 4 | 3 | 2 | 1 |
| (b) | 3 | 4 | 1 | 2 |
| (c) | 4 | 3 | 1 | 2 |
| (d) | 3 | 4 | 2 | 1 |

5. The given figure is that of a crystal showing some symmetry elements with axial angles $\alpha \neq \beta \neq \gamma \neq 90°$. The shape of the unit cell of the crystal is:

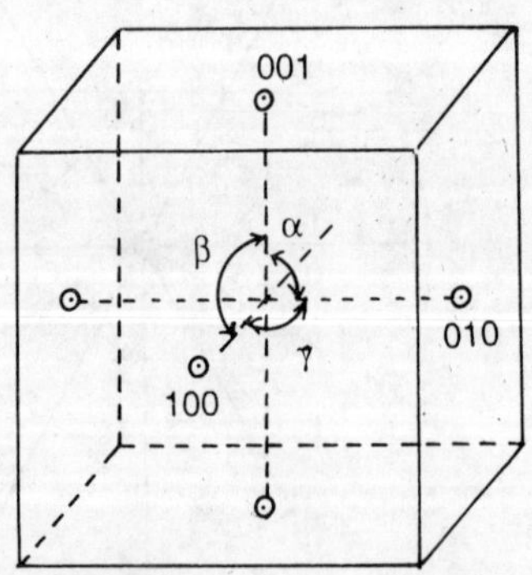


(a) general parallelepiped

(b) right prism on parallelogram

(c) rectangular parallelepiped

(d) square prism

6. Which one of the following is the correct sequence of the given minerals in increasing order of their silica content?

(a) Nepheline - leucite - hypersthene - orthoclase

(b) Nepheline - leucite - orthoclase - hypersthene

(c) Leucite - nepheline - orthoclase - hypersthene

(d) Leucite - nepheline - hypersthene - orthoclase

7. Match List I with List II and select the correct answer using the codes given below the lists:

| List I | List II |
|---|---|
| A. Carbonate | 1. Siderite |
| B. Phosphate | 2. Monazite |
| C. Oxide | 3. Periclase |
| D. Sulphide | 4. Sphalerite |

Codes :

| | A | B | C | D |
|---|---|---|---|---|
| (a) | 1 | 2 | 3 | 4 |
| (b) | 2 | 1 | 3 | 4 |
| (c) | 1 | 2 | 4 | 3 |
| (d) | 2 | 1 | 4 | 3 |

8. Match List I with List II and select the correct answer using the codes given below the lists:

| List I | List II |
|---|---|
| A. Chain silicate | 1. Quartz |
| B. Tektosilicate | 2. Beryl |
| C. Phyllosilicate | 3. Augite |
| D. Cyclosilicate | 4. Muscovite |

Codes:

| | A | B | C | D |
|---|---|---|---|---|
| (a) | 1 | 2 | 3 | 4 |
| (b) | 1 | 2 | 4 | 3 |
| (c) | 3 | 1 | 2 | 4 |
| (d) | 3 | 1 | 4 | 2 |

9. Which one of the following minerals is characterised chemically by high Si, $Fe^{2+}$, $Fe^{3+}$ and Na?

(a) Glaucophane (b) Arfvedsonite
(c) Riebeckite (d) Eckermanite

10. The triangular diagram shown in the given figure, represents the solid-solution in pyroxenes. The approximate position of hedenbergite is the point labelled.

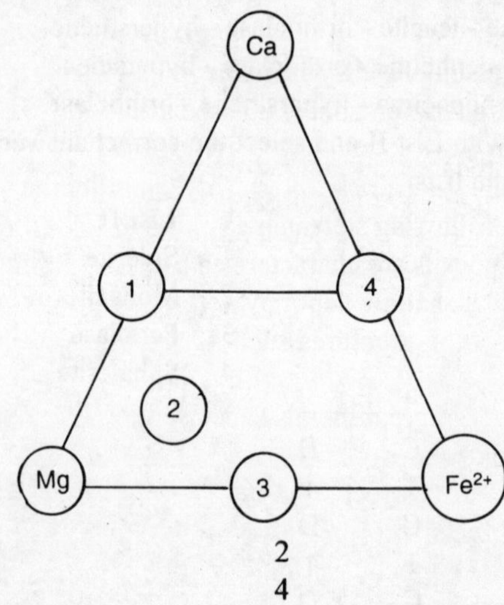


(a) 1 (b) 2
(c) 3 (d) 4

11. The given figure refers to quartz wedge in monochromatic yellow (580 mμ) light between crossed polars. Yellow colour band is seen at wavelength(s) (in unit of mμ)

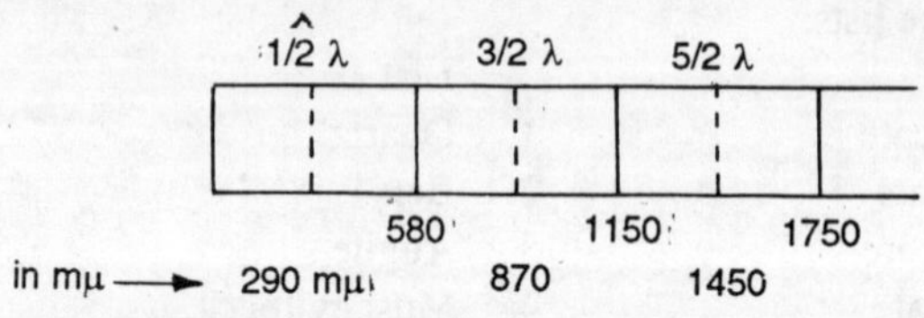


(a) 580 only (b) 290 only
(c) 290, 870 and 1450 (d) 580, 1150 and 1740

12. Interference colour of a mineral section is maximum in the plane:
(a) parallel to the optic axis.
(b) perpendicular to the optic axis.
(c) at 45° to the optic axis.
(d) at 60° to the optic axis.

13. Which one of the following pairs of metals is commonly associated with copper ores?
(a) Gold and molybdenum
(b) Tin and lead

(c) Manganese and chromium
(d) Platinum and nickel

14. The *minimum* metal concentration in an ore deposit which would ensure its profitable extraction is known as:
(a) specification (b) cut-off grade
(c) tenor (d) average grade

15. The mineral assemblage topaz, white mica and tourmaline, observed in the wall-rock alteration zones, is associated with:
(a) granite (b) basalt
(c) dolomite (d) amphibolite

16. Consider the following statements:
Epithermal deposits are characterised by:
1. deposition at shallow depth by nearly spent solution.
2. the presence of structures such as crustification, combs and drusy cavities.
3. the presence of ore minerals such as cinnabar, stibnite, realgar etc.
Of these statements:
(a) 1, 2 and 3 are correct
(b) 2 and 3 are correct
(c) 1 and 3 are correct
(d) 1 and 2 are correct

17. In hydrothermal processes of mineral deposit formation, the following stages are recognized:
1. Telethermal
2. Xenothermal
3. Epithermal
The correct sequence of these stages in ascending order of temperature and pressure is:
(a) 1, 3, 2
(b) 1, 2, 3
(c) 3, 1, 2
(d) 2, 1, 3

18. In a vertical section across a sulphide deposit, the site representing the best situation for supergene enrichment would be:
(a) above the water table
(b) below the water table
(c) in the oxidation zone
(d) in the zones rich in humus

19. Match List I (Class) with List II (Ore deposit) and select the correct answer using the codes given below the lists:

| List I | List II |
|---|---|
| A. Magmatic segregation | 1. Iron |
| B. Volcanic-exhalative | 2. Chromite |

C. Mechanical concentration 3. Copper
D. Sedimentary 4. Gold

Codes:

| | A | B | C | D |
|---|---|---|---|---|
| (a) | 1 | 2 | 3 | 4 |
| (b) | 1 | 2 | 4 | 3 |
| (c) | 2 | 3 | 4 | 1 |
| (d) | 2 | 3 | 1 | 4 |

20. Match List I with List II and select the correct answer using the codes given below the lists:

| List I | List II |
|---|---|
| A. Industrial variety of diamond | 1. Spodumene |
| B. Manufacture of fertilizer | 2. Crocidolite |
| C. Extraction of lithium | 3. Carbonado |
| D. Manufacture of asbestos | 4. Gypsum |

Codes:

| | A | B | C | D |
|---|---|---|---|---|
| (a) | 2 | 4 | 3 | 1 |
| (b) | 3 | 4 | 1 | 2 |
| (c) | 3 | 2 | 4 | 1 |
| (d) | 4 | 3 | 1 | 2 |

21. Match List I with List II and select the correct answer using the codes given below the lists:

| List I | List II |
|---|---|
| A. Boehmite | 1. Al O (OH) |
| B. Brucite | 2. Fe O (OH) |
| C. Geothite | 3. $Mg(OH)_2$ |
| D. Gibbsite | 4. $Al(OH)_3$ |

Codes:

| | A | B | C | D |
|---|---|---|---|---|
| (a) | 3 | 1 | 2 | 4 |
| (b) | 3 | 1 | 4 | 2 |
| (c) | 1 | 3 | 2 | 4 |

(d) A B C D
1 3 4 2

22. Titanium and vanadium bearing magnetite deposits are found in association with
(a) peridotite
(b) carbonatite
(c) gabbro-anorthosite
(d) adamellite

23. Match List I with List II and select the correct answer using the codes given below the lists:

| List I | List II |
|---|---|
| A. Chromium | 1. Chavara, Kerala |
| B. Copper | 2. Sukinda, Orissa |
| C. Lead-Copper | 3. Malanjkhand, M.P. |
| D. Thorium | 4. Sargipalli, Orissa |

Codes:
(a) A B C D
2 3 4 1
(b) A B C D
2 3 1 4
(c) A B C D
3 2 1 4
(d) A B C D
3 2 4 1

24. The precursor rock of East-Coast bauxite deposits is mainly:
(a) granite
(b) charnockite
(c) khondalite
(d) leptynite

25. Consider the following features:
1. Association with Bundelkhand granite massif.
2. Association with agglomeratic tuff of serpentine mass.
3. Instrusion into the Kaimur sandstone.
4. Traversed by a network of calcite veins.

Features of occurrence of diamond pipes at Majhgawan would include:
(a) 1, 3 and 4
(b) 2 and 3
(c) 2, 3 and 4
(d) 1 and 4

26. Which one of the following proximate contents of coal indicates the optimum by-product availability during coal carbonisation?
(a) Moisture (b) Volatile matter
(c) Fixed carbon (d) Ash

27. Match List I with List II and select the correct answer using the codes given below the lists:

| List I (Figures) | List II (Names) |
|---|---|
| A | 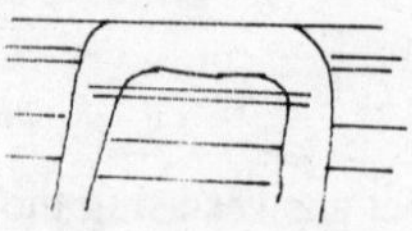 1. Sphenolith |
| B |  2. Cone sheet |
| C | 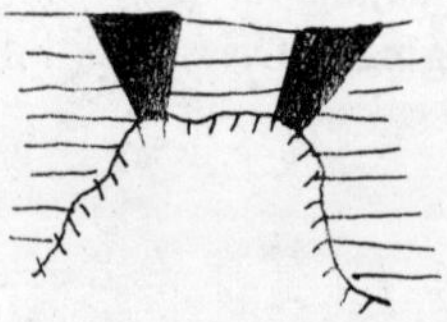 3. Ring dyke |
| D | 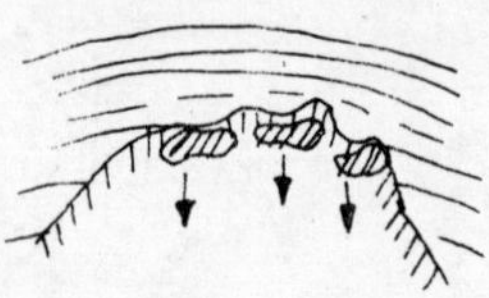 4. Diapiric intrusion |
| | 5. Pluton emplaced by magmatic stoping |

Codes:

| | A | B | C | D |
|---|---|---|---|---|
| (a) | 3 | 4 | 2 | 1 |
| (b) | 4 | 3 | 2 | 1 |
| (c) | 3 | 4 | 2 | 5 |
| (d) | 1 | 2 | 3 | 5 |

28. Which of the following pairs of correlations are correctly matched?
1. Lamprophyre............ Panidiomorphic texture
2. Granophyre................ Porphyritic texture with graphic growth
3. Andesite.................... Ophitic texture
4. Spilite........................ Pillow structure

Select the correct answer using the codes given below :
Codes:
(a) 1, 2 and 3
(b) 1, 2 and 4
(c) 2, 3 and 4
(d) 1, 3 and 4

29. Consider the following statements:
1. Migmatite is a structural term.
2. Komatiite is a compositional term.
3. Restite is a genetic term.
Of these statements
(a) 1 and 2 are correct
(b) 1, 2 and 3 are correct
(c) 2 and 3 are correct
(d) 1 and 3 are correct

30. Consider the following statements:
1. Granodiorite has no volcanic equivalent.
2. Volcanic equivalent of nepheline-syenite is phonolite.
3. Anorthosite has no volcanic equivalent.
4. Alnoite is hypabyssal igneous rock.
Of these statements
(a) 1, 2 and 3 are correct
(b) 1, 3 and 4 are correct
(c) 2, 3 and 4 are correct
(d) 1, 2 and 4 are correct

31. Match List I (Characteristics) with List II (Rocks) and select the correct answer using the codes given below the lists:

| | List I | | List II |
|---|---|---|---|
| A. | Mafic volcanic rocks containing 45-53 wt. % of $SiO_2$, plagioclase $An_{50+}$ Ca-rich clinopyroxene and glass | 1. | Gabbro |
| B. | Holocrystalline coarse to intermediate igneous rocks with dominant micro- or cryptoperthitic alkali feldspar | 2. | Basalt |
| C. | Phaneritic, alkali-rich $SiO_2$ oversaturated rock with one or two types feldspars, quartz, biotite and showing hypidiomorphic -granular texture | 3. | Syenite |

D. Coarse-grained basic intrusive rock with dominant plagioclase $An_{50+}$ clinopyroxene and subordinate olivine and showing hypidiomorphic granular texture — 4. Granite

Codes:

| | A | B | C | D |
|---|---|---|---|---|
| (a) | 3 | 2 | 1 | 4 |
| (b) | 3 | 2 | 4 | 1 |
| (c) | 2 | 3 | 4 | 1 |
| (d) | 2 | 3 | 1 | 4 |

32. Which one of the following pairs is *not* correctly matched?
(a) Eutectic crystallisation ............................ Isobaric invariant condition
(b) Cotectic crystallisation ............................ Isobaric univariant condition
(c) Binary minimum............. Isoberic invariant condition
(d) Peritectic crystallisation ......................... Isobaric univariant condition

33. Consider the following statements:
In the diopside-forsterite-quartz system, there are:
1. four binary eutectics.
2. three binary eutectics and one binary peritectic.
3. three bivariant fields separated by two univariant curves.
Of these statements:
(a) 2 alone is correct
(b) 2 and 3 are correct
(c) 1 and 3 are correct
(d) 1 alone is correct

34. Consider the following statements:
The course of crystallisation of a silicate melt is dependent upon:
1. maintenance of equilibrium of the early-formed crystals with the residual melt.
2. removal of the early-formed crystals from the melt.
3. the rate of solidification.
4. continuation of the unstable early-formed crystals as protected phases.

Of these statements:
(a) 1, 2 and 3 are correct
(b) 1, 3 and 4 are correct
(c) 2, 3 and 4 are correct
(d) 1, 2 and 4 are correct

35. The stage at which separation of pyrogenetic minerals takes place during the crystallisation of a magma is known as:
(a) pegmatitic (b) pneumatolytic
(c) orthomagmatic (d) hydrothermal

36. Which of the following criteria indicate small-scale assimilation effects resulting in contaminated magmatic rock?
1. Modified mineralogy at the interface of magmatic material and foreign rock.
2. Presence of xenoliths and embayed, corroded xenocrysts in magmatic material.
3. Sharp contact between parent magmatic material and foreign rock.
4. Contaminated rock plots on a straight line in variation diagram, connecting solid contaminant and parent magmatic material.

Select the correct answer using the codes given below:
Codes:
(a) 1, 2 and 4
(b) 2, 3 and 4
(c) 3 and 4
(d) 1, 3 and 4

37. Match List I and List II and select the correct answer using the codes given below the lists:

| List I *(Principal groups)* | List II *(Sedimentary rocks)* |
|---|---|
| A. Terrigenous clastic sediments | 1. Phosphorites |
| B. Biogenic, biochemical and organic sediments | 2. Ignimbrites |
| C. Chemical sediments | 3. Mudrocks |
| D. Volcaniclastic sediments | 4. Evaporites |

Codes:

| | A | B | C | D |
|---|---|---|---|---|
| (a) | 4 | 3 | 2 | 1 |
| (b) | 3 | 1 | 4 | 2 |
| (c) | 3 | 1 | 2 | 4 |
| (d) | 4 | 3 | 1 | 2 |

38. Which one of the following is the correct sequence of the decreasing (most stable to least stable) order of the chemical stability of the given minerals?
(a) Microcline - tourmaline - pyroxene - olivine
(b) Tourmaline - microcline - olivine - pyroxene
(c) Tourmaline - microcline - pyroxene - olivine
(d) Microcline - tourmaline - olivine - pyroxene

39. Which of the following criteria are used to describe a sedimentary facies?
1. The sedimentary rock itself
2. Associated igneous rocks
3. The imprints of depositional processes
4. The depositional environment as reflected in the rock association.

Select the correct answer using the codes given below:
Codes:
(a) 1, 2 and 3
(b) 2, 3 and 4
(c) 2 and 4
(d) 1, 3 and 4

40. Match List I with List II and select the correct answer using the codes given below the lists:

| | List I | | List II |
|---|---|---|---|
| A. | Organic structure | 1. | Tracks |
| B. | Solution structure | 2. | Stylolite |
| C. | Planar bedding structure | 3. | Laminations |
| D. | Bedding plane irregularities | 4. | Mudcracks |

Codes:

| | A | B | C | D |
|---|---|---|---|---|
| (a) | 1 | 2 | 4 | 3 |
| (b) | 1 | 2 | 3 | 4 |
| (c) | 2 | 1 | 3 | 4 |
| (d) | 2 | 1 | 4 | 3 |

41. Consider the following statements:
Wentworth's grade scale which is a systematic division of particle sizes into different grades,
1. provides standardised terminology for different particle sizes.
2. is applicable to all particle sizes starting from 1/1000 mm diameter.
3. is applicable only to the particles above 1/100 mm diameter.

Of these statements:
(a) 1 and 3 are correct
(b) 2 alone is correct

(c) 1 alone is correct
(d) 1 and 3 are correct

42. Which of the following reasons account for the fact that majority of preserved sedimentary rocks are marine?
1. Continental areas of sialic material constitute only 30% of the earth's surface.
2. Secular changes in climatic conditions.
3. Depression of continental blocks through time causing incursions of the oceans, that is epicontinental seas.
4. Stratigraphic sections on the continental blocks contain many unconformities, while deeper marine or oceanic sections are more complete.

Select the correct answer using the codes given below:

Codes:
(a) 1, 2 and 3
(b) 2, 3 and 4
(c) 1, 2 and 4
(d) 1, 3 and 4

43. Match List I (Characters) with List II (Names) and select the correct answer using the codes given below the lists:

| | List I | | List II |
|---|---|---|---|
| A. | Dark and generally poorly-sorted sandy rocks, typically forming thick and monotonous sequence of synorogenic association | 1. | Porcellanite |
| B. | Feldspathic sandstone commonly with a minimum of 25% alkali feldspar | 2. | Calcite |
| C. | Porous rock of low density will dull fractures, often containing fine clay or carbonate inclusions | 3. | Arkose |
| D. | Concretionary limestone containing calcite as a simple intergranular impregnation, but often forming almost pure limestone bodies within siltstone or sandstone hosts | 4. | Greywacke |

Codes:

| | A | B | C | D |
|---|---|---|---|---|
| (a) | 4 | 3 | 1 | 2 |

(b) A B C D
4 3 2 1

(c) A B C D
3 4 2 1

(d) A B C D
3 4 1 2

44. Match List I with List II and select the correct answer using the codes given below the lists:

| | List I | | List II |
|---|---|---|---|
| A. | Changes in sediments immediately after deposition and prior to metamorphic reconstitution | 1. | Halmyrolosis |
| B. | Precipitation of new minerals within the pore spaces of a sediment | 2. | Anadiagenesis |
| C. | Varied changes including compaction, secular dehydration etc. that take place in sediments during deep burial (1000 to 10,000 m) at temperatures reaching upto 100-200°C | 3. | Diagenesis |
| D. | The early changes that take place at the interface of the lithosphere and marine hydrosphere | 4. | Authigenesis |

Codes:

(a) A B C D
4 3 1 2

(b) A B C D
3 4 2 1

(c) A B C D
3 4 1 2

(d) A B C D
4 3 2 1

45. Though rarely observed, the most reliable evidence for meandering origin of a fluvial deposit is

(a) fining-upward sequence
(b) lateral accretion surface
(c) low sand/shale ratio
(d) wide dispersion of palaeocurrent

46. Consider the following factors:

1. Presence of fluid phase
2. Lower rate in the drop of temperature
3. Change in confining pressure
4. Penetrative deformation

Factors which control the higher rate of retrograde metamorphic reactions would include:

(a) 1, 2 and 3
(b) 1, 2 and 4
(c) 1, 3 and 4
(d) 2, 3 and 4

47. Consider the following statements:

1. Phyllonite is a product of cataclastic metamorphism.
2. Mylonite in a granitic terrain is a product of granitisation.
3. Fenite is a product of metasomatism.
4. Metamorphic rocks, developed with planar and linear structures are the products of dynamothermal metamorphism.

Of these statements:

(a) 1, 2 and 3 are correct
(b) 1, 2 and 4 are correct
(c) 1, 3 and 4 are correct
(d) 2, 3 and 4 are correct

48. Which one of the following pairs is not correctly matched?

(a) Staurolite.........Chloritoid or muscovite + quartz
(b) Sillimanite + biotite..........Chlorite + Magnetite
(c) Sillimanite + K-feldspar .......Sericite or muscovite
(d) Plagioclase.......... Zoisite or clilnozoisite + albite

49. Match List I (Mineralogical characteristics) with List II (Facies names) and select the correct answer using the codes given below the lists:

| List I | List II |
|---|---|
| A. Feldspar-free assemblages typified by jadeite-clinopyroxene + pyrope-rich garnet | 1. Greenschist |
| B. Augite + orthopyroxene + plagioclase and also Mg-Fe garnet | 2. Eclogite |
| C. Hornblende + plagioclase (An>20) | 3. Granulite |
| D. Albite + epidote + actinolite $\pm$ chlorite $\pm$ calcite | 4. Amphibolite |

Codes:

| | A | B | C | D |
|---|---|---|---|---|
| (a) | 3 | 2 | 1 | 4 |

|  | A | B | C | D |
|---|---|---|---|---|
| (b) | 2 | 3 | 1 | 4 |

|  | A | B | C | D |
|---|---|---|---|---|
| (c) | 2 | 3 | 4 | 1 |

|  | A | B | C | D |
|---|---|---|---|---|
| (d) | 3 | 2 | 4 | 1 |

50. Match List I (Characteristics) with List II (Names of texture/structure) and select the correct answer using the codes given below the lists:

| List I | List II |
|---|---|
| A. Exhibiting criss-cross divergent arrangement of minerals in a thermally metamorphosed rock | 1. Palimpsest |
| B. Intricately ingrown and inter-penetrating minerals usually with rod-like shape | 2. Helicitic |
| C. Bands of inclusions in a later-formed porphyroblasts | 3. Diablastic |
| D. Remnants of a previous structure/texture are preserved | 4. Decussate |

Codes:

|  | A | B | C | D |
|---|---|---|---|---|
| (a) | 3 | 4 | 1 | 2 |

|  | A | B | C | D |
|---|---|---|---|---|
| (b) | 4 | 3 | 2 | 1 |

| (c) | A | B | C | D |
|---|---|---|---|---|
|  | 4 | 3 | 1 | 2 |

| (d) | A | B | C | D |
|---|---|---|---|---|
|  | 3 | 4 | 2 | 1 |

51. Low-pressure metamorphic belts are common throughout the entire geological record, whereas high-pressure metamorphic belts are less known from older metamorphic terrains. Which of the following phenomena may explain this?

1. Change in geothemal gradient with timer.
2. Subdued orogenesis during the Precambrian.
3. Dominance of eugeosynclines during the Palaeozoic.
4. Variation in the rate of plate movements through time.

Select the correct answer using the codes given below:

Codes:

(a) 1, 2, 3 and 4
(b) 1, 3 and 4
(c) 2 and 3
(d) 1 and 4

52. A medium-to-coarse-grained granoblastic rock without granulite facies mineralogy and generally without planar and linear structures is:
    (a) Eclogite
    (b) Hornfels
    (c) Granofels
    (d) Leptynite
53. Factors which influence biostratigraphic zonation include:
    (a) extinction and ecological changes
    (b) ecological changes and migration
    (c) extinction and migration
    (d) extinction, ecological changes and migration.
54. The boundary between a shale unit that overlies a unit of interbedded limestone and shale is commonly placed at the:
    (a) top of the lowest limestone bed
    (b) top of the highest limestone bed
    (c) top of thickest limestone bed
    (d) base of the lowest shale bed
55. Type section of the Jurassic rocks is in
    (a) Kachch
    (b) Kumaon and Garhwal Himalayan
    (c) Antarctica
    (d) Jura mountains
56. Which one of the following Formation occurs in the Singbhum Group?
    (a) Banded iron ore Formation
    (b) Chaibasa Formation
    (c) Charbaoli Formation
    (d) Lohangi Formation
57. Charles D. Walcott in 1910 made a significant discovery of fossil fauna in the Middle Cambrian rocks of British Columbia. It is known as:
    (a) Ediacara fauna
    (b) Gunflint fauna
    (c) Cryptozoan fauna
    (d) Burgess shale fauna
58. The ideal section for the study of Permian-Triassic boundary in the Indian sub-continent is exposed in
    (a) Ladakh
    (b) Lahaul
    (c) Spiti
    (d) Kashmir
59. "Caenozoic" means
    (a) recent life
    (b) modern life

(c) ancient life
(d) earlier life

60. The Lameta beds are now well-known for their
(a) plant fossils
(b) dinosaurian eggs
(c) foraminiferal contents
(d) fish remains

61. The equivalent of the Haimantas in the Tethyan belt of the Garhwal-Kumaon Himalayas is known as:
(a) Shiala
(b) Martoli
(c) Garbyang
(d) Ralam

62. The Gondwana rocks of India can be correlated with the rocks of the other Gondwana Contents on the basis of their:
(a) lithology and faunal contents
(b) floral elements and lithology
(c) faunal contents and floral elements
(d) lithology, floral elements and faunal contents

63. The index fossil *Cardita* (Venericardia) *beaumonti* is connected with:
(a) Himalayan orogeny
(b) Cenomanian transgression
(c) end of Deccan Trap activity
(d) Palaeogene transgression

64. Which one of the following is the oldest?
(a) Ranibennur Group
(b) Chitradurga Group
(c) Sargur Group
(d) Bababudan Group

65. The lithological succession of the Kaimur Group is:
(a) Susnai Breccia
Bijaigarh Shales
Scarp sandstone and conglomerate
Dhandraul Quartzites
(b) Dhandraul Quartzites
Scarp sandstone and conglomerate
Bijaigrah Shales
Susnai Breccia
(c) Susnai Breccia
Scarp sandstone and conglomerate
Bijaigarh Shales
Dhandraul Quartzites

(d) Dhandraul Quartzites
Bijaigarh Shales
Scarp sandstone and conglomerate
Susnai Breccia

66. Consider the following statements:
1. First primate is found in Nagri Stage.
2. *Brahmapithecus* and *Gaindatherium* characterise the Chinji Stage.
3. *Macacus*, *Felis* and *Tetralophodon* characterise the Dhok Pathan Stage.
4. *Hipparian* first appeared in the Nagri Stage.

Of these statements
(a) 3 alone is correct
(b) 1 and 2 are correct
(c) 1, 2 and 3 are correct
(d) 2, 3 and 4 are correct

67. Consider the following statements:
A skeleton taken out of a grave is not a fossil because it is:
1. of recent origin.
2. in an unaltered condition.
3. not petrified.
4. preserved as a whole.

Of these statements
(a) 1 alone is correct
(b) 2 and 4 are correct
(c) 1, 2 and 3 are correct
(d) 1, 2, 3 and 4 are correct

68. Some wood and certain hard parts of animals seem to have been soaked in silica, as all of them now have pore spaces filled with silica and, in some cases, they are also surrounded by silica. This type of preservation is known as:
(a) leaching
(b) impregnation
(c) distillation
(d) permineralisation

69. Match List I with List II and select the correct answer using the codes given below the lists:

| List I | List II |
|---|---|
| A. *Paradoxides* | 1. Lias to U. Cretaceous |
| B. *Belemnites* | 2. M. Cambrian |
| C. *Phylloceras* | 3. Jurassic to Recent |
| D. *Cidaris* | 4. Triassic to Cretaceous |

Codes:

|  | A | B | C | D |
|---|---|---|---|---|
| (a) | 2 | 1 | 4 | 3 |

|  | A | B | C | D |
|---|---|---|---|---|
| (b) | 2 | 1 | 3 | 4 |

|  | A | B | C | D |
|---|---|---|---|---|
| (c) | 1 | 2 | 3 | 4 |

|  | A | B | C | D |
|---|---|---|---|---|
| (d) | 1 | 2 | 4 | 3 |

70. Archaeocyathids are index fossils of the:
    (a) Cambrian Period
    (b) Ordovician Period
    (c) Triassic Period
    (d) Tertiary Period
71. Genus *Monotis* present in Monotis shale of Triassic age is a:
    (a) bivalve
    (b) bryozoan
    (c) graptolite
    (d) foraminifer
72. Most of the echinoderms became extinct by the end of the Permian. Only one genus was solely responsible for the evolution of echinoderms in the Mesozoic Era. The genus question is:
    (a) *Echinocardium*
    (b) *Miocidaris*
    (c) *Proterocidaris*
    (d) *Archaeocidaris*
73. Lophophore is:
    (a) a thin double-layered membrane that lines the interior of the shell anteriorly from the body.
    (b) a variously shaped appendage attached to the anterior surface of the body and with the mouth at the middle in the plane of bilateral symmetry.
    (c) the organ by which the animals attach themselves to the substratum.
    (d) the mouth that opens directly into the digestive tube.
74. Which one of the following features in a fossil shell of pelecypod indicates the burrowing habit of the animal?
    (a) Adductor impression
    (b) Pallial sinus
    (c) Dentition
    (d) Ligamental groove
75. Consider the following statements:
    Heteromorphism in ammonites is believed to be the:
    1. result of degenerate population which were inadaptive.

2. representative of retrogressive trends in the evolution of ammonites leading to extinction.
3. form adapted to changing environment which under favourable conditions again showed normal coiling.

Of these statements:

(a) 1 and 3 are correct
(b) 1 and 2 are correct
(c) 3 alone is correct
(d) 2 and 3 are correct

76. In trilobites, the biramous appendages gave rise to two branches. The one used for walking is also known as:
(a) coxa
(b) semipodite
(c) outer ramus
(d) telopodite

77. *Tetragraptus* shows a geological range in time from :
(a) Lower Ordovician to Upper Ordovician
(b) Cambrian to Ordovician
(c) Upper Cambrian to Upper Ordovician
(d) Middle Cambrian to Upper Cambrian

78. Large-scale diversification and divergence of lineages is known as:
(a) parallel evolution
(b) adaptive radiation
(c) adaptive convergence
(d) phyletic speciation

79. *Glossopteris decipiens, Gangamopteris angustifolia, Gondwanidium validum* and *Noeggrathiopsis stoliczkana* are indicative of:
(a) Barran Measures
(b) Karharbari Series
(c) Raniganj Series
(d) Barakar Series

80. Foraminifera became important rock-forming elements for the first time during the:
(a) Carboniferous
(b) Permian
(c) Triassic
(d) Jurassic

**Directions**: The following thirteen items consist of two statements, one labelled the 'Assertion A' and the other labelled the 'Reason R'. You are to examine these two statements carefully and decide if the Assertion A and the Reason R are individually true and if so, whether the Reason is a correct explanation of the Assertion. Select your answers to these items using the codes given below the mark your answer sheet accordingly.

Codes:
(a) Both A and R are true and R is the correct explanation of A.
(b) Both A and R are true but R is not a correct explanation of A.
(c) A is true but R is false.
(d) A is false but R is true.

81. Consider the following figure:

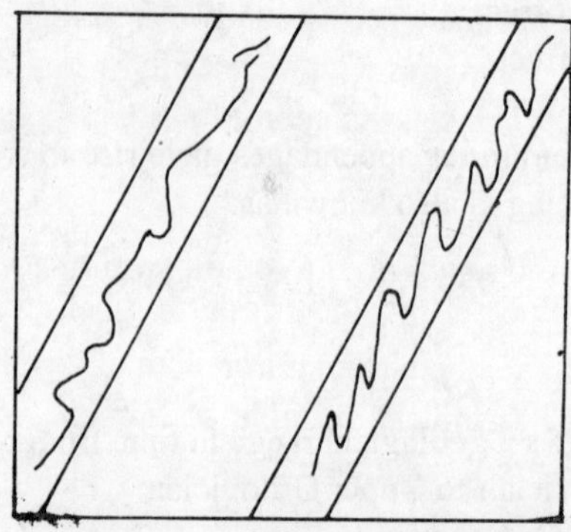

Assertion (A): The fold defined by the two limbs in the vertical section shown in the given figure is an overturned syncline.
Reason (R): The fold was formed by flexural slip mechanism.

82. Assertion (A) : An oblique-slip fault has vertical throw.
Reason (R) : An oblique-slip fault has strike-slip

83. Assertion (A) : Ultramafic rocks commonly form diapiric intrusions
Reason (R) : Density contrast plays an important role in the formation of diapirs.

84. Assertion (A) : Topaz has high hardness (8) and relatively high specific gravity (3.5).
Reason (R) : Topaz is a neosilicate and therefore has a dense structure.

85. Assertion (A) : Extinction angle of minerals never exceeds 45°.
Reason (R) : Extinction angle is the angle between a vibration direction and the crystallographic axis closest to it.

86. Assertion (A): The basal section of tourmaline is dichroic.
Reason (R) : The basal sectin of tourmaline is isotropic.

87. Assertion (A): Important petroleum deposits occur in Gujrat and Assam.
Reason (R): In Gujarat and Assam the geological-tectonic setting is the same for the occurrence of hydrocarbons.

88. Assertion (A): The formation of the core and the mantle was probably the first major differentiation process within the Earth.

Reason (R): Liquid immiscibility is an important magmatic differentiation process and it reflects separation of an initially homogeneous liquid into two compositionally distinct liquid phases.

89. Assertion (A): The less mature the sediment, less is the matrix and greater is the space it possesses.

Reason (R) : Primary porosity and permeability in sedimentary rocks increases with increasing textural maturity.

90. Assertion (A): The classification of particulate sediments according to grain size serves as a basis for genetic interpretation.

Reason (R): Grain size of a sediment is directly related to the nature and power of the transporting medium and indirectly to the environment of deposition.

91. Assertion (A): A typical example of isograd found in argillaceous rocks of moderate grade, separates a zone with assemblages containing muscovite and quartz from the one with sillimanite and microcline assemblages.

Reason (R): Any two zones, characterised by a particular rock type of differing mineralogy are represented by an isograd.

92. Assertion (A): The Deccan Trap activity is now considered to have lasted for not more than five million years.

Reason (R): Based on Planktic foraminifera, the Deccan Trap activity has been assigned a Maastrichtian age.

93. Assertion (A): Gastropods are largely used for identifying rocks of different ages.

Reason (R): Gastropods are found in sedimentary rocks of all ages from the Cambrian onwards.

94. The correct sequence of the decreasing order of abundance of the given elements in the Earth's crust is:
(a) Iron, oxygen, silicon, aluminium
(b) Oxygen, aluminium, silicon, iron
(c) Silicon, oxygen, iron, aluminium
(d) Oxygen, silicon, aluminium, iron

95. What is the correct chronological order (oldest to youngest) in which the following formed?
1. Atlantic ocean
2. Indian ocean
3. Pacific ocean
4. Mediterranean sea

Select the correct answer using the codes given below:
Codes:
(a) 3, 1, 4, 2
(b) 3, 1, 2, 4
(c) 1, 3, 2, 4
(d) 1, 3, 4, 2

96. When a collision between a continental plate and an oceanic plate takes place, the
(a) continental plate moves down
(b) oceanic plate moves down
(c) plates get sutured
(d) plates get sutured and the continental plate moves down

97. A seismogram records the following types of earthquake waves:
1. L waves
2. P waves
3. S waves

The correct chronological sequence of the arrival of these waves is:
(a) 2, 3, 1 (b) 2, 1, 3
(c) 3, 2, 1 (d) 3, 1, 2

98. Match List I with List II and select the correct answer using the codes given below the lists:

| List I | List II |
|---|---|
| A. Mid-oceanic ridge | 1. Alkali-rich basalt |
| B. Rift valley | 2. Potash-poor basalt |
| C. Island arc | 3. Andesite |
| D. Continent | 4. Plateau basalt |

Codes:

| | A | B | C | D |
|---|---|---|---|---|
| (a) | 2 | 1 | 4 | 3 |
| (b) | 2 | 1 | 3 | 4 |
| (c) | 1 | 2 | 3 | 4 |
| (d) | 1 | 2 | 4 | 3 |

99. Consider the following statements about Himalayan orogeny:
1. The orogeny resulted from the southward movement of the Indian plate.
2. The orogeny began about 25 million years ago.
3. The orogeny closed about one million years ago.
4. The Himalayas have a root of relatively light crustal rocks projecting into the denser mantle.

Of these statements
(a) 1 and 2 are correct
(b) 3 and 4 are correct
(c) 1 and 3 are correct
(d) 2 and 4 are correct

100. Consider the following statements:
The 'Gobi' desert of Mongolia owes its existence to:
1. the subtropical high-pressure belt where it falls.
2. its location in the continental interior.
3. the presence of high mountains across the prevailing winds.
Of these statements
(a) 1, 2 and 3 are correct
(b) 1 and 2 are correct
(c) 2 and 3 are correct
(d) 3 alone is correct

101. Consider the following pelagic deposits:
1. Diatom ooze
2. Globigerina ooze
3. Pteropod ooze
4. Radiolarian ooze
The correct sequence of the occurrence of these pelagic deposits with increasing depth of the ocean floor is:
(a) 3, 2, 4, 1
(b) 2, 3, 1, 4
(c) 3, 2, 1, 4
(d) 2, 3, 4, 1

102. Sheeting in granitic rocks is brought about by:
(a) thermal expansion
(b) dilation due to unloading
(c) diastrophism
(d) hydrolysis

103. In humid climate, the rate of weathering of the given rocks is such that:
(a) basalt > dunite > granite
(b) basalt > granite > dunite
(c) dunite > granite > basalt
(d) dunite > basalt > granite

104. Movement of individual rock masses along structural planes such as bedding, joint and fault on a slope is known as:
(a) debris slide
(b) slump
(c) rock slide
(d) debris fall

105. Which one of the following statements is true of braided streams?
   (a) Width of channel is less than its depth.
   (b) Width and depth of channel are equal.
   (c) Width of channel is greater than its depth.
   (d) Width and depth of a channel are unrelated.

106. Consider the following drainage patterns:
   1. Annular
   2. Trellis
   3. Deranged
   4. Dendritic

   Among these, those which reflect a strong structural control would include:
   (a) 1 and 2
   (b) 2 and 3
   (c) 3 and 4
   (d) 1, 2, 3 and 4

107. In a structural dome, sandstones dip 15° radially in all directions around intrusive rocks that are exposed in a centrally eroded topographic basin. Which one of the following landforms is most likely to be found on the sandstones?
   (a) Mesas
   (b) Hogbacks
   (c) Cuestas
   (d) Buttes

108. Which of the following statements are true of triangular facets that develop on dipping strata?
   1. Their lengths are inversely proportional to dip.
   2. Their apices point in the direction of dip.
   3. They develop due the formation of closely-spaced stream valleys.

   Select the correct answer using the codes given below:

   Codes:
   (a) 1, 2 and 3
   (b) 2 and 3
   (c) 1 and 2
   (d) 1 and 3

109. The bedding is ***not*** vertical along the hinge line in respect of:
   (a) reclined fold
   (b) recumbent fold
   (c) vertical fold
   (d) plunging upright fold

110. The principal axes of strain are direction in which:
   (a) there is maximum elongation
   (b) there is maximum shortening

(c) shear strain is maximum
(d) shear strain is zero

111. The stereographic projection of the two limbs of a fold shown in the given figure is that of a:

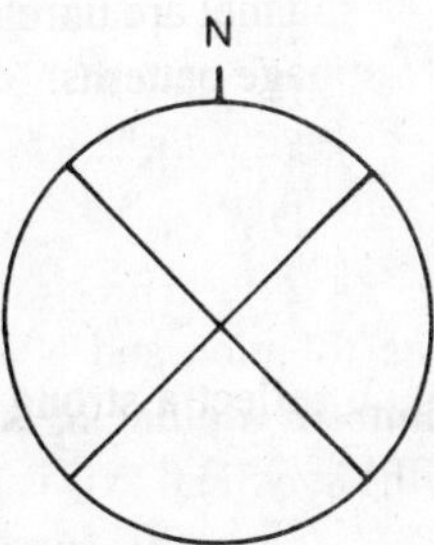


(a) non-plunging fold
(b) vertical fold
(c) recumbent fold
(d) reclined fold

112. The plan view shown in the given figure is that of a fault F-F intersecting a series of beds.

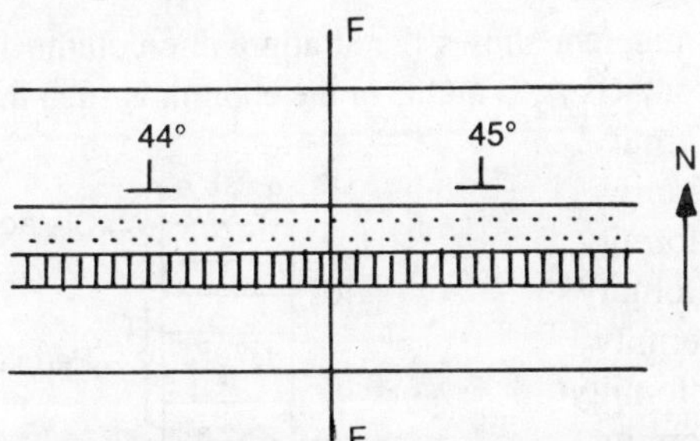


Which one of the following statement is true in this regard?
(a) The net slip is parallel to the bedding trace on the fault plane.
(b) The fault does not have any strike-slip.
(c) The fault does not have any dip-slip.
(d) The net-slip is equal to dip-slip.

113. Match List I with List II and select the correct answer using the codes given below the lists:

| | List I | | List II |
|---|---|---|---|
| A. | Cross-bedding | 1. | Dehydration |
| B. | Bifurcated vesicles in lavas | 2. | Igneous intrusion |
| C. | Graded bedding | 3. | Igneous extrusion |
| D. | Mud cracks | 4. | Sedimentation |
| | | 5. | Current direction |

Codes:

| | A | B | C | D |
|---|---|---|---|---|
| (a) | 5 | 2 | 4 | 1 |
| | A | B | C | D |
| (b) | 5 | 3 | 4 | 1 |
| | A | B | C | D |
| (c) | 3 | 2 | 1 | 4 |
| | A | B | C | D |
| (d) | 1 | 3 | 5 | 2 |

114. E-W-striking axial lane foliation and NW-SE-striking bedding are observed on the NE limb of a plunging syncline. The approximate direction of plunge of the synclinal axis is towards the:
(a) north (b) south
(c) east (d) west

115. The strike of tension joints formed due to a stress system with E-W oriented least principal stress axis will be:
(a) NE-SW
(b) NW-SE
(c) E-W
(d) N-S

116. The given diagram shows the stratigraphic column in an area. The nature of contacts A, B and C in the column is such that:

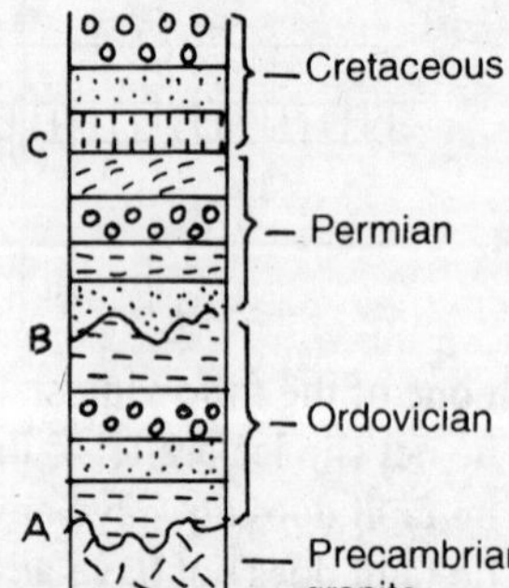


(a) A-Disconformity
B-Disconfirmity
C-Nonconformity
(b) A-Nonconformity
B-Disconformity
C- Paraconformity
(c) A-Disconformity
B-Paraconformity
C-Nonconformity
(d) A-Nonconformity
B-Paraconformity
C-Disconformity

117. The vertical section of a volcanic sequence showing pillow structures is shown in the given figure. The sequence is:

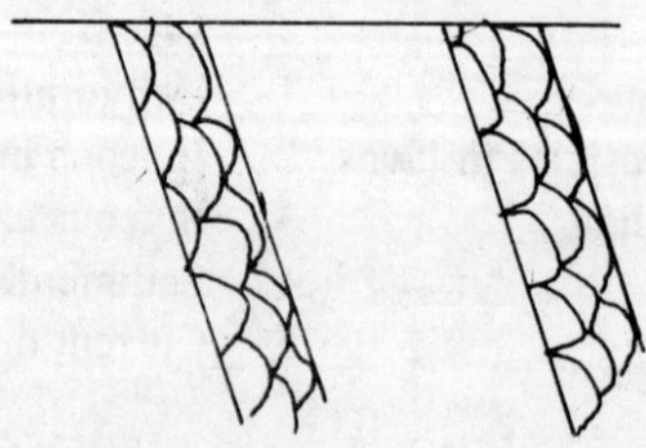

(a) normal
(b) overturned
(c) folded into an anticline
(d) folded into a syncline

118. Consider the following map:

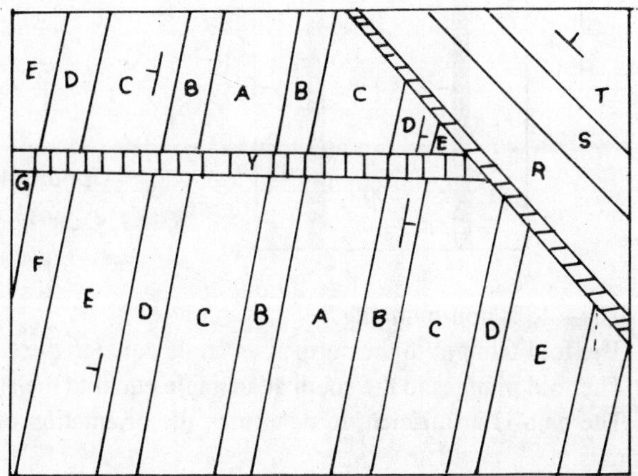


The correct sequence of events shown in the map is:

(a) Folding of A to G, faulting, intrusion of Y, deposition of R-T, intrusion of X.
(b) Folding of A to G, intrusion of Y, deposition of R-T, intrusion of X.
(c) Folding of A to G, deposition of R-T, intrusion of X and Y.
(d) Folding of A to G, faulting, deposition of R-T, intrusion of X and Y.

119. Consider the following statements regarding the relations between the outcrop of a sequence and ground slope:

'V' of the outcrop points:

1. downstream if the dip and slope are in the same direction, the dip being greater than the slope.
2. upstream if dip and slope are in the same direction, the dip being less than the slope.
3. upstream if the dip and slope are in opposite directions.

Of these statements:

(a) 1 and 2 are correct
(b) 1 and 3 are correct
(c) 2 and 3 are correct
(d) 1, 2 and 3 are correct

120. The given figure shows the outcrop of a folded bed on a uniformly sloping surface. Which one of the following statements is true in this regard?

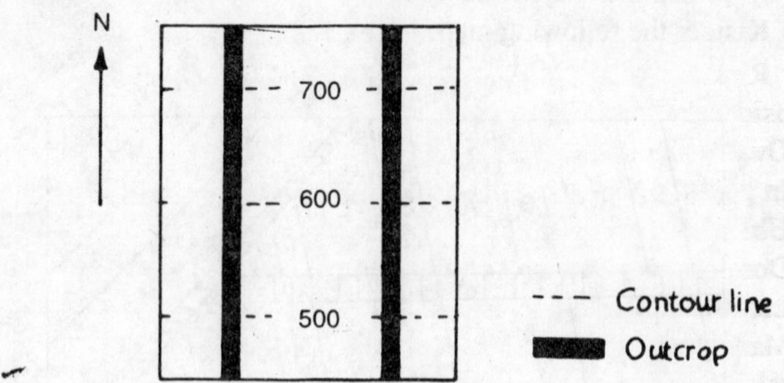


(a) The fold is non-plunging.
(b) The fold plunges to the north at an angle equal to the slope.
(c) The fold plunges to the south at an angle equal to the slope.
(d) The data is insufficient to determine the orientation of the fold axis.

## CIVIL SERVICES (PRELIMINARY) EXAMINATION - 1997 (KEY)

| | | | | | | | |
|---|---|---|---|---|---|---|---|
| 1-a | 2-b | 3-b | 4-d | 5-b | 6-d | 7-a | 8-d |
| 9-c | 10-d | 11-c | 12-b | 13-a | 14-c | 15-d | 16-a |
| 17-c | 18-b | 19-c | 20-b | 21-c | 22-b | 23-a | 24-c |
| 25-c | 26-c | 27-c | 28-b | 29-d | 30-c | 31-c | 32-d |
| 33-b | 34-a | 35-b | 36-b | 37-b | 38-a | 39-d | 40-b |
| 41-a | 42-b | 43-a | 44-b | 45-d | 46-a | 47-b | 48-a |
| 49-c | 50-b | 51-c | 52-c | 53-d | 54-b | 55-d | 56-b |
| 57-c | 58-c | 59-a | 60-b | 61-c | 62-d | 63-c | 64-c |
| 65-b | 66-c | 67-a | 68-b | 69-a | 70-d | 71-a | 72-c |
| 73-b | 74-c | 75-b | 76-d | 77-a | 78-b | 79-b | 80-c |
| 81-a | 82-a | 83-a | 84-c | 85-c | 86-a | 87-b | 88-b |
| 89-a | 90-a | 91-c | 92-c | 93-b | 94-d | 95-c | 96-b |
| 97-a | 98-b | 99-d | 100-b | 101-c | 102-b | 103-d | 104-c |
| 105-c | 106-a | 107-b | 108-a | 109-d | 110-a | 111-a | 112-a |
| 113-b | 114-c | 115-c | 116-b | 117-d | 118-a | 119-c | 120-d |

# 5. Civil Services (Preliminary)
## EXAMINATION 1996

1. Which one of the following rocks shows pillow structure?
   (a) Komatiite (b) Spilite
   (c) Rhyolite (d) Carbonatite
2. Consider the following features :
   1. Oval shape of alkali feldspar.
   2. In general, poorer in K-feldspar than average granite.
   3. Ubiquitous presence of two generations of quartz and alkali feldspar.
   4. Dominance of model $An_{40}$-plagioclase over alkali feldspar.
   5. Large size of alkali feldspars.
   6. Mantle of oligoclase on some of the alkali feldspar.

   Rapakivi granite is characterised by
   (a) 1, 2, 3 and 4 (b) 1, 3, 5 and 6
   (c) 1, 3, 4 and 5 (d) 2, 5 and 6
3. Which one of the following is thought to be confined in the Proterozoic?
   (a) Spinifex-textured komatiite
   (b) Massif-type anorthosites
   (c) Trondhjemitic lava flows
   (d) Layered ultramafic complexes
4. Which of the following pairs are correctly matched?
   1. Orthopyroxene +clinopyroxene ..... Harzburgite
   2. Olivine + clinopyroxene + orthopyroxene....Lherzolite
   3. Olivine + phlogopite + clinopyroxene ...Kimberlite
   4. Olivine + clinopyroxene .... Wehrlite

   Select the correct answer using the codes given below :
   Codes :
   (a) 2, 3 and 4 (b) 1, 2 and 3
   (c) 2 and 4 (d) 1, 3 and 4
5. Consider the following statements :
   1. A basic igneous rock where olivine shows reaction relation with Ca-poor pyroxene is a tholeiitic basalt.
   2. Kimberlite, derived from deeper mantle, shows fragmented texture.
   3. A rock with more than 85 percent albitic plagioclase is an anorthosite.
   4. Lamprophyres are basic to ultrabasic rocks containing phenocrysts of hydrous mafic minerals.

   Of these statements:
   (a) 1, 2 and 3 are correct
   (b) 1, 3 and 4 are correct
   (c) 2, 3 and 4 are correct
   (d) 1, 2 and 4 are correct

6. Sequence of crystallisation of liquid (D) in the anorthite-quartz system shown in the given figure, shows early separation of quartz followed by simultaneous crystallisation of anorthite and quartz at point, E. Applying the Phase Rule, P + F = C + 2, one can infer that the degree of freedom and the variance of the system at E would be:

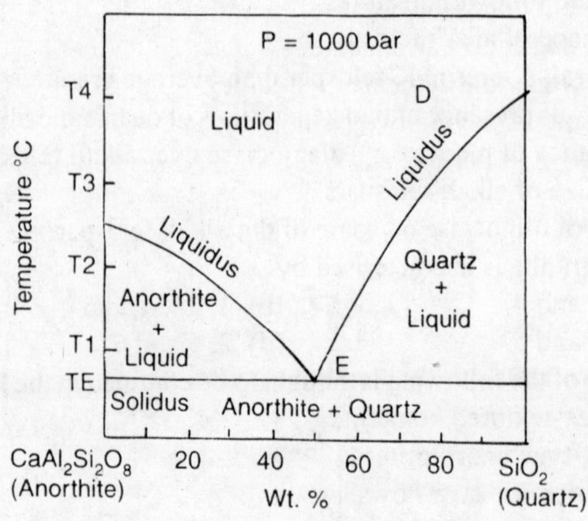


(a) one and univariant (b) two and univariant
(c) zero and univariant (d) two and divariant

7. Which one of the following is the correct order of crystallisation of minerals as per Bowmen's reaction principle?
(a) Mg-pyroxene—hornblende-biotite—orthoclase
(b) Alivine—pyroxene—orthoclase—hornblende
(c) Fe-pyroxene—plagioclase—Mg-pyroxene—quartz
(d) Olivine—alkali feldspar-plagioclase -quartz

8. After the fractional crystallisation of alkali basalt magma, the residual liquid will produce:
(a) quartz and K-feldspar rich rock
(b) quartz and feldspathoid rich rock
(c) feldspar and feldspathoid rich rock
(d) quartz and muscovite rich rock

9. A granitic differentiate *cannot* be generated by:
(a) fractional crystallisation of alkali basalt
(b) fractional crystallisation of tholeiitic basalt
(c) melting of the crust
(d) fractional crystallisation of calc-alkali basalt

10. Match List I with List II and select the correct answer using the codes given below the lists :

| List I | List II |
|---|---|
| A. Rudaceous rock containing only/dominantly one kind of material | 1. Polymodal |
| B. Detrital rock consisting of fragments of many different materials | 2. Lutite |
| C. Sandstone which is slightly flexible in thin slabs because of ball and socket arrangement of the grains and cement | 3. Polymict |
| | 4. Oligomict |
| D. Sediment displaying two or more maxima in its particle size | 5. Itacolumite |

Codes :

| | A | B | C | D | | A | B | C | D |
|---|---|---|---|---|---|---|---|---|---|
| (a) | 4 | 1 | 2 | 3 | (b) | 4 | 3 | 5 | 1 |
| (c) | 2 | 1 | 5 | 4 | (d) | 2 | 3 | 4 | 1 |

11. Which one of the following is the main basis of classification of sedimentary rocks?
    (a) Process of formation of sedimentary material
    (b) Source of sedimentary material
    (c) Environment of deposition
    (d) Texture
12. Most of the current structures in sandy sediments have been formed by flows that are:
    (a) laminar and rapid
    (b) laminar and tranquil
    (c) turbulent and rapid
    (d) turbulent and tranquil
13. Consider the following statements :
    1. Laterite is a sedimentary rock of residual origin.
    2. A sandstone containing more than 25% feldspar is an arkose.
    3. 'Epiclastic' sediments are derived from erosion of pre-existing clastic sedimentary rocks.
    4. Chert is a chemically precipitated rock.

    Of these statements:
    (a) 1, 2 and 3 are correct
    (b) 1, 3 and 4 are correct
    (c) 2, 3 and 4 are correct
    (d) 1, 2 and 4 are correct
14. Pebbles in imbricated sediments generally lie with their long axes parallel to the flow direction and plungin.
    (a) gently downcurrent
    (b) steeply upcurrent
    (c) gently upcurrent
    (d) steeply downcurrent
15. Which one of the following structures shows *minimum* directional variability?
    (a) Parting lineation
    (b) Ripple mark
    (c) Planar-tabular cross-bedding
    (d) Trough cross-bedding

16. Which one of the following detrital constituents of sedimentary rocks provides the best evidence for reconstructing the provenance composition?
(a) Quartz (b) Feldspar
(c) Heavy minerals (d) Rock fragments

17. Match the diagenetic fabrics labelled A, B, C and D with the figures labelled 1, 2, 3 and 4 and select the correct answer using the codes given below the lists of fabrics :

| *Diagenetic fabrics* | *Figures* |
|---|---|
| A . Pressure solution fabric | 1. |
| B. Cementation and cavity-filling fabric | 2. |
| C. Neomorphic fabric | 3. |
| D. Replacement fabric | 4. |

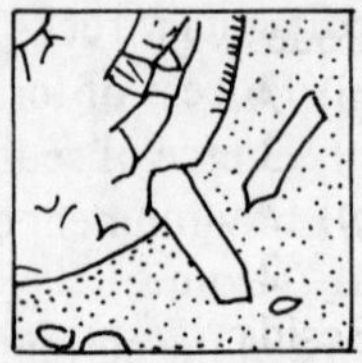

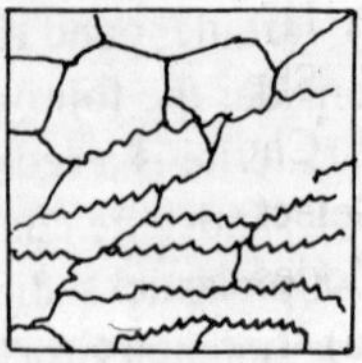

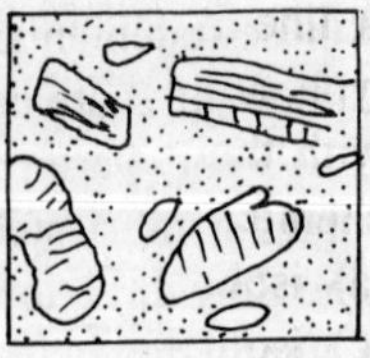

Codes:

| | A | B | C | D | | A | B | C | D |
|---|---|---|---|---|---|---|---|---|---|
| (a) | 4 | 2 | 3 | 1 | (b) | 3 | 2 | 4 | 1 |
| (c) | 3 | 1 | 4 | 2 | (d) | 4 | 1 | 3 | 2 |

18. Consider the following minerals :

| | |
|---|---|
| 1. Zircon | 2. Muscovite |
| 3. Orthoclase | 4. Plagioclase |
| 5. Biotite | |

The correct sequence of the decreasing order of the chemical stability of these minerals in a sedimentational motif is:

(a) 1, 2, 3, 4, 5
(b) 1, 2, 4, 3, 5
(c) 2, 1, 3, 5, 4
(d) 2, 1, 5, 3, 4

19. Match List I with List II and select the correct answer using the codes given below the lists :

| List I | List II |
|---|---|
| A. Reef | 1. Fresh water carbonate |
| B. Allochems | 2. Evaporitic carbonate |
| C. Tufa | 3. Bioherm |
| D. Caliche | 4. Framework carbonate |

Codes :

| | A | B | C | D | | A | B | C | D |
|---|---|---|---|---|---|---|---|---|---|
| (a) | 3 | 1 | 4 | 2 | (b) | 2 | 4 | 1 | 3 |
| (c) | 2 | 1 | 3 | 4 | (d) | 3 | 4 | 1 | 2 |

20. Besides temperature, which of the following physical variables govern the process of regional metamorphism?

1. Lithostatic pressure
2. Hydrostatic pressure
3. Shearing stress
4. Chemical activity of percolating fluids

Select the correct answer using the codes given below :

Codes :

(a) 1, 2 and 3 (b) 2, 3 and 4
(c) 2 and 4 (d) 1, 3 and 4

21. A line that distinguishes two metamorphic zones in a particular rock type, and that which represents the intersection of the ground surface with the pressure/temperature plane of univariant equilibrium for a chemical reaction, is designated as an:

(a) isoline (b) isograd
(c) isotherm (d) isozone

22. Which of the following index minerals can appear in different grades of metamorphosed argillaceous rocks in Barrovian type of metamorphism?

| | |
|---|---|
| 1. Chlorite | 2. Tremolite |
| 3. Almandine garnet | 4. Staurolite |
| 5. Diopside | |

Select the correct answer using the codes given below .
Codes :

(a) 1, 2, 4 and 5 (b) 2, 3 and 4
(c) 2, 3 and 5 (d) 1, 3 and 4

23. Rocks formed by metasomatism at igneous contacts is defined as:
(a) skarns (b) mylonite
(c) hornfels (d) aureole

24. Granoblastic texture in a metamorphic rock is developed by:
(a) high stress and low temperature.
(b) low lithostatic pressure and temperature.
(c) high lithostatic pressure and temperature.
(d) high stress and temperature.

25. Development of foliation in a metamorphic rock is mainly a function of:
(a) stress and temperature.
(b) lithostatic pressure and temperature.
(c) temperature and chemically active fluids.
(d) chemically active fluids and stress.

26. Tektites are formed by:
(a) carbonic metamorphism
(b) complex organic processes
(c) impact of meteorites
(d) differentiation

27. Consider the following diagnostic characteristics :
1. Dominant granulite paragenesis.
2. Dominant amphibolite paragenesis.
3. Usually with mesoperthite or albitelabradorite in the absence of K-feldspar.
4. Consisting exclusively of hypersthene granite only.
5. Sagenetic exsolutions in biotite.
6. Sphene, ilmenite and rutile - usual accessories in the rocks.

Diagnostic characteristics of Indian Charnockite suite of rocks would include:
(a) 1, 3, 5 and 6 (b) 1, 2, 4 and 6
(c) 2, 3, 4 and 5 (d) 2, 3, 4 and 6

28. Plate movements take place at the junctions of:
(a) sialic and simatic layers
(b) crust and sub-crust
(c) crust and the mantle
(d) lithosphere and asthenosphere

29. As one moves north from the Ganga plains towards the Central Himalayas, the thickness of the continental crust will:
(a) decrease

(b) increase
(c) change but not systematically
(d) not change

30. The break-up of Gondwanaland is believed to have started in the:
(a) Permian
(b) Carboniferous
(c) Jurassic
(d) Cretaceous

31. Consider the following statements :
Magnetic 'stripes'
1. are symmetrical on either side of the mid-oceanic ridge.
2. are caused by the extrusions and intrusions of basalt.
3. provide a magnetic record of ocean-floor spreading.
Of these statements
(a) 1, 2 and 3 are correct
(b) 1 and 2 are correct
(c) 2 and 3 are correct
(d) 1 and 3 are correct

32. Measurements of gravitational attraction at Kaliana and Kalianpur in Northern India led to the development of:
(a) triangulation technique.
(b) refinement of the Cavendish experiment.
(c) fixation of North Star as the reference point.
(d) concept of isostasy.

33. Earthquakes caused by strike slip faulting originate at a depth of:
(a) less than 15 km
(b) 15 to 20 km
(c) 20 to 50 km
(d) more than 50 km

34. Eugeosyncline is recognised by the association of :
(a) limestones and serpentines.
(b) quartzites and limestones.
(c) thin cover of sediments and absence of volcanic activity.
(d) thick cover of sediments and abundant eruptives.

35. In a mountain glacier valley, flowage of ice mass takes place by:
(a) plastic deformation
(b) plastic deformation and slippage
(c) plastic deformation, slippage and overthrust
(d) slippage and overthrust

36. Match List I with List II and select the correct answer using the codes given below the lists :

| List I *(Zone of deposition)* | List II *(Type of deposit)* |
|---|---|
| A. Littoral zone | 1. Oolite sand |
| B. Neritic zone | 2. Coral reef |
| C. Bathyal zone | 3. Coral mud |
| D. Abyssal zone | 4. Globigerina ooze |

Codes :

| (a) A | B | C | D | (b) A | B | C | D |
|---|---|---|---|---|---|---|---|
| 1 | 2 | 3 | 4 | 2 | 1 | 4 | 3 |
| (c) A | B | C | D | (d) A | B | C | D |
| 1 | 3 | 2 | 4 | 2 | 3 | 4 | 1 |

37. The mineral stability series in a weathering process is as follows :

| | |
|---|---|
| Olivine | Calcic plagioclase |
| Augite | Calcic-alkali plagioclase |
| Hornblende | Alkalic-calcic plagioclase |
| Biotite | Alkalic plagioclase |
| | Potash feldspar |
| | Muscovite |
| | Quartz |

According to this mineral stability series in weathering:

(a) biotite is formed from hornblende during weathering
(b) biotite is more stable than hornblende
(c) hornblende is formed from biotite
(d) hornblende is more stable than biotite

38. The given figure shows the crescentic sand-dune of Barchan type, a feature found in desert. The predominant wind blows in the direction labelled:

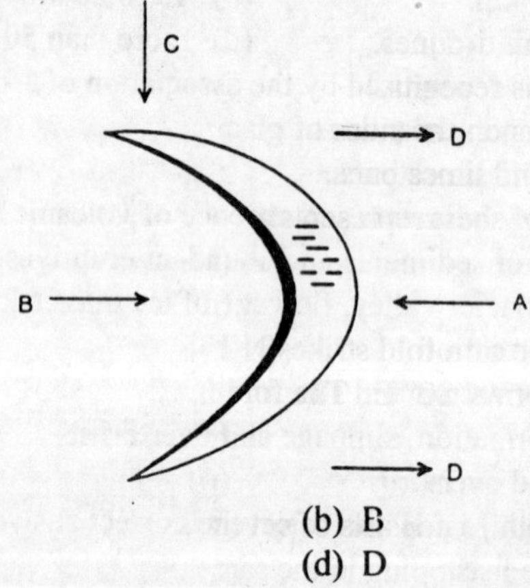


(a) A (b) B
(c) C (d) D

39. There is a systematic sequence of youth, maturity and old age in the development of landforms. Differences in landforms are explained in terms of differences in structures, processes and stage. Which one of the following concepts fits well with the above statement?

(a) Dynamic equilibrium
(b) Epeirogeny
(c) Geomorphic cycle
(d) Morphogenetic region

40. Regionwise analysis showing percentage distribution of different drainage patterns is presented in the following table :

| Region | Dendritic | Trellis | Rectangular |
|---|---|---|---|
| Region-I | 30% | 30% | 40% |
| Region-II | 70% | 20% | 10% |
| Region-III | 30% | 40% | 30% |
| Region-IV | 50% | 20% | 30% |

Based on the above data, the two regions that show pronounced structural control include:
(a) I and II (b) I and III
(c) II and III (d) II and IV

41. The playa lakes develop on a:
(a) desert basin (b) stony desert
(c) semi-arid basin (d) desert mountain

42. Which one of the following is the most suitable condition/set of conditions for the maximum development of karsts?
(a) Corrasion.
(b) High porosity and permeability.
(c) Soluble rocks with extensive jointing.
(d) Arid and semi-arid climate.

43. 'Fosses' are:
(a) a type of sand-dunes.
(b) flood-plain deposits.
(c) depressions on the sides of glacial troughs.
(d) a type of off-shore bars.

44. Valley-in-valley best represents:
(a) rejuvenation (b) meandering
(c) erosion (d) deposition

45. The axial plane of a fold strikes N 170° and dips 20°/N 80°. Its hinge line plunges 20°/N 80° E. The fold is:
(a) reclined (b) upright
(c) reclined antiform (d) reclined synform

46. Imbricate structure consists of several:
(a) gravity faults dipping in the same direction.
(b) gravity faults dipping in different directions.
(c) thrust faults dipping in the same direction.
(d) thrust faults dipping in different directions.

47. Cross beds and graded beds do not occur together, because they represent:
(a) contrasted lithology
(b) contrasted ages
(c) contrasted depositional conditions
(d) both contrasted lithology and age

48. Match List I with List II and select the correct answer using the codes given below the lists :

| | List I | | List II |
|---|---|---|---|
| A. | Columnar jointing | 1. | Tension fractures related to faulting |
| B. | Feather joints | 2. | Tension joints |
| C. | Joints perpendicular to the fold axes | 3. | Joints parallel to the axial planes of folds |
| D. | Release joints | 4. | Common in orogenic belts |

Codes:

| | A | B | C | D | | A | B | C | D |
|---|---|---|---|---|---|---|---|---|---|
| (a) | 2 | 3 | 4 | 1 | (b) | 3 | 1 | 2 | 4 |
| (c) | 2 | 1 | 4 | 3 | (d) | 1 | 2 | 4 | 3 |

49. Which among the following is not suitable for determining top and bottom in layered rocks?

(a) Wave ripple (b) Syngenetic concretion
(c) Flow cast (d) Mud-cracks

50. The sequence of events as interpreted from the geological map in the given diagram is :

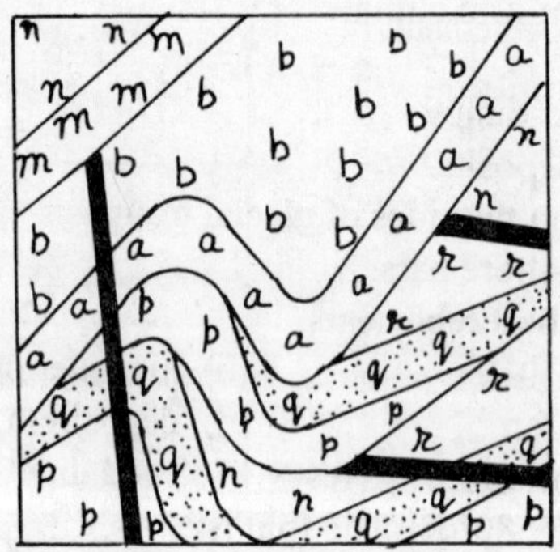

(a) Deposition of beds m, n ↑ youngest
—Unconformity—
Dyke intrusion
Folding
Deposition of beds a, b
—Unconformity —
Faulting
Dyke intrusion
Deposition of beds p, q, r ⊥ oldest

(b) Deposition of beds m, n ↑ youngest
—Unconformity—
Dyke intrusion
Folding
Deposition of beds a, b
—Unconformity—
Dyke intrusion
Deposition of beds p, q, r ⊥ oldest

(c) Deposition of beds p, q, r ↑ youngest
—Unconformity—
Dyke intrusion
Folding
Deposition of beds a, b
—Unconformity—
Dyke intrusion
Faulting
Deposition of beds p, q, r ⊥ oldest

(d) Deposition of beds m, n ↑ youngest
—Unconformity—
Dyke intrusion
Folding
Faulting
Deposition of beds a, b
—Unconformity—
Dyke intrusion
Deposition of beds p, q, r ⊥ oldest

51. Horizontal Jurassic beds lie over horizontal Devonian beds; the contact is also horizontal. The contact represents:
 (a) not an unconformity
 (b) a nonconformity
 (c) a disconformity
 (d) a paraconformity

52. 'Windows' are most common in:
 (a) Deccan Traps of Western India
 (b) Gondwanas of Eastern India
 (c) Precambrians of Southern India
 (d) Tertiaries of the Himalayas

53. The toposheet 'X' lies east of the toposheet 550/4 and south east of toposheet 550/3 as shown in the given diagram. The number of toposheet 'X' is:

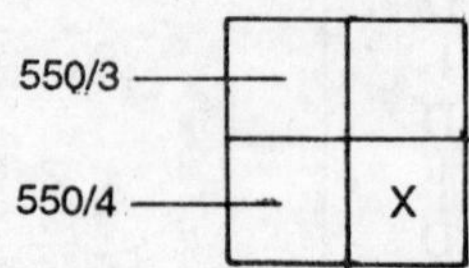


(a) 550/5 (b) 550/8
(c) 550/7 (d) 550/9

54. A contact between shale and sandstone (C-C) is shown in the given diagram. Which one of the following interpretations drawn from the diagram is correct?

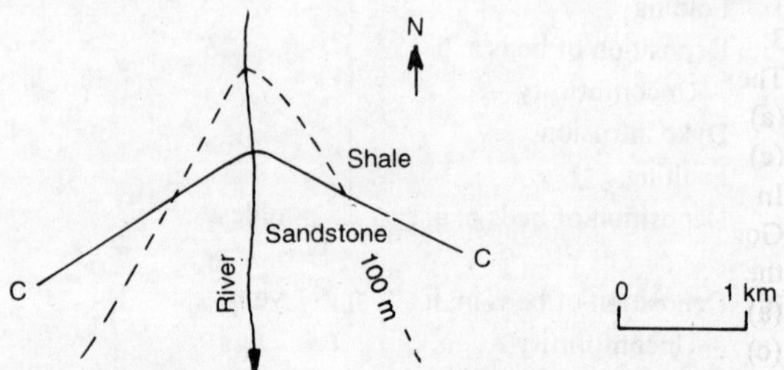


(a) The rocks dip north (b) The rocks dip south
(c) The rocks are horizontal (d) The rocks are vertical

55. An essential condition for the recognition of a formation is that it must be:
(a) a sedimentary unit (b) fossiliferous
(c) of the same age throughout (d) mappable

56. If a lithostratigraphic formation extends laterally for a considerable distance, then its age from one end to the other:
(a) will always be the same.
(b) will always be different.
(c) may be different but the difference will increase gradually.
(d) cannot be unequivocally ascertained.

57. Match List I with List II and select the correct answer using the codes given below the lists :

| List I | List II |
|---|---|
| A. Riphean | 1. Palaeogene |
| B. Cenomanian | 2. Proterozoic |
| C. Thanetian | 3. Neogene |
| D. Burdigalian | 4. Cretaceous |

Codes:

| | | | |
|---|---|---|---|
| (a) A | B | C | D |
| 1 | 4 | 3 | 2 |
| (b) A | B | C | D |
| 4 | 2 | 3 | 1 |
| (c) A | B | C | D |
| 2 | 4 | 3 | 1 |
| (d) A | B | C | D |
| 2 | 4 | 1 | 3 |

58. The time span of Proterozoic Eon ranges from:
(a) 570 Ma to 493 Ma (b) 1500 Ma to 570 Ma
(c) 2500 Ma to 570 Ma (d) 4000 Ma to 2500 Ma

59. Consider the following :
1. Series 2. Zone
3. System 4. Stage
Their correct sequence in the ascending order is:
(a) 2, 4, 1, 3 (b) 1, 2, 3, 4
(c) 3, 1, 4, 2 (d) 2, 1, 4, 3

60. In the statement "Indian sub-continent which formed a part of the Gondwana Supercontinent separated out around **X** million years before the present", 'X' stands for:
(a) 180 (b) 135
(c) 65 (d) 35

61. In the statement, "Cenozoic period of the earth's history began around **X** million years from the present", X' stands for
(a) 5 (b) 35
(c) 65 (d) 135

62. Consider the following formations:
1. Muth Quartzite
2. Syringothyris Limestone
3. Fenestella Shale
The correct stratigraphic order of these formations is:
(a) 1, 3, 2 (b) 2, 1, 3
(c) 3, 2, 1 (d) 1, 2, 3

63. The age of the Panjal Volcanic Succession of Kashmir is:
(a) Jurassic to Cretaceous (b) Permian to Triassic
(c) Devonian to Carboniferous (d) Silurian to Devonian

64. Match List I with List II and select the correct answer using the codes given below the lists :

| List I | List II |
|---|---|
| A. Copper and iron | 1. Damuda Group |
| B. Coal | 2. Singbhum-Gangapur Group |
| C. Petroleum | 3. Khondalite of Eastern Ghat |
| D. Bauxite | 4. Barail Group |

Codes :

| | A | B | C | D | | A | B | C | D |
|---|---|---|---|---|---|---|---|---|---|
| (a) | 2 | 1 | 4 | 3 | (b) | 2 | 4 | 1 | 3 |
| (c) | 3 | 1 | 2 | 4 | (d) | 1 | 3 | 4 | 2 |

65. Consider the following of the Vindhyan Supergroup:
    1. Rewa Group
    2. Semri Group
    3. Bhander Group
    4. Kaimur Group

    The correct sequence of these Vindhyan rocks in the ascending order is:
    (a) 2, 4, 3, 1 (b) 4, 2, 3, 1
    (c) 2, 4, 1, 3 (d) 4, 2, 1, 3
66. Vindhyans are:
    (a) older than 700 M.Y. (b) older than 1800 M.Y.
    (c) younger than 600 M.Y. (d) younger than 700 M.Y.
67. Consider the following statements :
    1. The Upper Vindhyan succession is three times larger in thickness than the Lower Vindhyans.
    2. The Upper Vindhyans have a predominant lithology of Shales and Limestones.
    3. The basal rocks of the Vindhyans rest unconformably over the Bijawar Group.

    Of these statements
    (a) 1 and 2 are correct (b) 1 and 3 are correct
    (c) 1, 2 and 3 are correct (d) 2 and 3 are correct
68. Consider the following statements :
    1. Gondwana sequence represents a short episode of continental sedimentation.
    2. Gondwana sequence has Equisetaceous plants in the Lower, Cycads and Conifers in the Upper and Ferns common to both.
    3. The Gondwana sedimentation confining to the river valleys has contemporaneous relationship with faulting.

    Of these statements
    (a) 1, 2 and 3 are correct (b) 2 and 3 are correct
    (c) 1 and 3 are correct (d) 1 and 2 are correct
69. Presence of undecomposed feldspars in Talchir Formation indicates:
    (a) humid climate (b) semi-arid climate
    (c) cold climate (d) arid climate
70. The occurrence of sandstone beds, grits, conglomerates, silt and clay beds and the availability of a rich and varied mammalian fauna in the Siwalick deposits provide evidence of:
    (a) the prevalence of the glacial conditions.
    (b) a warm humid climate and a shallow marine environment.

(c) a warm humid climate and fluviatile environment.
(d) a cold humid climate and a shallow marine environment.

71. Consider the following Proboscideans which occur in the Siwalik rocks:

1. *Listriodon* 2. *Stegodon*
3. *Elephas* 4. *Mastodon*

Their correct chronological order is:

(a) 1, 4, 2, 3 (b) 1, 4, 3, 2
(c) 4, 1, 2, 3 (d) 4, 1, 3, 2

72. An index fossil may be defined as the one which has a:

(a) limited geographic distribution and shows a short range in time.
(b) wide geographic distribution and a well extended stratigraphic range.
(c) limited geographic distribution and a wide range in time.
(d) wide geographic distribution and a limited stratigraphic range.

73. The process of fossilisation is influenced by environmental factors. Which one of the following provides a more favourable condition for the preservation of organisms?

(a) Oxygenated environment.
(b) Burial in coarse-grained sediments.
(c) Burial in fine-grained sediments.
(d) Embedment in hard rocks.

74. When the individual corallites in a compound coral are free and diverge from one another, the corallum is termed as:

(a) dendroid (b) fasciculate
(c) massive (d) basaltiform

75. Which one of the following characters help to differentiate *Gangamopteris* from *Glossopteris*?

(a) Presence of midrib (b) Absence of midrib
(c) Presence of reticulate venation (d) Presence of parallel venation

76. Two valves of a lamellibranch shell are referred to as:

(a) Larger and smaller (b) Right and left
(c) Dorsal and ventral (d) Anterior and Posterior

77. Branchiopods are:

(a) bottom dwelling (benthonic) marine animals.
(b) bottom dwelling (benthonic) fresh water and marine animals.
(c) free-swimming (nektonic) fresh water and marine animals.
(d) free floating (planktonic) marine animals.

78. The two valves of a brachiopod shell are:

(a) equal in size and inequilateral.
(b) unequal in size and equilateral.
(c) unequal in size and inequilateral.
(d) equal in size and equilateral.

79. The larger valve with a coral-like conical structure and the smaller valve with a toothed lid is found in:

(a) *Ostrea* (b) *Hippurites*
(c) *Gryphea* (d) *Exogyra*

80. Which one of the following is common to both Lower and Upper Gondwana deposits?
(a) *Glossopteris* (b) *Ptilophyllum*
(c) *Sphenophyllum* (d) None of the above

81. In a brachiopod shell, the plane of symmetry passes:
(a) between the two valves.
(b) across the two valves from umbo to the opposite margin.
(c) across the two valves from side to side.
(d) between/across none of the above.

82. Match List I with List II and select the correct answer using the codes given below the lists :

| **List I** *(Name of the mineral)* | **List II** *(Symmetry characters)* |
|---|---|
| A. Vesuvianite | 1. $a \neq b \neq c$ ; $\alpha = \beta = \gamma = 90°$ |
| B. Calcite | 2. $a = b \neq c$ ; $\alpha = \beta = \gamma = 90°$ |
| C. Sulphur | 3. $a \neq b \neq c$ ; $\alpha = \beta = 90°$ ; $\beta \neq 90°$ |
| D. Albite | 4. $a \neq b \neq c$ ; $\alpha \neq \beta \neq \gamma \neq 90°$ |
| | 5. $a = b \neq c$; $\alpha = \beta = 90°$; $\gamma = 120°$ |

Codes:

| | A | B | C | D | | A | B | C | D |
|---|---|---|---|---|---|---|---|---|---|
| (a) | 3 | 2 | 1 | 4 | (b) | 2 | 5 | 1 | 4 |
| (c) | 4 | 2 | 3 | 1 | (d) | 2 | 5 | 3 | 1 |

83. Match List I (Stereographic projections) with List II (Crystal system) and select the correct answer using the codes given below the lists :

**List I** **List II**

(A)

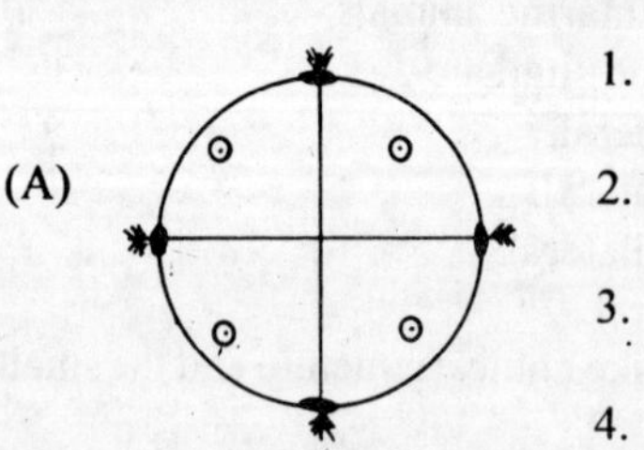

1. Orthorhombic
2. Tetragonal
3. Hexagonal
4. Isometric

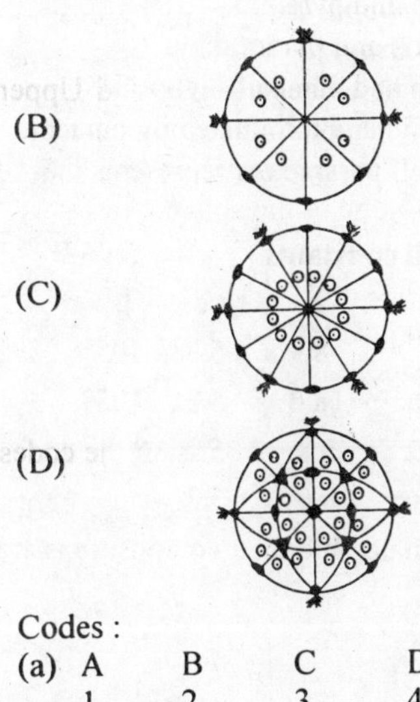


Codes :

| | | | | |
|---|---|---|---|---|
| (a) | A | B | C | D |
| | 1 | 2 | 3 | 4 |
| (b) | A | B | C | D |
| | 2 | 1 | 3 | 4 |
| (c) | A | B | C | D |
| | 1 | 2 | 4 | 3 |
| (d) | A | B | C | D |
| | 2 | 1 | 4 | 3 |

84. Which of the following pairs of twinnings and twin planes are correctly matched?

1. Manebach Law ........................... (001)
2. Brazil Law .................................. (1120)
3. Baveno Law ............................... (021)
4. Albite Law ................................. (100)

Select the correct answer using the codes given below :

Codes :

(a) 2, 3 and 4
(b) 1, 3 and 4
(c) 1, 2 and 4
(d) 1, 2 and 3

85. The crystal shown in the given figure represents combinations of:

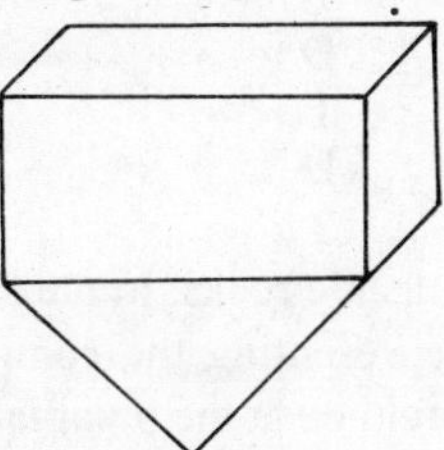

(a) open forms, terminated by pedion and pyramid.
(b) closed forms, terminated by pedion and pyramid.
(c) open forms, terminated by pedion and pinacoid.
(d) open forms, modified by dipyramid and terminated by pinacoid.

86. Which one of the following sets of cell parameters represents that of tremolite ?

| **Crystal class** | **Cell constants** | | | |
|---|---|---|---|---|
| | a | b | c | β |
| (a) 2/m | 9.7 | 8.9 | 5.3 | 105° |
| (b) 2/m | 9.8 | 18.0 | 5.3 | 105° |
| (c) 2/m 2/m 2/m | 18.2 | 8.8 | 5.2 | |
| (d) 2/m 2/m 2/m | 18.6 | 18.0 | 5.3 | |

87. Which of the following pairs of carbonates and their compositions are correctly matched?
1. Smithsonite .................. $ZnCO_3$
2. Witherite ...................... $BaCO_3$
3. Magnesite ..................... Ca (Mg, Fe) $CO_3 2H_2O$
4. Cerrusite ..................... $PbCO_3$

Select the correct answer using the codes given below :
Codes :
(a) 2, 3 and 4 (b) 1, 2 and 3
(c) 1, 2 and 4 (d) 1, 3 and 4

88. Match List I (Polymorphs of quartz) with List II (Temperature range, °C; stability) and select the correct answer using the codes given below the lists :

| **List I** | **List II** |
|---|---|
| A. Tridymite | 1. 0 to 573 |
| B. β-quartz | 2. 573 to 867 |
| C. Cristobalite | 3. 867 to 1470 |
| D. α-quartz | 4. 1470 to 1783 |

Codes :

| | A | B | C | D |
|---|---|---|---|---|
| (a) | 1 | 2 | 3 | 4 |
| (b) | 2 | 3 | 4 | 1 |
| (c) | 3 | 2 | 4 | 1 |
| (d) | 3 | 4 | 1 | 2 |

89. Sanidine is found in volcanic rocks. In the triangular diagram (shown in the given figure) representing the composition of feldspars, the location of sanidine would be at the point labelled

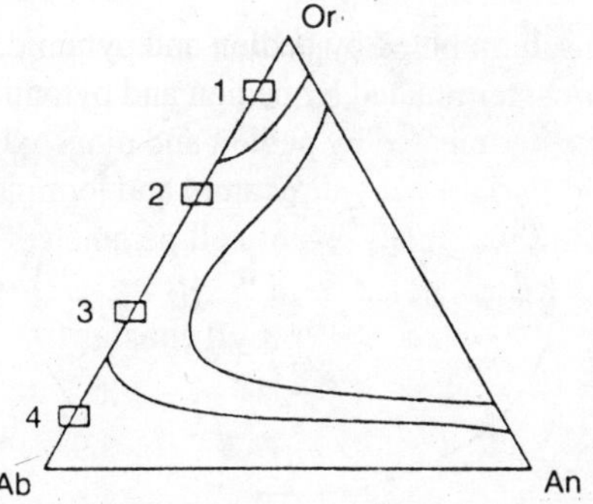


(a) 1
(b) 2
(c) 3
(d) 4

90. Monoclinic pyroxenes are considered to be members of the four component system, diopside-hedenbergite-enstatite-ferrosilite. In the phase diagram shown in the given figure, sub-calcic augite will fall in the field labelled.

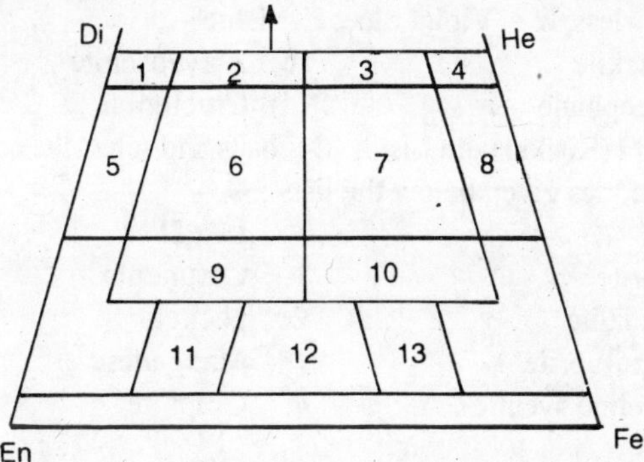


(a) 6
(b) 9
(c) 10
(d) 11

91. Which one of the following pairs of garnets is non-isometric above 800°C and may invert near that temperature to an isotropic form or else remain birefringent with inversion twinning?
(a) Grossularite-spessartite
(b) Pyrope-almandine
(c) Andradite-melanite
(d) Uvarovite-schorlomite

92. The given figure shows an optical indicatrix that illustrates the optical features of a crystal. Which one of the following minerals shows such an optical behaviour?

C-axis and Optic axis

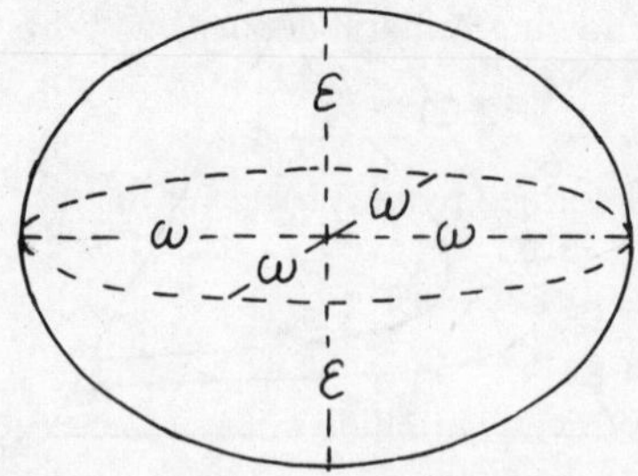


(a) Calcite (b) Olivine
(c) Gypsum (d) Quartz

93. A biaxial mineral that shows parallel or symmetrical extinction on faces in zones parallel to at least one crystallographic direction is:
(a) Triclinic (b) Monoclinic
(c) Orthorhombic (d) Pseudo-orthorhombic

94. Which one of the following minerals shows the scheme of pleochroism, X = Colourless; Y = Violet blue; Z = Blue?
(a) Riebeckite (b) Arfvedsonite
(c) Glaucophane (d) Crocidolite

95. Match List I (Rocks) with List II (Deposits and select the correct answer using the codes given below the lists :

| List I | List II |
|---|---|
| A. Kodurite | 1. Corundum |
| B. Khondalite | 2. Nickel |
| C. Gabbro-norite | 3. Manganese |
| D. Nepheline syenite | 4. Graphite |

Codes :

| | A | B | C | D |
|---|---|---|---|---|
| (a) | 3 | 1 | 4 | 2 |
| (b) | 2 | 3 | 4 | 1 |
| (c) | 3 | 4 | 2 | 1 |
| (d) | 2 | 4 | 1 | 3 |

96. Consider the following processes :
1. Dissemination
2. Residual liquid segregation
3. Segregation
4. Immiscible liquid segregation

The late magmatic deposits are formed by the processes listed at:
(a) 2 and 4 (b) 3 and 4

(c) 1 and 2 (d) 1 and 3

97. Consider the following mineral deposits :

1. Iron 2. Manganese
3. Copper 4. Aluminium

Those which are formed by the process of residual concentration would include:

(a) 1 and 4 (b) 1, 2 and 3
(c) 1, 2 and 4 (d) 2 and 3

98. Consider the following sulphide minerals :

1. Sphalerite
2. Chalcopyrite
3. Galena

The usual sequence of crystallisation of these sulphide minerals in hydrothermal deposits is:

(a) 3, 1, 2 (b) 1, 3, 2
(c) 1, 2, 3 (d) 2, 1, 3

99. What is the correct sequence of the following rock-forming silicates in order of *decreasing* stability in the formation of clay deposits over granitic terrains?

1. Quartz
2. Potash feldspar
3. Muscovite

Select the correct answer using the codes given below :

Codes :

(a) 1, 2, 3 (b) 1, 3, 2
(c) 2, 3, 1 (d) 2, 1, 3

100. Lindgren's classification of ore deposits is based on

(a) structural parameters
(b) association of ore minerals
(c) location of the deposits
(d) depth-temperature parameters

101. A common primary texture formed on account of rapid flocculation of materials in hydrothermal cavity-filled deposit is called:

(a) Ribbon banding
(b) Colloform banding
(c) Crustiform banding
(d) Rhythmic microbanding

102. Match List I (Deposit) with List II (Location/stratigraphic position) and select the correct answer using the codes given below the lists :

| List I | List II |
|---|---|
| A. Khetri copper | 1. Bastar district, MP |
| B. Zawar Pb-Zn | 2. Ambadongar, Gujarat |

| | | | |
|---|---|---|---|
| C. | Fluorite | 3. | Delhi Group |
| D. | Largest hematite deposit (Indian) | 4. | Aravalli Supergroup |

Codes :

| | A | B | C | D |
|---|---|---|---|---|
| (a) | 3 | 4 | 1 | 2 |
| (b) | 4 | 3 | 2 | 1 |
| (c) | 3 | 4 | 2 | 1 |
| (d) | 4 | 3 | 1 | 2 |

103. Match List I (Location) with List II (Mineral deposit) and select the correct answer using the codes given below the lists :

| List I | | List II | |
|---|---|---|---|
| A. | Sonapahar | 1. | Copper-lead |
| B. | Salem | 2. | Kyanite |
| C. | Sargipalli | 3. | Chromium |
| D. | Lapsaburu | 4. | Sillimanite |
| | | 5. | Magnesite |

Code :

| | A | B | C | D |
|---|---|---|---|---|
| (a) | 5 | 4 | 3 | 1 |
| (b) | 5 | 4 | 1 | 2 |
| (c) | 4 | 5 | 2 | 3 |
| (d) | 4 | 5 | 1 | 2 |

104. Which of the following are true for gold mineralisation in Kolar area?

1. Hypothermal
2. Telethermal
3. Localisation; structural and stratigraphic
4. En Echelon drag folds are characteristic
5. Auriferous quartz veins in hornblende schists
6. Charnockite, main host rock

Select the correct answer using the codes given below :

Codes :

(a) 1, 3, 4 and 5

(b) 2, 3, 4 and 6

(c) 1, 4, 5 and 6

(d) 2, 3, 4 and 5

105. Consider the following copper ore minerals :
 1. Tetrahedrite
 2. Malachite
 3. Chalcopyrite
 4. Cuprite

 The correct sequence of these minerals in order of *ascending* zones of secondary sulphide enrichment is:

 (a) 3, 1, 2, 4 (b) 3, 1, 4, 2
 (c) 1, 3, 4, 2 (d) 1, 3, 2, 4

106. Which one of the following is the correct meaning of the term 'blue dust'?
 (a) Coal ash
 (b) Weathered copper ore
 (c) Powdery iron ore
 (d) Dust generated during diamond cutting

107. Which one of the following raw materials is used in the manufacture of 'Sorel cement'?
 (a) Kyanite
 (b) Gypsum
 (c) Bauxite
 (d) Magnesite

108. To which one of the following ages do the oil-bearing rock formations of Cambay basin belong?
 (a) Cambrian — Ordovician
 (b) Jurassic — Triassic
 (c) Oligocene — Miocene
 (d) Cretaceous — Eocene

**Directions :** The following 12 (twelve) items consist of two statements, one labelled the 'Assertion A' and the other labelled the 'Reason R'. You are to examine these two statements carefully and decide if the Assertion A and the Reason R are individually true and if so, whether the Reason is a correct explanation of the Assertion. Select your answers to these items using the codes given below and mark your answer sheet accordingly.

Codes :
 (a) Both A and R are true and R is the correct explanation of A.
 (b) Both A and R are true but R is not a correct explanation of A.
 (c) A is true but R is false.
 (d) A is false but R is true.

109. Assertion (A) : The Miller index for a unit pyramid in the orthorhombic system is (111).

 Reason (R) : The unit pyramid in the orthorhombic system meets all the axes at the same distance.

110. Assertion (A) : Light is said to be polarized when the wave motion is confined to vibrations in a single plane perpendicular to the direction of propagation.

Reason (R) : Light is a wave motion with vibrations taking place in all the directions perpendicular to the direction of propagation.

111. Assertion (A) : Petroleum occurs in Gujarat, Assam and Bombay off-shore regions.

Reason (R) : Tertiary Formations occur in Assam, Gujarat and Bombay off-shore regions in an undisturbed state.

112. Assertion (A) : Quartz-feldspar (usually in 30:70 ratio) intergrowth is often found in granitic rocks and looks like ancient cuneiform writings (graphic intergrowth).

Reason (R) : The constant ratio and intergrowth of crystals suggests crystallisation of a melt of the two components in eutectic proportions, while the shape is controlled by the atomic structure of the minerals.

113. Assertion (A) : In a normal zoned plagioclase crystal, the core is more calcic than the rim.

Reason (R) : The more calcic the plagioclase, the higher the temperature of formation.

114. Assertion (A) : Garnet, feldspar, cordierite, andalusite, staurolite and chloritoid tend to form porphyroblasts, whereas quartz, carbonates and micas tend to be matrix-formers.

Reason (R) : Porphyroblasts are the result of the association of one mineral having small N/G with one or more minerals having a large N/G, where N (nucleation) = number of new crystals per unit time and G (growth) = increase in size per unit time.

115. Assertion (A) : In arid regions river valleys have steep sides.

Reason (R) : There is little moisture in the soil to assist downslope movement.

116. Assertion (A) : Drainage patterns are very useful in the interpretation of geomorphic features.

Reason (R) : Drainage patterns reflect the influence of initial slopes, inequalities in rock, hardness, structural controls and recent diastrophism.

117. Assertion (A) : A *uvala* is similar to a *polje* in its origin and extent.

Reason (R) : A *uvala* is formed by roof collapse and a *polje* is formed by solutional modifications of structurally disturbed zones.

118. Assertion (A) : The time classification is primarily dependent on the distribution of organisms and the first appearance or

extinction of the faunal and floral elements.

Reason (R) : Time divisions and time-rock divisions have the same names and only the suffix change denotes the difference.

119. Assertion (A) : From the Early Gondwana sedimentation there has been a perceptible change in the climate from glacial to warm and humid condition, represented by glacial tillite followed by coal bearing deposits.

Reason (R) : India was a part of the Gondwana Supercontinent in Late Carboniferous times, which was located close to the South Pole, but moved gradually northwards in the Permian.

120. Assertion (A) : The type of fossilisation depends on several factors and it is not necessary that only hard parts can get preserved during fossilisation.

Reason (R) : Animals with hard parts were much more abundant as compared to soft-bodied animals and therefore they had a greater chance of preservation.

## CIVIL SERVICES (PRELIMINARY) EXAMINATION - 1996 (KEY)

| | | | | | | | |
|---|---|---|---|---|---|---|---|
| 1. (b) | 2. (b) | 3. (d) | 4. (d) | 5. (d) | 6. (c) | 7. (a) | 8. (d) |
| 9. (b) | 10. (a) | 11. (d) | 12. (d) | 13. (a) | 14. (a) | 15. (b) | 16. (c) |
| 17. (b) | 18. (a) | 19. (d) | 20. (b) | 21. (b) | 22. (d) | 23. (d) | 24. (d) |
| 25. (d) | 26. (c) | 27. (d) | 28. (d) | 29. (b) | 30. (a) | 31. (a) | 32. (b) |
| 33. (a) | 34. (d) | 35. (d) | 36. (c) | 37. (b) | 38. (a) | 39. (c) | 40. (b) |
| 41. (a) | 42. (c) | 43. (a) | 44. (a) | 45. (a) | 46. (b) | 47. (a) | 48. (c) |
| 49. (a) | 50. (a) | 51. (d) | 52. (d) | 53. (a) | 54. (a) | 55. (d) | 56. (a) |
| 57. (c) | 58. (d) | 59. (c) | 60. (b) | 61. (c) | 62. (d) | 63. (b) | 64. (a) |
| 65. (c) | 66. (a) | 67. (a) | 68. (b) | 69. (c) | 70. (c) | 71. (a) | 72. (d) |
| 73. (c) | 74. (b) | 75. (b) | 76. (b) | 77. (c) | 78. (b) | 79. (a) | 80. (d) |
| 81. (a) | 82. (b) | 83. (a) | 84. (c) | 85. (d) | 86. (c) | 87. (c) | 88. (b) |
| 89. (a) | 90. (c) | 91. (a) | 92. (b) | 93. (b) | 94. (b) | 95. (c) | 96. (a) |
| 97. (c) | 98. (a) | 99. (a) | 100. (c) | 101. (a) | 102. (c) | 103. (a) | 104. (a) |
| 105. (d) | 106. (c) | 107. (d) | 108. (a) | 109. (b) | 110. (c) | 111. (c) | 112. (a) |
| 113. (b) | 114. (a) | 115. (b) | 116. (a) | 117. (b) | 118. (b) | 119. (a) | 120. (c) |

## 1. Civil Services (Preliminary) Examination, 1994

1. In advanced stage of migmatisation when the palaeosome and neosome can no longer be identified separately, the resulting texture is called
   a) nebulitic  b) stromatic  c) agmatitic  d) schlieren
2. The appearance of cordierite in regionally metamorphosed rocks is controlled by
   a) FeO/MgO ratio of the rock  b) $Al_2O_3$/MgO ratio of the rock
   c) $Al_2O_3$/FeO ratio of the rock  d) $Fe_2O_3$ content of the rock
3. According to Kelvin, the age of the Earth is
   a) 20 to 40 million years  b) 500 million years
   c) 1000 million years  d) 4500 million years
4. In Bikini and other similar atolls of the Pacific ocean, the 'Moho' lies at a depth of
   a) 40 km  b) 30 km  c) 20 km  d) 12 km
5. Match List I with List II and select the correct answer using the codes given below the lists :

| List I | List II |
|---|---|
| A. Gloup | 1. Waves |
| B. Blow-cut | 2. Plate tectonics |
| C. Cone of depression | 3. Wind |
| D. Triple junction | 4. Ground water |

Codes : a) A—1 B—3 C—4 D—2
b) A—1 B—3 C—2 D—4
c) A—3 B—1 C—2 D—4
d) A—3 B—1 C—4 D—2

6. The Earth's crust is the thickest at
   a) New Delhi  b) Dacca  c) Islamabad  d) Kathmandu
7. Match List I with List II and select the correct answer using the codes given below the lists :

| List I (*Plateau basalts*) | List II (*Age*) |
|---|---|
| A. Deccan Traps, India | 1. Miocene to Recent |
| B. Siberia, Russia | 2. Upper Cretaceous-Palaeocene |
| C. Columbia & Snake river, USA | 3. Early Jurassic |
| D. Lake Superior | 4. Precambrian |

Codes : a) A—2 B—3 C—1 D—4
b) A—2 B—1 C—3 D—4
c) A—1 B—3 C—4 D—2
d) A—1 B—4 C—3 D—2

8. Consider the following statements about island arcs
1. They have their convex sides directed towards the ocean.
2. They form a line of active or recently extinct volcanoes.
3. They coincide with a belt of positive gravity anomalies.
4. Deep-focus earthquakes occur beneath the continental side of the island arcs.

Of these statements
a) 2, 3 and 4 are correct b) 1, 2 and 3 are correct
c) 1, 2 and 4 are correct d) 1, 3 and 4 are correct

9. Consider the following shoreline features
1. Stacks 2. Bays and headlands 3. Sea arches

The correct sequence in which these features are formed by marine erosion is
a) 1, 3, 2 b) 2, 3, 1 c) 2, 1, 3 d) 1, 2, 3

10. Which one of the following drainage patterns is characterised by the tributaries flowing parallel to the main stream?
a) Dendritic b) Trellis
c) Both dendritic and trellis d) Radial

11. Which of the following are characteristic of the mature stage of fluvial cycle?
1. Maximum possible relief 2. Well integrated drainage
3. Broad stream divides

Select the correct answer using the codes given below
Codes : a) 1, 2 and 3 b) 1 and 2 c) 2 and 3 d) 1 and 3

12. The given figure is the topographical map of an area; the values of the contours are given in metres. Which one of the following geological structures is indicated by the topographical contours?

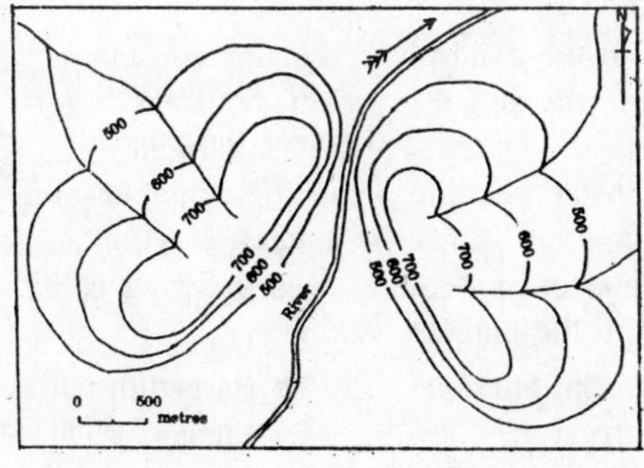

a) Plunging synform b) Nonplunging synform
c) Nonplunging antiform d) Plunging antiform

13. Some viscous lavas when poured out form mushroom-like features in the crater. These are called
a) plug domes b) cumulo-domes c) cinder cones d) dike ridges

14. Hums are
a) residual hills in an aeolian terrain
b) residual hills in a karst terrain
c) hillocks of residual soil
d) monadnocks in an alluvial terrain

15. If a stream is actively incising its valley and also eroding it laterally, thereby shifting its meander belt liberally, then which one of the following types of features will be formed in due course?
a) Paired terraces b) Non-paired terraces
c) Backswamp deposits d) Alluvial fans

16. When a body undergoes a permanent deformation without visible fractures due to the forces that are applied, then the deformation is known as
a) plastic b) elastic c) rupture d) clastic

17. A band of pelite exists between two psammitic bands of rocks which in turn have granite rocks on both sides. During folding
a) all the bands will appear to be similarly folded
b) the pelite will show more intense folding than the other two bands
c) the lower band will be folded most intensely
d) the upper band will be the most intensely folded

18. What is the relationship between the stratigraphic throw and the vertical throw of an inclined normal fault if the beds are horizontal?
a) Stratigraphic throw is greater than the vertical throw
b) Stratigraphic throw is smaller than the vertical throw
c) Both stratigraphic throw and the vertical throw are equal
d) There is no relationship between the two

19. An ideal concentric fold is the one in which the thickness of the beds measured
a) parallel to the axial plane remains constant
b) parallel to the isogons remains constant
c) normal to the bedding planes is variable
d) normal to the bedding planes remains constant

20. The given diagram shows the outcrops of folded beds, dips on the two limbs and cross bedding seen at two localities. The structure represented in the diagram is
a) an antiformal anticline b) a synformal syncline
c) an antiformal syncline d) a synformal anticline

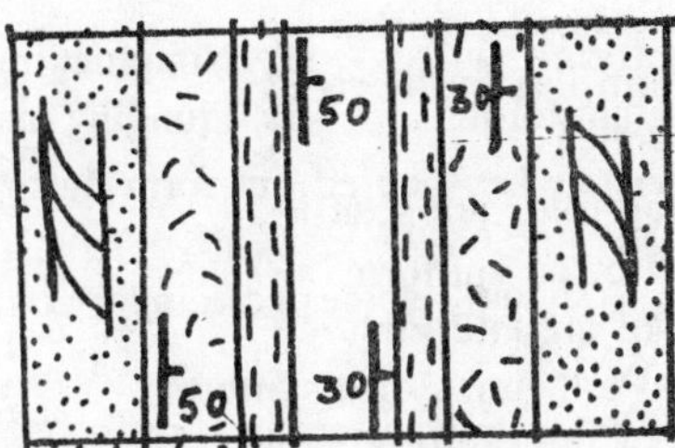


21. Supratenuous fold is a structure which is formed by
   a) compression in the upper layers of the crust
   b) gravitational sliding on the earth's surface
   c) magmatic intrusion
   d) differential compaction of sediments laid over an uneven surface
22. Which one of the following sets of features is likely to be produced during penecontemporaneous deformation process?
   a) Strain-slip cleavage, nonconformity, graded bedding
   b) Slump balls, overturned current bedding, flame structures
   c) Cut-and-fill structure, disconformity, ripple marks
   d) Thrusts, flute casts, slaty cleavage
23. Beds show folds with axial plane schistosity, striking NNE-SSW and vertical; the schistosity contains a mineral lineation which is parallel to the strike of the schistosity. The intersection lineation on the schistosity varies in orientation in all possible manner. This means that there is
   a) superposed folding
   b) heterogeneous shear on the schistosity
   c) flattening on the schistosity d) tangential longitudinal strain
24. Consider the following statements
   In the case of salt diapirs (salt domes), the movement upward of the salt may be influenced by
   1. fracturing and faulting of the cover
   2. yielding of the cover and flow of salt
   3. variation of density between the cover and salt
   4. solid flowage under the influence of gravitational forces

   Of these statements
   a) 1, 2 and 3 are correct b) 2, 3 and 4 are correct
   c) 1, 3 and 4 are correct d) 1, 2 and 4 are correct
25. The given figure shows the relationship between bedding and the axial plane schistosity in a cross-section. The relationship suggests

   Bedding

   Schistosity

   a) that a synform lies to the right

b) that a synform lies to the left
c) that an antiform lies to the right
d) none of the above

26. Right-handed drag folds are seen in
a) the right limb of an antiform
b) the left limb of an antiform
c) both right and left limbs of a synform
d) the left limb of the synform alone

27. The given figure illustrates an outcrop pattern on a flat ground. Which one of the following sets of events is arranged in the correct chronological order?

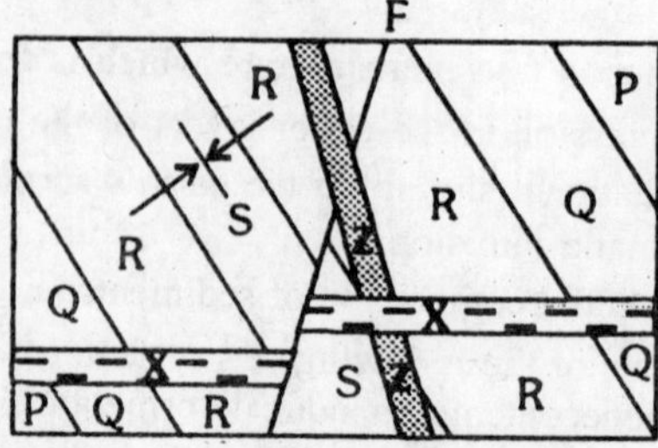


a) Intrusion X → faulting → folding → intrusion Z
b) Intrusion Z → folding → intrusion X → faulting
c) Folding → intrusion Z → faulting → intrusion X
d) Folding → intrusion X → faulting → intrusion Z

28. Abney Hand Level consists of which of the following?
1. A sighting tube  2. A graduated semicircular arc
3. A circular level  4. A freely swinging pendulum
Select the correct answer using the codes given below :
Codes : a) 1, 3 and 4 b) 1 and 3 c) 1 and 2 d) 2, 3 and 4

29. In which of the following crystal systems do all the crystallographic axes intersect at right angles?
1. Isometric 2. Hexagonal 3. Orthorhombic 4. Tetragonal
Select the correct answer using the codes given below :
Codes : a) 1 and 2 b) 1 and 3 c) 1, 3 and 4 d) 2, 3 and 4

30. A face with Miller indices (211) will intersect the a-axis at which one of the following values in relation to the fundamental distance?
a) Four times b) Three times c) Two times d) One-half times

31. Match List I with List II and select the correct answer using the codes given below the lists :

| List I (*Twinning*) | | List II (*Mineral*) |
|---|---|---|
| A. | 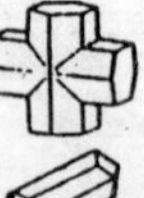 | 1. Staurolite |
| B. |  | 2. Rutile |
| C. | 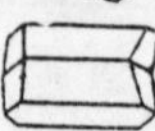 | 3. Plagioclase |

D. 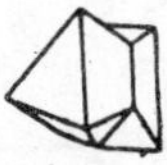 4. Spinel

Codes : a) A—1 B—2 C—3 D—4
b) A—1 B—3 C—4 D—2
c) A—2 B—1 C—3 D—4
d) A—2 B—1 C—4 D—3

32. Which of the following characters define an Octahedron belonging to the Isometric system?
1. Solid with equilateral triangular faces
2. Each face cuts all the three axes
3. 8-faced solid each face having indices {111}
Select the correct answer using the codes gives below
Codes : a) 1 and 2 b) 2 and 3 c) 1 and 3 d) 1, 2 and 3

33. Consider the following minerals
1. Hypersthene 2. Kyanite 3. Quartz 4. Anthophyllite
The correct sequence of the increasing order of the specific gravities of these minerals is
a) 1, 2, 3, 4 b) 1, 2, 4, 3 c) 3, 4, 1, 2 d) 3, 1, 4, 2

34. Consider the following statements :
Perthite is a product of
1. mixing of two feldspars 2. ex-solution
3. eutectic crystallisation 4. replacement phenomenon
Of these statements
a) 1 and 2 are correct b) 1, 2 and 3 are correct
c) 1, 3 and 4 are correct d) 2, 3 and 4 are correct

35. Match List I with List II and select the correct answer using the codes given below the lists :

| List I | List II |
|---|---|
| A. Pyroxene group | 1. Hyalophane |
| B. Amphibole group of minerals | 2. Oligoclase |
| C. Feldspar of $BaAl_2Si_2O_8$ composition | 3. Inosilicate |
| D. Feldspar of $Ab_{80}An_{20}$ composition | 4. Celsian |
| | 5. Chain silicate |

Codes : a) A—5 B—2 C—4 D—1
b) A—3 B—2 C—4 D—1
c) A—3 B—5 C—4 D—2
d) A—4 B—5 C—3 D—2

36. The given figure shows a plagioclase grain having
a) reverse zoning
b) oscillatory zoning

c) normal zoning

d) no zoning

37. Match List I with List II and select the correct answer using the codes given below the lists :

| List I (*Mineral*) | List II (*Chemical composition*) |
|---|---|
| A. Muscovite | 1. $K(Li_2AlSi_4O_{10})(OH)_2$ |
| B. Biotite | 2. $KMg_3(AlSi_3O_{10})(OH)_2$ |
| C. Phlogopite | 3. $K(MgFe)_3(AlSi_3O_{10})(OH)_2$ |
| D. Lepidolite | 4. $K(Al_2AlSi_3O_{10})(OH)_2$ |

Codes :
a) A—4 B—3 C—1 D—2
b) A—3 B—4 C—1 D—2
c) A—3 B—4 C—2 D—1
d) A—4 B—3 C—2 D—1

38. Match List I with List II and select the correct answer using the codes given below the lists :

| List I | List II |
|---|---|
| A. Garnet containing manganese & aluminium | 1. Almandine |
| B. Chrome-bearing garnet | 2. Spessartite |
| C. Lithium-bearing mica | 3. Spodumene |
| D. Lithium-bearing pyroxene | 4. Uvarovite |
| | 5. Zinnwaldite |

Codes :
a) A—2 B—1 C—5 D—3
b) A—3 B—2 C—1 D—5
c) A—3 B—4 C—1 D—5
d) A—2 B—4 C—5 D—3

39. Which one of the following is a set of physical and optical properties of more common amphiboles?

a) Colourless, H-4, first order polarisation colours, extinction angle greater than 20 degrees

b) Dark coloured, H-6, pleochroic, two sets of cleavage at 56 ∧ 124 degrees, extinction along the acute bisectrix of cleavages

c) Dark coloured, H-6, with perfect basal cleavage like micas, extinction angle greater than 20 degrees

d) Dark coloured, H-6, pleochroic, basal sections with two sets of cleavages at right angles, extinction angle greater than 30 degrees

40. Match List I with List II and select the correct answer using the codes given below the lists :

| List I | List II |
|---|---|
| A. Sillimanite | 1. Relative refringence |
| B. Becke test | 2. Isotropic |
| C. Uniaxial negative | 3. Parallel extinction |
| D. Basal section of uniaxial mineral | 4. Calcite |

Codes :

| | | | | |
|---|---|---|---|---|
| a) | A—3 | B—1 | C—4 | D—2 |
| b) | A—1 | B—3 | C—4 | D—2 |
| c) | A—3 | B—1 | C—2 | D—4 |
| d) | A—1 | B—4 | C—3 | D—2 |

41. Which of the following properties characterize isotropic substances?
   1. Light is transmitted with the same velocity in all directions.
   2. Wave surface is a sphere and the wave front a circle.
   3. R.I. varies with direction of transmission of light.
   4. They are singly refracting.

   Select the correct answer using the codes given below :
   Codes :
   a) 1, 2 and 4 b) 2 and 3
   c) 2 and 4 d) 1, 3 and 4

42. The common gangue minerals found to be associated with chromite deposits would include
   a) quartz-microcline b) oligoclase-orthoclase
   c) biotite-orthoclase d) olivine-serpentine

43. Consider the following minerals
   1. Chalcopyrite 2. Pyrite
   3. Galena 4. Bornite

   The correct paragenetic order in a hydrothermal cavity-filled deposit of these minerals is
   a) 2, 1, 3, 4 b) 1, 2, 3, 4
   c) 1, 2, 4, 3 d) 2, 1, 4, 3

44. Match List I with List II and select the correct answer using the codes given below the lists :

| List I *(Features)* | List II *(Deposits/Minerals)* |
|---|---|
| A. Fissures with parallel strike but intersecting dips in ore deposit | 1. Stockworks |
| B. Unsupported residual nuclei of a mineral or structure in ore mineral | 2. Chalcopyrite |
| C. Mineral developed by secondary enrichment process in copper deposit | 3. Replacement deposit |
| D. Interlacing network of ore-bearing veinlets of small dimensions (in a mass of rock), veinlets consisting of comb, crustification and druses | 4. Conjugated |
| | 5. Chalcocite |

Codes :

| | | | | |
|---|---|---|---|---|
| a) | A—3 | B—4 | C—5 | D—2 |

b) A—4 B—3 C—5 D—1
c) A—3 B—4 C—2 D—1
d) A—4 B—3 C—2 D—1

45. Consider the following processes

1. Evaporation
2. Oxidation
3. Leaching
4. Injection

The supergene enrichment zone is developed by

a) 1 and 4
b) 1 and 2
c) 2 and 3
d) 3 and 4

46. In a layered sequence of dolomite (bottom), shale (middle) and sandstone (top), the ideal site of hydrothermal ore localisation is

a) within the shale
b) within the sandstone
c) at the junction of dolomite and shale
d) at the junction of shale and sandstone

47 Which of the following elements form hypothermal ore deposits?

1. Mercury
2. Gold
3. Tin
4. Tungsten

Select the correct answer using the codes given below :

Codes :

a) 2, 3 and 4
b) 1, 2 and 3
c) 1, 2 and 4
d) 1, 3 and 4

48. The main host rock of lead-zinc mineralisation at Zawar, Rajasthan, is

a) phyllite of Aravalli group
b) dolomite of Delhi group
c) dolomite of Aravalli group
d) amphibolite of banded gneissic complex

49. Which one of the following minerals is used to manufacture special (SOREL) cement?

a) Gypsum
b) Kyanite
c) Magnesite
d) Mica

50. Books of muscovite mica are concentrated in which one of the following parts of pegmatites?

a) Core
b) Outer fringe
c) Inner zone
d) Unzoned pegmatites

51. Match List I with List II and select the correct answer using the codes given below the lists :

| List I (*Mineralization*) | List II (*Deposit*) |
|---|---|
| A. Sulphide mineralization in garnetiferous chlorite schist of Daling formation | 1. Mosabani |

| | |
|---|---|
| B. Galena-chalcopyrite-pyrite bodies in mica-schists and phyllites in the western part of Singhbhum-Gangpur shear zone | 2. Dariba-Rajpura |
| C. Lead-zinc sulphide bodies within graphitic schists along the contact between schists and tremolite marble in Bhilwara Group | 3. Sandur |
| D. Copper sulphide and sphalerite in schist belt in shield area | 4. Sargipalli |
| | 5. Rangpo |

Codes :

a) A—5 B—4 C—2 D—3
b) A—4 B—5 C—2 D—1
c) A—5 B—4 C—1 D—2
d) A—4 B—5 C—3 D—1

52. Match List I with List II and select the correct answer using the codes given below the lists :

| List I (*Deposit*) | List II (*Location*) |
|---|---|
| A. Lignite | 1. Donimalai |
| B. Iron ore | 2. Kanjamalai |
| C. Chromite | 3. Palana |
| D. Magnetite | 4. Sukinda |
| | 5. Malanjhkhand |

Codes :

a) A—3 B—1 C—4 D—2
b) A—5 B—2 C—1 D—4
c) A—3 B—1 C—5 D—2
d) A—5 B—2 C—4 D—3

53. Which of the following pairs are correctly matched?

1. Zinc — Anglesite
2. Lead — Cerussite
3. Copper — Covellite
4. Manganese — Braunite

Select the correct answer using the codes given below

Codes :

a) 1, 3 and 4
b) 2, 3 and 4
c) 1, 2 and 4
d) 1, 2 and 3

54. The maximum coal reserves of Gondwana period are contained in

a) Barakar of Lower Gondwana
b) Jabalpur of Upper Gondwana
c) Talchir of Lower Gondwana
d) Kamthi of Lower Gondwana

55. Which of the following conditions support the formation of coal?
    1. Heavy rainfall
    2. Low relief of hinterland
    3. High relief of hinterland
    4. Warm and humid climate

    Select the correct answer using the codes given below

    Codes :

    a) 1 and 2 b) 3 and 4 c) 1 and 4 d) 2 and 4
56. Trace fossils would include
    a) burrow, skin impression and trail
    b) burrow, boring and trail
    c) burrow, skin impression and boring
    d) skin impression, boring and trail
57. "What is not preserved as fossil is conserved after translation or transformation." Which one of the following serves as a good example of the above statement?
    a) Mould
    b) Petrified wood
    c) Coal
    d) Desiccated worm
58. Which one of the following groups of organisms is found to be highly useful in the correlation of fossiliferous rocks?
    a) Planktons
    b) Sessile benthos
    c) Vagrant-benthos
    d) Land-living organisms
59. The correct order of taxonomic sequence is
    a) phylum, order, family, class, genus, species
    b) phylum, family, class, order, genus, species
    c) phylum, class, order, family, genus, species
    d) phylum, order, class, family, genus, species
60. Consider the following events
    1. Great decline of graptolites and trilobites
    2. Abundance of planktonic foraminifera
    3. Abundance and extinction of Archaeocyathids
    4. Abundance and extinction of fusilinids

    a) 1, 2, 3, 4
    b) 2, 4, 1, 3
    c) 3, 1, 4, 2
    d) 3, 2, 1, 4
61. Which one of the following is mainly responsible for getting fossil brachiopods mostly in closed valve condition?
    a) Articulating teeth and sockets
    b) Shell symmetry
    c) Muscle system
    d) Tight-fit of the valves
62. Consider the following lamellibranch genera
    1. Pecten
    2. Exogyra
    3. Ostraea
    4. Alectryonia
    5. Spondylus and
    6. Gryphaea

    The morphological feature common to all these genera is that all of them

a) have equilateral shells b) are inequivalved
c) are monomyarians d) have dysodont dentition

63. Shells of early Palaeozoic lamellibranchs are not generally found preserved except in the form of moulds and casts. This is because
a) they lacked durable shells
b) their shells were mostly chitinous
c) their tests were mostly calcareous
d) their shells were aragonitic

64. Aristotle's lantern is
a) a masticatory apparatus in echinoids
b) an organ for locomotion in echinoids
c) used by echinoids for generating currents
d) used by echinoids as a defence mechanism against predators

65. Gastropods live in
a) marine shallow water only b) fresh water only
c) fresh and brackish waters
d) marine, fresh and brackish waters and on land

66. The shell of a cephalopod which is involute with a sharp venter is known as
a) sphaerocone b) serpenticone
c) oxycone d) phragmocone

67. Which one of the following plant genus is found both in Lower and Upper Gondwana formations?
a) Gondwanidium b) Williamsonia
c) Alethopteris d) Nilssonia

68. Which one of the following assemblages of genera is known to be confined to Nagri Formation of the Siwaliks?
a) Listriodon, Amphicyon, Tetrabelodon
b) Mastodon, Hipparian, Prostegodon
c) Aceratherium, Tetrabelodon, Telematodon
d) Camelus, Hyoboops, Leptobos

69. Not only the largest dinosaurs but also the largest known land animals found in the Jurassic and the Cretaceous rocks are called
a) Sauropods b) Erinthopods
c) Plesiosaurus d) Tyrannosaurus

70. Consider the following statements about Eohippus
1. It was the earliest and simplest form of horse found in the Eocene.
2. It had three toes in both front and hind feet.
3. It had the size of a sheep or a small pony.
4. It had bigger and complex teeth.

Of these statements

a) 2, 3 and 4 are correct b) 1 and 4 are correct
c) 2 and 3 are correct d) 1 alone is correct

71. Match List I with List II and select the correct answer using the codes given below the lists :

| List I (*Group*) | List II (*Formation*) |
|---|---|
| A. Papaghani | 1. Irlakonda quartzite |
| B. Cheyair | 2. Bairankonda quartzite |
| C. Nallamalai | 3. Pulivendla quartzite |
| D. Kistna | 4. Gulcheru quartzite |

Codes :

| | | | | |
|---|---|---|---|---|
| a) | A—4 | B—3 | C—2 | D—1 |
| b) | A—4 | B—2 | C—3 | D—1 |
| c) | A—2 | B—3 | C—1 | D—4 |
| d) | A—2 | B—1 | C—3 | D—4 |

72. The age of the oldest pelagic sediments on the present sea floor is

a) Late Jurassic b) Early Pleistocene
c) Eocambrian d) Ordovician

73. Isolated deposits of inland seas of Precambrian era are found in Western Rajasthan giving rise to

a) shales b) limestones
c) phosphorites d) bauxite

74. Consider the following statements
Devonian period is marked by

1. septate corals such as Phillipsastrea
2. dominance of fishes
3. land flora with a turf-like growth of Psilophyton and Rhynia
4. primitive mammals

Of these statements

a) 1, 2 and 4 are correct b) 1, 3 and 4 are correct
c) 2, 3 and 4 are correct d) 1, 2 and 3 are correct

75. The only formation of Gondwana that contains primary carbonate is represented by

a) Kota Formation b) Raniganj Formation
c) Jabalpur Formation d) Parsora Formation

76. Which of the following statements regarding the Deccan Traps and associated sedimentary formations are correct?

1. Lameta beds occur below the traps.
2. Few intertrappeans have yielded Clypeaster, Venus and the molluscs.

3. Intertrappean beds near Bombay have yielded fossil frogs.
4. The maximum thickness of the traps is about 2000 metres.

Select the correct answer using the codes given below

Codes :

a) 2, 3 and 4 b) 1, 2 and 3
c) 1, 3 and 4 d) 2 and 4

77. Epeiric seas were likely to have persisted during the sedimentation of

a) Vindhyans of Son valley b) Tertiaries of Assam
c) Jurassic of Kutch d) Cretaceous of South India

78. Which of the following formations belong to Mio-pliocene and are correlated with each other?

1. Cuddalore, Warkalli and Rajamahendry sandstone
2. Murree, Pegu and Manchhar
3. Lilang, Kuling and Irrawaddy
4. Siwalik, Irrawaddy and Manchhar

Select the correct answer using the codes given below

Codes :

a) 1, 2 and 3 b) 2, 3 and 4
c) 1 and 4 d) 4 alone

79. The commonly accepted view with regard to the age of the Tal Formation of Lower Himalaya is

a) Permian b) Late Devonian
c) Early Cambrian d) Jurassic to Cretaceous

80. Dharwar Super Group in Karnataka includes rocks showing pillow structure. What does it indicate?

a) Rocks were igneous in origin
b) Submarine origin of the volcanic lava flows
c) Rocks were sedimentary in origin
d) None of the above

81. Consider the following statements

The Dharwar succession of Karnataka is divided into Lower and Upper on the basis of

1. fossil evidence 2. geochronological data
3. lithostratigraphy 4. unconformity

Of these statements

a) 1 and 2 are correct b) 3 and 4 are correct
c) 1 and 3 are correct d) 2 and 4 are correct

82. The two diamond bearing conglomerate horizons of Central India are associated with

1. Panna shale 2. Ganurgarh shale

3. Upper Rewa sandstone
4. Jhiri shale
5. Dhandraul quartzite

The correct stratigraphic position of these horizons is

a) between 1 and 5 and 2 and 3
b) between 1 and 4 and 3 and 4
c) between 2 and 5 and 1 and 3
d) between 1 and 3 and 4 and 5

83. The following are the names of zones and beds taken at random from the Triassic succession of Spiti

1. Daonella limestone
2. Otoceras zone
3. Megalodon limestone
4. Hedenstroemia bed
5. Monotis shale

The correct order according to their stratigraphic positions is

a) 2, 4, 1, 5, 3
b) 1, 2, 4, 5, 3
c) 2, 4, 3, 5, 1
d) 1, 2, 4, 3, 5

84. The first ice advancement in Kashmir took place during

a) Dhok Pathan Formation
b) Pinjor Formation
c) Boulder conglomerate Formation
d) Tatrot Formation

Directions : The following eleven items consist of two statements, one labelled the 'Assertion' and the other labelled the 'Reason'. You are to examine these two statements carefully and decide if the Assertion and the Reason are individually true and if so, whether the Reason is a correct explanation of the Assertion. Select your answers to these items using the codes given below and mark your answer sheet accordingly.

CODES :

a) Both Assertion and Reason are true and Reason is the correct explanation of Assertion.
b) Both Assertion and Reason are true but Reason is NOT a correct explanation of Assertion.
c) Assertion is true but Reason is false.
d) Assertion is false but Reason is true.

85. Assertion : Dinosaurs and a few other reptiles became extinct by the end of the Cretaceous.
Reason : Mammals evolved rapidly during the Tertiary Era.

86. Assertion : Appreciable oil reserves are found in the Digboi area of Assam.
Reason : In Digboi area, the oil and gas traps are salt domes of Permocarboniferous age.

87. Assertion : Kyanite has a hardness of 4 to 5 parallel to the elongation of the crystal, and a hardness of 6 to 7 normal to the elongation.

Reason : Moh's scale of hardness is based on the relative resistance of minerals to scratches on their surfaces.

88. Assertion : The cleavage of halite may be termed as (100) or cubic and halite is also described as a mineral having three directions of cleavage at 90°.

Reason : The angular relationship between cleavage planes is constant in any given mineral in as much as cleavage planes always parallel an actual or possible crystal face and are therefore subject to the law of constancy of interfacial angles.

89. Assertion : Crenulation cleavage is found only when puckers are present in the rock.

Reason : Crenulation cleavage planes separate microlithons within which folded planar fabric is present.

90. Assertion : The strike of the axial plane of an inclined, plunging isoclinal fold is parallel to the direction of plunge of the fold axis.

Reason : The strikes of the two limbs of such a fold are parallel.

91. Assertion : The landforms of arid regions display greater angularity.

Reason : The downslope movement of material is insignificant in arid regions.

92. Assertion : The basaltic crust of the ocean floors is characterised by 'zebra-like' stripes of magnetic anomalies.

Reason : These magnetic anomalies are due to the magnetic declination.

93. Assertion : The abundance of feldspars in arkoses suggests their deposition under warm and humid climatic conditions.

Reason : Decomposition of feldspars is inhibited under arid or very cold climate.

94. Assertion : The size of fine sedimentary particles can be determined on the basis of their settling velocity.

Reason : According to Stoke's law, settling velocity varies with the specific gravity of the particles.

95. Assertion : In general, magmas are about 10% lighter than their equivalent solidified rock at the same temperature. Hence minerals are more likely to sink in a crystallising magma.

Reason : A process in which continuous plating of crystals from an ascending magma on to the walls of the magma conduit takes place is called "flow crystallisation" which is responsible for the absence of cumulates in layered complexes.

96. Match List I with List II and select the correct answer using the

codes given below the lists :

| List I (*Features in volcanic rocks*) | List II (*Texture*) |
|---|---|
| A. A felt-like interweaving of feldspar microlites | 1. Intergranular |
| B. Triangular or polygonal areas between feldspar laths are filled with glass or its alteration product or cryptocrystalline material | 2. Hyalopilitic |
| C. The interspaces between feldspar laths are filled with granules of pyroxene, olivine, and in some, magnetite | 3. Interserial |
| D. A felt of crystals is set in a residuum of glass | 4. Pilotaxitic |

Codes :

a) A—4 B—3 C—1 D—2
b) A—3 B—4 C—2 D—1
c) A—4 B—3 C—2 D—1
d) A—3 B—4 C—1 D—2

97. Which of the following statements are correct?
1. Phonolite is the volcanic equivalent of nepheline syenite.
2. Rhyolite is the volcanic equivalent of granite.
3. Trachyte is the volcanic equivalent of gabbro.
4. Andesite is the volcanic equivalent of diorite.

Codes :

a) 1, 3 and 4 b) 1, 2 and 3
c) 1, 2 and 4 d) 2, 3 and 4

98. Troctolite is a variety of gabbro containing essentially
a) olivine, orthopyroxene and labradorite-bytownite plagioclase
b) orthopyroxene and labradorite-bytownite plagioclase
c) olivine and orthopyroxene
d) olivine and labradorite-bytownite plagioclase

99. Consider the following statements

Spilites are
1. soda rich basalts 2. potash rich basalts
3. usually associated with quartz keratophyres
4. usually associated with andesites

Of these statements

a) 1 and 3 are correct b) 1 and 4 are correct
c) 2 and 3 are correct d) 2 and 4 are correct

100. Consider the following rocks

1. Granite 2. Basalt 3. Anorthosite 4. Diorite

The correct sequence of the decreasing order of the anorthite content in plagioclase in these rocks is

a) 3, 2, 4, 1 b) 2, 3, 4, 1 c) 3, 2, 1, 4 d) 2, 3, 1, 4

101. The combination of leucite and olivine is stable at

a) high temperature and high pressure
b) low temperature and low pressure
c) high temperature and low pressure
d) low temperature and high pressure

102. The equilibrium diagram of the system Albite-Anorthite is shown in the following figure. If the water pressure is increased the freezing point of

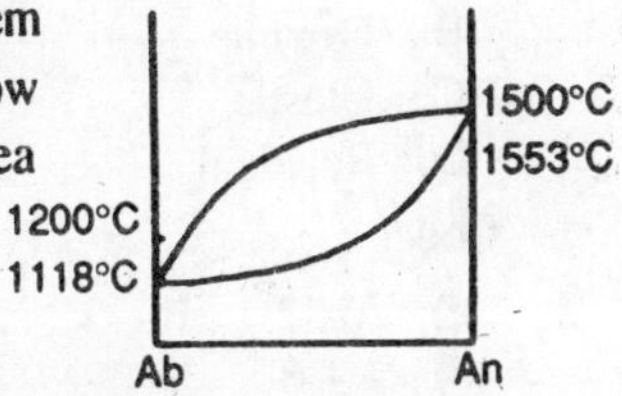

a) albite alone will be lowered
b) anorthite alone will be lowered
c) both albite and anorthite will be lowered
d) both albite and anorthite will be raised

103. Eutectic crystallisation takes place when there is

a) no miscibility of the solid phases
b) complete miscibility of the solid phases
c) partial miscibility between the solid and the liquid phase

104. Consider the following statements

Viscosity of magma increases with

1. increase in temperature 2. increase in silica content
3. decrease in temperature 4. decrease in silica content

Of these statements

a) 1 and 2 are correct b) 2 and 4 are correct
c) 3 and 4 are correct d) 2 and 3 are correct

105. Which of the following features characterise the island-arc magmas?

1. High level magma chambers and high volatile contents
2. Highly porphyritic nature except for those of tholeiitic series
3. Low level magma chambers and low volatile contents
4. Caldera is never produced
5. All major ferromagnesian minerals tend to be relatively Mg-rich
6. Highly calcic phenocryst phase display oscillatory zoning

Select the correct answer using the codes given below :

Codes :

a) 1, 3, 4 and 5 b) 1, 2, 5 and 6
c) 2, 3, 4 and 6 d) 1, 2, 4 and 5

106. The formation of large volumes of andesitic magma is attributed to
   a) mixing of basaltic and rhyolitic magma
   b) partial melting of upper mantle rocks under high pressure
   c) fractional crystallisation of basaltic magma
   d) contamination of basaltic magma by assimilation of crustal rocks

107. Match List I with List II and select the correct answer using the codes given below the lists :

| List I (*Origin*) | List II (*Sediment*) |
|---|---|
| A. Chemical | 1. Graywacke |
| B. Organic | 2. Terra Rosa |
| C. Clastic | 3. Stalactite |
| D. Residual | 4. Guano |

Codes :

| | | | | |
|---|---|---|---|---|
| a) | A—2 | B—3 | C—4 | D—1 |
| b) | A—3 | B—2 | C—1 | D—4 |
| c) | A—2 | B—3 | C—1 | D—4 |
| d) | A—3 | B—4 | C—1 | D—2 |

108. In modern oceans, the removal of dissolved silica is principally carried out by
   a) radiolaria
   b) silicoflagellates
   c) diatoms
   d) sponges

109. The contact between products of normal and catastrophic sedimentation is likely to be
   a) gradational
   b) abrupt but conformable
   c) intercalated
   d) abrupt, but unconformable

110. Consider the following geometries of the sand bodies
   1. Blanket
   2. Tabular
   3. Prism
   4. Shoestring

   The correct sequence of the geometry of these sand bodies, based on decreasing order of the width-thickness ratio is
   a) 2, 3, 4, 1
   b) 1, 2, 3, 4
   c) 1, 3, 2, 4
   d) 2, 4, 3, 1

111. Which of the factors given below favour the development of authigenic overgrowths during cementation?
   1. Precipitation of cement in optical continuity with the other crystal
   2. Both the crystals having same mineral composition
   3. Mostly, neomorphism of the primary matrix
   4. Commonly, developme t in relation to single crystal grains

   Select the correct answer using the codes given below

   Codes :

a) 1, 2 and 4 b) 2 and 3
c) 2 and 4 d) 1, 3 and 4

112. The observation that softer minerals of a clastic sediment show angularity whereas the harder minerals are well rounded is indicative of
a) dump deposition b) hydraulic jump
c) single source d) multiple sources

113. Which of the following geological processes are responsible for the abundance of matrix in greywackes?
1. Deposition in deep marine environment
2. Deposition from a fraction load
3. Diagenetic alteration of sand-sized grains
4. Compaction of pelitic fragments

Select the correct answer using the codes given below

Codes :
a) 1 and 3 b) 1 and 4
c) 2 and 3 d) 3 and 4

114. A limestone deposit formed by sedentary organisms in a stratiform manner enclosed in rocks of different lithologic character is known as
a) bioherm b) biostrome
c) pelagic limestone d) calcarenite

115. In retrogressive metamorphism, mineralogical changes take place due to
a) readjustment of high grade metamorphic minerals to falling temperature
b) increase in temperature and pressure due to invading magma
c) extreme pressure-temperature conditions at great depth
d) recrystallisation at high temperature and low pressure

116. Kyanite is a characteristic mineral phase of argillaceous sediments which have undergone
a) high pressure metamorphism
b) high temperature metamorphism
c) low pressure metamorphism
d) low temperature metamorphism

117. With increasing metamorphism, staurolite normally forms only in rocks containing
a) garnet b) chloritoid
c) biotite d) prehnite

118. The metasomatic alteration of a granite with the help of superheated steam and fluorine results in the formation of
a) muscovite granite b) greisen

c) tourmaline granite d) schorl rock

119. Which one of the following mineral assemblages is expected at the contact of marble and amphibolite?
   a) Diopside - grossularite - idocrase
   b) Andalusite - cordierite - almandine
   c) Quartz - calcite - tremolite
   d) Grossularite - plagioclase - wollastonite

120. Crystalloblastic series is best defined as the arrangement of metamorphic minerals
   a) in order of increasing grade
   b) according to size
   c) in order of decreasing force of crystallisation
   d) in order of increasing force of crystallisation

## ANSWERS

| | | | | | |
|---|---|---|---|---|---|
| 1. d | 2. b | 3. a | 4. d | 5. d | 6. d |
| 7. b | 8. c | 9. b | 10. b | 11. b | 12. d |
| 13. b | 14. b | 15. c | 16. a | 17. b | 18. c |
| 19. a | 20. d | 21. d | 22. b | 23. b | 24. c |
| 25. a | 26. c | 27. d | 28. c | 29. c | 30. c |
| 31. a | 32. d | 33. c | 34. b | 35. c | 36. b |
| 37. d | 38. d | 39. b | 40. a | 41. a | 42. d |
| 43. d | 44. d | 45. c | 46. a | 47. d | 48. c |
| 49. c | 50. c | 51. a | 52. a | 53. b | 54. a |
| 55. c | 56. b | 57. b | 58. b | 59. c | 60. a |
| 61. a | 62. b | 63. b | 64. a | 65. d | 66. b |
| 67. a | 68. b | 69. a | 70. d | 71. a | 72. a |
| 73. c | 74. d | 75. c | 76. c | 77. c | 78. c |
| 79. d | 80. b | 81. d | 82. a | 83. a | 84. c |
| 85. b | 86. c | 87. b | 88. a | 89. a | 90. a |
| 91. a | 92. a | 93. a | 94. a | 95. b | 96. a |
| 97. c | 98. b | 99. b | 100. a | 101. c | 102. d |
| 103. b | 104. d | 105. b | 106. d | 107. d | 108. c |
| 109. d | 110. b | 111. d | 112. a | 113. b | 114. b |
| 115. a | 116. a | 117. b | 118. b | 119. c | 120. b |

## 2. Civil Services (Preliminary) Examination, 1993

1. Which one of the following criteria is MOST suitable for determining the internal constitution of the earth?
   a) Composition of Volcanic emanations
   b) Composition of meteorites
   c) Refracted seismic waves d) Reflected seismic waves
2. Match List I with List II and select the the correct answer using the codes given below the lists

| List I | List II |
|---|---|
| A. Zone of divergence | 1. Melange |
| B. Zone of convergence | 2. Deep-sea trench |
| C. Transform fault | 3. Mid-oceanic ridges |
| D. Ocean-Continent collision | 4. Hot spot |
| | 5. Shallow focus earthquakes |

Codes :

| | | | | |
|---|---|---|---|---|
| a) | A—2 | B—3 | C—4 | D—1 |
| b) | A—3 | B—2 | C—5 | D—1 |
| c) | A—5 | B—3 | C—4 | D—2 |
| d) | A—3 | B—4 | C—2 | D—5 |

3. Uranium-238 is more abundant than uranium-235 in the earth because
   a) Uranium-235 has a shorter half-life than uranium-238
   b) Uranium-238 does not undergo fission while uranium-235 does
   c) Uranium-235 was present in low amounts at the time of the evolution of the solar system
   d) Uranium-235 reacts more readily with the radiation from cosmic rays
4. The deep foci earthquakes are practically restricted to the
   a) Himalayan belt b) Mediterranean-Alpine belt
   c) Mid-oceanic ridge of the atlantic
   d) Deep-sea trench areas of the Pacific
5. Geosynclines with initial infilling of volcanic activity is described as
   a) Taphrogeosyncline b) Zeugogeosyncline
   c) Eugeosyncline d) Miogeosyncline
6. Which one of the following is the correct sequence of events leading to volcanic activity of the central type?
   a) Formation of magma — overheating of the crust — release of dissolved gases — injection of feeders — explosive eruption
   b) Overheating of the crust — formation of magma — injection of feeders — release of dissolved gases — explosive eruption
   c) Release of dissolved gases — formation of magma — overheating

of the crust — injection of feeders — explosive eruption

d) Injection of feeders — overheating of the crust — formation of magma — release of dissolved gases — explosive eruption

7. Which among the following is the correct sequence of minerals of increasing stability resulting from continued weathering of rocks?
   a) Biotite, K-feldspar, muscovite, quartz
   b) Olivine, muscovite, hornblende, biotite
   c) Augite, quartz, biotite, muscovite
   d) Hornblende, muscovite, K-feldspar, quartz

8. Match List I with List II and select the correct answer using the codes given below the lists :

| List I | List II |
|---|---|
| A. Bars | 1. Wave erosion |
| B. Natural arches | 2. Glaciated coastline |
| C. Coastal equilibrium | 3. Blowholes |
| D. Fjords | 4. Longshore drift |
| | 5. Straight coastline |

Codes :

| | | | | |
|---|---|---|---|---|
| a) | A—4 | B—1 | C—5 | D—2 |
| b) | A—1 | B—3 | C—5 | D—4 |
| c) | A—3 | B—1 | C—2 | D—5 |
| d) | A—2 | B—5 | C—4 | D—1 |

9. What is the correct sequence in which the following features develop?
   1. Karst Window
   2. Natural Tunnel
   3. Natural Bridge.

   Select the correct answer from the codes given below :

   Codes :

   a) 1, 2, 3 b) 2, 1, 3 c) 3, 2, 1 d) 3, 1, 2

10. The following geomorphic agencies move different materials by different methods
    1. Running water
    2. Waves and currents .
    3. Wind
    4. Glacier

    "Traction" is present in respect of

    a) 1 and 3
    b) 2, 3 and 4
    c) 2 and 4
    d) 1, 2, 3 and 4

11. Physiological drought, in which moisture is NOT available to the plants, occurs in
    a) Polar deserts
    b) Topographic deserts
    c) Tropical deserts
    d) Temperature deserts

12. Which one of the following drainage patterns is illustrated by the above figure?

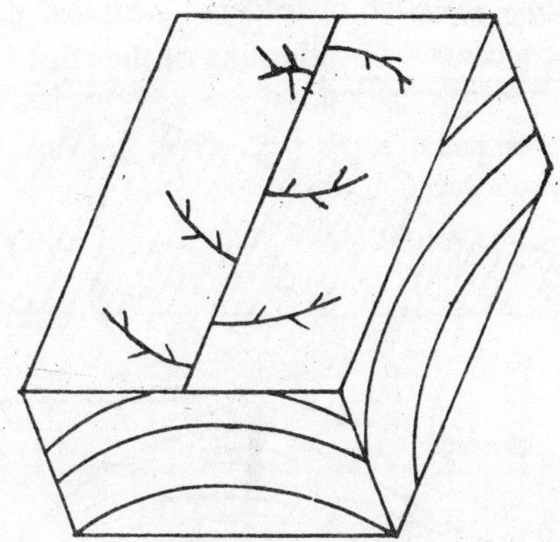

a) Trellis b) Radial c) Dendritic d) Annular

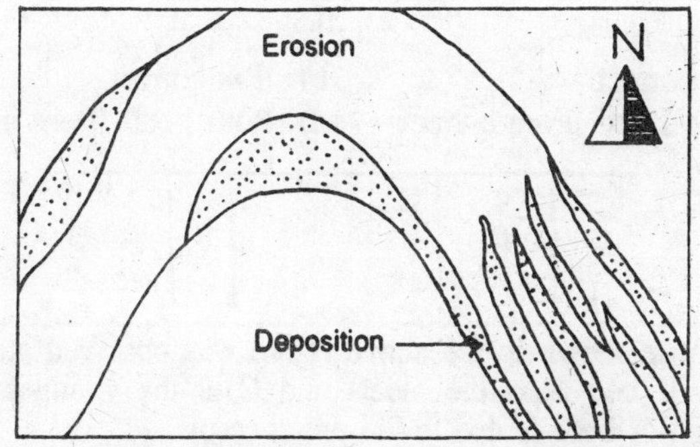


13. As shown in the above figure, the stream erodes its bank on the curve and deposits same material at some distance to assume a braided pattern. The direction of flow of water will be
    a) N to S
    b) NE to SW
    c) E to W
    d) NW to SE
14. Which of the following are saline takes?
    1. Chilka lake
    2. Sambhar lake
    3. Pulicat lake.

    Select the correct answer using the codes given below :

    Codes :

    a) 1 and 2 b) 1 and 3 c) 2 and 3 d) 1, 2 and 3
15. Horns are residual landforms produced by the action of
    a) Streams
    b) Wind
    c) Glaciers
    d) Sub-surface water
16. In the diagrams of a sand dune (I) and a drumlin (II), shown above the arrows indicate the directions of movement of geological agent.

Of these two diagrams

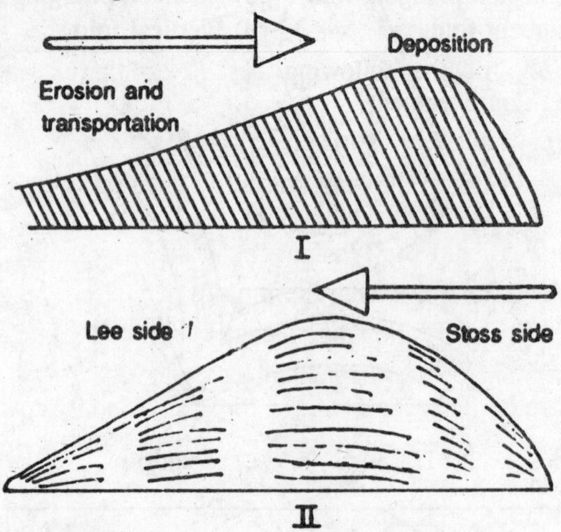


a) I is correct  b) II is correct
c) Both I and II are correct  d) Both I and II are incorrect

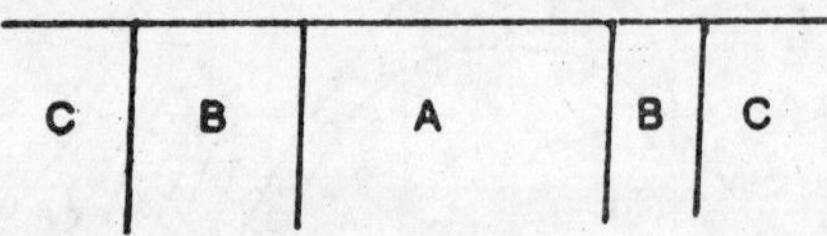


17. The profile shown in the above figure was observed along a road cutting. If bed A is the oldest and C is the youngest, then the repetition of beds is due to the presence of
    a) An anticline  b) A syncline
    c) A normal fault  d) A reverse fault

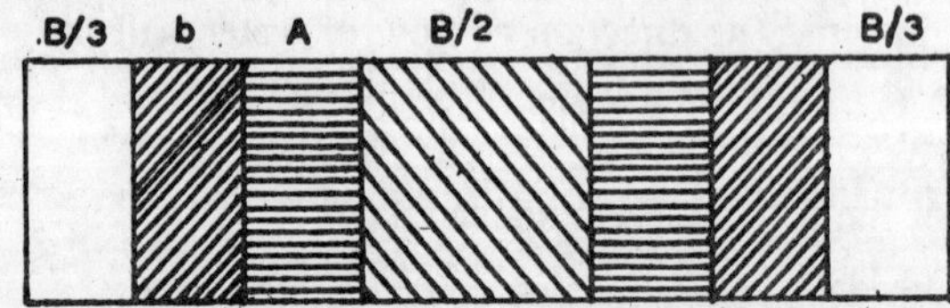


18. The above figure illustrates an outcrop pattern on flat topography. Macrocephalites in Bed B/3, Crioceras in Bed A, Peltoceras in Bed B and Scaphites in Bed B/2 are found. Repetition of beds in the map is
    a) Because of anticlinal folding
    b) Not explainable from the available data
    c) Because of synclinal folding
    d) Because of faulting
19. The 100 m, 2090 m and 300 m stratum contours of a folded surface when projected on the map appear as a single curved line. This type

of structural contour map is typical of

a) upright non-plunging fold b) inclined plunging fold
c) recumbent fold d) vertical fold

20. Which one of the following best explains the nature of the Narmada-Son lineament?

a) Normal fault b) Thrust c) Wrench fault d) Shear zone

21. Which one of the following crustal dynamics is best suited for the formation of a normal fault?

a) Compression b) Tension c) Lateral spreading
d) Down-slope movement due to gravity

22. Which one of the following characterises a convolute bedding?

a) Crumpling of the lamination increases downward initially and gradually dies out
b) The affected bed shows external irregularities in its thickness
c) Laminations are not ruptured and can be traced across several undulations
d) Laminations are ruptured and cannot be traced across adjacent unduiations.

23. Which one of the following statements is correct?

a) The lower surface of lava flows is vesicular, ropy, spiny and weathered
b) Pyroclastic deposits grade downward into laminated and finer materials
c) The flat part marks the lower surface of incomplete amygdule
d) The mafic minerals are coarser in the lower part and the feldspathic minerals are coarser in the upper part of a sill

24. A small outcrop of Gondwana rocks surrounded by Archaean rocks is observed in an area. The correct interpretation of the outcrop is

a) Window b) Klippe c) Inlier d) Outlier

25. Which one of the following is the most important characteristic of an unconformity?

a) Truncation of sequences b) Time-gap
c) Presence of conglomerate d) Tectonic break

26. The R.F. of a given map of 2 cm to 1 km is

a) 1 : 50000 b) 1 : 5000 c) 1 : 500 d) 1 : 100000

27. Consider the following pairs :

| *Factors and situations* | *Methods of mapping* |
|---|---|
| 1. Large-scale topographical map given. Time is not an important consideration. Most accurate result desired. | Exposure mapping |
| 2. Small-scale map given. Limited time. Highly inclined bedded rocks. Exploratory surveys. | Contact mapping |
| 3 Small-scale map given. Less detailed | Traverse mapping |

work. Recomnaissance studies.

Of these pairs

a) 1, 2 and 3 are correctly matched
b) 1 and 2 are correctly matched
c) 1 alone is correctly matched
d) 2 and 3 are correctly matched

28. Match List I and List II and select the correct answer using the codes given below the lists :

| List I<br>(*Forms : Monoclinic System*) | List II<br>(*Name*) |
|---|---|
| A. An open form {001} of two faces | 1. Prism |
| B. An open form {010} of two faces | 2. Orthopinacoid |
| C. An open form {100} of two faces | 3. Hemi-Bipyramids |
| D. An open form of four faces, the indices of the unit form are {110} | 4. Basal pinacoid |
| | 5. Clino-pinacoid |

Codes :

a) A—4 B—5 C—2 D—1
b) A—5 B—4 C—1 D—3
c) A—3 B—1 C—2 D—4
d) A—2 B—3 C—4 D—5

29. Match List I and List II and select the correct answer using the codes given below the lists :

| List I<br>(*Type of twins*) | List II<br>(*Minerals*) |
|---|---|
| A. Geniculate twin | 1. Gypsum |
| B. Maltese cross-twin | 2. Microcline |
| C. Pericline twin | 4. Staurolite |
| D. Swall-tail twin | 5. Ilmenite |

Codes :

a) A—4 B—2 C—3 D—5
b) A—3 B—4 C—2 D—1
c) A—2 B—5 4—2 D—3
d) A—1 B—3 C—2 D—4

30. In the crystals of orthorhombic system, which one of the following terms is used for a face that makes intercepts on the vertical axis and one of the horizontal axes?

a) Dome b) Pyramid c) Prism d) Pinacoid

31. The plagioclase series regularly varying in composition from Na-rich to Ca-rich members can be arranged as

a) Albite-andesine-oligoclase-labradorite-bytownite-anorthite
b) Albite-oligoclase-labradorite-andesine-bytownite-anorthite
c) Albite-oligoclase-andesine-labradorite-bytownite-anorthite
d) Albite-oligoclase-andesine-bytownite-labradorite-anorthite

32. With increasing An content, which of the following physical-optical properties increase in plagioclase feldspars?

1. Specific gravity
2. Cleavage angle 010 ∧ 001
3. Refractive indices
4. Birefringence
5. Optic axial angle.

Select the correct answer using the codes given below :

Codes :

a) 1, 2 and 4
b) 1 and 3
c) 1, 3 and 5
d) 2, 3, 4 & 5

33. Which one of the following pairs of garnets does NOT show solid solution?

a) Pyrope-Spessartite
b) Almandine-Pyrope
c) Almandine-Spessartite
d) Grossularite-Andradite

34. If in a uniaxial –ve crystal, light travels normal to the optic axis, the extraordinary-ray will vibrate parallel to

a) a-axis
b) b-axis
c) c-axis
d) Normal to principal section

35. To measure accurately the retardation of an anisotropic mineral, the accessory used is

a) Gypsum plate
b) Berek compensator
c) Quartz wedge
d) Optical goniometer

36. Consider the following statements regarding the colours of a mineral :

They depend upon the

1. Orientation of the mineral
2. Thickness of the mineral slide
3. Birefringence of the mineral
4. Absorption of the mineral.

Of these statements

a) 1 and 2 are correct
b) 2 and 3 are correct
c) 1, 2 and 3 are correct
d) 1, 3 and 4 are correct

37. Consider the following characteristics :

1. High relief
2. Pale yellow to brown pleochroism
3. Parallel extinction
4. Weak birefringence.

Which one of the following minerals has the above-mentioned characteristics?

a) Biotite
b) Hornblende
c) Olivine
d) Staurolite

38. Consider the following minerals :

1. Olivine
2. Pyrrhotite
3. Chromite

The correct paragenetic sequence of these minerals in an early magmatic deposit is

a) 1, 2, 3 b) 1, 3, 2 c) 3, 2, 1 d) 2, 3, 1

39. Oxidation of a Pb-Zn-Pyrite deposit will result in the
   a) Formation of placer deposit of pyrite
   b) Liberation of metals like Pb, Zn, and Fe
   c) Formation of residual deposit of zincite
   d) Formation of the deposit containing sphalerite and galena
40. When a portion of the country rock is traversed by a number of very closely spaced and mutually interlinked cracks, mineralisation may occur conveniently along the available openings leading to the development of mineral deposits known as
   a) Stockworks b) Ladder veins
   c) Saddle reefs d) Gash veins
41. Which one of the following types of deposits is formed by high-temperature fluids, in low pressure environment and a shallow depth?
   a) Telethermal b) Xenothermal
   c) Hypothermal d) Epithermal
42. Among the following sulphide ores of copper, the copper content is more in
   a) Chalcopyrite b) Chalcocite
   c) Covellite d) Bornite
43. Prospecting for minerals in sea-beds is for polymetallic deposits of
   a) Sn, W, Ni b) Ag, Zn, Cu
   c) Cu, Zn, Ni d) Cu, Ag, Ni
44. Sulphide mineralisation in Delhi Supergroup of rocks is localised along the
   a) Primary stratification planes
   b) Shear zones c) NE-SW tectonic lineament
   d) axial trace of third generation of folds ($F_3$)
45. Which one of the following lithologic assemblages best represents the host rocks of uranium mineralization along the Singhbhum shear zone in Bihar?
   a) Quartzite-chlorite schist b) Chlorite schist-biotite schist
   c) Felspathic schist-chlorite schist
   d) Felspathic schist-biotite schist
46. The principal byproduct metals in the ores of Zawar mines include
   a) Mo, Cd and Ag b) Cd, Ag and W
   c) Ag, W and Au d) W, Au and Co
47. Which one of the following metal associations is found in the Sargipalli deposits of Orissa?
   a) Lead and zinc b) Lead and copper
   c) Zinc and copper d) Lead, zinc and copper

48. Arrange the following lithounits of the Indian residual bauxite deposits in the correct sequence from top to bottom :

1. Ferruginous laterite
2. Lithomarge
3. Ferruginous bauxite
4. Bauxite

Select the correct answer using the codes given below :

Codes :

a) 1, 2, 3, 4 b) 2, 3, 4, 1 c) 4, 2, 3, 1 d) 1, 3, 4, 2

49. Which one of the following processes has been responsible for the occurrence of diamond in the Vindhyan Supergroup at Rewa, M.P.?

a) Magmatic segregation
b) Metamorphic segregation
c) Hydrothermal concentration
d) Residual segregation connected with surfacial processes

50. Which of the following statements are applicable to workable coal deposits?

1. The fuel ratio is high in anthracite and low in lignite.
2. Presence of sulphur in coals does not normally alter the quality and usability of coals.
3. High ash content may place a high rank coal in a low grade.
4. Absence of sulphur normally does not affect the quality of coals.

Of these statements

a) 1 and 3 are correct
b) 2 and 4 are correct
c) 3 and 4 are correct
a) 1 and 2 are correct

51. Which one of the following is a concordant intrusion forming structural basin?

a) Sill b) Phacolith c) Lopolith d) Laccolith

52. Which of the following pairs are correctly matched?

1. Trachyte Sheeting
2. Rhyolite Flow bands
3. Basalt Columnar joints
4. Granite Schlieren

Select the correct answer using the codes given below :

Codes :

a) 2, 3 and 4
b) 1, 3 and 4
c) 1, 2 and 3
d) 1, 2 and 4

53. Match List I with List II and select the correct answer using the codes given below the lists :

| List I (*Texture*) | List II (*Rock showing the texture*) |
|---|---|
| A. Allotriomorphic | 1. Trachyte |
| B. Flow | 2. Obsidian |
| C. Vitrophyric | 3. Dolerite |
| D. Ophitic | 4. Aplite |
| | 5. Lamprophyte |

Codes :

a) A—1 B—2 C—3 D—4
b) A—4 B—1 C—2 D—3
c) A—3 B—4 C—5 D—2
d) A—2 B—3 C—4 D—5

54. Explosive eruptions often involve rhyolitic magmas. This is because of their

a) High silica content b) High water content
c) Low volatile solubility d) Deep-seated nature

55. Match List I and List II and select the correct answer using the codes given below the lists :

| List I | List II |
|---|---|
| A. Dunite | 1. Melanocratic dyke rock |
| B. Felsite | 2. Monomineralic rock |
| C. Ijolite | 3. Leucocratic hypobyssal rock |
| D. Lamprophyte | 4. Undersaturated rock |
| | 5. Oversaturated rock |

Codes :

a) A—2 B—1 C—3 D—4
b) A—5 B—3 C—4 D—2
c) A—1 B—4 C—3 D—5
d) A—2 B—3 C—4 D—1

56. Under deep-seated conditions and at high temperatures of crystallisation of a mafic igneous rock, the stable pyroxene mineral pair is

a) Augite-hypersthene b) Ferroaugite-pigeonite
c) Augite-pigeonite d) Ferroaugite-hypersthene

57. ˊhe coexistence of talc, magnesite and serpentine in a kimberlitic groundmass suggests that the assemblage formed

a) Near the surface
b) Under P-T conditions existing in the upper mantle
c) Under P-T conditions existing in the lower crust
d) Under P-T conditions where eclogite will be stable

58. In a binary system showing an eutectic relation between two components A and B

a) A and B can mix only eutectic proportion below the solidus
b) A and B can mix in all proportions below the solidus
c) A and B can mix in all proportions above the liquidus
d) A and B can mix in the only eutectic proportion above the liquidus

59. Consider the following minerals :

1. Biotite 2. Diopside 3. Olivine 4. Hornblende

The correct order in which these minerals crystallise from the magma is

a) 3, 2, 4, 1 b) 4, 3, 1, 2
c) 2, 1, 3, 4 d) 1, 4, 2, 3

60. The occurrence of doliomorph signifies
a) Complete reaction b) Incomplete reaction
c) Congruent melting d) No reaction

61. Alkaline rocks showing banded structure are formed by the
a) Differentiation of a basic alkaline magma
b) Metasomatic alternation of certain foliated country rocks
c) Assimilation of limestones by a basaltic magma
d) Action of volatile constituents of a magma

62. Under mantle conditions olivine (rich in $Mg_2SiO_4$) would
a) Coexist with plagioclase (rich in anorthite molecule) below 60 km
b) Not coexist with anorthite-rich plagioclase below 60 km
c) Remain in equilibrium throughout the P-T range of the mantle with anorthite-rich plagioclase
d) Not remain in equilibrium with plagioclase under any of the P-T conditions.

63. Consider the following statements regarding the granite series. Tonalite-Granodiorite-Granite-Alkali granite :
1. There is a progressive increase in $SiO_2$ and $K_2O$.
2. There is a progressive decrease in $TiO_2$, MgO and CaO.
3. There is a progressive increase in $SiO_2$ and $Al_2O_3$.
4. There is a progressive decrease in $K_2O$, MgO and $Na_2Ø$.

Among these, those characteristics which can be attributed to the given granite series include
a) 1 and 3 b) 2 and 3 c) 1 and 3 d) 3 and 4

64. Match List I with List II and select the correct answer using the codes given below the lists :

| List I *(Rock type)* | List II *(Origin)* |
|---|---|
| A. Guano | 1. Chemical |
| B. Graywacke | 2. Residual |
| C. Chert and flint | 3. Clastic |
| D. Laterite | 4. Organic |
| | 5. Bioclastic |

Codes :

| | | | | |
|---|---|---|---|---|
| a) | A—4 | B—3 | C—1 | D—2 |
| b) | A—3 | B—4 | C—2 | D—5 |
| c) | A—2 | B—1 | C—3 | D—4 |
| d) | A—5 | B—2 | C—4 | D—3 |

65. Small ripples are converted into large ripples ('dunes') when
   a) Both grain size and flow velocity increase in lower flow regime
   b) Both grain size and flow velocity increase in higher flow regime
   c) Grain size increases but flow velocity decreases in lower flow regime
   d) Grain size decreases but flow velocity increases in higher flow regime
66. Theoretically the mean size clastic sediments
   a) Has no effect on porosity but affects permeability
   b) Has no effect on either porosity or permeability
   c) Has no effect on permeability but affects porosity
   d) Affects both porosity and permeability
67. Which one of the following diagrams of sedimentary structures correctly depicts the direction of current flow?

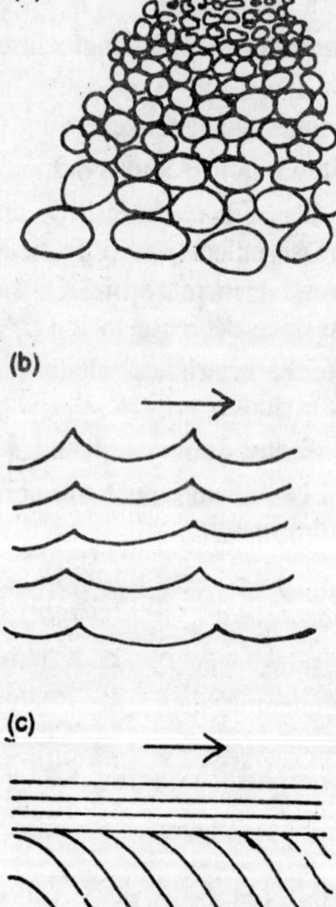

(d)

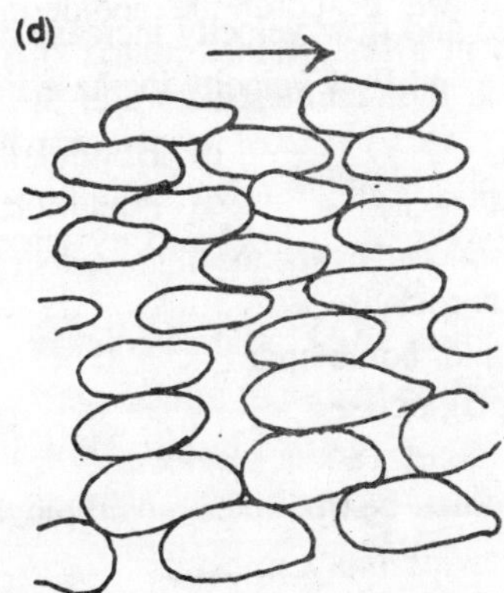

68. A heavy mineral assemblage of rounded grains of tourmaline, zircon and rutile suggests a provenance consisting of
    a) Pegmatites  b) Acid igneous rocks
    c) Basic igneous rocks  d) Reworked sedimentary rocks
69. Glauconite occurs exclusively in the sediments formed in
    a) Freshwater environment  b) Lacustrine environment
    c) Marine environment  d) Glacial environment
70. In arid and semi-arid regions, limestones precipitated from evaporating soil mixture as surface or near-surface crusts are known as
    a) Sparry  b) Calcrete  c) Micrite  d) Vadose
71. The presence of omphacite and pyrope in metamorphic rocks indicates that the rocks were formed under
    a) Low P - low T conditions  b) Low P - high T conditions
    c) High P - high T conditions  d) High P - low T conditions
72. Match List I with List II and select the correct answer using the codes given below the lists :

| List I (*Type of metamorphism*) | List II (*Rock Type*) |
|---|---|
| A. Contact metamorphism | 1. Glauco-phane schist |
| B. Cataclastic metamorphism | 2. Phyllite |
| C. Regional metamorphism | 3. Hornfels |
| D. Burial metamorphism | 4. Phyllonite |
| | 5. Greisen |

Codes :

| | | | | |
|---|---|---|---|---|
| a) | A—3 | B—4 | C—5 | D—2 |
| b) | A—2 | B—5 | C—4 | D—3 |
| c) | A—4 | B—3 | C—1 | D—2 |
| d) | A—3 | B—4 | C—2 | D—1 |

73. Coarse porphyroblasts of cordierite and/or andalusite (chaistolite variety) enclosed in a fine-grained matrix of graphite, sericite and quartz suggest that the texture/structure of the rock is

a) Maculose
b) Hornfelsic
c) Blastoporphyritic
d) Poikiloblastic

74. The assemblage produced by the medium grade regional metamorphism of a basic igneous rock is

a) Garnet, anorthite, hornblende
b) Hornblende, plagioclase
c) Garnet, mica, quartz, sodic plagioclase
d) Sillimanite, garnet, quartz, mica, sodic plagioclase

75. Consider the following index minerals :

1. Chlorite 2. Garnet 3. Kyanite 4. Biotite

The correct order of increasing grade of metamorphism of these minerals is

a) 4, 3, 2, 1
b) 2, 1, 3, 4
c) 1, 4, 2, 3
d) 3, 2, 4, 1

76. Which one of the following processes leads to the development of alternate light and dark coloured banding seen in some metamorphic rocks?

a) Fractional crystallization
b) Metamorphic differentiation
c) Anatexis
d) Gravitational segregation

77. Which one of the following evidences supports a metamorphic origin of granite?

a) Chilled and sharp contacts of granite
b) Apophyses of granite in the adjacent rocks
c) Basic front and graded contact of granite
d) Xenoliths of limestone in granite

78. Melting can occur in metamorphic rocks

a) Only if they come into contact with magmas
b) At high temperatures, even if they do not come into contact with magma
c) When magmatic rocks are metamorphosed
d) When the rocks contain large amounts of water

79. The phenomenon that strata characterized by similar fossils occupy corresponding positions in different stratigraphic sections, is known as

a) Correlation b) Taxonomy c) Homotaxis d) Heredity

80. Match List I and II and select the correct answer using the codes given below the lists :

| List I | List II |
|---|---|
| A. Sausar Series | 1. Cuddapah |

B. Semri Series — 2. Archaean
C. Cheyair Series — 3. Vindhyan
D. Barakar Series — 4. Damuda

Codes :

a) A—3 B—4 C—2 D—1
b) A—4 B—1 C—3 D—2
c) A—1 B—2 C—4 D—3
d) A—2 B—3 C—1 D—4

81. The Aravalli mountain chain was formed as a result of intense orogenic movements at the
a) Close of the Dharwar era
b) Beginning of the Gondwana era
c) Close of the Mesozoic era
d) Beginning of the Cenozoic era

82. Upper Palaeozoic mountain building movement (Orogeny) is known as
a) Hercynian Orogeny
b) Caledonian Orogeny
c) Alpine-Himalayan Orogeny
d) Huronian Orogeny

83. In the recent past, there has been a controversy over the age of
a) Zewan beds
b) Krol Formation
c) Lameta beds
d) Niniyur beds

84. The Cretaceous rocks of Trichinopoly can be ascribed to
a) Marine transgressional environment
b) Eugeosynclinal environment
c) Mio-geosynclinal environment
d) Fluviatile environment

85. The youngest volcanic activity in the Indian subcontinent, is in the
a) Sylhet Trap
b) Rajmahal Trap
c) Deccan Trap
d) Panjal Trap

86. Which one of the following pairs is correctly matched?
a) Spiti shales : Triassic
b) Zewan series : Cambrian
c) Ranikot Series : Palaeocene
d) Baripada beds : Oligocene

87. The age of the Lower Karewa beds is
a) Holocene
b) Pleistocene
c) Pliocene
d) Plio-pleistocene

88. The great reptiles amongst the land animals and the ammonites amongst the marine ones disappeared towards the end of the
a) Triassic Period
b) Cretaceous Period
c) Jurassic Period
d) Permian Period

89. The Permian succession of Salt Range is classified as part of
a) Productus Limestone
b) Zewan Formation
c) Kuling Shales
d) Katrol Series

90. Consider the following statements regarding the study of fossil flora and fauna :

It is useful in

1. Age determination
2. Correlation
3. Palaeoenvironmental reconstructions

Of these statements

a) 1 and 2 are correct b) 2 and 3 are correct
c) 1 and 3 are correct d) 1, 2 and 3 are correct

91. The stratigraphic position of Lameta beds is

a) Intertrappean b) Infratrappean
c) Epitrappean d) Post-trappean

92. The Barakar Formation belongs to the

a) Lower Permian b) Triassic
c) Jurassic d) Eocene

93. The history of Tethys is very well known from

a) Cambrian Period onwards b) Triassic Period onwards
c) Devonian Period onwards d) Jurassic Period onwards

94. Which one of the following is useful in the correlation of coal and oil-bearing strata, in tracing patterns of climatic fluctuations and migrating vegetational belts?

a) Foraminifera b) Plant fossils
c) Spores and pollen d) Vertebrate fossils

95. Which one of the following groups of animals has the capacity to leave large number of fossils in deep-sea sediments?

a) Foraminifera b) Graptolites
c) Ammonites d) Trilobites

96. The process of fossilization take place in the

a) Presence of oxygen b) Absence of oxygen
c) Presence of nitrogen d) Presence of carbon dioxide

97. Trilobites are found only in the rocks belonging to

a) Archaeozoic Era b) Proterozoic Era
c) Palaeozoic Era d) Mesozoic Era

98. Consider the following index genera :

1. Nummulites
2. Monograptus
3. Ceratites
4. Paradoxites

The correct chronological sequence of these genera is

a) 1, 2, 3, 4 b) 1, 3, 4, 2
c) 4, 2, 3, 1 d) 4, 3, 2, 1

99. Which one of the following fossils is highly preferred as index fossil of the Ordovician Period?

a) Didymograptus b) Paradoxides
c) Productus d) Fenestella

100. A valve of a lamellibranch shell with two muscle scars of unequal shape and size is described as

a) Anisomyarian b) Heteromyarian
c) Monomyarian d) Isomyarian

101. Match List I with List II and select the correct answer using the codes given below the lists :

| List I | List II |
|---|---|
| A. Dimorphism | 1. Tabulate coral |
| B. Sponges | 2. Micraster |
| C. Halysites | 3. Foraminifera |
| D. Fasciole | 4. Porifera |

Codes :

a) A—1 B—2 C—3 D—4
b) A—2 B—3 C—4 D—1
c) A—3 B—4 C—1 D—2
d) A—4 B—3 C—2 D—1

102. Which one of the following brachiopods possesses concavo-convex shells?

a) Atrypa b) Rhynchonella
c) Spirifer d) Productus

103. The facial suture which cuts cephalon margin in front of genal angle is known as

a) Opisthoparian b) Proparian
c) Hypoparian d) Gonatoparian

104. Two gastropod genera Turritella and Cerithium can be distinguished from each other on the basis of

a) Spiral angle b) Shape of aperture
c) Surface ornamentation
d) Presence and/or absence of columella

105. Rounded saddles and denticulate lobes in ammonoids characterise suture lines of

a) Goniatitic type b) Ceratitic type
c) Ammonite type d) Nautiloid type

106. Among the Siwalik mammals which one of the following does NOT belong to the group?

a) Sus b) Rhinoceros
c) Giraffa d) Dinotherium

107. Which one of the following echinoid genera is regular?

a) Cidaris b) Micraster
c) Stigmatopygus d) Holectypus

108. Consider the following stages in the development and evolution of a mammal :

1. Mesohippus 2. Pliohippus
3. Merychippus 4. Eohippus

The correct chronological order of development is

a) 1, 2, 3, 4 b) 4, 1, 3, 2
c) 3, 4, 2, 1 d) 2, 3, 1, 4

The following 12 items consist of two statements, one labelled the 'Assertion' and the other labelled the 'Reason'. You are to examine these two statements carefully and decide if the Assertion and the Reason are individually true and if so, whether the Reason is a correct explanation of the Assertion. Select your answers to these 12 items using the codes given below and mark your answer sheet accordingly :

CODES :

a) Both Assertion and Reason are true and Reason is the correct explanation of Assertion.
b) Both Assertion and Reason are true and Reason is NOT a correct explanation of Assertion.
c) Assertion is true but Reason is false.
d) Assertion is false but Reason is true.

109. Assertion : The mountains do not deflect the plumb-line as much as they should.
Reason : The crust is underlain by lighter material and in the mountains there are roots of denser material.

110. Assertion : Most of the earth's topography is younger than Tertiary and most of it is not older than Pleistocene.
Reason : Cenozoic rocks conceal most of the older rocks.

111. Assertion : Anticlines are eroded to from valleys more commonly.
Reason : The anticlines are more stable along their axes.

112. Assertion : Sanidine is found only in volcanic rocks.
Reason : Sanidine is a high temperature form of orthoclase.

113. Assertion : Graded bedding is characteristic of many greywackes.
Reason : Greywackes are mostly of marine origin.

114. Assertion : Different crystals of any one mineral have different symmetry characters.
Reason : The law of constancy of interfacial angles governs symmetry characters in crystals.

115. Assertion : No plane that is a symmetry plane in the individual crystal can become a twinning place in the compound crystal.
Reason : If such a plane were assumed as twinning plane, the resulting crystal would be identical with the individual crystal.

116. Assertion : In an uniaxial mineral, birefringence is the greatest

when light travels parallel to the optic axis.

Reason : Birefringence is the difference between refractive indices of a mineral.

117. Assertion : Calcite, in thin section, exhibits twinkling when the stage of the microscope is rotated.

Reason : Refractive index of calcite for ordinary ray is much higher than that of balsam, but it is lower for the extraordinary ray.

118. Assertion : Barite deposits occur in the Cuddappah Supergroup of Andhra Pradesh.

Reason : They are confined to the Cumbum limestones.

119. Assertion : The Saurashtra bauxite is ideally suited for chemical and refractory industries.

Reason : It contains 60 percent alumina, about 2 percent each of $Fe_2O_3$ and $SiO_2$ and about 1.3 per cent $TiO_2$.

120. Assertion : Petroleum occurs in the offshore regions of India.

Reason : Petroleum occurrence is common in fluviatile environment.

## ANSWERS

| | | | | | |
|---|---|---|---|---|---|
| 1. d | 2. b | 3. a | 4. d | 5. c | 6. a |
| 7. a | 8. a | 9. a | 10. a | 11. a | 12. a |
| 13. b | 14. d | 15. c | 16. c | 17. a | 18. c |
| 19. a | 20. c | 21. b | 22. a | 23. d | 24. d |
| 25. b | 26. a | 27. b | 28. a | 29. b | 30. a |
| 31. c | 32. c | 33. d | 34. c | 35. b | 36. c |
| 37. d | 38. c | 39. b | 40. a | 41. b | 42. b |
| 43. c | 44. b | 45. c | 46. a | 47. a | 48. d |
| 49. d | 50. a | 51. c | 52. a | 53. b | 54. a |
| 55. d | 56. a | 57. b | 58. c | 59. a | 60. b |
| 61. b | 62. c | 63. a | 64. a | 65. b | 66. d |
| 67. d | 68. d | 69. c | 70. b | 71. c | 72. d |
| 73. a | 74. b | 75. c | 76. b | 77. c | 78. b |
| 79. c | 80. d | 81. a | 82. a | 83. c | 84. a |
| 85. a | 86. c | 87. d | 88. b | 89. a | 90. d |
| 91. b | 92. a | 93. c | 94. c | 95. a | 96. b |
| 97. c | 98. c | 99. a | 100. a | 101. c | 102. d |
| 103. a | 104. a | 105. c | 106. d | 107. a | 108. b |
| 109. a | 110. c | 111. c | 112. d | 113. a | 114. b |
| 115. a | 116. b | 117. a | 118. c | 119. a | 120. c |

## 3. Civil Services (Preliminary) Examination, 1992

1. Which of the following is an artificial diamond
   a) Bort  b) Carbonado
   c) Graphite  d) Stishovite

2. Match the following :

| List I | List II |
|---|---|
| 1. Geniculate Twins | (a) Gypsum |
| 2. Carlsbad Law | (b) Albite |
| 3. Tailsmooth Twins | (c) Rutile |
| 4. Dauphine Law | (d) Quartz |

   a) 1—(c) 2—(b) 3—(a) 4—(d)
   b) 1—(b) 2—(c) 3—(d) 4—(a)
   c) 1—(d) 2—(c) 3—(b) 4—(a)
   d) 1—(a) 2—(b) 3—(c) 4—(d)

3. Match the following :

| List I | List II |
|---|---|
| 1. Serir | A. Wind |
| 2. Uvala | B. Glacier |
| 3. Valley Flat | C. Running water |
| 4. Cavern | D. Underground water |

   a) A—4 B—1 C—3 D—2
   b) A—3 B—2 C—1 D—4
   c) A—1 B—2 C—3 D—4
   d) A—2 B—1 C—3 D—4

4. Kimberlites are associated with
   a) Subduction zones  b) Cratons
   c) MORS  d) Transform junctions

5. The gasteropod showing sinistral coiling is
   a) Physa  b) Murex  c) Planorbis  d) Turritella

6. Which of the following are transition beds between cretaceous and eocene?
   a) Montian  b) Danian  c) Thanetian  d) Neocomian

7. Match the following :

| List I | List II |
|---|---|
| 1. Trigonia | A. Schizodont |
| 2. Pecten | B. Isodont |
| 3. Meretrix | C. Heterodont |
| 4. Mytilus | D. Dysodont |

   a) A—1 B—2 C—3 D—4
   b) A—2 B—1 C—4 D—3

c) A—1 B—3 C—2 D—4
d) A—4 B—2 C—1 D—3

8. Which of the following belong to reworked sediments?
   a) Glauconite and chert
   b) Brown Hornblende and Tourmaline
   c) Glauconite and Micas
   d) Zircon and Tourmaline
9. Smaltite is an ore mineral of
   a) Cobalt
   b) Manganese
   c) Copper
   d) Molybdenum
10. Along the fault place the horizontal and vertical separation are known as
    a) Heave and Throw
    b) Stratigraphic throw and throw
    c) Plunge and Heave
    d) Plunge and Throw
11. The mineral which crystallises just before quartz according to the reaction series is
    a) Hornblende
    b) Pyroxene
    c) Mica
    d) Feldspar
12. In a glaciated landscape the features encountered from top to bottom would be
    a) Cirque, Hanging Valley, Outwash Plain, Moraine
    b) Hanging Valley, Cirque, Outwash Plain, Moraine
    c) Outwash Plain, Hanging Valley, Cirque, Moraine
    d) Moraine, Outwash Plain, Hanging Valley, Cirque
13. The Aluminium Silicate formed due to low grade Metamorphism of pelitic rocks is
    a) Andalusite
    b) Cordierite
    c) Sillimanite
    d) Kyanite
14. The mineral assemblage sillimanite + graphite is typical to
    a) Khondalite
    b) Charnockite
    c) Leptynite
    d) Augen Gneiss
15. The 3 factors playing the dominant role in metamorphism are
    1. Pressure
    2. Temperature
    3. Volatiles
    4. Colloids
    a) 1, 2 and 3 are correct
    b) 1 and 3 are correct
    c) 2 and 3 are correct
    d) All the above are correct
16. The deposits which are not in situ are
    a) Talus/scree
    b) Bauxite
    c) Laterite
    d) Monazite sand
17. Match the following :

| List I | List II |
|---|---|
| 1. Diastem | A. Marine formations |

2. Angular unconformity B. Angular disposition of beds
3. Non-conformity C. Volcanic and Sedimentary rock
4. Parallel unconformity D. Parallel disposition

a) A—1 B—3 C—2 D—4
b) A—2 B—3 C—3 D—4
c) A—1 B—2 C—3 D—4
d) A—2 B—1 C—4 D—3

18. The by-product obtained in Kolar Gold Field is
a) Silver b) Molybdenum
c) Arsenic d) Copper

19. Mound-like structures of organic origin are
a) Bioherms b) Biostromes
c) Lithostromes d) None of the above

20. The typical hypothermal ore deposit is that of
a) Gold b) Copper
c) Mercury d) Manganese

21. The factors responsible for the rise of Magma are
a) Release of pressure and radioactive heat
b) Release of pressure only c) Radioactive heat only
d) None of the above

22. The role played by volatiles in Magma
a) Release of volatiles leads to high viscosity
b) Release of volatiles leads to low viscosity
c) Release of volatiles has no effect on viscosity
d) Addition of volatiles increases viscosity

23. The nature of Magma below Island Arcs is
a) Rhyolitic b) Andesitic
c) Granitic d) Basaltic

24. Choose the correct code w.r.t. Island Arcs
1. They lie on the continental side of trenches
2. They are associated with Andesitic volcanism
3. They are associated with negative gravity anomalies
4. They are zones of low heat flow

a) 1, 2, 4 b) 1, 2, 3 c) 1, 2 d) 2, 4

25. Trigonal symmetry is characterised by
a) No planes, 4 axis and no centre
b) 4 planes, 4 axis and no centre
c) No planes, 4 axis and a centre
d) No planes, no axis and a centre

26. Choose the correct code w.r.t. loess
1. Wind blown deposit
2. Deposit shows vertical permeability

3. Deposit is buff coloured
4. Deposit shows horizontal permeability

a) 1, 2, 3 b) 1, 2, 4 c) 1, 3, 4 d) 1, 2, 3, 4

27. The stratigraphic horizon in Gondwana system showing mixed facies is
    a) Talchir series b) Umia series
    c) Panchet series d) Ranigunj series
28. Inherited structures unaffected by metamorphism are termed
    a) Skaarn b) Palimpsest
    c) Hornfels d) Decussate fabric
29. The crustal thickness is maximum under
    a) Island arcs b) Subduction zones
    c) Folded Mountains d) Shield Areas
30. The oldest man is
    a) Ramapithecus b) Neanderthal man
    c) Cro-Magnon man d) Aegyptopethecus
31. Sharp narrow valleys are usually attributed to
    a) Fluvial agencies b) Tectonic factors
    c) Glacial erosion d) Volcanic eruption
32. Tors are
    a) Huge monolithic pillars of rock due to sub-surface weathering
    b) Residual boulders due to stream erosion
    c) Enlarged craters of a volcano
    d) None of the above
33. Foraminifers having a small test and large proloculus are described as
    a) Microspheric b) Megalospheric
    c) Flagellate d) Uninucleate
34. Match the following :

| List I | List II |
|---|---|
| A. Uniserial foraminifera | 1. Nodosaria |
| B. Blind Trilobite | 2. Agnostus |
| C. Simple suture line | 3. Nautilus |
| D. Uncoiled Ammonite | 4. Scaphites |

a) A—1 B—2 C—3 D—4
b) A—2 B—1 C—4 D—3
c) A—4 B—1 C—2 D—3
d) A—3 B—2 C—4 D—1

35. Cuspate spits in a foreland beach are due to
    a) Marine Transgression b) Marine Regression
    c) Both Transgression and Regression

d) None

36. Exfoliation domes are due to
a) Solar weathering
b) Chemical weathering
c) Glacial action
d) River erosion

37. The age of Tipam series of Assam is
a) Pilocene
b) Oligocene
c) Miocene
d) Eocene

38. The basic parameters for defining a sedimentary facies are
a) Composition and geometry
b) Environment and geometry
c) Only geometry
d) Only composition

39. Choose the correct sequence of orogenies
a) Acadian, Caledonia, Hercynian, Himalayan
b) Caledonian, Acadian, Hercynian, Himalayan
c) Hercynian, Caledonian, Acadian, Himalayan
d) Acadian, Hercynian, Himalayan, Caledonian

40. Wackes are characterised by
a) Higher proportion of matrix
b) Lower proportion of matrix
c) Absence of matrix
d) Equal amounts matrix and cement

41. The assemblage sphenophyllum pacopteris, schizoneura characterises
a) Lower Gondwana flora
b) Upper Gondwana flora
c) Middle Gondwana flora
d) None of the above

42. The lithified portions of alluvial fans are termed as
a) Bajada
b) Fanglomerate
c) Agglomerate
d) Orthoconglomerate

43. The first elephant stegodon appears for the first time in
a) Dhok pathan stage
b) Nagri stage
c) Pinjor stage
d) Kamlial stage

44. In spirifer, the hinge line is
a) Straight and long
b) Straight and small
c) Curved and long
d) Curved and small

45. The concordant structure formed by lavas with a sagged base and of great size is termed
a) Lopolith
b) Laccolith
c) Phacolith
d) Batholith

46. Colloform structure pertains to
a) Liesegang banding
b) Igneous lamination
c) Organic structure
d) None

47. Liquidus in a solid solution series is
a) Lower melting point curve
b) Upper melting point curve
c) Upper freezing point curve
d) Lower freezing point curve

48. The eutectic proportion of Anorthite and Olivine is

a) 70 : 30 b) 45 : 55 c) 42 : 58 d) 30 : 70

49. The Nummulitic limestones of Kashmir belong to
   a) Eocene
   b) Jurassic
   c) Miocene
   d) Pleistocene
50. $Fe_3Al_2(SiO_4)_3$ is the composition of
   a) Spessartite
   b) Almandine
   c) Uvarovite
   d) Andradite
51. Tadapatri shales and Nagari quartzites belong to
   a) Papaghani series
   b) Cheyair series
   c) Nallamallai series
   d) Srisailan series
52. One of the most important groups of Devonian vertebrates are
   a) Bony fishes
   b) Dinosaurs
   c) Flying reptiles
   d) Horses
53. Cement grade limestone is found in the stratigraphic horizon of
   a) Rohtas formation
   b) Rewa formation
   c) Panna formation
   d) Kheinjua formation
54. Kioto or Megalodon limestones of Spiti Synclinorium belong to
   a) Rhaetic b) Danian c) Montian d) Ladinic
55. Choose the correct sequence
   1. Rehmani beds
   2. Ceratite beds
   3. Belemenite beds
   4. Umia beds
   a) 4, 3, 2, 1
   b) 1, 2, 3, 4
   c) 2, 1, 3, 4
   d) 2, 1, 4, 3
56. Match the following :

| List I | List II |
|---|---|
| A. Deltidium | 1. Brachiopod |
| B. Rhabdosome | 2. Graptolite |
| C. Calyx | 3. Coral |
| D. Septum | 4. Ammonoid |

   a) A—2 B—1 C—3 D—4
   b) A—1 B—2 C—3 D—4
   c) A—1 B—2 C—4 D—3
   d) A—2 B—1 C—4 D—3
57. A good example of a solitary coral is one of the following
   a) Calceola
   b) Zaphrentis
   c) Xylodes
   d) Heriophyllum
58. Podiform chromite occurs in
   a) Cratons
   b) Subduction zones
   c) Oceanic Ridges
   d) Transform faults
59. The igneous rock containing sodic feldspars and belonging to sub-aqueous environment is known as

a) Spillite
b) Alkali syenite
c) Nordmarkite
d) Trachyte

60. Hypersolvus granites are those characterised by
a) Absence of perthitic feldspar
b) Presence of a single perthitic feldspar
c) Absence of microcline
d) Absence of orthoclase

61. Consider the following :
1. Sharp contact with country rock
2. Uniform chemical composition
3. Corroded borders
4. Dykes and sills
Of the above, which constitute evidence for igneous activity
a) 1 & 4
b) 1, 2 & 4
c) 2, 3, & 4
d) 1, 2, 3 & 4

62. The igneous rock with calcic composition and presence of rare earth minerals is
a) Komatiite
b) Carbonatite
c) Enderbite
d) Ijolite

63. The volcanic equivalent of Nepheline syenite is
a) Trachyte
b) Phonolite
c) Latite
d) Dacite

64. Olenus beds are equivalent to
a) Bala
b) Tremdec
c) Moric
d) Carnic

65. Match the following :

| List I | List II |
|---|---|
| A. Kyanite | 1. Refractory |
| B. Mica | 2. Insulator |
| C. Barytes | 3. Oil drilling |
| D. Gypsum | 4. Cement |

a) A—1 B—2 C—3 D—4
b) A—3 B—2 C—1 D—4
c) A—2 B—3 C—1 D—4
d) A—4 B—3 C—2 D—1

66. The assemblage glaucophone, lawsonite and aragonite is typical to
a) Blue-Schist facies
b) Ecologite facies
c) Pyroxene granulite facies
d) Green-Schist facies

67. Omphacite and pyrope characterise
a) Hornfels
b) Ecologite
c) Kimberlite
d) Granulite

68. The typical fossil of intertrappean beds is
a) Lingula
b) Cardita Beaumoniti
c) Physa
d) Nucula

69. Porphyroblastic minerals of metamorphic rocks are due to
   a) Solid diffusion b) Metasomatism
   c) Granitisation d) Shearing stress
70. In a rock containing Olivine and Hornblende being subject to weathering which of the following would be true?
   a) Olivine is lost before Hornblende
   b) Hornblende is lost before Olivine
   c) Both are lost simultaneously
   d) The order of loss cannot be predicted
71. Consider the following :
   1. Actinolite is a monoclinic amphibole
   2. Actinolite is optically negative
   3. Actinolite is essentially a metamorphic mineral
   4. Actinolite typically occurs in the blue-schist facies
   The correct code is
   a) 1 & 4 b) 2 & 4 c) 1, 3, 4 d) 1, 2, 3
72. The Magnesian Mica phlogophite typically occurs in
   a) Komatiites b) Kimberlites
   c) Dunites d) Monzonites
73. Which of the following is non-volcanic
   a) Cirque b) Caldera c) Crater d) Pipe
74. Mica deposits in India occur in
   a) A.P., Bihar, Rajasthan b) A.P., Bihar, Gujarat
   c) Bihar, Gujarat, Rajasthan d) Bihar, Gujarat, Maharashtra
75. The beds of which can be used as a marker bed
   a) Bentonite b) Coal
   c) Gypsum d) Limestone
76. Consider the following :
   1. P-waves travel with the greatest velocity
   2. P-waves have long wave lengths
   3. P-waves are transmitted through all media
   4. P-waves vibrate perpendicular to the direction of propagation
   a) 1, 2, 4 b) 2, 3, 4 c) 1, 3 d) 1, 2, 3, 4
77. Choose the correct stratigraphic succession of Gondwana group
   a) Damuda, Panchet, Mahadeva
   b) Mahadeva, Damuda, Panchet
   c) Panchet, Mahadeva, Damuda
   d) Mahadeva, Panchet, Damuda.
78. The assemblage Albite, Epidote, Chlorite characterises
   a) Green-schist facies b) Pyroxene Hornfels facies
   c) Granulite facies d) Amphibolites facies

79. The Montmorillonitic clay used in oil drilling is
a) Bentonite b) Fuller's earth
c) Barium mud d) None of the above

80. Identify the fold type

a) Recumbent fold b) Isoclinal fold
c) Box fold d) Cylindrical fold

81. Consider the following structure

In the above structure, the axial planes of the syncline is between
a) A & B b) B & C
c) A & C d) Cannot be said

82. According to Goldich mineral stability series
a) $Na_2O$ is less stable than $K_2O$
b) $Na_2O$ is more stable than $K_2O$
c) Both are equally stable d) None of the above is correct

83. According to the concept of Metamorphic facies, the maximum number of stable phases in metamorphic reactions is
a) 4 b) 6 c) 8 d) 10

84. The maximum number of symmetry elements in a stereographic projection belong to
a) Hexagonal crystal b) Cubic crystal
c) Monoclinic crystal d) Orthorhombic crystal

85. The reflection plane of 001 is one of the following
a) $0\,\bar{0}\,1$ b) $0\,0\,\bar{1}$ c) $\bar{0}\,0\,1$ d) $0\,1\,0$

86. The horizontal axis occur in crystals of
a) Hexagonal system b) Monoclinic system
c) Triclinic system d) Cubic system

87. If a network of parallel streams are developed along strike and dip of the formations, the drainage pattern is referred as
a) Braided b) Annular c) Trellis d) Dendritic

88. Consider the following :

1. Nickpoints
2. Meandering
3. Graded stream

The correct chronology of landforms from youth to old age are

a) 1, 2, 3 b) 1, 3, 2 c) 3, 2, 1 d) 2, 3, 1

89. In a structure resembling a plunging syncline, if the stratigraphic order is not known it is termed as

a) Antiform b) Synform
c) Cylindrical fold d) Plunging fold

90. The illustration below depicts

a) Replacement b) Dissemination
c) Cavity filling d) Magmatic injection

91. An example of a strategic mineral in India is one of the following

a) Sulphur b) Copper
c) Graphite d) Chromium

92. Equus and Rhinoceras characterising Pliocene-lower Pleistocene occur in which stratigraphic horizon of Siwaliks

a) Kámlial b) Pinjor
c) Dhok Pathan d) Nagri

93. Match the following :

| List I | List II |
|---|---|
| A. Chromite | 1. Oxide |
| B. Chlorite | 2. Silicate |
| C. Barytes | 3. Sulphate |
| D. Chalcopyrite | 4. Sulphide |

a) A—1 B—3 C—2 D—4
b) A—2 B—3 C—4 D—1
c) A—4 B—3 C—2 D—1
d) A—1 B—2 C—3 D—4

94. The variety of Sandstone usually derived from a granitic terrain is

a) Subgraywacke b) Arkose
c) Conglomerate d) Lithic arenite

95. The molecular proportion of minerals derived from the chemical analysis is used in

a) Normative classification b) Mineralogical classification
c) Chemical classification d) Textural classification

96. Analogous chemical formulae and similar structures characterise

a) Isomorphs b) Polymorphs
c) Pseudomorphs d) Isotopes

97. In a vertical section where a bed is striking North-East the correct dip of the bed will be

a) 10° NW
b) 45° NW-SE
c) 0° SE
d) 90° N-S

98. Eastern Ghats constitute
a) A sill
b) A dyke
c) A lopolith
d) A batholith

99. The basic difference between a shear fracture and a fault is
a) Amount of shear
b) Amount of stress
c) Amount of brecciation
d) None of the above

100. Orpiment is an Ore mineral of
a) Arsenic
b) Tungsten
c) Rubidium
d) Molybdenum

## ASSERTION & REASON

CODES :

a) Both Assertion & Reason are true and Reason is the explanation of Assertion.

b) Both Assertion & Reason are true and Reason is NOT the explanation of Assertion.

c) Assertion is true and Reason is false.

d) Assertion is false and Reason is true.

101. Assertion : In hornfelsic fabric, there is no preferred orientation of metamorphic minerals.
Reason : Hornfelsic fabric is indicative of lack of high pressure.

102. Assertion : Foraminifers are important faunal groups for biostratigraphic zonation of tertiary.
Reason : Alveolina oblonga is the zone fossil of Upper Eocene.

103. Assertion : Banded Iron ores are bedded type of deposits.
Reason : Banded Iron ores are sedimentary in origin.

104. Assertion : The loci of thickest accumulation of sediments are Geosynclines.
Reason : Geosynclines are deeply subsiding basins.

105. Assertion : Pigeonite, i.e., the lime-poor pyroxene is only found in saturated basaltic lavas.
Reason : Pigeonite is unstable in high pressures.

106. Assertion : Two member ring silicates are unlikely.
Reason : Two member rings necessitate the close approach of two highly charged silicon ions.

107. Assertion : Joints are genetically related to folds and faults.
Reason : Many joints are symmetrically disposed about folds and faults.

108. Assertion : Oceanic crust shows symmetry in composition and age w.r.t. MOR.

Reason : Basalt Volcanism and accretion of crust is bilaterally symmetrical.

109. Assertion : Pyroclasts erupted from continental side of Andesite line are more voluminous.

Reason : Volcanoes on continental side of Andesite line are supplied with high pressure gas.

110. Assertion : Upper Vindhyans are floodplain and deltaic deposits.

Reason : Cross-bedding, ripple marks are structures of upper Vindhyans.

111. Assertion : Gondwanaland as a whole experienced glaciation during paleozoic.

Reason : Gondwana continents were placed near to south pole in paleozoic.

112. Assertion : L-waves are responsible for most of the destructive force in earthquake activity.

Reason : They develop in immediate vicinity of epicentre and vibrate vertically.

113. Assertion : Deformed bodies show anisotropic fabric.

Reason : Fabric elements of deformed bodies show preferred orientation.

114. Assertion : Pseudomorphs are false forms.

Reason : Their external shapes are not consistent with the internal arrangement of units.

115. Assertion : Coral reefs are stereophytic.

Reason : Stereophytes are built up from the beginning as solid masses.

116. Assertion : Dissolved metallic sulphates deposit metals in zone of enrichment.

Reason : Zone of enrichment is a reducing zone.

117. Assertion : Ventral valve of fixed pelcypods is thicker than the dorsal valve.

Reason : Ventral valve is the fixed valve.

118. Assertion : Binary components do not form solid solutions.

Reason : Binary components are not miscible in each other.

119. Assertion : Stress minerals are typical to epizone.

Reason : Recrystallisation is weak while pressure of load greater in epizone.

120. Assertion : Cores of eroded orogenic belts contain high grade metamorphic rocks.

Reason : Katazonal conditions prevail in cores of orogens.

## ANSWERS

| | | | | | |
|---|---|---|---|---|---|
| 1. b | 2. a | 3. c | 4. a | 5. a | 6. b |
| 7. a | 8. a | 9. a | 10. a | 11. c | 12. a |
| 13. b | 14. a | 15. a | 16. a | 17. c | 18. b |
| 19. a | 20. a | 21. a | 22. a | 23. b | 24. b |
| 25. a | 26. a | 27. b | 28. b | 29. c | 30. a |
| 31. b | 32. a | 33. b | 34. a | 35. b | 36. a |
| 37. c | 38. a | 39. a | 40. a | 41. a | 42. b |
| 43. a | 44. a | 45. a | 46. a | 47. c | 48. a |
| 49. a | 50. b | 51. b | 52. a | 53. a | 54. a |
| 55. c | 56. b | 57. b | 58. b | 59. a | 60. b |
| 61. b | 62. b | 63. b | 64. b | 65. a | 66. a |
| 67. b | 68. c | 69. a | 70. a | 71. d | 72. b |
| 73. a | 74. a | 75. a | 76. c | 77. a | 78. a |
| 79. b | 80. a | 81. a | 82. a | 83. b | 84. b |
| 85. b | 86. a | 87. c | 88. b | 89. b | 90. b |
| 91. a | 92. b | 93. d | 94. b | 95. a | 96. a |
| 97. b | 98. a | 99. a | 100. a | 101. a | 102. b |
| 103. a | 104. a | 105. a | 106. a | 107. a | 108. a |
| 109. a | 110. a | 111. a | 112. a | 113. a | 114. a |
| 115. b | 116. b | 117. b | 118. a | 119. a | 120. a |

## 4. Civil Services (Preliminary) Examination, 1991

1. Annular stream patterns are associated with
   a) Basins  b) Domes
   c) Anticlines  d) Anticlinoria
2. Consider the following :
   1. Contact metamorphism  2. Plutonic metamorphism
   3. Thermal metamorphism  4. Dynamothermal metamorphism

   Uniform pressure and heat are the main factors in
   a) 1, 4  b) 1 only  c) 1, 3, 4  d) 1, 2, 3
3. Folds whose profiles are essentially semi-circular are termed as
   a) Similar folds  b) Parallel folds
   c) Cylindrical folds  d) Non-cylindrical folds
4. Match the following :

| List I | List II |
|---|---|
| A. Phaneric texture | 1. Dolerite |
| B. Diabasic texture | 2. Gabbro |
| C. Holocrystalline texture | 3. Spillite |
| D. Pillow structure | 4. Granite |

   a) A—2 B—1 C—4 D—3
   b) A—4 B—1 C—2 D—3
   c) A—1 B—2 C—4 D—3
   d) A—1 B—2 C—3 D—4
5. Consider the following :
   1. Size  2. Weight
   3. Hardness  4. Distance of transport

   The degree of rounding depend upon
   a) 1, 3, 4  b) 1, 2, 3  c) 2, 3  d) 1, 2, 3, 4
6. A texture in which phenocrysts form clusters has been termed as
   a) Seriate texture  b) Spherulitic texture
   c) Glomeroporphyrtic texture  d) Panidiomorphic texture
7. Cardita Beaumonti is considered to be the index fossil of Danian for the Indian subcontinent. The age of the pelicypod cardita ranges from
   a) Jurassic to present day  b) Triassic to present day
   c) Cretaceous to present day  d) Miocene to present day
8. Match the following :

| List I | List II |
|---|---|
| A. Kiruna Magnetite | 1. Gravitative accumulation |
| B. Bushveld chromite | 2. Dissemination |
| C. Insizwa Nickel | 3. Immiscible liquid segregation |
| D. Kimberlite Diamonds | 4. Residual liquid injection |

a) A—4 B—1 C—3 D—2
b) A—2 B—1 C—4 D—3
c) A—4 B—1 C—2 D—3
d) A—1 B—2 C—3 D—4

9. Match the following :

| List I | List II |
|---|---|
| A. Pyrrhotite | 1. Weakly magnetic |
| B. Haematite | 2. Strongly magnetic |
| C. Quartz | 3. Moderately magnetic |
| D. Tourmaline | 4. Non-magnetic |

a) A—2 B—4 C—1 D—3
b) A—2 B—3 C—4 D—1
c) A—1 B—2 C—3 D—4
d) A—3 B—4 C—2 D—1

10. A community of inter-related organisms inhabiting an area is called as

a) Biotope
b) Thanatocoenose
c) Biocoenose
d) Thanatope

11. Homotaxial Infratrappeans are

1. Bagh beds
2. Kateru beds
3. Duddukuru beds
4. Hippuritic limestone

The correct code is

a) 1, 3 b) 1, 2, 3 c) 1, 3, 4 d) 1 only

12. The most widespread basaltic volcanism occurring on the earth's crust today is associated with

a) Transform faults
b) Continental Rift valleys
c) MORS
d) Benioff zones

13. The composition of earth's crust is intermediate between

a) Granite and Basalt
b) Granite and Rhyolite
c) Granite and Gabbro
d) Granite and Andesite

14. Consider the following :

1. Deccan traps
2. Sylhet traps
3. Khewra traps
4. Panjal traps

The correct chronology is given by

a) 1, 2, 4, 3 b) 1, 2, 3, 4 c) 1, 4, 3, 2 d) 4, 3, 2, 1

15. According to Sarkar the early proterozoic - middle proterozoic boundary occurs at

a) 1400 M.Y.
b) 1600 M.Y.
c) 900 M.Y.
d) 2500 M.Y.

16. Consider the following :

1. Garnets
2. Plagioclase

3. Orthopyroxenes 4. Amphiboles

Of the above, isomorphs are given by

a) 1, 2 b) 1, 2, 3 c) 1, 2, 4 d) 1, 2, 3, 4

17. Sideronitic texture is a feature pertaining to
    a) Immiscible liquid segregation
    b) Magmatic injection
    c) Early gravitative segregation
    d) Late magmatic segregation
18. Globular gasteropod among the following is
    a) Helix b) Turbo c) Natica d) Trochus
19. "Pebble" grade size according to the Wentworth-Udden scale is
    a) 256-128 mm b) 64-32 mm
    c) 32-64 mm d) 64-4 mm
20. Gravity faults form under
    a) tensile stress regime b) compressive stress regime
    c) Shear stress regime d) None of the above
21. Rocks with equidimensional minerals formed under regional metamorphism of slates are
    a) Hornfels b) Skaarns
    c) Eclogites d) Amphibolites
22. The Turonian marine transgression in Narmada Valley is represented by
    a) Bagh beds b) Lameta beds
    c) Nimar Sandstone d) Nodular Limestone
23. Benioff zones are associated with
    a) Andesite Volcanism b) Calc-alkaline-plutonism
    c) Calc-alkaline Volcanism/Plutonism
    d) Rhyolite Volcanism
24. Fall of Meteorites from time to time is explained by
    a) Tidal hypothesis b) Protoplanet hypothesis
    c) Dust Cloud hypothesis d) Planetismal hypothesis
25. According to Schurmann's rule of less soluble sulphides, copper can replace
    a) Zinc, Nickel, Cobalt, Mercury
    b) Zinc, Nickel, Cobalt, Bismuth
    c) Zinc, Nickel, Cobalt, Gold
    d) Mercury, Gold, Zinc, Cobalt
26. The most obvious evidence for continent-reassembly is
    a) Morphological similarity of coastlines
    b) Glacial deposits distribution
    c) Paleomagnetic data d) Sea floor spreading

27. Match the following :

| List I | List II |
|---|---|
| A. Orthis | 1. Straight hinge line |
| B. Productus | 2. Curved hinge line |
| C. Terebratula | 3. Free form |
| D. Spirifer | 4. Delthyrium |

a) A—3 B—4 C—2 D—1
b) A—4 B—3 C—1 D—2
c) A—4 B—3 C—2 D—1
d) A—1 B—2 C—3 D—4

28. Sediment filled eroded channels preserved in stratal sequences typical of many deltaic sequences are called
a) Sole marks b) Wash-outs
c) Sand dikes d) Flute casts

29. The relations between the three sets of joints in "Mural jointing" is given by
a) Tow joint sets are horizontal while one is vertical
b) Two joint sets are vertical and one is horizontal
c) All three are horizontal d) All three are vertical

30. Match the following :

| List I | List II |
|---|---|
| A. Quartz | 1. Trachytes |
| B. Sanidine | 2. Phonolites |
| C. Nepheline | 3. Syenites |
| D. Alkali Feldspar | 4. Granites |

a) A—4 B—1 C—2 D—3
b) A—4 B—2 C—1 D—3
c) A—4 B—3 C—1 D—2
d) A—1 B—2 C—3 D—4

31. The paleozoic intertrappeans in Panjal traps are called
a) Kateru beds b) Gangamopteris beds
c) Karewas d) Lavender clays

32. The Alpine-Himalayan belt of earthquakes is
a) Shallow focus b) Deep focus
c) Intermediate focus d) None of the above

33. A segment of the earth's crust exhibiting similar geological and geophysical properties is called
a) Shield b) Platform c) Orogen d) Crustal type

34. Consider the following :
1. Thorium-Lead 2. Uranium-Lead
3. K-Ar

Of the above, which series has the least half-life

a) 1 b) 3 c) 2 d) 1, 2

35. Match the following :

| List I | List II |
|---|---|
| A. Aravalli | 1. N-S |
| B. Satpura | 2. E-W |
| C. Eastern Ghat | 3. NE-SW |
| D. Western Ghat | 4. NNE-SSW |

a) A—3 B—2 C—4 D—1
b) A—2 B—3 C—4 D—1
c) A—1 B—2 C—3 D—4
d) A—4 B—3 C—2 D—1

36. The boundary between basalts of oceanic crust with their Island Arcs and Andesite-Dacite-Rhyolite Volcanic rocks or Circum-Pacific belt is

a) Benioff zone
b) Ring of fire
c) Andesite line
d) Steinmann's trinity

37. Consider the following :

1. Gangamopteris beds
2. Zewan beds
3. Panjal traps
4. Muth quartzite
5. Syringothyris limestone

The correct succession of paleozoic rocks of Kashmir is given by

a) 4, 5, 3, 1, 2
b) 5, 4, 1, 3, 2
c) 1, 2, 3, 4, 5
d) 3, 4, 5, 1, 2

38. Meteorites containing 50% silicates are

a) Tektites
b) Chondrites
c) Achondrites
d) Pallasites

39. Hydrothermal deposits formed between surface and a depth of 1500 mts. are termed

a) Epithermal
b) Mesothermal
c) Telethermal
d) Hypothermal

40. Consider the following :

1. Si : Al ratio of Micas is 3 : 1
2. All Micas are optically negative
3. All Micas show (100) perfect cleavage
4. All are monoclinic

The correct code is

a) 1, 2, 4 b) 1, 3, 4 c) 1, 2, 3, 4 d) 1, 2

41. Griffthides belongs to

a) Cambrian
b) Ordovician
c) Silurian
d) Carboniferrous

42. Match the following :

| List I | List II |
|---|---|
| A. Autoclastic | 1. Loess |
| B. Anemoclastic | 2. Glacial-till |
| C. Bioclastic | 3. Coral-reef |
| D. Atmoclastic | 4. Bauxite |

a) A—2 B—1 C—3 D—4
b) A—3 B—2 C—1 D—4
c) A—2 B—1 C—4 D—3
d) A—1 B—2 C—3 D—4

43. Consider the following :
1. Recumbent fold
2. Thrust fault
3. Net slip greater than a mile
4. Active hanging wall

Of the above, characters true for Nappes are
a) 1 only b) 1, 3 c) 1, 2, 4 d) 1, 2, 3, 4

44. Earth's crust reaches its maximum thickness under
a) Shield areas
b) Archean orogens
c) Phanerozoic orogens
d) None of the above

45. The Twin-Sun theory of the origin of earth was advanced by
a) Fred Hoyle
b) Jeans and Jeffrey's
c) Kuiper
d) Von Schmidt

46. Consider the following :
1. Transform motion between offset ridges
2. Original offsets of MOR
3. Transcurrent motion beyond offset ridges
4. Strike-slip faults

Of the above, characters true of transform faults are
a) 3, 4. b) 1, 2 c) 1, 2, 4 d) 1, 2, 3, 4

47. Relict Oceanic crust is represented by
a) Tholeiite
b) Ophiolite
c) Tektite
d) None

48. Polarity reversals of earth occur once in every
a) 10 M.Y.
b) 100 M.Y.
c) 1 M.Y.
d) 11 M.Y.

49. Contemporary Volcanism and Sedimentation is characteristic of
a) Eugeosyncline
b) Miogeosyncline
c) Parageosyncline
d) Tephrageosyncline

50. When axial planes of folds dip directly down the axial surface they are termed as
a) Homoclines
b) Generative folds
c) Accordion folds
d) Reclined folds

51. Match the following :

| List I | List II |
|---|---|
| A. Joint system | 1. Persistent joint |
| B. Joint set | 2. Intersecting joint sets |
| C. Conjugate joints | 3. Parallel joints |
| D. Master joints | 4. Joints under same stress |

a) A—2 B—3 C—4 D—1
b) A—3 B—2 C—4 D—1
c) A—4 B—3 C—2 D—1
d) A—1 B—2 C—3 D—4

52. Consider the following :

1. Hornblende
2. Muscovite
3. Hypersthene
4. Spinel

Of the above, the group of saturated minerals is given by

a) 1, 2, 3, 4
b) 2, 3, 4
c) 1, 2, 3
d) 1, 2, 3, 4

53. Straight hinge-line without ears is represented by the lamellibranch

a) Pecten b) Exogyra c) Arca d) Unio

54. Match the following :

| List I | List II |
|---|---|
| A. Graded bedding | 1. Near shore |
| B. Current bedding | 2. Eugeosynclinal |
| C. Torrential bedding | 3. Alluvial fans |
| D. Herring-Bone bedding | 4. Tidal flats |

a) A—2 B—1 C—3 D—4
b) A—4 B—1 C—2 D—3
c) A—2 B—1 C—4 D—3
d) A—3 B—4 C—2 D—1

55. A cavity lined with minerals identical with those of the host rock is called

a) Geode b) Agate c) Druse d) Septarian

56. Consider the following :

1. Biotite
2. Staurolite
3. Almandine
4. Chlorite

The correct sequence of minerals with increasing grade of metamorphism is given by

a) 1, 2, 3, 4
b) 4, 1, 3, 2
c) 4, 1, 2, 3
d) 4, 3, 2, 1

57. Match the following :

List I List II

A. Pleuromya — 1. Dysodont
B. Nucula — 2. Schizodont
C. Trigonia — 3. Desmodont
D. Mytilus — 4. Taxodont

a) A—3 B—1 C—2 D—4
b) A—4 B—3 C—2 D—1
c) A—3 B—4 C—2 D—1
d) A—4 B—3 C—1 D—2

58. Pink interference colours are typical to
a) Garnet b) Augite c) Sphene d) Aegirine

59. Match the following :

| List I | List II |
|---|---|
| A. Carlsbad law | 1. Geniculate twins |
| B. Rutile law | 2. Quartz |
| C. Dauphine law | 3. Orthoclase |
| D. Albite law | 4. Albite |

a) A—3 B—1 C—2 D—4
b) A—3 B—2 C—1 D—4
c) A—1 B—2 C—3 D—4
d) A—2 B—3 C—1 D—4

60. Consider the following :
1. Delhi 2. Aravalli 3. Banded Gneissic complex 4. Rialo
The correct Archean succession of Rajasthan is given by
a) 3, 2, 4, 1 b) 3, 4, 2, 1 c) 1, 2, 3, 4 d) 4, 2, 1, 3

61. Vindhyans occupy the time period between
a) 1400-900 M.Y. b) 900-570 M.Y.
b) 1600-900 M.Y. d) 2500-1600 M.Y.

62. Country rocks for Bauxite in Katni area are
a) Vindhyans b) Gondwanas
c) Deccan traps 4) Precambrians

63. Tectonic disturbance is associated with
a) Orthoconglomerate b) Paraconglomerate
c) Petromict conglomerate d) Intraformational conglomerate

64. The basic difference between Gabbros and Basalts is
a) Gabbros are feldspar rich, while Basalts are feldspar poor
b) Gabbros are feldspar poor, while Basalts are feldspar rich
c) Gabbro has plagioclase while Basalt lacks plagioclase
d) Gabbro lacks plagioclase while Basalt has plagioclase

65. The end product of fractional crystallization of a Basaltic magma is
a) Gabbro b) Granite c) Dolerite d) Diorite

66. The most common minerals of banded iron ore are
 a) Magnetite and Haematite b) Haematite and Jasper
 c) Haematite and Chert d) Magnetite and Chert
67. Consider the following :
 1. Uranium 2. Copper 3. Apatite
 The correct sequence of paragenesis in Singhbhum Copper Thrust is given by
 a) 1, 2, 3 b) 3, 2, 1 c) 3, 1, 2 d) 2, 1, 3
68. In a plunging fold where both bedding and cleavage dip in the same direction but cleavage shows gentile dips than bedding, the beds are
 a) Rightside up b) Overturned
 c) Cannot be determined d) Inclined
69. Consider the following :
 1. Boudinage 2. Foliation 3. Fold axis 4. Mullion structure
 The penetrative fabric element is
 a) 2 b) 3 c) 1 d) 4
70. Match the following :

| List I | List II |
|---|---|
| A. Ribbon lake | 1. Fluvial |
| B. Bolson | 2. Glacial |
| C. Rapid | 3. Arid |
| D. Chimney | 4. Karst |
| | 5. Marine |

 a) A—2 B—3 C—1 D—4
 b) A—1 B—3 C—5 D—4
 c) A—2 B—3 C—1 D—5
 d) A—1 B—2 C—3 D—4
71. Back-arc basins are
 a) Andean Geosynclines b) Mediterranean Geosynclines
 c) Himalayan Geosynclines d) Japan sea Geosynclines
72. Largest negative gravity anomalies are recorded at
 a) MOR b) Trenches
 c) Island Arcs d) Continental rift valleys
73. Surface manifestations of mantle plumes are
 a) MOR/Arc b) Seamount/MOR
 c) Seamount/Volcanic Islands d) None of the above
74. Oceanic ridges are called oceanic rises only when they are
 a) Spreading centres b) Extinct
 c) Bounded by Volcanoes d) Seismically active
75. Continental drift is set into motion by
 a) Transform faults b) MOR c) Subduction zones d) Island arcs

76. Brachiopods are the typical faunal groups of
   a) Paleozoic/Mesozoic b) Mesozoic/Tertiary
   c) Tertiary/quarternary d) Only Paleozoic

77. Consider the following :
   1. Zaphrentis 2. Calceola
   3. Cystiphyllum 4. Columnaria

   The correct chronological sequence is given by
   a) 1, 2, 3, 4 b) 4, 3, 2, 1
   c) 3, 4, 1, 2 d) 2, 1, 3, 4

78. Match the following :

| List I | List II |
|---|---|
| A. Hipparion | 1. Dhok Pathan |
| B. Equus Sivalensis | 2. Tatrot |
| C. Dryopithecus | 3. Pinjor |
| D. Stegodon | 4. Chinji |
| | 5. Kamlial |

   a) A—5 B—3 C—4 D—1
   b) A—5 B—2 C—4 D—1
   c) A—4 B—2 C—5 D—1
   d) A—1 B—2 C—3 D—4

79. The coal fields of upper Assam are associated with
   a) Tippam b) Barail c) Jaintia d) Surma

80. Match the following :

| List I | List II |
|---|---|
| A. Archean Proterozoic boundary | 1. 600 M.Y. |
| B. Lower/middle proterozoic | 2. 1600 M.Y. |
| C. Upper proterozoic | 3. 2500 M.Y. |
| D. Beginning of Phanerozoic | 4. 2500-1600 M.Y. |

   a) A—3 B—4 C—2 D—1
   b) A—3 B—2 C—4 D—1
   c) A—4 B—3 C—2 D—1
   d) A—4 B—2 C—3 D—1

81. Consider the following :
   1. Garnet 2. Alkali feldspar
   3. Kyanite 4. Talc

   Of the above, stress minerals are given by
   a) 1, 3 b) 1, 2, 3 c) 2, 3 d) 2, 3, 4

82. Angular unconformity in Cuddapah basin occurs between
   a) Chitravati and Nallamallai b) Papaghani and Chitravati
   c) Srisailam and Chitravati d) Papaghani and Srisailam

83. According to the doctrine of isostacy elevated mountain ranges should show

a) Positive gravity anomolies b) Negative gravity anomolies
c) Positive magnetic anomolies d) Negative magnetic anomolies

84. Match the following :

| List I | List II |
|---|---|
| A. Scalene triangle | 1. Trisoctahedron |
| B. Isoscles triangle | 2. Hexoctahedron |
| C. Equilateral triangle | 3. Octahedron |
| D. Trapezium | 4. Trapezohedron |

a) A—2 B—1 C—3 D—4
b) A—1 B—2 C—3 D—4
c) A—3 B—1 C—2 D—4
d) A—1 B—2 C—3 D—4

85. Bronzite is a variety of

a) Hypersthene b) Diopside
c) Enstatite d) Augite

86. Zoned olivines have rims of

a) Forsterite b) Tephroite c) Iddingsite d) Fayalite

87. X = Pink, Y = Yellow, Z = Green is the pleochroic formula of

a) Jadiete b) Aegirine c) Pigeonite d) Hypersthene

88. The chemical composition Ca Si Ti 05 belongs to

a) Topaz b) Peruskite c) Sphene d) Mullite

89. Suture zones are demarcated by

a) Green schist b) Blue schist
c) Amphibolite d) Eclogite

90. Blanket sands are associated with

a) Quartz-arenites b) Arkoses
c) Graywackes d) Calc-arenites

91. Rocks in which modal olivine exceeds 90% are termed

a) Kimberlite b) Saxonite
c) Websterite d) Dunite

92. The eutectic proportion 42 : 58 belongs to

a) Orthoclase : Albite b) Orthoclase : Quartz
c) Diopside : Enstatite d) Anorthite : Olivine

93. The two leading solvents in supergene enrichment are

a) Ferric sulphate / Ferrous sulphate
b) Ferrous sulphate / Sulphuric acid
c) Ferric sulphate / Sulphuric acid
d) Nitric acid / Sulphuric acid

94. Gold in Wynaad Gold field is associated with
   a) Amphibolites
   b) Pegmatites
   c) Hornblende-schists
   d) Iron Pyrites

95. Consider the following :
   1. Manganese, Copper, Cobalt
   2. Depth between 3500-600 mts
   3. Volcano-chemical nature
   4. Non-depletable

   Of the above, the characters true for polymetallic nodules are given by
   a) 1, 2, 4
   b) 1, 3, 4
   c) 1, 2, 3
   d) 1, 2, 3, 4

96. The biggest linear basin of the Dharwar type in India is
   a) Sandur
   b) Shimoga
   c) Chitradurga
   d) Singhbhum

97. Magmatic carbonates are termed
   a) Carbonatites
   b) Kimberlites
   c) Komatiite
   d) Harzburgite

98. Porphyritic granites rich in orthoclase are called
   a) Adamellites
   b) Aplites
   c) Rapakivi granites
   d) Monzonites

99. Skew twins are associated with
   a) Columbite
   b) Fluorite
   c) Calcite
   d) Staurolite

100. Channel bars indicate
   a) Graded streams
   b) Valley flats
   c) Braided streams
   d) Flood plains

## ASSERTION & REASON

CODES :

a) Both Assertion & Reason are true and Reason is the explanation of Assertion.

b) Both Assertion & Reason are true and Reason is NOT explanation of Assertion.

c) Assertion is true and Reason is false.

d) Assertion is false and Reason is true.

101. Assertion : Cross-bedding and graded bedding never occur together.

   Reason : Cross-bedding forms in shallow waters while graded bedding forms in deep waters.

102. Assertion : The separation of Australia, Africa, India in early to middle Mesozoic led to evolution of Marsupials in Gondwanaland continents.

   Reason : Continental break up/drift may lead to evolution of specialised groups due to adaptive pressure.

103. Assertion : Transform faults are aligned parallel to motion of bounding plates.

Reason : Plate motions are normal to ridge-axis.

104. Assertion : Recorded earthquakes defined on a World-map mark outlines of lithospheric plates.

Reason : Hypocentre distribution is concentrated along plate margins.

105. Assertion : Opal rarely occurs in older sandstones.

Reason : Devitrifies into chalcedony.

106. Assertion : Hydroxyl-bearing minerals like hornblende and biotite are typical to plutonic igneous rocks.

Reason : Hydroxyl-bearing minerals tend to break down in a volcanic environment.

107. Assertion : In Cuddapah, every younger series overlaps the older.

Reason : Shape and depth of basin varied from time to time

108. Assertion : Lower triassic witnessed dry and arid climate.

Reason : Panchet series, of lower triassic are red beds.

109. Assertion : Barrowian Isogrades are also Isotherms.

Reason : Barrowian Isogrades indicate changes in degree of metamorphism corresponding to changes in temperature.

110. Assertion : Rhyolites and Trachytes show flow structures in great perfection.

Reason : Flow structure develops well in viscous lavas.

111. Assertion : Niger delta is post drift feature.

Reason : Niger delta overlaps Africa and South America fit.

112. Assertion : Quartz is piezoelectric mineral.

Reason : Quartz develops an electric charge when subject to stress.

113. Assertion : Best source rocks for oil/gas are fine-grained rocks.

Reason : Coarse grained sediments permit Winnowing and oxidation of hydrocarbons.

114. Assertion : Prograde metamorphism may lead to a hydrothermal system.

Reason : Prograde metamorphism involves progressive expulsion of water and volatiles.

115. Assertion : Longitudinal sections of tetragonal and hexagonal crystals are anisotropic.

Reason : Longitudinal section of Hexagonal and Tetragonal crystals show straight extinction.

116. Assertion : Hypocentres of deep-focus earthquakes are as deep as

700 km.

Reason : Lithospheric plates plunge till 700 km deep into mantle.

117. Assertion : Geologically younger sedimentary rocks display greater variety of heavy minerals.

Reason : Heavy minerals have been probably removed from geologically older sediments by differential solution.

118. Assertion : P-waves or body waves behave anomalously in a region of impending earthquake.

Reason : Their velocity approaches the velocity of S-waves.

119. Assertion : Rift valleys of MOR are regions of highest heat flow on the earth.

Reason : They are on ascending limbs of convection cells.

120. Assertion : Irregular echinoids are characterised by a bilateral symmetry.

Reason : The radial symmetry so typical of regular echinoids evolved into the bilateral symmetry of irregular echinoids.

## ANSWERS

| | | | | | |
|---|---|---|---|---|---|
| 1. b | 2. d | 3. c | 4. b | 5. d | 6. c |
| 7. b | 8. a | 9. b | 10. c | 11. b | 12. c |
| 13. a | 14. d | 15. b | 16. b | 17. b | 18. a |
| 19. d | 20. a | 21. a | 22. b | 23. c | 24. d |
| 25. b | 26. a | 27. c | 28. b | 29. b | 30. a |
| 31. b | 32. c | 33. d | 34. a | 35. a | 36. c |
| 37. a | 38. d | 39. a | 40. a | 41. d | 42. d |
| 43. d | 44. c | 45. a | 46. d | 47. b | 48. c |
| 49. a | 50. d | 51. a | 52. c | 53. c | 54. a |
| 55. c | 56. b | 57. c | 58. c | 59. a | 60. a |
| 61. b | 62. b | 63. d | 64. a | 65. b | 66. c |
| 67. c | 68. b | 69. a | 70. c | 71. d | 72. b |
| 73. c | 74. d | 75. b | 76. a | 77. b | 78. c |
| 79. b | 80. a | 81. a | 82. a | 83. b | 84. a |
| 85. a | 86. d | 87. d | 88. c | 89. b | 90. a |
| 91. d | 92. a | 93. c | 94. d | 95. d | 96. a |
| 97. a | 98. c | 99. d | 100. c | 101. a | 102. a |
| 103. a | 104. a | 105. a | 106. a | 107. a | 108. a |
| 109. a | 110. a | 111. a | 112. a | 113. a | 114. a |
| 115. b | 116. b | 117. a | 118. c | 119. a | 120. a |

## 1. Civil Services (Preliminary) Examination, 1990

1. The present day atmosphere and oceans are a result of
   a) Cosmic precipitations
   b) Volcanic exhaltations
   c) Primordial ocean and atmosphere evolution
   d) None of the above

2. When two continental plates weld, it is described as
   a) Suture zone
   b) Andesite line
   c) Island arc
   d) Subduction zone

3. The oldest sediments are of the age
   a) 100 m.a.
   b) 200 m.a.
   c) 2000 m.a.
   d) 400 m.a.

4. Calc-alkali volcanoes occurring above Benioff zones are called
   a) Andesite volcanoes
   b) Island arcs
   c) Guyots
   d) MORs

5. Lithosphere comprises of
   a) LVZ and Mantle
   b) LVZ and upper Mantle
   c) LVZ and upper core, including mantle
   d) Only the Crust

6. Choose the correct sequence from the following
   a) Valley—Gully—Canyon—Gorge
   b) Gully—Valley—Gorge—Canyon
   c) Gully—Valley—Canyon—Gorge
   d) Valley—Gully—Gorge—Canyon

7. Identify the drainage pattern given in the diagram

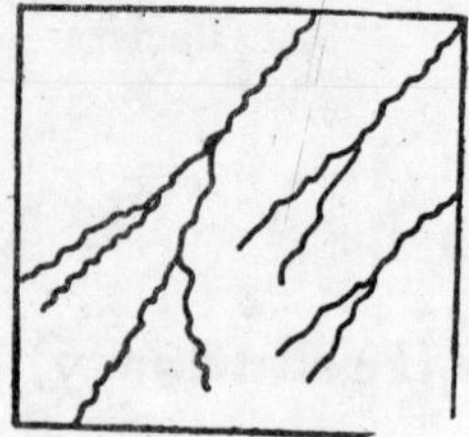

a) Trellis
b) Dendritic
c) Anastomosing
d) Rectangular

8. Which of the following affords geological evidence for the existence of Gondwana land ?
a) Fauna and flora
b) Geometric fit
c) Glaciation
d) Volcanism

9. Choose the correct sequence from top of surface to the bottom of a water-table
a) Percolated zone–Capillary zone—Saturated zone
b) Capillary zone—Percolated zone—Saturated zone
c) Saturated zone—Percolated zone—Capillary zone
d) Capillary zone—Saturated zone—Percolated zone

10. Consider the following w.r.t. Ventifacts
(i) They are a result of wind abrasion
(ii) Strong winds are necessary for their formation
(iii) Presence of vegetation promotes ventifact formation
(iv) They are associated with loess
a) (i) and (ii) are correct
b) (i), (ii) and (iii) are correct
c) (ii) and (iii) correct
d) All of the above are correct

11. When a beach occupies a swamp, it reflects
a) Progradation
b) Transgression
c) Offlap
d) Eustatic change

12. Match the following and choose the correct answer

| | | | |
|---|---|---|---|
| A. | Crag | 1. | Sea |
| B. | Loess | 2. | Wind |
| C. | Nick point | 3. | Glaciers |
| D. | Shingle | 4. | Stream |

a) A—3 B—2 C—4 D—1
b) A—4 B—2 C—1 D—3
c) A—1 B—2 C—3 D—4
d) A—4 B—3 C—2 D—1

13. Caliche deposits are formed due to
    a) Colloidal action
    b) Evaporation
    c) Organic precipitation
    d) Biochemical action
14. Red beds are formed due to
    a) Oxidation
    b) Iron oxide precipitation
    c) Reduction
    d) Organic action
15. Extinction of trilobites is due to the fact that
    a) They were eaten up by sharks
    b) The waters became toxic
    c) They reached acme of evolution
    d) They failed to adaptively radiate
16. Codsider the following data and answer tbe question following it

| *Drainage density* | *Slope* | *Average Annual Rainfall* |
|---|---|---|
| 1. 7.2 | 9 | 900 cm |
| 2. 8.0 | 10 | 1000 cm |
| 3. 9.0 | 7 | 600 cm |
| 4. 9.0 | 8 | 900 cm |

The maximum erosion will occur at
a) 1 and 2
b) 2 and 4
c) 3 and 4
d) 1 and 4

17. Identify the fault shown in the diagram

a) Normal fault
b) Reverse fault
c) Strike-slip fault
d) Transform fault

18. In a structural map, the dip joints are represented by the symbol

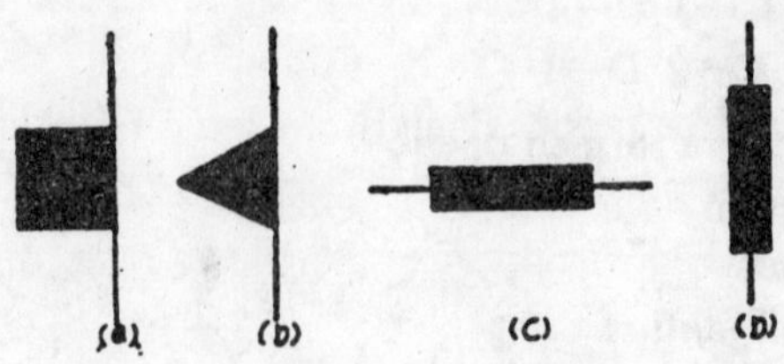


19. Stratigraphic throw equals vertical throw when the beds are
a) Vertical
b) Horizontal
c) Oblique
d) None of the above

20. On a level surface a bed is encountered at 100 m at A, 120 m at B and 150 m north of A. The dip of the bed is
a) North-east
b) North-West
c) North
d) South-east

21. Match the following and choose the correct answer

| | |
|---|---|
| A. Horizontal axial plane | 1. Recumbent fold |
| B. Circular outcrop | 2. Cyllindrical fold |
| C. Equal limbs | 3. Symmetrical fold |
| D. Convergence of dips | 4. Plunging fold |

a) A—1 B—2 C—3 D—4
b) A—4 B—3 C—2 D—1
c) A—3 B—2 C—1 D—4
d) A—1 B—3 C—2 D—4

22. Consider the following illustration

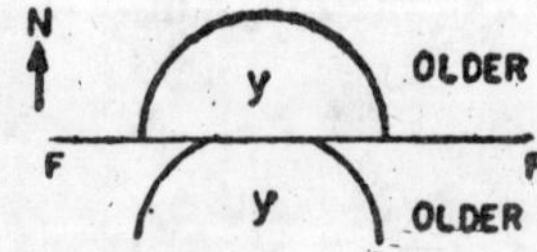

a) Northern block is upthrown and moved down
b) Southern block is upthrown and moved down
c) Southern block is downthrown and moved up
d) Northern block is downthrown and moved up

23. In the plane of Lineation if foliation occurs
a) It will be parallel to lineation
b) It will be perpendicular to lineation
c) It will be not be perpendicular, but inclined to lineation
d) It will be randomly oriented will respect to lineation

24. If slickensides and fault plane dip in opposing direction, the fault may be a
a) Reverse fault
b) Normal fault
c) Strike fault
d) Strike-slip fault

25. With reference to the diagram given below, if a river flows in the N-S direction, it will dip

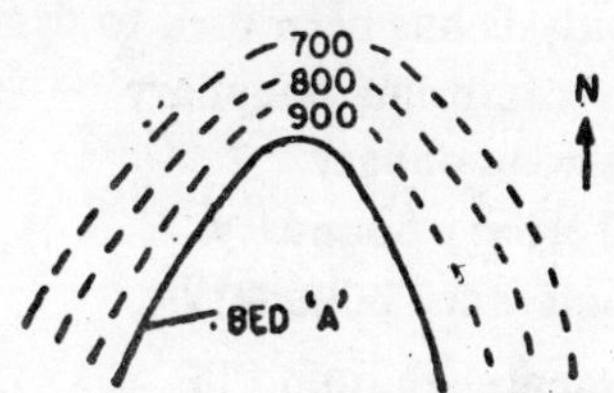


a) Upstream
b) Downstream
c) Oliquely to bed A
d) Parallel to bed A

26. A bed strikes N 45° E in an outcrop. If in the vertical section it strickes N 45° W and if the dip is measured in the vertical section
a) It will show no dip
b) It will show maximum dip
c) It shows dip less than the maximum dip
d) It shows two different dips

27. Which one of the following can be used as a marker bed ?
a) Red Bed
b) Bentonite bed
c) Laminated shale
d) Conglomerate bed

28. The term used to denote the stratigraphic horizon where a species reaches its acme its
   a) Acme zone
   b) Biozone
   c) Epibole
   d) Teilzone

29. Pick out the correct group of coequals
   a) Cuddapah, Chattisgarh, Bijawer
   b) Semris, Chattisgarh,Sullavais
   c) Cuddapah, Pakhals and Chattisgarh
   d) Semris, Pakhals and Gwaliors

30. The present seas essentially in their present character evolved in
   a) Paleozoic
   b) Mesozoic
   c) Precambrian
   d) Tertiary

31. Pollen-grain analysis has been used to demarcate
   a) Cambrian-Precambrian boundary
   b) Permo-Triassic boundary
   c) Cretaceous-Tertiary boundary
   d) Neogene-Quaternary boundary

32. The oldest sediments are found in
   a) Sargur schists
   b) Kolar schists
   c) Dharwar schists
   d) Older metamorphic group

23 The largest time period among the following is
   a) Cambrian
   b) Tertiary
   c) Cretaceous
   d) Triassic

34. The geological period which records both continental and marine facies in India is
   a) Triassic
   b) Jurassic
   c) Permian
   d) Eocene

35. Pick out the correctly matched pair from the following

a) Zewan beds——Triassic

b) Gangamopteris beds——Cambrian

c) Karewa beds——Carboniferous

d) Bagh beds——Triassic

36. The strata which record the complete evolutionary history of a species is termed as

a) Biolith

b) Biocoenose

c) Thanatocoenose

d) None of the above

37. In Siwaliks, the absence of Proboscidea and the presence of primates is recorded in

a) Kamlial stage

b) Nagri stage

c) Dhok Pathan stage

d) Chinji stage

38. The fossil group of the lower palaeozoic is

a) Graptolites, echinoids, trilobites

b) Branchiopods, graptolites and trilobites

c) Gasteropods, graptolites and trilobites

d) Corals, graptolites and trilobites

39. The body chamber filled with gas (air) in cephalopods is called

a) Phragmacone

b) Gyrocone

c) Septa

d) Siphuncle

40. The Gondwana flora belonging to Parsora stage is

a) Schizoneura

b) Phyllotheca

c) Noegeropthiosis

d) Gangamopteris

41. The Brachial-skeleton forms a loop in one of the following brachiopods

a) Rhynconella

b) Terebratula

c) Terebratulina

d) Spirifer

42. Which of the following gasteropods show both dextral and sinistral colling ?

a) Physa

b) Murex

c) Fusina

d) Helix

43. The earliest trilobites were characterized by

a) A large pygidium

b) Primitive condition of eyes

c) Absence of genal angles

d) Presence of genal angles

44. The main part of the echinoid test containing numerous plates is described as

a) Peristome

b) Apical disc

c) Corona

d) None of the above

45. The cretaceous period is marked by

a) Disappearance of certain vertebrates

b) Appearance of flowering plants

c) Disappearance of flowering plants

d) None of the above

46. When the last whorl is so large that it covers all the other whorls, the condition is said to be

a) Convolute

b) Planispiral

c) Globular

d) Helicoidal

4 . Absence of brachial skeleton and presence of pedicle opening on the ventral valve is typical of

a) Atremata

b) Neotremata

c) Protremata

d) Teleotremata

48. Madreporic plate is a/an

a) Genital plate

b) Occular plate

c) Ambulacral plate

d) Inter-ambulacral plate

49. Lepidodendron flora is

a) absent in lower Gondwanas of India
b) present in lower Gondwanas of India
c) absent in Gondwanas of India as a whole
d) recorded in the upper Gondwanas of India

50. If Andesite undergoes weathering it will give rise to
a) Graywacke
b) Orthoquartzite
c) Arkose
d) Calcarenite

51. Consider the following statements
(i) Ripple marks are current produced structures
(ii) They are typical of geosynclinal facies
(iii) They are true subaerial structures
(iv) They accur along with groove casts
a) only (i) and (iii) are correct
b) only (ii) and (iv) are correct
c) (i) (iii) and (iv) are correct
d) all are correct

52. Feldspar is absent in
a) Gondite
b) Kodurite
c) Khondalite
d) Charnockite

Q. 53. and Q. 54. are based on the diagram below which represents the fundamental traingle used to classify sedimentary rocks

53. In the traingle, the Point A represents

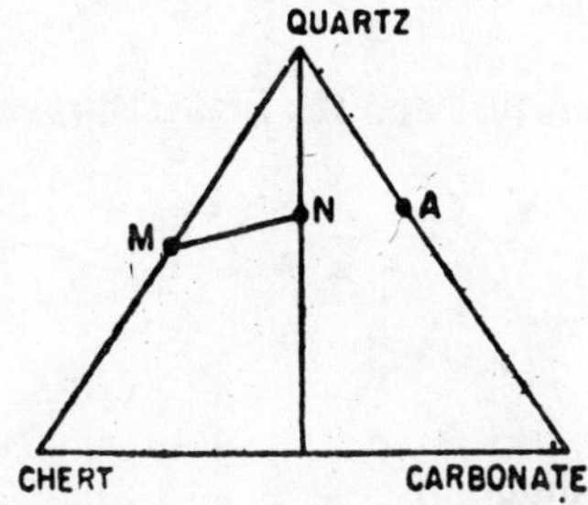


a) Carbonaceous sand
b) Carbonaceous sand plus some chert

c) Cherty limestone

d) None of the above

54. The line NM represents

a) Chert, quartz plus some carbonate

b) Subequal amounts of quartz and chert

c) Equal amounts of quartz and carbonate

d) Subequal amounts of chert and carbonate

55. Spandite a peculiar type of feldspar occurs in the hybrid rock called

a) Khondalite

b) Kodurite

c) Gondite

d) Charnockite

56. The process responsible for rims of Augite enclosing olivine crystals is

a) Fractional crystallization

b) Gravity settling

c) Assimilation

d) Magmatic mixing

57. Eclogites are distinguished from the amphibolites on the basis of

a) Pyroxene

b) Amphibole

c) Garnet

d) Plagioclase

58. Perthite, quartz, sillimanite, & pyroxene are present in

a) Khondalite

b) Charnockite

c) Eclogite

d) Amphibolite

59. The highest temperature and the highest pressure facies among the following is

a) Granulite

b) Eclogite

c) Blueschist

d) Sanidinite

60. Evaporites are made up of

a) Dolomite, gypsum and anhydrite

b) Gypsum, halite and anhydrite

c) Limestone, gypsum and anhydrite

d) Dolomite, gypsum and limestone

61. The correct sequence of assemblage from the inner aureole to the outer aureole is

a) Biotite-schist ---→ Andalusite schist ---→ Spotted slate ---→Injection hornfels

b) Andalusite schist ---→ Injection hornfels ---→ Biotite Schist ---→ Spotted slate

c) Injection hornfels ---→ Biotite schist ---→ Andalusite schist ---→ spotted slate

d) Andalusite schist --→ Biotite schist ---→ Spotted slate ---→ Injection hornfels

62. The transition from low-pressure facies to high-pressure facies is marked by a change from

a) Heulandite to Lawsonite

b) Lawsonite to Heulandite

c) Lawsonite to Sanidinite

d) None of the above

63. If cherty limestone undergoes weathering it will give rise to

a) Chert and limestone fragments equally

b) Subequal amounts of chert and limestone

c) Only chert fragments

d) Only limestone fragments

64. Bedded chert is formed due to

a) Diagenetic replacement

b) Direct deposition of colloidal silica

c) Chemical precipitation of silica

d) Organic precipitation of silica

65. Presence of incipient crystals and the dominant temperature effects is a characteristic feature of

a) Granulose fabric

b) Hornfels fabric

c) Schistose fabric

d) Gneissose fabric

66. Fennitisation is a process connected with

a) Kimberlites

b) Carbonatites

c) Komatiites

d) Meteorites

67. Choose the correct pair from the following

a) Groove castes---Paleocurrent direction

b) Graded bedding——Subaerial structure
c) Flute casts——Top of beds
d) Ripple marks——Subaerial structure

68. Dolstone comprises of
a) Dolomite and limestone plus evaporite
b) Only calcium carbonate
c) Only magnesium carbonate
d) Both calcium and magnesium carbonates

69. Meta-aluminous condition is represented by
a) $Al_2O_3 > K_2O + Na_2O + CaO$
b) $Al_2O_3 > K_2O + Na_2O$
c) $Al_2O_3 < K_2O + Na_2O$
d) $Al_2O_3 > Na_2O + CaO$

70. Consider the following statements
(i) Quartz is typical of undersaturated rocks
(ii) Olivine is typical of oversaturated rocks
(iii) Feldspars are typical of alkaline rocks
a) All are correct
b) (i) and (ii) are correct
c) (ii) and (iii) are correct
d) Only (iii) is correct

71. The most important volatile of magmas is
a) Hydrogen
b) Nitrogen
c) Oxygen
d) Water

72. The correct sequence in stratified igneous complexes is
a) Magnetite ——→ olivine ——→ augite ——→ hornblende
b) Olivine ——→ magnetite ——→ augite ——→ hornblende
c) Augite ——→ olivine ——→ magnetite ——→ hornblende
d) Magnetite ——→ olivine ——→ hornblende ——→ augite

73. The lithological succession of Gondwanas is
a) Sand ——→ Shale ——→ Coal
b) Sand ——→ Coal ——→ Shale
c) Coal ——→ Sand ——→ Shale
d) Shale ——→ Sand ——→ Coal

74. The typical breakdown products of high-grade metamorphic assemblage to low-grade metamorphic assemblage are
a) Quartz-Albite

b) Albite-Epidote
c) Epidote-Quartz
d) Epidote-Calcite

75. Match the following and choose the correct answer

| | | | |
|---|---|---|---|
| A. | Sanidine | 1. | Bytownite |
| B. | $Ab_{30}$—$An_{70}$ | 2. | Orthoclase |
| C. | (010) twin lamellae | 3. | Labradorite |
| D. | Optically negative | 4. | Albite |

a) A—2 B—3 C—4 D—1
b) A—2 B—3 C—1 D—4
c) A—3 B—2 C—4 D—1
d) A—1 B—2 C—3 D—4

76. The pleochroic absorption of Hornblende is represented by

a) X>Y>Z
b) Y>X>Z
c) Z>Y>X
d) Z>X>Y

77. Consider the following statements

(i) Triclinic sections are only pinacoidal
(ii) They have only one plane of symmetry
(iii) They have no axis of symmetry
(iv) They lack a centre of symmetry

a) (i) and (iii) are correct
b) (i), (iii) and (iv) are correct
c) (ii) and (iii) are correct
d) All are correct

78. Match the following and choose the correct answer

| | | | |
|---|---|---|---|
| A. | Unique 4-fold axis | (1) | Tourmaline |
| B. | Unique 2-fold b-axis | (2) | Apophyllite |
| C. | Unique 6-fold axis | (3) | Sphene |
| D. | Only centre | (4) | Albite |

a) A—2 B—3 C—1 D—4
b) A—2 B—4 C—3 D—1
c) A—1 B—2 C—3 D—4
d) A—4 B—3 C—2 D—1

79. Arrange the following in the increasing order of refractive index

a) Albite —→ fluorspar —→ garnet —→ diamond

b) Fluorspar ⟶ albite ⟶ garnet ⟶ diamond
c) Albite ⟶ garnet ⟶ fluorspar ⟶ diamond
d) Albite ⟶ garnet ⟶ diamond ⟶ fluorspar

80 Match the following and choose the correct answer

| | | | |
|---|---|---|---|
| A. | Uvarovite | 1. | Variety of Grossular |
| B. | Andradite | 2. | Chromium-calcium garnet |
| C. | Cinnamon stone | 3. | Iron-aluminium silicate |
| D. | Almandine | 4. | Calcium-iron silicate |

a) A—2 B—3 C—1 D—4
b) A—2 B—4 C—1 D—3
c) A—1 B—2 C—3 D—4
d) A—4 B—3 C—2 D—1

81. The sharpest transition or inversion point occurs between
a) α-quartz and β-quartz
b) Tridymite and cristobalite
c) Quartz and tridymite
d) Quartz and cristobalite

82. The most common alteration product of plagioclase feldspars is
a) Epidote
b) Scapolite
c) Chlorite
d) Mica

83. The representative formula for the Mica group as a whole is
a) $X_4\ Y_{8-8}\ Z_8\ O_{20}(OH, F)_4$
b) $X_2\ Y_{4-6}\ Z_8\ O_{20}(OH, F)_2$
c) $X_2\ Y_{4-6}\ Z_8\ O_{20}(OH, F)_4$
d) $X_{2-4}\ Y_{4-6}\ Z_8\ O_{10}(OH, F)_4$

84. Match the following and choose the correct answer

| | | | |
|---|---|---|---|
| A. | Hypersthene | 1. | Resinous |
| B. | Zinc blende | 2. | Adamantine |
| C. | Brucite | 3. | Pearly |
| D. | Diamond | 4. | Metallic |

a) A—4 B—1 C—3 D—2
b) A—1 B—2 C—3 D—4
c) A—1 B—3 C—4 D—2
d) A—1 B—3 C—2 D—4

85 Electric charges developed on opposite crystal faces due to lack of stress is described as
a) Piezoelectricity

b) Pyroelectricity

c) Phosphorescence

d) Fluorescence

86. Maximum interference colours are shown by uniaxial minerals in longitudinal sections because

a) One vibration direction coincides with one crystallographic direction

b) Two vibration directions coincide with three crystallographic directions

c) Two vibration directions coincide with two crystallographic directions

d) Three vibration directions coincide with two crystallographic directions

87. Consider the following statements

(i) Prism angle of amphibole is 120°

(ii) Orthopinacoid is not present in pyroxenes

(iii) Crystals of amphiboles terminate with three faces

(iv) Pyroxenes are 8-sided in transverse sections

a) (i), (iii) and (iv) are correct

b) (i), (ii) and (iv) are correct

c) (ii) and (iv) are correct

d) All are correct

88. Epidote shows an elongation along b-axis because it crystallizes in

a) Orthorhombic system

b) Monoclinic system

c) Triclinic system

d) Tetragonal system

89. Consider the following statements

(i) Amorphous crystals lack internal structure

(ii) They are isotropic

(iii) They exhibit anamalous interference colours

(iv) They are delinquescent

a) Only (i) and (ii) are correct

b) (i), (ii) and (iv) are correct

c) (ii) and (iv) are correct

d) (i), (ii) and (iii) are correct

90. The following metals are found as we proceed away from the intrusive Batholith outwards in that order

a) Copper—Iron—Zinc—Tin
b) Iron—Tin—Copper—Zinc
c) Tin—Iron—Copper—Zinc
d) Copper—Tin—Iron—Zinc

91 Diamonds are found in
a) Andhra Pradesh and Madhya Pradesh
b) Madhya Pradesh and Maharashtra
c) Madhya Pradesh and Gujarat
d) Only Madhya Pradesh

92. Gold in the KGF is localized at the contact of
a) Massive and fibrous amphibolite
b) Quartz and amphibolite
c) Fibrous amphibolite and granular amphibolite
d) Amphibolite and granite

93. Orpiment is a mineral of
a) Antimony
b) Molybdenum
c) Arsenic
d) Gallium

94. Consider the following statements
(i) Mesothermal deposits form farthest from the intrusive
(ii) They are associated with high to medium temperature gangue
(iii) They are usually intermingled with xenothermal deposits
(iv) They form at a lower temperatures than hypothermal deposits
a) (ii) and (iv) are correct
b) (i), (ii) and (iv) are correct
c) (ii), (iii) and (iv) are correct
d) All are correct

95. Pick the group belonging to Damodar valley coal fields
a) Ranigunj, Bokaro, Giridih
b) Ranigunj, Talcher, Giridih
c) Ranigunj, Bokaro, Singrauli
d) Singrauli, Bokaro, Giridih

96. The age of Lignite deposits of Palana in Rajasthan is
a) Miocene
b) Eocene
c) Oligocene

d) Pliocene

97. Consider the following statements for the Supergene Enrichment Process

(i) Shallow occurrence of pyrite is necessary for supergene enrichment

(ii) Primary sulphides should occur below watertable

(iii) Carbonates are found in enriched zone

(iv) Protore is found associated with gossans

a) (i) and (ii) are correct

b) (i), (ii) and (iii) are correct

c) (ii), (iii) and (iv) are correct

d) All are correct

98. Match the following and choose the correct answer

| | | | |
|---|---|---|---|
| A. | Lead-zinc | 1. | Precambrian Dharwars |
| B. | Magnetite | 2. | Vindhyan system |
| C. | Pyrite | 3. | Gangpur series |
| D. | Lignite | 4. | Cuddalore sandstone |

a) A—1 B—2 C—3 D—4

b) A—3 B—1 C—2 D—4

c) A—3 B—1 C—4 D—2

d) A—4 B—3 C—2 D—1

99. Match the following and choose the correct answer

| | | | |
|---|---|---|---|
| A, | Neyveli lignite | 1. | Miocene |
| B. | Iron deposits in Ranigunj coal field | 2. | Damuda series |
| C. | Oligocene coal | 3. | A.P. coals |
| D. | Kamtee sandstones | 4. | Barails |

a) A—1 B—2 C—4 D—3

b) A—2 B—1 C—3 D—4

c) A—4 B—3 C—2 D—1

d) A—1 B—4 C—2 D—3

100. If a tourmaline plate is placed over a dot the observer sees two dots, this is due to

a) Refraction

b) Reflection

c) Double refraction

d) Polarisation

101. Match the following and choose the correct answer

| | | | |
|---|---|---|---|
| A. | Dauphine law | 1. | Tetragonal system |
| B. | Pericline law | 2. | Quartz |

C. Iron-cross twin 3. Albite
D. Geniculate twins 4. Pyrite

a) A—2 B—3 C—4 D—1
b) A—1 B—2 C—3 D—4
c) A—4 B—3 C—2 D—1
d) A—2 B—3 C—1 D—4

Questions 102—111 are based on Reasoning and Assertion, in which two statements would be given, which are followed by five alternatives. Only one of them is fully correct. Mark

a) When both Assertion and Reason are correct and the Reason explains the Assertion
b) When both Assertion and Reason are correct but the Reason does not support the Assertion
c) When Assertion is correct and Reason is incorrect
d) When Reason is correct and Assertion is false
e) When both Assertion and Reason are false

102. Assertion : It is believed that oil migrated into Tippam formation from the underlying Surmas
Reason : Surmas are found to be devoid of micro-fauna and micro-flora

103. Assertion : Gouge becomes silicified in the vicinity of fault to develop into pseudotachylite
Reason : Local frictional heat is responsible for silicification and development of pseudotachylite

104. Assertion : Misfit rivers are those that occupy lesser areas in their drainage channels
Reason : Misfit rivers are formed due to stream piracy

105. Assertion : Sphericity is said to be a characteristic of source rock weathering
Reason : Transportation has little effect on sphericity

106. Assertion : When ordinary light strikes a tourmaline plate, it vibrates in two directions within the same plane
Reason : Light becomes plane polarised upon striking the tourmaline plate

107. Assertion : Fossil life is not found in Archaens
Reason : Archaen seas were toxic

108. Assertion : Cordierite is found in Norites
Reason : Cordierite is formed from pelitic assemblages

109. Assertion : Symbols such as $\sqrt{2}a$, $\alpha$ or 2a 1.736 ........b, c are impossible
Reason : According to the Law of Rational Indices, the intercepts that any crystal face makes on the crystallographic axis are either infinite or small rational multiples of intercepts made by unit forms

110. **Assertion :** Corresponding Interfacial angles are constant for all crystals of a given mineral

**Reason :** The atomic structure of the crystal for any one mineral is fixed

111. **Assertion :** Coral reefs are formed in shallow waters

**Reason :** Subsidence of reef keeps pace with the building of the reef

## ANSWERS

| | | | |
|---|---|---|---|
| 1. b | 2. a | 3. b | 4. b |
| 5. b | 6. b | 7. a | 8. a |
| 9. a | 10. a | 11. b | 12. a |
| 13. b | 14. b | 15. d | 16. b |
| 17. c | 18. c | 19. b | 20. a |
| 21. a | 22. a | 23. c | 24. a |
| 25. c | 26. a | 27. b | 28. c |
| 29. b | 30. a | 31. b | 32. a |
| 33. a | 34. b | 35. a | 36. b |
| 37. b | 38. b | 39. a | 40. c |
| 41. a | 42. a | 43. b | 44. c |
| 45. a | 46. a | 47. b | 48. a |
| 49. c | 50. a | 51. a | 52. a |
| 53. a | 54. a | 55. b | 56. a |
| 57. a | 58. b | 59. b | 60. b |
| 61. a | 62 a | 63. b | 64. b |
| 65. b | 66. b | 67. d | 68. d |
| 69. b | 70. d | 71. d | 72. a |
| 73. a | 74. a | 75. a | 76. c |
| 77. a | 78. a | 79. b | 80. b |
| 81. a | 82. b | 83. c | 84. a |
| 85. b | 86. c | 87. a | 88. b |
| 89. a | 90. c | 91. a | 92. a |
| 93. c | 94. a | 95. b | 96. b |
| 97. a | 98. b | 99. a | 100. c |
| 101. a | 102. b | 103. a | 104. d |
| 105. c | 106. a | 107. c | 108. b |
| 109. a | 110. a | 111. b | |

## 2. Civil Services (Preliminary) Examiuation, 1989

1. The approximate density of the earth is
   a) 5.5
   b) 5.8
   c) 5.1
   d) 5.2

2. Plate motions are recognised by
   a) Their relative motion to one another
   b) Their motion with respect to fixed points
   c) Their motion with respect to poles
   d) All of the above

3. The competency of a stream means
   a) Its ability to carry huge boulders
   b) Its ability to carry large amount of load
   c) Its ability to cause widespread erosion
   d) None of the above

4. Flat-topped seamounts are known as
   a) Submarine volcanoes
   b) Guyots
   c) Groynes
   d) Terraces

5. Which of the following is the best method for determining the age of rocks ?
   a) Palaeomagnetic evidences
   b) Structural features of rocks
   c) Radioactive methods
   d) Seismic refraction profiling

6. Which of the following is a fresh water lake ?
   a) Wular
   b) Chilka
   c) Pulicat
   d) Dal

7. Ria Coast is
   a) A faulting scarp along the coast
   b) A topographic high near the coast
   c) A submerged fjord

d) A coastline running parallel to the adjacent highlands

8. The past geologic processes can be explained by the processes taking place presently by
   a) Werner's concept
   b) Charles Lyell's theory
   c) Uniformitarianism
   d) Law of faunal succession

9. In the following figure indicate the direction of the wind

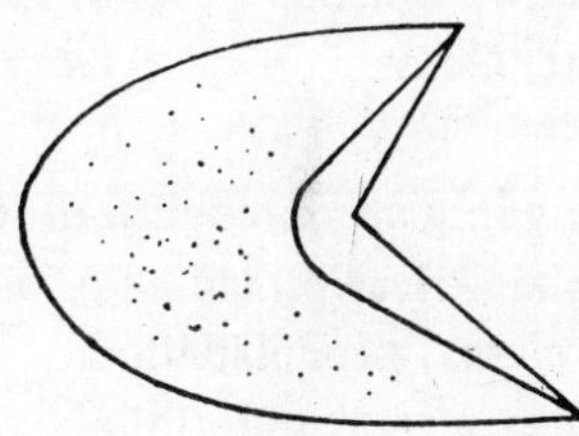

   a) E to W
   b) W to E
   c) N to S
   d) S to N

10. The figure given below shows that it is

   a) An end moraine
   b) A lateral moraine
   c) A medial moraine
   d) Not a moraine

11. From the figure given below, it can be seen that

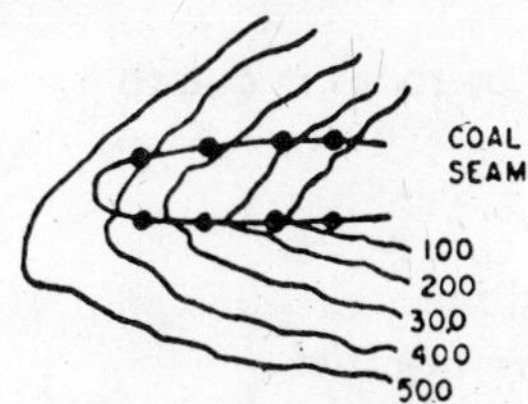


   a) The coal seam is dripping higher than the river beds
   b) The river beds are dipping higher than the coal seam
   c) Both are dipping equally
   d) None of the above

12. Calcium Carbonate is generally formed due to
   a) Evaporation of water

b) Precipitation
c) Loss of carbondioxide
d) All of the above
e) None of the above

13. The clinometer compass can be used to find the structural trend of rocks containing the following minerals
a) Magnetite, Pyrite, Sphalerite
b) Chromite, Magnetite, Galena
c) Galena, Sphalerite, Gold
d) None of the above

14. Arrange the following in a chronological order
a) Waterfalls, peneplain, river piracy
b) Waterfalls, river piracy, peneplanation
c) Peneplanation, river piracy, waterfalls
d) River piracy, peneplanation, waterfalls

15. Which of the following is associated with subduction zones ?
a) Blueschists
b) Amphibolites
c) Eclogites
d) Greywackes

16. In a road cutting the following sequence of beds are encountered

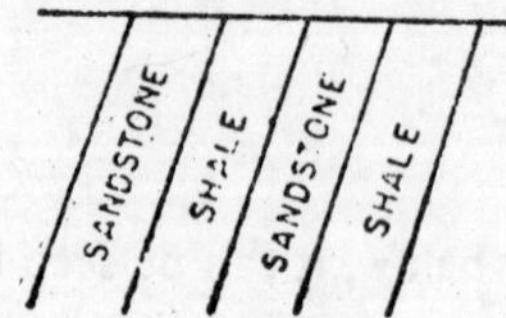


The above sequence of rocks is due to
a) Folding
b) Faulting
c) Folding and Faulting
d) Presence of Unconfirmities

17. In the case of a vertical fault, it is difficult to measure the slip because
a) The HW is missing
b) The FW is missing
c) The HW and FW are missing
d) The vertical fault has no slip

18. The joints that are perpendicular to the axial planes of folds are described as

a) Cross joints
b) Strike joints
c) Dip joints
d) Diagonal joints

19. The figure given below is a/an

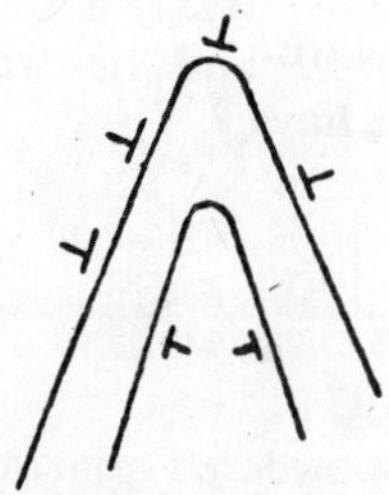

a) Plunging anticlinorium
b) Overturned fold
c) Doubly plunging anticline
d) Plunging synclinorium

20. The stages involved in the formation of a Nappe are

a) Horizontal beds—–folding—–recumbent fold—–forward movement—–nappe

b) Horizontal beds—–forward movement—–folding—–recumbent fold—–nappe

c) Horizontal beds—–recumbent fold—–forward movement —–folding—–nappe

d) None of the above

21. Which of the following minerals exhibit four sets of joints ?

a) Galena
b) Sphalerite
c) Leucite
d) Wavellite

22. From the attitudes given below, the following can be inferred

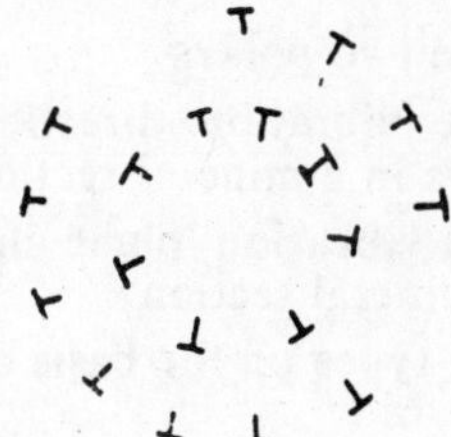

a) Basin
b). Syncline
c) Trough

d) Lake

23. An overturned fold is one in which
   a) One limb is overturned with respect to the other
   b) Both the limbs are turned through angles more than 90°
   c) The AP is vertical and both the limbs dip uniformly
   d) One limb is twice the size of the other limb

24. A bed is stricking N 30° W and its true dip is 30°. How many apparent dips can it have ?
   a) 2
   b) 3
   c) 1
   d) 4

25. The specific gravity of a powdered mineral can be determined with the help of a
   a) Chemical balance
   b) Pycnometer
   c) Jolly's spring balance
   d) Walker's Steel Yard

26. Which of the following exhibit pyroelectricity ?
   a) Hemihedral crystals
   b) Tetartohedral crystals
   c) Holohedral crystals
   d) None of the above

27. In Crystallograph, the interfacial angle is
   a) The angle between the two faces of a crystal
   b) The angle between the normals of to two faces of a crystal
   c) The angle between the two edges of a crystal
   d) None of the above

28. Extinction angle is
   a) The angle between the light and dark positions of a mineral
   b) The angle between the two polars
   c) The angle between the vibration directions of the ordinary and extraordinary rays in a mineral section
   d) The angles between a vibration plane and crystallographic direction in a given mineral section

29. Crystals are divided into types on the basis of
   a) Disposition of axis
   b) Symmetry elements
   c) Presence or absence of centre
   d) Nature of the crystals

30. Match the following and choose the correct answer

| *Mineral* | *Hardnesis* |
|---|---|
| a) Corundum | (i) 9.0 |
| b) Galena | (ii) 2.5 |
| c) Fluorite | (iii) 7.0 |

a) A—(iii) B—(i) C—(ii)
b) A—(i) B—(iii) C—(ii)
c) A—(ii) B—(iii) C—(i)
d) None of the above

31. Polysynthetic twinning may be described as
a) A type of repeated twinning in which the successive composition surfaces remain parallel
b) A type of repeated twinning in which the successive composition surfaces does not remain parallel
c) A type of twinning in which the individual immitates a higher symmetry
d) None of the above

32. Match the following and choose the correct answer

| | |
|---|---|
| A. Cross-hatched twinning | 1. Microcline |
| B. Cyclic twinning | 2. Aragonite |
| C. Polysynthetic twinning | 3. Plagioclase |
| D. Penetration twins | 4. Pyrite |

a) A—1 B—2 C—3 D—4
b) A—1 B—2 C—4 D—3
c) A—1 B—4 C—2 D—3
d) A—3 B—2 C—1 D—4

33. Which of the following mineral is biaxial and exhibits pseudo-hexagonal symmetry ?
a) Biotite
b) Olivine
c) Mica
d) Microcline

34. Which the help of a Berek's Compensator, the following can be determined ?
(i) Thickness of the mineral plate
(ii) Retardation
(iii) Identification of the mineral
a) (i) and (ii) only
b) (ii) only
c) (i), (ii) and (iii) only
d) (ii) and (iii) only

35. The minerals of which of the following series show isomorphic series
    a) Olivine
    b) Silica
    c) Zeolite
    d) Felspathoid
36. Which of the following is a freshwater fossil ?
    a) Physa
    b) Cerathium
    c) Nucula
    d) Ammonite
37. Arrange the following in the order of increasing age and complexity

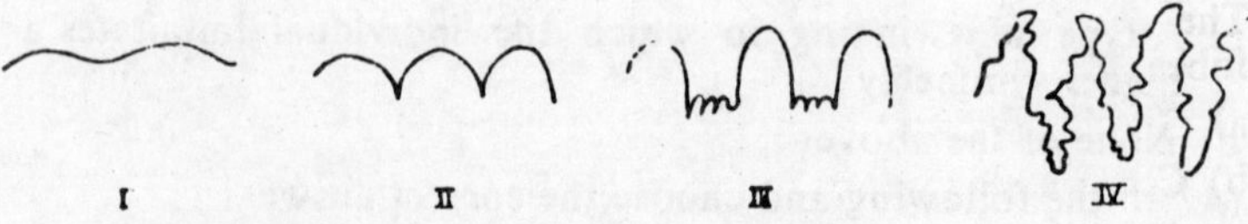


    a) I, II, III, IV
    b) I, II. IV, III
    c) IV, III, II, I
    d) III, IV, II, I
38. Which of the following trilobites has high accentuation of eyes ?
    a) Agnostus
    b) Acidaspis
    c) Paradoxides
    d) Microdiscus
39. Which of the following has fused genital plates ?
    a) Micraster
    b) Clipeaster
    c) Holaster
    d) Hemiaster
40. Match the following and choose the correct answer

| | |
|---|---|
| A. Calceola | 1. Gasteropoda |
| B. Conularia | 2. Coral |
| C. Spondylus | 3. Lamellibranch |
| D. Crania | 4. Brachiopod |

    a) A—2 B—1 C—3 D—4

b) A—2 B—4 C—1 D—3

c) A—4 B—1 C—3 D—2

d) A—2 B—3 C—1 D—4

41. The characteristic feature of Ostrea is

a) Equivalved nature

b) Inequivalved nature

c) Characteristic dentition

d) None of the above

42 Which of the following organisms does NOT from coral reefs ?

a) Ammonites

b) Corals

c) Brachiopods

d) Bryozoans

43. The Cephalopod showing rounded saddles and crenulated lobes is

a) Goniatite

b) Ceratite

c) Ammonite

d) Nautilus

44. Arrange the following fossils in the order of their increasing age

a) Agnostus—Tetragraptus—Productus—Goniatite

b) Agnostus—Productus—Tetragraptus—Goniatite

c) Agnostus—Tetragraptus—Goniatite—Productus

d) Productus—Agnostus—Tetragraptus—Goniatite

45. Find the odd man out

a) Lamination

b) Slaty cleavage

c) Schistosity

d) Foliation

46. Graywacke is differentiated from lithic arenites on the basis of

a) Silica content

b) Feldspar content

c) Percentage of matrix

d) Presence or absence of cement

47. Flysch sequences are characteristically formed in

a) Eugeosynclines

b) Miogeosynclines

c) Orogenic troughs

d) None of the above

48. The most abundant sedimentary rock found in the Earth's crust is

a) Shale

b) Sandstone

c) Limestone

d) None of the above

49. Stylolites are formed due to

a) Pressure solution and temperature effects

b) Pressure and solution effects

c) Pressure effects only

d) Solution effects only

50. Stromatolites are

a) Green algae

b) Blue algae

c) Organo-sedimentary structures

d) Sedimentary structures

51. Which of the following conditions satisfy the formation of oscillatory ripple marks ?

a) It has pointed crests and broad troughs

b) It is formed in the shallow waters

c) As the depth increases the wavelength of the ripple marks increases

d) All of the above

52. Cross-bedding which is truncated at the top and bottom is found in

a) Lacustrine environments

b) Marine environments

c) Deltaic environments

d) None of the above

53. Which of the following conglomerates is subjected to least erosion ?

a) Beach conglomerates

b) Conglomerates transported along with a host of angular rocks

c) Conglomerates transported by a stream

d) None of the above

54. Arrange the following in the order of increasing grade of metamorphism

a) Phyllite, khondalite, staurolite schist, sillimanite schist

b) Phyllite, staurolite schist, sillimanite schist, khondalite

c) Phyllite, sillimanite schist, staurolite schist, khondalite
d) Khondalite, phyllite, staurolite schist, sillimanite schist

55. Pyrometamorphism is a result of
a) Inclusion of xenoliths in basalts
b) Contact of magma with granitic rocks
c) Contact of magma with gabbroic rocks
d) Contact of magma with granitic and gabbroic rocks

56. Khondalite is composed of
a) Sillimanite schist with garnet and apatite
b) Sillimanite schist with no garnet and apatite
c) Hypersthene granite with graphite
d) None of the above

57. In which of the following rocks mica is absent ?
a) Garnet—sillimanite graphite gneiss
b) Garnet—sillimanite mica schist
c) Peridotites
d) None of the above

58. Which of the following is NOT a texture ?
a) Crystalloblastic
b) Diablastic
c) Xenoblastic
d) Cataclastic

59. Which of the following facies has the least pressure effects ?
a) Amphibolite
b) Eclogite
c) Sanidinite
d) Zeolite

60. The most optimum condition for the formation of a porphyritic texture is
a) Differential cooling under different environments
b) Cooling uniformly under the same conditions
c) Cooling differently under same environments
d) Differential cooling under the same environment

61. A sill a
a) Horizontal igneous intrusive
b) Vertical igneous intrusive
c) Concave igneous plution
d) None of the above

62. The characteristic texture of lamprophyres is

a) Panidiomorphic
b) Hypidiomorphic
c) Allotriomorphic
d) Xenomorphic

63. Rapakivi texture consists of

a) An intergrowth of K-feldspar and plagioclase
b) An intergrowth texture in which plagioclase is enclosed in K-feldspar
c) Large oval K-feldspar mantled by rims of plagioclase in a fine-grained groundmass
d) A parallel arrangement of feldspars in a fine-grained groundmass

64. Match the following and choose the correct answer

| | |
|---|---|
| A. Phaccolith | 1. A vertical igneous intrusion |
| B. Laccolith | 2. A concave igneous plution with a sagging base |
| C. Lopolith | 3. Dome-shaped igneous plutons |
| D. Dyke | 4. Convex igneous bodies showing effects |

a) A—4 B—2 C—3 D—1
b) A—2 B—3 C—1 D—4
c) A—4 B—3 C—2 D—1
d) A—3 B—2 C—4 D—1

65. Match the following and choose the correct answer

| | |
|---|---|
| A. Perthite texture | 1. An intergrowth texture in which plagioclase contains inclusions of K-feldspar |
| B. Antiperthite texture | 2. An intergrowth texture in which K-felspar contains inclusions of plagioclase |
| C. Porphyritic texture | 3. Texture resulting from simultaneous crystallization of two minerals |
| D. Intergrowth texture | 4. Large crystals are set in a glassy groundmass |
| | 5. Numerous small crystals are uniformly scattered through a large crystal |

a) A—2 B—1 C—5 D—3
b) A—2 B—1 C—4 D—3
c) A—1 B—2 C—5 D—3
d) A—1 B—2 C—4 D—3

66. Which of the following rocks do not have a volcanic equivalent ?
a) Granite
b) Dunite
c) Syenite
d) Gabbro

67 Diorite is composed of
a) Calcic plagioclase, hornblende and quartz
b) Calcic plagioclase, biotite and quartz
c) Sodic plagioclase and hornblende
d) Quartz, potash feldspar and mafics

68. Match the following and choose the correct answer

| | |
|---|---|
| A. Ropy lava | 1. Smooth lava surface |
| B. Pillow lava | 2. Spilites |
| C. Siliceous sinter | 3. Crater |
| D. Block lava | 4. Rough lava surface |

a) A—1 B—2 C—3 D—4
b) A—2 B—1 C—3 D—4
c) A—4 B—2 C—3 D—1
d) A—1 B—3 C—2 D—4

69. The Hypothermal deposits are formed at the temperatures of
a) 100—300°C
b) 300—500°C
c) 500—700°C
d) 700—900°C

70. From the figure given name below, the mineral that was formed last
a) Bornite
b) Chalcocite
c) Chalcopyite
d) Pyrite

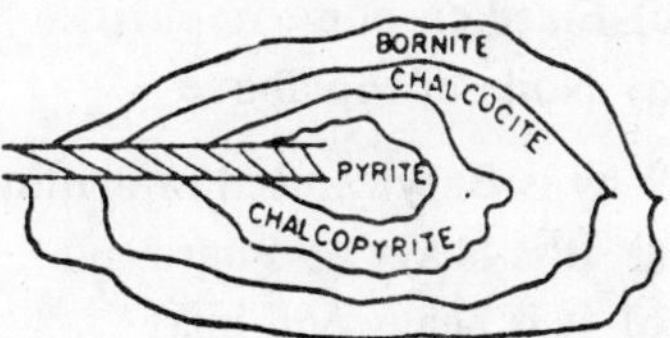


71. From the figure in Q. 70, the minerals are formed by which process ?
a) Replacement
b) Cavity filling
c) Mechanical concentration
d) Residual concentration

72. Arrange the following minerals in the order of their increasing iron content

a) Limonite<goethite<haematite<magnetite

b) Goethite<limonite<magnetite<haematite

c) Goethite<limonite<haematite<magnetite

d) Limonite<goethite<magnetite<haematite

73. Phosphatic deposits are formed under

a) Stable shelves

b) Unstable shelves

c) Lacustrine environments

d) Deltaic environments

74. Why is mica used in the electrical industry ?

1. It has high dielectric constant
2. It is a poor conductor of electricity
3. It has shining metallic lustre
4. It is an easily cleavable mineral

a) 1 and 2 only

b) 1, 2 and 3 only

c) 1, 2 and 4 only

d) 3 and 4 only

75. Which of the following is replacing chromite as a refractory mineral in modern steel industries ?

a) Dolomite and coke

b) Limestone and coke

c) Dolomite and magnesite

d) Limestone and magnesite

76. What is the fuel ratio ?

a) Fixed carbon÷volatile matter

b) Volatile matter÷fixed carbon

c) Fixed carbon÷impurities

d) None of the above

77. Why is Barytes used in drilling ?

a) It is heavy and hard

b) It is heavy and inert

c) It is soft and combustible

d) It is heavy and porous

78. In the Digboi oilfields, the oil is obtained from

a) Tippam formation

b) Barail formation

c) Surma formation

d) Sylhet formation

79. Bombay high is the highest producing oilfield in India. Which of the following is true of this oilfield ?

a) We find large gravity anomalies in this area

b) We find large seismic anomalies in this area

c) It is a topographic high having large gravity anomalies

d) It is a topographic high having large seismic anomalies

80. Which of the following is the thickest coal seam in India ?

a) Jharia

b) Karanpura

c) Singrauli

d) Bokaro

81. The conditions for the formation of placer deposits are

1. High relief of the source rocks

2. Deep weathering of sediments

3. Concentration of deposits by mechanical processes

a) 1 only

b) 1 and 2 only

c) 2 and 3 only

d) All of the above

82. Most of the manganese deposits in India are obtained from

a) Sausar series

b) Sakoli series

c) Kolhan series

d) Rialo series

83. Match the following and choose the correct answer

| | |
|---|---|
| A. Kyanite | 1. Drilling mud |
| B. Gypsum | 2. Plaster |
| C. Barytes | 3. Refractory |
| D. Vermicullite | 4. Insulator |

a) A—3 B—2 C—4 D—1

b) A—3 B—1 C—4 D—2

c) A—3 B—2 C—1 D—4

d) A—4 B—3 C—1 D—2

84. The famous Makrana marble is obtained from

a) Alwar group

b) Delhi group

c) Rialo series

d) Sausar group

85. Gondwana rocks range in age from

a) U. Carboniferous to M. Cretaceous
b) U. Carboniferous to M. Jurassic
c) M. Carboniferous to U. Cretaceous
d) M. Carboniferous to M. Jurassic

86. Which of the following are the equivalents of the lower Vindhyans ?
a) Cuddapahs
b) Kurnools
c) Chattisgarh group
d) None of the above

87. When did the earliest marine transgression take place on the western border of the Indian Peninsula ?
a) Traissic
b) Jurassic
c) Cretaceous
d) Carboniferous

88. Arrange the following Gondwana flora in the order of their increasing age
a) Glossopteris—Schizoneura—Williamsonia
b) Glossopteris—Williamsonia—Schizoneura
c) Schizoneura—Williamsonia—Glossopteris
d) Williamsonia—Schizoneura—Glossopteris

89. Sivamalai Rocks are mainly composed of
a) Charnokites
b) Anorthosites
c) Alkaline rocks
d) Granites

90. The age of the deccan traps is
a) Uppermost cretaceous to early eocene
b) Uppermost cretaceous to lower oligocene
c) Middle cretaceous to eocene
d) Cretaceous to lower oligocene

91. Which of the following is a freshwater formation ?
a) Barakar
b) Ranigunj
c) Jabalpur
d) Karharbari

92. The age of the Syringothyris limestones is
a) U. Carboniferous
b) M. Carboniferous

c) Jurassic

d) Cretaceous

93. In the Gondwana system, where does the Permian-Triassic break lie ?

a) Between Panchet and Mahadeva

b) Between Panchet and Raniguni

c) Between Barakar and Panchet

d) In the middle of the Panchet

94. Which of the following is the correct sequence of the Cretaceous rocks of Tiruchinapalli ?

a) Niniyur —Ariyalur—Tiruchinapalli—Uttatur

b) Uttatur—Tiruchinapalli—Ariyalur—Niniyur

c) Uttatur—Ariyalur—Tiruchinapalli—Uttatur

d) Ariyalur—Niniyur—Tiruchinapalli—Uttatur

95. Pick out the correct sequence of the Jurassic rocks from the following

a) Umia—Katrol—Chari—Patcham

b) Patcham—Chari—Katrol—Umia

c) Umia—Chari—Patcham—Katrol

d) Patcham—Katrol—Chari—Umia

96. The equivalents of the Cuddalore sandstones in A P. are

a) Rajahmundry sandstone

b) Tirupati sandstone

c) Banganapalli sandstone

d) Srisailam quartzite

97. Pleochroism is

a) Exhibition of a changing colouration on the rotation of a microscopic stage

b) Change in the quantity and quality of light due to the unequal absorption by a mineral

c) Changing series of colours exhibited by a mineral due to the interference of light from the grain boundaries

d) None of the above

98. Paleocurrent analysis can be used for

1. Determining the direction of transport of grains

2. The direction of younging of the sediments

3. Determining the palaeoslopes

a) 1 only

b) 1 and 2 only

c) 2 and 3 only

d) 1 and 3 only

e) All of the above

**99.** A metamorphic facies may be described as

a) A set of mineral assemblages at equilibrium with one another

b) A group of rocks of different chemical composition characterized by a definite set of minerals

c) A group of rocks of similar chemical composition with similar set of minerals

d) None of the above

100. Assertion : The shape of the earth is flattened near the poles and bulged near the equator

Reason : The above shape is due to the rotation of the Earth

a) Both A and R are correct and R is the correct explanation or A

b) Both A and R are correct but R is NOT the correct explanation of A

c) A is correct, R is false

d) A is false, R is correct

## ANSWERS

| | | | |
|---|---|---|---|
| 1. a | 2. c | 3. b | 4. b |
| 5. c | 6. a | 7. a | 8. c |
| 9. b | 10. a | 11. b | 12. d |
| 13. c | 14. b | 15. a | 16. c |
| 17. c | 18. a | 19. c | 20. a |
| 21. b | 22. a | 23. b | 24. b |
| 25. b | 26. b | 27. b | 28. d |
| 29. b | 30. a | 31. a | 32. a |
| 33. c | 34. a | 35. a | 36. a |
| 37. a | 38. c | 39. d | 40. a |
| 41. a | 42. a | 43. b | 44. c |
| 45. a | 46. c | 47. c | 48. a |
| 49. b | 50. c | 51. d | 52. c |
| 53. b | 54. b | 55. a | 56. a |
| 57. d | 58. d | 59. d | 60. c |
| 61. a | 62. a | 63. c | 64. c |
| 65. b | 66. b | 67. a | 68. a |
| 69. d | 70. a | 71. a | 72. d |
| 73. a | 74. c | 75. b | 76. a |
| 77. d | 78. a | 79. c | 80. c |
| 81. d | 82. a | 83. c | 84. c |
| 85. a | 86. b | 87. b | 88. a |
| 89. c | 90. a | 91. a | 92. b |
| 93. d | 94. c | 95. b | 96 a |
| 97. b | 98. e | 99. b | 100. a |

## 3. Civil Services (Preliminary) Examinations 1988

1. The age of the oldest sea floor is about
   a) 100 MY
   b) 200 MY
   c) 300 MY
   d) 50 MY
2. According to the theory of Plate
   a) MORs mark the boundaries of the plates
   b) MORs and Continental Mountain chains mark the boundaries of the plates
   c) Island arcs and mountain chains form the boundaries of plates
   d) MORs, Island arcs, Trenches and mountain chains mark the boundaries of of the plates
3. Match the following and choose the correct answer

| | |
|---|---|
| A. 1950 | 1. Assam |
| B. 1967 | 2. Kangra |
| C. 1934 | 3. Koyna |
| D. 1905 | 4. Bihar |

   a) A—4 B—3 C—2 D—1
   b) A—1 B—3 C—4 D—2
   c) A—4 B—3 C—1 D—2
   d) A—1 A—4 C—3 D—2
4. Mohorovicic discontinuity is a seismic interface between
   a) Crust and Mantle
   b) Mantle and Core
   c) Outer Core and Inner Core
   d) Upper Mantle and Lower Mantle
5. The portion of the geosyncline where thick, poorly sorted sediments generally associated with volcanics are found are known as
   a) Eugeosynclines
   b) Miogeosynclines
   c) Zeugogeosynclines
   d) Taphrogeosynclines
6. Earthquakes originate beneath Island arcs at great depths. Which of the following statements is correct ?
   a) Island arcs are found not far from the continent
   b) Earthquake zones have a steep dip beyond the arcs towards the trenches

c) Earthquakes originating along the Island arcs have deep focii

d) All of the above

e) None of the above

7. Andesite is a typical volcanic rock of the Island arcs because

a) It is formed from a predominantly acidic magmas

b) It is formed by the partial mixing of the oceanic and continental crust

c) It is generated by the collision of two oceanic plates

d) None of the above

8. Which of the following minerals in a sedimentary rock indicates rapid uplift ?

a) Augite

b) Quartz

c) Felspar

d) Zircon

9. Bajada is

a) A series of coalescing alluvial fans formed at the piedmont slopes

b) Flat-toped hills formed by fluvial erosion

c) A shallow basin surrounded by mountains from all the sides

d) None of the above

10. Sets of landforms which indicates fall in the sealevel are

a) Wave-cut benches, beaches, ridges

b) Sea caves, Estuaries, Lagoons

c) Flood plains, extensive beaches, alluial fans

d) None of the above

11. Geological work of groundwater results in

(1) Corrosion

(2) Abrasion

(3) Cementation

a) 1 only

b) 1 and 2 only

c) 1 and 3 only

d) 1, 2 and 3

12. The Mineral stability series is given below : Choose the statement that is wrong

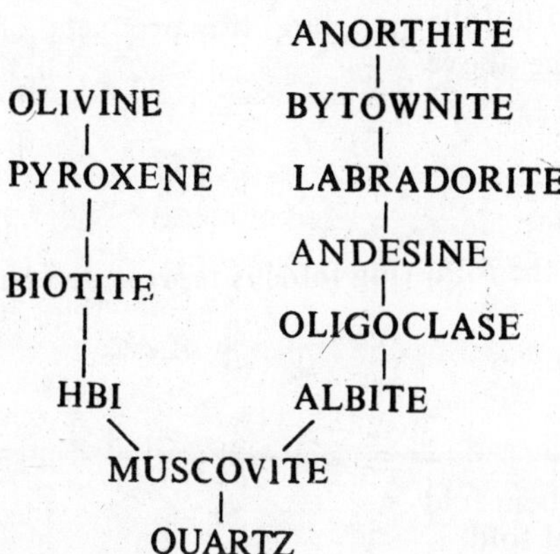


a) Quartz and muscovite are the predominant constituents of the weathered residues
b) Olivine weathers more easily than pyroxene in a rock containing both these minerals
c) Olivine weathers to pyroxene and pyroxene weathers to biotite
d) None of the above

13. Radial pattern of drainage is found in
a) Volcanoes
b) Hard rocky areas
c) Uplifted areas
d) Glaciers

14. The drainage pattern produced by parallel faulting and alternate beds of resistant and non-resistant rocks is
a) Trellis
b) Parallel
c) Radial
d) Dendritic

15. The most appropriate term for a ridge dipping gently on one side and steeply on the other is
a) Hogbacks
b) Mesas
c) Buttes
d) Cuestas

16. Karst topography is found dominantly in terrains of
a) Limestones
b) Sandstones
c) Granites
d) Metamorphic rocks

17. Which of the following are the products of plastic deforma-tion ?
    a) Folds
    b) Faults
    c) Joints
    d) Cleavage planes

18. Which of the following folds is represented below ?

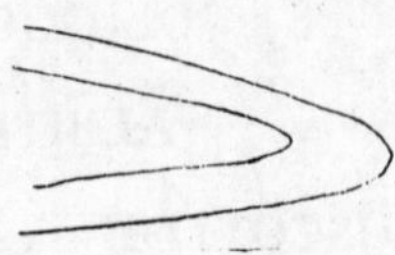

    a) Recumbent fold
    b) Vertical fold
    c) Isoclinal fold
    d) Any of the above

19. If drag folds are seen only in the plan and not in a section, then the major fold is
    a) Plunging
    b) Non-Plunging
    c) Doubly plunging
    d) Overturned

20. The stratigraphic order of three formations is given below. Identify the fold

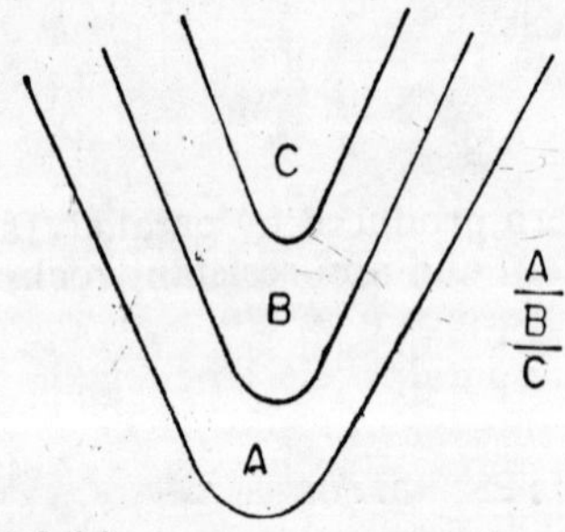


    a) Overturned fold
    b) Isoclinal fold
    c) Synformal Anticline
    d) Antiformal syncline

21. Hade of a fault is 45°, then
    a) Dip=Hade=Heave=Throw
    b) Dip=Hade=Hade≠Throw
    c) Dip≠Hade≠Hade=Throw
    d) Dip≠Hade=Hade≠Throw

22. Study the diagram given below and choose the most appro-priate answer.

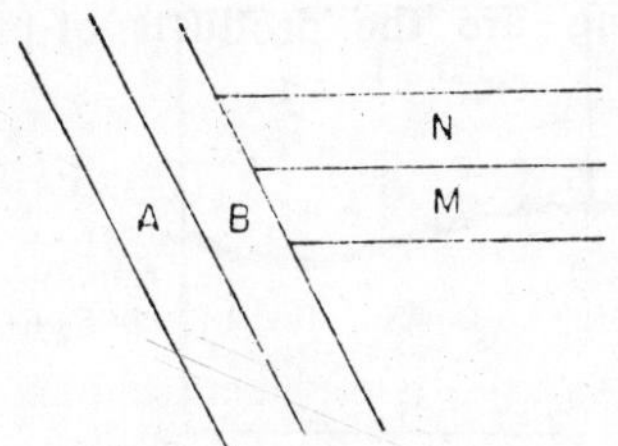


a) A and B are older than M and N
b) B is older than M
c) A is older than B
d) M is older than A

23. The figure given below depicts an overturned anticline. Which of the following statements is true ?

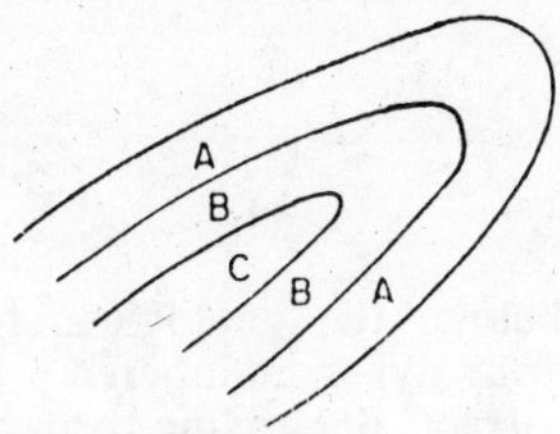


a) The strata on the left limb is overturned
b) The strata on the left limb is normal
c) B is older than C
d) A is older than C

24. **Two sets of beds are shown below. They are separated by an unconfirmity. A contains Ceratites fossils and B consists of Paradoxides. Then which of the following statements is true**

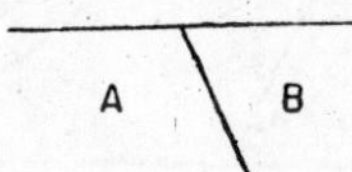


a) A is older than B
b) B is older than A
c) Both A and B are of same age
d) The information is insufficient

25. The Profile diagram of a Thrust fold is given below. Which of the following is a Window ?

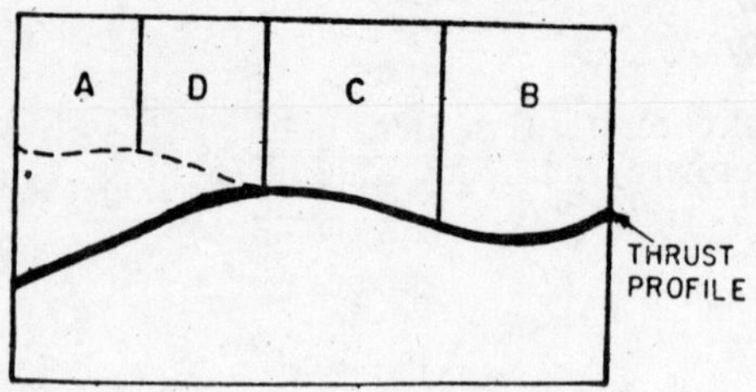


a) A
b) B
c) C
d) D

26. On a map scale 1:25000, 1km distance is represented by
a) 2.5 cm
b) 5.0 cm
c) 4 cm
d) 10 cm

27. In the map given below, the geological formations are seen to be intruded by four dykes numbered 1 to 4. Which of the following is the correct decreasing order of the age of the dykes ?

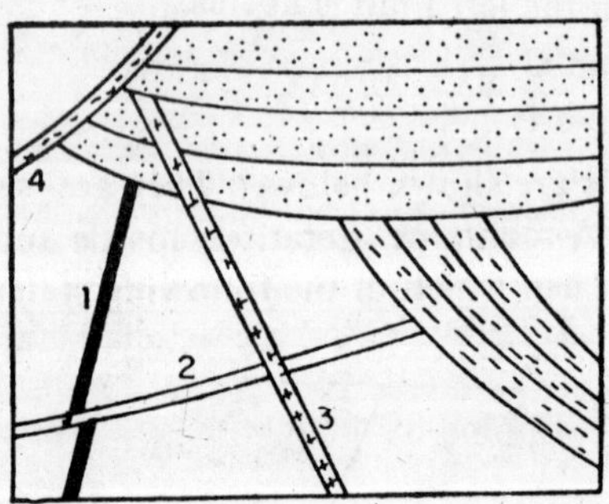


a) 1, 2, 3, 4
b) 2, 1, 3, 4
c) 3, 1, 4, 2
d) 4, 3, 2, 1

28. The strike of a bed using a compass was found to be N 10°E. If the magnetic north were 10° E of true north. The true N reading is
a) Due North
b) N 20°E

c) N 10°W
d) N 20°W

29. The contour map of an area marked N is given and the contour values are marked. The least slope is found in the direction

a) N
b) S
c) E
d) W

30. A bed is situated on a valley floor. Its contours are given. Then the bed is

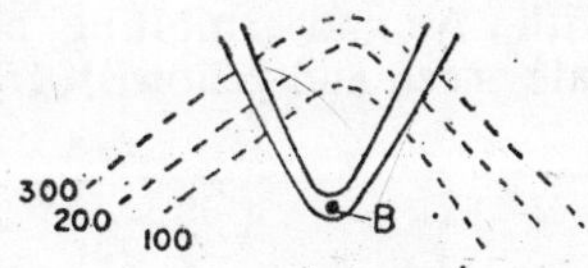


a) Horizontal
b) Vertical
c) Dips upstream
d) Dips downstream

31. Three axes of four-fold symmetry is shown by
a) Olivine
b) Galena
c) Barytes
d) Hornblende

32. Face which lie in the zone $1\bar{1}0$, 101, 011 is
a) 112
b) 221
c) 100
d) 321

33. Cross-hatched twinning of microcline is due to a wedge-shaped twinning on
a) Albite + Pericline law
b) Albite + Carlsbad law
c) Albite + Manebach law
d) Carlsbad + Pericline law

34. Which of the following minerals indicate highest pressure during metamorphism ?
a) Chlorite
b) Biotite
c) Hornblende

d) Garnet

35. The relative abundance of elements in the crust is

a) Si>Al>Fe
b) Si>Al>Fe
c) Al>Si>Fe
d) Fe>Si>Al

36. Which of the following is the blue variety of Corundum ?

a) Saphhire
b) Sapphirine
c) Ruby
d) Amethyst

37. An amphibole with z∧c and exhibiting pleochroism in shades of light green, pale green and yellowish green is

a) Tremolite
b) Actinolite
c) Hornblende
d) None of the above

38. Chiastolite is a variety of

a) Andalusite
b) Diopside
c) Sillimanite
d) Hornblende

39 A Bertrand Lens is used to study

a) Refractive index
b) Interference figures
c) Optic sign
d) Optical anomalies

40. If the R.I. of a certain material in a mineral section were 1.64. Quartz would show

a) High positive relief
b) High negative relief
c) Low positive relief
d) Low negative relief

41. The extinction shown by a basal section of pyroxenes is

a) Straight
b) Symmetrical
c) Inclined
d) None of the above

42. Extinction angle is the angle between

a) The vibration direction and the crystallographic planes

b) The light and the dark positions in a mineral
c) The vibration directions of the two polars
d) None of the above

43. Interference phenomenon depends upon
I. Refractive index of the mineral
II. Thickness of the mineral plate
III. Composition of the minerals

a) I only
b) I and II only
c) I and III only
d) All of the above

44. When a mineral shows straight extinction, it belongs to
a) Isometric and Tetragonal systems
b) Tetragonal and Triclinic systems
c) Tetragonal and Monoclinic systems
d) Monoclinic and Triclinic system

45. In a crossed dispersion, the b-axis is
a) Perpendicular to the vibration plane
b) Perpendicular to the wave normal
c) Parallel to the vibration plane
d) None of the above

46. A mineral exhibits 55° cleavage angle and 18° extinction angle and is optically positive and biaxial. It would be
a) Anthophyllite
b) Cummingtonite
c) Diopside
d) Tremolite

47. Manganese nodules of the ocean floor are important because beside manganese they also contain
a) Ni, Co, Fe, Mo
b) Ni, Co, Fe, Cr
c) Ni, Co, Mg, S, Ag
d) Ni, Co, Fe, Au, Ag, Hg

48. Epigenetic deposits are formed
a) Before the source rocks
b) At the same time as the source rocks
c) After the deposition of source rocks
d) None of the above

49. Martitization generally takes place when
   a) Pyrite is oxidized to limonite
   b) Hematite is oxidized to goethite
   c) Magnetite is oxidized to hematite
   d) None of the above

50. Gossans are products of the oxidation of
   a) Oxides
   b) Sulphides
   c) Oxides and sulphides
   d) Oxides, Sulphides and Chlorides

51. When euhedral crystals are there in the matrix the texture developed is described as
   a) Idiomorphic
   b) Porphyritic
   c) Poikilitic
   d) Ophitic

52 Development of a graphic texture is due to
   a) Simultaneous crystallization of two minerals
   b) Exsolution of two minerals
   c) Unequal crystallization of two minerals
   d) None of the above

53. Manganese is used in the manufacture of

   1. Wet batteries    2. Chemicals    3. Dry batteries

   a) 1 and 2 only
   b) 2 and 3 only
   c) 1 and 3 only
   d) All of the above

54. Which of the following is represented by the diagram given below ?

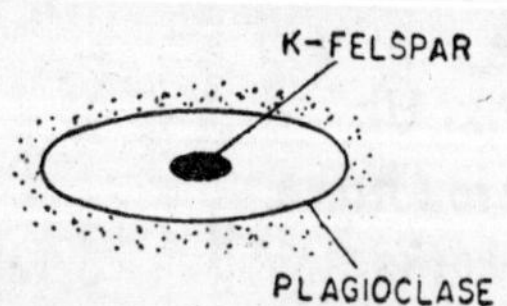


   a) Rapakivi texture
   b) Orbicular structure
   c) Graphic texture

d) Mirmekite structure

55. The reservoir of the Bombay high oilfield is in
   a) Limestones
   b) Sandstones
   c) Shales
   d) None of the above

56. Which of the following is NOT a place for the occurrence of diamond ?
   a) Golconda
   b) Rewa
   c) Panna
   d) Siwaliks

57. Which of the following is used in steel hardening ?
   a) Manganese
   b) Chromium
   c) Nickel
   d) Zinc

58. Indian iron and manganese deposits are mostly found in
   a) Palaeozoics
   b) Precambrians
   c) Mesozoics
   d) Cenozoic

59. East Coast bauxite is mostly found in
   a) Andhra Pradesh
   b) Orissa
   c) Andhra Pradesh and Orissa
   d) Andhra Pradesh, Orissa and Tamil Nadu

60. Assay value of gold is calculated in
   a) Percentage
   b) gm/tonne
   c) Parts per million
   d) Any of the above

61. Which of the following is the correct sequence of Kaimurs ?
   a) Dhandraul quartzites-Bijaigarh shales Scarp sandstones and Lts Susnai breccia
   b) Dhandraul quartzites-Scarp sandstones and limestones Bijaigarh shales-Susnai breccia
   c) Bijaigarh shales-Dhandraul quartzites-Scarp sandstones and

limestones-Susnai breccia

d) None of the above

62. Choose the correct stratigraphic sequence from the following
a) Nallamalai/Papaghani/Cheyair/Kistna
b) Nallamalai/Cheyair/Papaghani/Kistna
c) Cheyair/Nallamalai/Papaghani/Kistna
d) Papaghani/Cheyair/Nallamalai/Kistna

63. The most extensive felsic magmatism during the Vindhyan times was
a) Malani volcanics
b) Dras volcanics
c) Bhimtal volcanics
d) Deccan volcanics

64. Hipparian had
a) Two toes
b) Three toes
c) Four toes
d) One toe

65. Rajamundry Sandstones and Cuddalore Sandstones are equivalents of
a) Siwaliks
b) Vindhyans
c) Karewas
d) Kurnools

66. The Pre-Cambrian—Cambrian boundary in the Kashmir is located
a) Above Lolab Formation
b) Below Lolab formation
c) In the middle of Lolab formation
d) None of the above

67. Which of the following does not belong to the Puranas ?
a) Bhander
b) Gwalior
c) Bijawer
d) Jodhpur Sandstone

68. The formation Coral Limestone is found in
a) Lower Baghs
b) Upper Baghs

c) Badhora beds
d) None of the above

69. The Ranigunj formation is rich in
   a) Glossopteris indica, Vertebraria indica and Schizoneura
   b) Glossopteris indica, Vertebraria indica and Williamsonia
   c) Glossopteris indica, Vertebraria indica and Ptillophyllum
   d) None of the above

70. Magnesite deposits are commonly associated with
   a) Sandstones and quartzites
   b) Peridotites or limestones/dolomites
   c) Ultrabasic rocks only
   d) Granites and granodiorites

71. The most favourable sites for placer deposits are
   a) Upper reaches
   b) Lower reaches
   c) Middle reaches
   d) None of the above

72. Match the following and choose the correct answer

| | |
|---|---|
| A. Bassein | 1. Ferrous minerals |
| B. Naumundi | 2. Non-ferrous minerals |
| C. Ghatsila | 3. Fuel |
| D. Singrauli | 4. Coal |

   a) A—3 B—1 C—2 D—4
   b) A—3 B—2 C—1 D—4
   c) A—2 B—1 C—3 D—4
   d) A—1 B—2 C—4 D—3

73. Which of the following are the calcareous rocks associated with thermal metamorphism ?
   a) Skarn
   b) Calcirudites
   c) Carbonatites
   d) Marbles

74. Thermal metamorphism of argillaceous rocks produces
   a) Shale
   b) Slate
   c) Hornfelses
   d) Phyllite

75. Match the following and choose the correct answer

A. Anorthosites — 1. Omphacite+Garnet

B. Blueschists — 2. Quartz+felspar+hypersthene

C. Charnockites — 3. Lawsonite+glaucophane+jadeite

D. Eclogites — 4. Labradorite

a) A—4 B—2 C—3 D—1
b) A—4 B—3 C—2 D—1
c) A—4 B—3 C—1 D—2
d) A—3 B—2 C—4 D—1

76. Rocks predominantly composed of amphibole and plagioclase are
a) Amphibolites
b) Anorthosites
c) Harzburgites
d) Lherzolites

77. Match the following and choose the correct answer

| *Dentition* | *Species* |
|---|---|
| A Taxodont | 1. Spondylus |
| B Isodont | 2. Ctenodonta |
| C. Heterodont | 3. Mytilus |
| D. Dyssodont | 4. Venus |

a) A—4 B—3 C—2 D—1
b) A—4 B—2 C—3 D—1
c) A—2 B—3 C—4 D—1
d) A—2 B—1 C—4 D—3

78. Aristotle's lantern and Perignathic girdle are not found in
a) Holectypina
b) Spatangina
c) Holaster
d) Micraster

79. Match the following and choose the correct answer

A. Agnostus — 1. Apical disc.

B. Cardiaster — 2. Dextral coiling

C. Murex — 3. Blind fossil

D. Nautiloid — 4. Primitive sutures

a) A—2 B—1 C—4 D—3
b) A—2 B—1 C—3 D—4
c) A—3 B—1 C—2 D—4
d) A—3 B—4 C—2 D—1

80. Agnostus has

a) Three eyes

b) Two eyes

c) Compound eyes

d) No eyes

81. The term 'Bonanza' is associated with

a) Gold

b) Copper

c) Iron

d) Manganese

82. The most common mode of alteration of peridotites and dunites is

a) Kaolinization

b) Sericitization

c) Scapolitization

d) Chloritization

83. 'Fenite' is the product of

a) Alkali metasomatism

b) Lime metasomatism

c) Sedimentation

d) High grade metamorphism

84. Gneiss is the most abundant rock on the surface of the earth because

a) It has been preserved under favourable conditions

b) The conditions for the formation of gneiss were more favourable than for the formation of other rocks

c) Gneisses have not been subjected to large-scale weathering

d) None of the above

85 A weathered rock with dominant angular grains may be described as

a) Conglomerate

b) Breccia

c) Grit

d) Cobble

86. Incongruent melting yields a liquid and a solid from an original solution. The liquid formed has

a) Same composition as the original solid

b) Different composition as the original solid

c) Same composition as the newly formed solid

d) None of the above

87. Presence of water vapour and gases in a magma
   a) Decreases its melting point
   b) Increases its melting point
   c) Decreases its viscocity
   d) Increases its viscocity

88. Match the following and choose the correct answer

| | |
|---|---|
| A. Eras | 1. Group |
| B. Epoch | 2. System |
| C. Age | 3. Stage |

   a) A—3 B—2 C—1
   b) A—1 B—2 C—3
   c) A—2 B—3 C—1
   d) A—2 B—1 C—3

89. The main boundary fault lies between
   a) Vindhyans and the archaens
   b) Vindhyans and the younger rocks
   c) Siwaliks and the younger rocks
   d) Siwaliks and the older rocks

90. Lamellibranchs are
   a) Smooth on branchial sides and rough on convex sides
   b) Smooth on convex sides and rough on branchial sides
   c) Rough on both the sides
   d) None of the above

91. The Rajmahal volcanics is the earliest manifestation of volcanic activity of
   a) Deccan traps
   b) Dras volcanics
   c) Sylhet volcanics
   d) None of the above

92. Narimannam is associated with
   a) Cauvery
   b) Godavari
   c) Damodar basin
   d) Cachar basin

93. Which of the following are intrusive igneous rocks of calcareous composition ?
   a) Skarn
   b) Marble
   c) Dolomites

d) Carbonatites

94. Arenaceous rocks range in size from
a) 1/16—2 mm
b) 2 mm—64 mm
c) 64—256 mm
d) 1/16—1/256 mm

95. Glacially formed deposits are found in
a) Talchir
b) Damuda
c) Panchet
d) Mahadev

96. If a crystal shows regular growth outwards from the centre it is known as
a) Accretion
b) Incretion
c) Intercretion
d) Secretion
e) None of the above

97. Assemblages of strata representing deposition during distinct time units are called
a) Time units
b) Time-rock units
c) Rock units
d) None of the above

98. Assertion : Nautilus Danicus is an index fossil
Reason : It has been used in the correlation of cretaceous rocks of India
a) Both A and R are correct and R is the correct explanation of A
b) Both A and R are correct but R does not explains A
c) A is correct R is false
d) R is correct A is false

## ANSWERS

| | | | |
|---|---|---|---|
| 1. b | 2. d | 3. c | 4. a |
| 5. a | 6. d | 7. b | 8. c |
| 9. a | 10. b | 11. b | 12. c |
| 13. a | 14. a | 15. d | 16. a |
| 17. a | 18. a | 19. c | 20. c |

| | | | |
|---|---|---|---|
| 21. b | 22. d | 23. b | 24. b |
| 25. d | 26. a | 27. a | 28. b |
| 29. a | 30. c | 31. b | 32. a |
| 33. a | 34. d | 35. a | 36. a |
| 37. c | 38. a | 39. b | 40. c |
| 41. a | 42. a | 43. b | 44. c |
| 45. b | 46. b | 47. a | 48. c |
| 49. c | 50. b | 51. a | 52. a |
| 53. b | 54. a | 55. a | 56. d |
| 57. a | 58. b | 59. c | 60. b |
| 61. b | 62. d | 63. a | 64. b |
| 65. a | 66. c | 67. d | 68. b |
| 69. a | 70. b | 71. c | 72. b |
| 73. a | 74. c | 75. b | 76. a |
| 77. d | 78. b | 79. c | 80. d |
| 81. a | 82. b | 83. a | 84. b |
| 85. b | 86. b | 87. c | 88. b |
| 89. c | 90. b | 91. a | 92. a |
| 93. d | 94. a | 95. a | 96. b |
| 97. a | 98. b | | |

## 4. Civil Services (Preliminary) Examination, 1987

N.B. : Choose the Answer which is most appropriate.

1. Arrange the following in the correct sequence
   a) Sedimentation—Geosynclinal subsidence—Lateral compression—Orogenic upliftment
   b) Geosynclinal subsidence—Sedimentation—Lateral compression—Orogenic upliftment
   c) Geosynclinal subsidence—Lateral compression--Sedimentation—Orogenic upliftment
   d) Lateral compression—Sedimentation—Geosynclinal subsidence--Orogenic upliftment

2. Which of the following is a living fossil ?
   a) Productus
   b) Nucula
   c) Arca
   d) Lingula

3. Which of the following can be described as a fossil ?
   a) Human skeleton of about 5000 years back
   b) Lamellibranch remains preserved in Pliestocene deposits and belonging to existing genus
   c) Concretion of Cretaceous age
   d) Dendrites

4. Which of the following is NOT a lower Gondwana fossil ?
   a) Glossopteris
   b) Gangamopteris
   c) Vertebraria
   d) Ptillopbylium

5. The plant remains of the Gondwanas are found as
   a) Petrified deposits
   b) Carbonized deposits
   c) Phosphatic deposits
   d) None of the above

6. The lower Gondwana fossils are found in the Extra Peninsula in association with
   a) Chail
   b) Zeewan beds
   c) Karewas
   d) Infra—Krol

7. The Kurnool Group of A.P. is equivalent of
   a) Semri series
   b) Upper Vindhyans
   c) Dharwars
   d) Cuddapahs
8. The rock succession of Khondalite is found in which of the following Precambrian provinces ?
   a) Dharwar
   b) Eastern Ghat
   c) Singhbhum
   d) Orissa
9. Which of the following has richest fossils among the Siwaliks ?
   a) Nagri
   b) Dhokpathan
   c) Kamlial
   d) Chinji
10. The Hipparian fossils appeared first is
    a) Pinjor
    b) Chinji
    c) Nagri
    d) Kamlial
11. The Siwaliks are mostly
    a) Fluviatile deposits
    b) Glacial deposits
    c) Marine deposits
    d) Intermontane deposits
12. Gangamopteris can be differentiated from Glossopteris by
    a) Presence of venation
    b) Absence of midrib
    c) Acute leaf margin
    d) Size of the leaves
13. Cleavage is the property of
    a) Crystalline substances
    b) Amorphous substances
    c) Non-crystalline solid substanccs
    d) Partly soluble or colloidal substances
14. The mineral Quartz exhibits
    a) Nesosilicate structure
    b) Phyllosilicate structure
    c) Ionosilicate structure
    d) Three-dimensional network

15. In a three-dimensional network, each tetrahedron shares
   a) One oxygen atom
   b) Two oxygen atoms
   c) Three oxygen atoms
   d) Four oxygen atoms

16. Match the following and choose the correct answer from the codes given below

| A | B |
|---|---|
| (a) Pyroxene | 1. Labradorite |
| (b) Mica | 2. Anthophyllite |
| (c) Felspar | 3. Augite |
| (d) Amphibole | 4. Opal |
| (e) Silica | 5. Phlogopite |

   a) a—3 b—4 c—1 d—5 e—2
   b) a—3 b—4 c—2 d—5 e—1
   c) a—3 b—5 c—1 d—2 e—4
   d) a—1 b—5 c—3 d—2 e—4

17. Which of the following is wrongly matched ?

| *Twin Plane* | *Twin* |
|---|---|
| a) 001 | Manebach |
| b) 001 | Carlsbad |
| c) 010 | Carlsbad |
| d) 021 | Baveno |

18. Arrange the following in the correct order of their hardness
   a) Talc<Felspar<Fluorite<Quartz
   b) Talc<Quartz<Felspar<Fluorite
   c) Talc<Fluorite<Felspar<Quartz
   d) Fluorite<Talc<Felspar<Quartz

19. A set of beds is given. The dip directions are marked. The plunge direction is given. Where would the vertical dip be measured ?

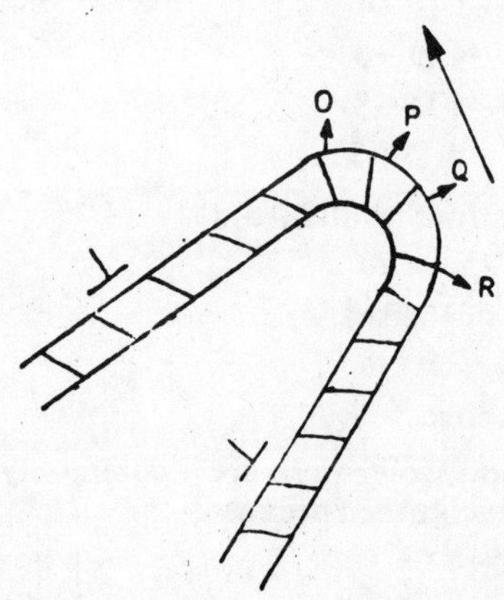

a) O
b) P
c) Q
d) R

20. Identify the fold given in the diagram

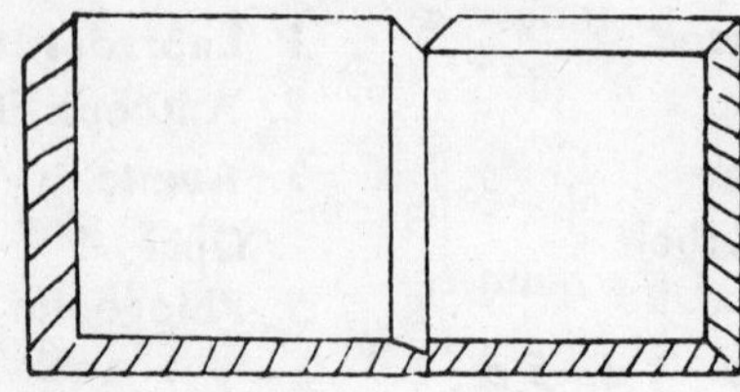

a) Non-cyllindrical fold
b) Cyllindrical fold
c) Non-planar non-cyllindrical fold
d) Planar and cyllindrical fold

1. Plumose markings are a feature of

a) Joints
b) Faults
c) Folds
d) Disconfirmities

22. Match the following and choose the correct answer from the code given below

| | |
|---|---|
| A. Angular Unconfirmity | 1. The beds show parallelism |
| B. Disconfirmity | 2. Beds above and below the UC are non-parallel |
| C. Non-confirmity | 3. The beds are parallel but the contact is distinctly local |
| D. Local unconfimity | 4. The lower plutonic rocks are overlain by sedimentary rocks |

a) A—1 B—2 C—3 D—4
b) A—2 B—1 C—4 D—3
c) A—2 B—1 C—4 D—3
d) A—1 B—1 C—4 D—3

23. An irregular coastline results from

a) Emergence of the sea
b) Submergence of the sea
c) Faulting
d) None of the above

24. Earthquakes and volcanics are commonly found to be associated with each other because

a) They lie at the same regions of the earth
b) They lie on the active plate margins
c) They have similar origin
d) All of the above
e) None of the above

25. The movement of the plates results from .
a) Mantle drag
b) Ridge push
c) Slab pull
d) Convection of the mantle

26. The most reliable method of determining the age of the crust is
a) The accumulation of sediments in a sequental order
b) The rate of cooling of the oceans and seas
c) Radioactive methods of dating
d) The measurement of the salinity of sea water

27. Correlation of rocks is done mainly to
a) Study the sequential order of the formation of beds
b) Study the evolution of the stratiform rocks
c) Link the rocks having the same fauna and flora
d) Determine the time equivalence

28. In a biostratigraphic unit the nomenclature takes into account
a) The locality in which the horizon is best developed
b) Fossil assemblage of the horizon
c) The age in which the horizon is best developed
d) The chemical and mineral similiarities of the formation

29. According to the continental drift hypothesis, which of the following continent was surrounded by rest of the continents ?
a) Asia
b) Africa
c) North America
d) Antartica

30. Which of the following shows reversed zoning ?

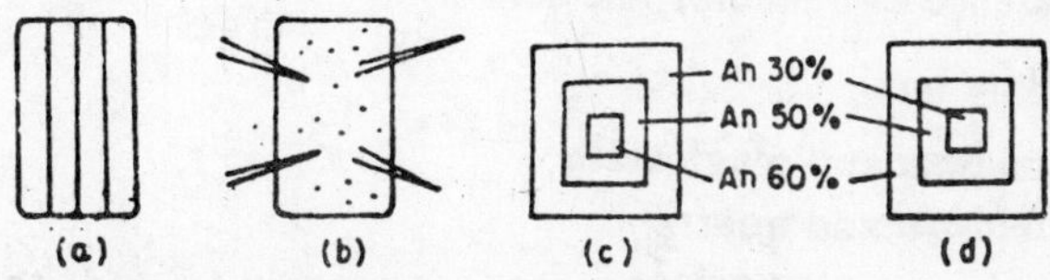


31. Which of the following textures has NOT resulted from rapid cooling at chilled margins ?

a) Hypogranular
b) Intersertal
c) Granophyric
d) Vitrophyric

32. Which of the following does not correspond with the others given in the series ?
a) Hornfelses
b) Slate
c) Schist
d) Phyllite

33. Which one of the following correctly represent the discontinuous reaction series ?
a) Pyroxene-olivine-biotite-hornblende
b) Olivine-pyroxene-biotite-hornblende
c) Olivine-pyroxene-hornblende-biotite
d) Hornblende-pyroxene-olivine-biotite

34. A 'doliomorphic' rock
a) Has both igneous and metamorphic minerals
b) Has both light and dark-coloured constituents
c) Has mineral constituents which are related to the reaction series
d) None of the above

35. The volcanic eruptions during the Dharwar times is reflected by the presence of
a) Ash beds
b) Lava flows
c) Volcanic rocks
d) Pillow lavas

36. The Precambrian can be differentiated from the rest of the formations by
a) Presence of igneous rocks
b) Absence of primitive life and presence of highly metamorphosed rocks
c) Intense orogenic activities
d) Presence of metamorphic series

37. A perthite has
a) K-feldspar and plagioclase
b) K-feldspar and quartz
c) Quartz and plagioclase
d) Albite and anorthite

38. Which one of the following series is NOT found in common rock associations ?
    a) Granite-granodiorite-monzonite
    b) Gabbro-syenite-spilite
    c) Syenite-nepheline-syenite-monzonite
    d) Basalt-Andesite-volcanic tuffs

39. Sometimes a mineral is unable to withstand high temperature metamorphism and becomes unstable and reverts to low-temperature metamorphic assemblage. This process is called
    a) Palingenesis
    b) Diapthoresis
    c) Anatexis
    d) Metasomatism

40. Large dunes which are aligned in the general direction of the winds are called
    a) Barchans
    b) Transverse dunes
    c) Parabolic dunes
    d) Longitudinal dunes

41. Loess is
    a) Desert sand deposit
    b) Desert silt and clay deposit
    c) Alluvial sand deposit
    d) Alluvial silt and clay deposit

42. The deposits such as sand, clay and silt formed at the mouth of the rivers due to floods are termed as
    a) Alluvium
    b) Colluvium
    c) Interalluvium
    d) Proluvium

43. Landforms are a function of
    a) Geologic structure
    b) Rate of erosion
    c) Plate motions
    d) Igneous and metamorphic processes

44. If the scale is taken as 1="1000,' what will be the thickness of a line drawn with a pencil ?
    a) 100"
    b) 500"
    c) 100"
    d) 50"

45. The lakes formed in the cut-off meanders are called
   a) Kayals
   b) Playas
   c) Oxbow lakes
   d) Tarns

46. 'Phanerozoic' is the term used to designate
   a) Azoic age
   b) Proterozoic age
   c) Archaen age
   d) Cambrian to recent

47. Glacial-formed deposits are heterogenous, ill-sorted and angular bccause
   a) The particles are not subjected to any great degree of attrition amongst themselves
   b) The particles suffer little grinding
   c) The particles are not subjected to much transportation
   d) All of the above

48. Which one of the following statements correctly designate that the core of the earth is made of solid metals ?
   1. A large number of mateors have abundant metals, especially iron in their composition
   2. P and S waves do not pass through solids
   3. The average density of the earth is 5.2 whereas the average density of the crust is only 2.4
   a) 1 and 2 only
   b) 1 and 3 only
   c) 2 and 3 only
   d) 1, 2 and 3 only

49. India is word's first in the resources of one atomic mineral. The element recovered from it is
   a) Uranium
   b) Plutonium
   c) Thorium
   d) Rubidium

50. Marine manganese nodules are formed in which of the following ways
   a) Deposition by seas into the ocean bottoms
   b) Deposition by currents from various sources
   c) Insitu precipitation from the oceanic sediments of various types
   d) A combination of all the above

51. Lignite deposits of India are associated with
   a) Rajmahal series
   b) Jabalpur series
   c) Barakar series
   d) Cuddalore sandstone

52. The following diagram designates

   a) Stylolite
   b) Convolute bedding
   c) Bioherm
   d) Stromatolite

53. Strike-slip faults of very deep origin are termed as
   a) Transcurrent faults
   b) Strike-faults
   c) Dip slip faults
   d) Oblique faults

54. Omphacite and garnet are the chief constituents of
   a) Anorthosites
   b) Granulites
   c) Eclogites
   d) Leptynites

55. The movement of the transform faults are
   a) Parallel to the line of the fault
   b) Perpendicular to the line of the fault
   c) Inclined to the line of the fault
   d) None of the above

56. The metamorphic facies which is characteristic of low geothermal gradients and found associated with geosynclines, is referred as
   a) Blueschist facies
   b) Glaucophane schist facies
   c) Eclogite facies
   d) Sanidinite facies

57. The deposition of calcite around spherulitic layers is called
   a) Oomicrite
   b) Oosparite

c) Intersparite

d) Pelsparite

58. Identify the given structures and choose the correct answer from the given codes

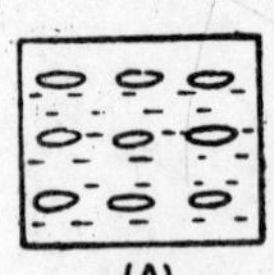
(A)

(B)

(C)

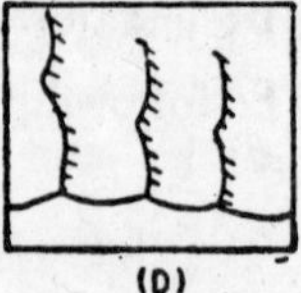
(D)

a) A—Lenticular bedding B—Climbing ripples C—Convolute bedding D—Parallel ripples

b) A—Lenticular bedding B—Convolute bedding C—Climbing ripples D—Parallel ripples

c) A—Parallel ripples B—Climbing ripples C—Lenticular bedding D—Convolute bedding

d) A—Convolute bedding B—Climbing ripples C—Parallel ripples D—Lenticular bedding

59. Although diamond and graphite have the same chemical composition, they differ from one another because

a) They are formed at different conditions

b) They are formed from different sources

c) They have different lattices

d) They have different isotopes

60. The Pb-Zn deposits of Rajasthan are associated with

a) Quartzites

b) Mica-dolerites

c) Archaen schists

d) Limestones

61. Which of the following is NOT a monominerallic ultrabasic rock ?

a) Pyroxenite

b) Anorthosite

c) Dunite

d) Leptynite

62. Coal is found in association with

a) Sphalerite

b) Pyrrhotite

c) Graphite

d) Pyrite

63. Gold occurs as
   1. Gold tellurides
   2. Gold-quartz lodes
   3. Gold-quartz sulphide lodes

   a) 1 only
   b) 1 and 2 only
   c) 2 and 3 only
   d) 1, 2 and 3

64. The Great Boundary Fault lies between
   a) U. Vindhyans and Precambrian rocks
   b) U. Vindhyans and Cambrian rocks
   c) Siwaliks and younger rocks
   d) Siwaliks and older rocks

65. Match the following and choose the correct answer

| | |
|---|---|
| (A) Chalcopyrite | (1) Kheonjhar, Orissa |
| (B) Chromite | (2) Singhbhum, Bihar |
| (C) Pyrolusite | (3) Ghatsila |
| (D) Sillimanite | (4) Khasi Hills |

   a) A—2 B—1 C—4 D—3
   b) A—2 B—4 C—3 D—1
   c) A—3 B—2 C—1 D—4
   d) A—3 B—1 C—2 D—4

66. Match the following and choose the correct answer

| | |
|---|---|
| (A) Canyon | (1) Alteration product |
| (B) Chlorite | (2) Erosional ridge |
| (C) Bird's foot | (3) Matured river valley |
| (D) Cuesta | (4) Matured delta |

   a) A—3 B—1 C—4 D—2
   b) A—3 B—1 C—2 D—4
   c) A—2 B—1 C—4 D—3
   d) A—2 B—1 C—3 D—4

67. Match the following and choose the correct answer

| *Drainage Pattern* | *Structure* |
|---|---|
| (A) Anastomosing | (1) Glacial deposits |
| (B) Dendritic | (2) Tectonically disturbed rocks |
| (C) Trellis | (3) Domes |
| (D) Radial | (4) Uniformly dipping beds |

a) A—4 B—1 C—2 D—3
b) A—2 B—1 C—4 D—3
c) A—2 B—1 C—3 D—4
d) A—3 B—2 C—1 D—3

68. Glacial deposits are
a) Homogenous and well sorted
b) Heterogenous and unsorted
c) Homogenous and unsorted
d) Heterogenous and well sorted

69. Arrange the following in the increasing order of their R.I.
Hbl, Qtz, Fluorite, Zircon
a) Qtz<Hbl<Fluorite<Zircon
b) Fluorite>Qtz<Hbl<Zircon
c) Hbl<Fluorite<Qtz<Zircon
d) Zircon<Fluorite<Hbl<Qtz

70. If the contour lines run parallel to a bedding, then it is assumed that the beds are
a) Horizontal
b) Vertical
c) Inclined
d) Steeply dipping

71. Identify the rocks marked in the given map

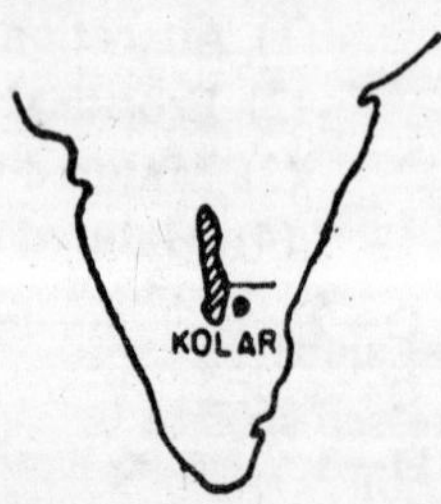


a) Peninsular Gneiss
b) Champion Gneiss
c) Kolar Schist Belt
d) Clospet Granite

72. Isostasy denotes
a) Balance reached between the areas which have suffered intensive seismic activity
b) Balance reached between granitic and basaltic masses
c) Equilibrium condition existing between various rock formations

d) None of the above

73. Match the following and choose the correct answer

| | |
|---|---|
| (A) Acidic | (1) Gabbro |
| (B) Basic | (2) Peridotite |
| (C) Intermediate | (3) Granite |
| (D) Ultrabasic | (4) Monzonite |

a) A—3 B—1 C—2 D—4
b) A—3 B—2 C—4 D—1
c) A—3 B—1 C—4 D—2
d) A—4 B—1 C—3 D—2

74. Which of the following can be used as a way-up criteria ?
a) Raindrops—Concave upwards
b) Mudcracks—Convex upwards
c) Ripple marks—Concave upwards
d) Stromatolites—Convex upwards

75. The suture lines in Cephalopods evolved in the order
a) Nautiloid—Ceratite—Ammonite—Goniatite
b) Goniatite—Nautiloid—Ceratite—Ammonite
c) Nautiloid—Goniatite—Ceratite—Ammonite
d) Ceratite—Nautiloid—Goniatite—Ammonite

76. Gypsum deposits are found associate with
a) Jodhpur formation
b) Nagaur formation
c) Nagthat formation
d) Bikaner formation

77. Burrowing lamellibranchs have
a) Large siphuncle tube
b) Well-developed pallial sinus
c) Elongated and compressed shells
d) Well-developed dentition

78. In the Regular echinoids
a) Mouth is at the centre of the apical disc
b) Mouth lies outside the apical disc
c) New cells are added from the peristome
d) Madreporic plates are absent

79. Which of the following is characteristic of Echinoids ?
a) Ambulacra are larger than interambulacra
b) Ambulacra are smaller than interambulacra

c) Ambulacra are perforated
d) Interambulacra are perforated

80. The chief characteristic of deposits occurring in suspension is
a) Absence of current bedding/ripple marks
b) Fine-grained size
c) Black colour of the sediments
d) Abundance of fossils

81. The basal section of microcline shows
a) Symmetrical extinction
b) Inclined extinction
c) Parallel extinction
d) Isotropism

82. Twinkling in Calcite is
a) An anomalous optical property
b) A property resulting from R.I. differences
c) A property resulting from differences in relief
d) None of the above

83. Banded Hematite Quartzites are formed as a result of
a) Simultaneous precipitation of silica and iron oxides
b) Alternate precipitation of silica and iron oxides
c) Intrusion of quartz into iron oxides
d) Intrusion of iron oxides into quartz

84. Ripple marks and current bedding are features associated with ———environments
a) Abyssal plains
b) Continental slope
c) Littoral and neritic
d) Continental

85. A clinometer is used to determine
a) Strike
b) Dip
c) Bearing
d) Slope

86. Under crossed nicols leucite exhibits
a) Optical positive character
b) Optical negative character
c) Symmetrical extinction
d) Isotropism

87. The only symmetry element that is present in the Axinite type of crystals is

a) Plane of symmetry
b) Axes of symmetry
c) Centre of symmetry
d) Axes of rotary inversion

88. Fossils which have short geologic range and wide aerial distribution are described as
a) Range fossils
b) Index fossils
c) Living fossils
d) None of the above

89. Oxidation is more when
a) There is low erosion and high fluctuation of GW
b) Gradient is more and erosion is rapid
c) Gradient is low and erosion is slow
d) There is high erosion and low fluctuation of GW

90. The economic deposits associated with Vindhyan rocks is
a) Barytes
b) Limestones
c) Diamonds
d) Gold

91. The movement of the groundwater is effected by
a) Presence of pore spaces in the rocks
b) Presence of permeable formations
c) Fluid potentiality associated with it
d) Gravitational pull of the Earth

92. Manganese deposits of India are of
a) Igneous origin
b) Sedimentary origin
c) Metamorphic origin
d) Volcanic origin

93. Match the following and choose the correct answer

| | | | |
|---|---|---|---|
| A. | Chalcopyrite | 1. | $AS_2S_3$ |
| B. | Realgar | 2. | $CuFeS_2$ |
| C. | Sphalerite | 3. | $ZnCO_3$ |
| D. | Chromite | 4. | $Cr_2O_3$ |
| | | 5. | $FeCr_2O_4$ |

a) A—2 B—1 C—3 D—5
b) A—2 B—1 C—3 D—4
c) A—2 B—4 C—3 D—5

d) A—5 B—1 C—2 D—4

94. Alluvial terraces are

a) Extensive deltas formed by flood plains
b) Ridge-like areas on the coasts
c) Uplifted areas formed by streams
d) None of the above

95. The Miller symbol for the indices $2a_1\ 2a_2\ \bar{a}_3 \alpha C$ is

a) $11\bar{2}0$
b) $11\bar{3}0$
c) $22\bar{1}0$
d) $22\bar{3}0$

96. Which of the following forms have 24 faces ?

(i) Trapezohedron
(ii) Tetrahexahedron
(iii) Scalenohedron
(iv) Hexoctahedron

a) (i) only
b) (i) and (ii) only
c) (i) and (iii) only
d) (i), (ii) and (iii) only

97. In the structural maps, foliation is represented as

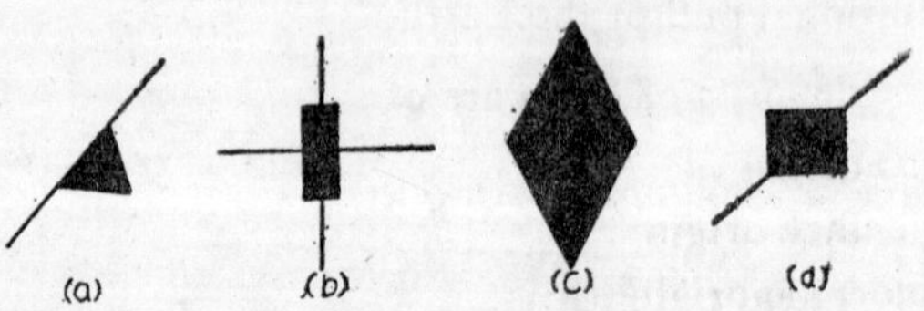


98. Match the following and choose the correct answer

| | |
|---|---|
| (i) Thrust faults | (a) High stress energy |
| (ii) Gravity faults | (b) Intermediate stress energy |
| (iii) Hinge faults | (c) Low stress energy |

a) (i)—a (ii)—c (iii)—b
b) (i)—c (ii)—a (iii)—b
c) (i)—c (ii)—b (iii)—a
d) (i)—a (ii)—b (iii)—c

99. The largest deposits of magnesite are found in

a) Andhra Pradesh

b) Karnataka
c) Bihar
d) Tamil Nadu

100. Majghawan Pipes are associated with
a) Kimberlites
b) Mica Peridotites
c) Khondalites
d) Charnockites

101. Pick out the correct metamorphic sequence from the following
a) Slate—Phyllite—Schist—Gneiss—Granulite
b) Slate—Schist—Phyllite—Gneiss—Granulite
c) Granulite—Slate—Schist—Phyllite—Gneiss
d) Phyllite—Schist—Gneiss—Slate—Granulite

102. Glass is formed due to
a) Slow cooling of the magma
b) Rapid cooling of the magma
c) Gradual cooling of the magma
d) None of the above

103. High degree of wall-rock alteration occurs when
a) Granite intrudes shale
b) Ultrabasic rocks intrude shale
c) Ultrabasic rocks intrude dolerite
d) None of the above

104. Which is the correct sequence of rocks found in a layered complex ?
a) Anorthosite—Dunite—Peridotite—Pyroxenite
b) Pyroxenite—Dunite—Anorthosite—Peridotite
c) Dunite—Pyroxenite—Peridotite—Anorthosite
d) None of the above

105. Tectonic earthquakes are generated due to
a) Release of energy (shear)
b) Volcanic eruptions
c) Crumpling of crustal blocks
d) Thrust faulting

106. Quartz exhibits undulose extinction due to
a) Deformation stress
b) Twinning
c) Cleavage
d) Impurities

107. Bauxite is
   a) Aluminium oxide
   b) Hydrous aluminium oxide
   c) A complex aluminium salt
   d) Hydrous aluminium chloride

## ANSWERS

| | | | |
|---|---|---|---|
| 1. b | 2. d | 3. a | 4. d |
| 5. b | 6. d | 7. a | 8. b |
| 9. b | 10. b | 11. a | 12. b |
| 13. a | 14. d | 15. d | 16. c |
| 17. b | 18. c | 19. b | 20. d |
| 21. a | 22. c | 23. c | 24. b |
| 25. d | 26. c | 27. d | 28. b |
| 29. b | 30. c | 31. b | 32. a |
| 33. b | 34. c | 35. b | 36. b |
| 37. a | 38. b | 39. b | 40. d |
| 41. b | 42. a | 43. a | 44. c |
| 45. c | 46, d | 47. c | 48. b |
| 49. c | 50. c | 51. d | 52. d |
| 53. a | 54. c | 55. a | 56. a |
| 57. b | 58. a | 59. c | 60. a |
| 61. d | 62. d | 63. d | 64. a |
| 65. d | 66. a | 67. b | 68. b |
| 69. b | 70. a | 71. b | 72. b |
| 73. c | 74. d | 75. c | 76. a |
| 77. c | 78. a | 79. b | 80. b |
| 81. b | 82. c | 83. b | 84. c |
| 85. c | 86. d | 87. c | 88. b |
| 89. a | 90. c | 91. b | 92. c |
| 93. a | 94. c | 95. b | 96. b |
| 97. d | 98. d | 99. d | 100. a |
| 101. a | 102. b | 103. d | 104. a |
| 105. a | 106. a | 107. b | |

---

## 5. UPSC Geologists Examination 1989

### (A) PAPER—I

(General Geology, Geomorphology, Structural Geology, Stratigraphy and Palaeotology)

1. Which of the following explains the expansion of the Universe ?
   a) Planetesimal hypothesis
   b) Nebular hypothesis
   c) Big Bang theory
   d) Novo hypothesis

2. Which of the following is an aseismic ridge ?
   a) Mid Atlantic ridge
   b) Pacific rise
   c) Lomonosov ridge
   d) Indian oceanic ridge

3. Thermal Remnant Magnetism is acquired when a mineral is cooled
   a) Above the Curie temperature
   b) Below the Curie temperature
   c) At the Curie temperature
   d) Randomly

4. Isostatic compensation is attained at
   a) MOHO
   b) Asthenosphere
   c) Stratosphere
   d) Mesosphere

5. According to the Vine-Mathew's hypothesis, the polarity is reversed every
   a) 10 m.y.
   b) 2 m.y.
   c) 1 m.y.
   d) ½ m.y.

6. Decay of $K^{40}$ gives the daughter products
   a) $Ca^{40}-Ar^{40}$
   b) $K^{39}-Ar^{40}$
   c) $K^{39}-Ar^{39}$
   d) $Ca^{40}-K^{39}$

7. Dating of Quarternary rocks is done by using
   a) K—Ar
   b) Rb—Sr
   c) $C^{12}/C^{14}$
   d) $U^{238}/Pb^{207}$

8. During the Precambrian times the rate of production of radiogenic heat was almost 7 times as at present. This was mainly attributed to
   a) $K^{40}$
   b) $U^{238}$
   c) Cosmic rays
   d) Solar radiations

9. The crust at young orogenic mountain belts like Himalayas is
   a) As thick us the average thickness of the crust
   b) ½ the average thickness of the crust
   c) Twice the average thickness of the crust
   d) 5 times the average thickness of the crust

10. The volcanism found at the Island Arcs is
   a) Andesitic
   b) Basaltic
   c) Tholeiitic
   d) Rhyolitic

11. Sir Edward Bullard made an attempt to match the continental outlines with the help of a computer at a depth of
   a) 500 m
   b) 1000 m
   c) 2000 m
   d) 50 m

12. Which of the following denote the inclined zones along which the continents get subducted ?
   a) Trenches
   b) Benioff zones
   c) Island arcs
   d) Andesite line

13. Geosynclinal sedimentation involves
   a) Thick sedimentary sequences of deep water
   b) Thick sedimentary sequences of shallow water
   c) Thin sedimentary sequences of shallow water
   d) None of the above

14. Choose the correct order of the following glaciations

a) Riss-Wurm-Mindel-Gunz
b) Mindel-Wurm-Riss-Gunz
c) Gunz-Riss-Mindel-Wurm
d) Wurm-Riss-Mindel-Gunz

15. According to the Paleaontologic and paleomagnetic evidences, the Tatrot-Pinjor boundary is marked by
a) Gauss-Gilbert transition
b) Matuyama-Brunhes transition
c) Brunhes-Gauss transition
d) Top of the Brunhes

16. The MORs reveal a pattern for
a) Volcanism
b) Ocean floor spreading
c) Seismicity
d) Subduction of lithosphere

17. According to the plate tectonics theory in about 100 m.a.
a) The African ocean will open up and the African rift would form a new ocean
b) India would collide with Antartica
c) Australia would split into microcontinents
d) All of the above

18. Match the following and choose the correct answer

| *Volcanism* | *Volcanic activity* |
|---|---|
| A. Hawaiin | 1. Violent |
| B. Stombolian | 2. Strong explosion |
| C. Vesuvian | 3. Silent explosive |
| D. Volcanic | 4. Quiet |

a) A—4 B—1 C—3 D—2
b) A—4 B—3 C—2 D—1
c) A—4 B—3 C—1 D—2
d) A—3 B—4 C—2 D—1

19. Which of the following is an igneous pluton which is upfaulted by arcuate faults ?
a) Bysmalith
b) Conolith
c) Laccolith
d) Lopolith

20. Identify the features shown in the map

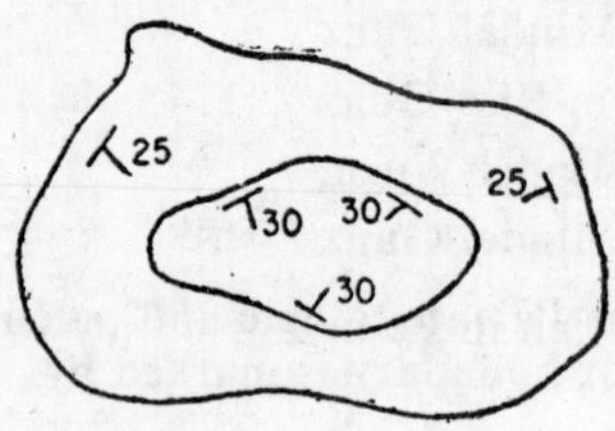


a) Ring dykes
b) Laccolith
c) Lopoliths
d) Batholith

21. Sabal major is associated with
a) Subathu
b) Dagshai
c) Kasauli
d) Garbyang

22. Nummulites fossils are found in
a) Subathu
b) Dagshai
c) Kasauli
d) Shali

23. Cumbum Phyllites form a part of
a) Cuddapahs
b) Kurnools
c) Vindhyans
d) Bhimas

24. Which of the following formation is diamond-bearing ?
a) Vaimpalli
b) Banganapalli
c) Agglomeratic slates
d) Spiti shales

25. The Pliocene-Pliestocene boundary in the Siwaliks can be demarcated by the presence of mammalian fossils
a) Equus-Elephas
b) Dinotheres-Primates
c) Moeritheres-Stegodon
d) None of the above

26. A bar that connects an island with another island or an island with the mainland is called

a) Tumbolo
b) Spit
c) Hook
d) Groyne

27. Soils which are rich in Fe or Al or both are known as
a) Seicozems
b) Pedelfers
c) Chernozems
d) Loams

28. Which of the following is true of black cotton soils ?
(i) They are formed at high temperatures
(ii) They are rich in clay minerals
(iii) They are formed only by the Deccan Traps
(iv) They are succeptible to cracking on exposure to sunlight
a) (i), (iii) and (iv) only
b) (i) and (iii) only
c) (i), (ii) and (iv) only
d) All of the above

29. A glaciated valley which is completely submerged by a stream is described as
a) Ria
b) Fjord
c) Drumlin
d) Hornito

30. Cup-shaped smooth mounds in the glaciated regions are known as
a) Drumlins
b) Kettles
c) Eskers
d) Cirques

31. Knob and Kettle topography is associated with
a) Glacial regions
b) Fluvial regions
c) Karst regions
d) Arid regions

32. Inselbergs are formed due to
a) Alluvial action
b) Glacial action
c) Solution action
d) Wind action

33. 'Bolson' is

a) A basin whose base shows sagging as a result of piling sediments
b) A basin with centripetal dips and rimmed by mountains
c) A basin with centrifugal dips and rimmed by mountains
d) None of the above

34. An 'arete' is

a) A ridge formed by the stream action
b) A depositional structure of glaciers
c) A ridge formed by the cutting of two cirques
d) A cup-shaped hollow in glaciated terrains

35. Indus is a/an

a) Insequent stream
b) Obsequent stream
c) Subsequent stream
d) Antecedent stream

36. A stream which flows in the same direction of the trunk stream is known as

a) Obsequent
b) Resequent
c) Insequent
d) Subsequent

37. The lateral shifting in the channel of a meandering river is shown by

a) Meander bars
b) Channel bars
c) Point bars
d) None of the above

38. Blind valleys are associated with

a) Aeolian erosion
b) Alluvial erosion
c) Glacial erosion
d) Karst erosion

39. Eurydesma is indicative of

a) Cold climates
b) Warm climates
c) Arid climates
d) Humid climates

40. The periods Carboniferous-Permian marks the extinction of

a) Foraminifers
b) Trilobites
c) Graptolites
d) Corals

41. Lamellibranchs are classified on the basis of
a) Dentition
b) Adductor impressions
c) Number of gills
d) Presence or absence of pallial sinus

42. The Heterodont dentition is exhibited by
a) Venus
b) Pleuromya
c) Mytilus
d) Heliotype

43. The Tiruchirapalli formation extends from
a) Aptian to Albian
b) Turonian to Senomanian
c) Danian to Albian
d) None of the above

44. Arrange the following formations of Assam in the proper sequence
1. Surma
2. Barail
3. Dihing
4. Tippam
a) 4123
b) 4132
c) 4321
d) 2341

45. The Mesozoic era is described as the age of
a) Reptiles
b) Mammals
c) Amphibians
d) Birds

46, Devonian is described as the age of
a) Reptiles
b) Fishes
c) Amphibians
d) Sharks

47. Graptolites were
a) Planktonic
b) Neritic
c) Sessile
d) Benthonic

48. Physa (Bullinas) principi found in the Intertrappean beds is associated with
a) Panjal traps
b) Deccan traps
c) Baflaiz volcanics
d) Dras volcanics

49. Lingula is a
a) Long range form
b) Short range form
c) Index fossil
d) Extinct form

50. Otoceras woodwardi has an important bearing in defining the boundary between
a) Permian and Carboniferous
b) Silurian and Devonian
c) Permian and Triassic
d) Triassic and Jurassic

51. Paradoxides is an index fossil of
a) Cambrian
b) Ordovician
c) Silurian
d) Devonian

52. Which of the following belongs to the Silurian ?
a) Tetragraptus
b) Didymograptus
c) Monograptus
d) Dictyonema

53. The age of the Equus sivanensis is
a) Pliocene
b) Pliestocene
c) Miocene
d) Eocene

54. Ramapithecus, a descendent of man, has been described from the Siwalik formations of
    a) Kumaon Himalyas
    b) Spiti valley
    c) Arunachal Pradesh
    d) Ladakh

55. Sivapithecus is a/an
    a) Primate
    b) Rodent
    c) Perrisodactyl
    d) Artiodactyl

56. A megalospheric form is one in which
    a) The proloculus is larger than the shell size
    b) The proloculus is smaller than the shell size
    c) Both proloculus and the shell size are large
    d) Both proloculus and the shell size is small

57. Foraminifers became adapted to the benthonic mode of life in
    a) Jurassic
    b) Cretaceous
    c) Tertiary
    d) Cambrian

58. Globotruncana is
    a) Planktonic
    b) Benthonic
    c) Argillaceous
    d) Arenaceous

59. Fussilinid is a characteristic foraminifer of
    a) Permian
    b) Jurassic
    c) Triassic
    d) Cretaceous

60. Monotis salinaria is a characteristic fossil of
    a) Triassic
    b) Jurassic
    c) Cretaceous
    d) Eocene

61. Cardita beaumonti is an index fossil for

a) Cambrian
b) Carboniferous
c) Permian
d) Cretaceous

62. An index fossil has
a) Short stratigraphic range and wide geographic distribution
b) Short geographic range and wide stratigraphic distribution
c) Wide stratigraphic and geographic distribution
d) None of the above

63. Which of the following conditions is essential for the formation of fossils ?
a) Burial of fossils by natural agencies
b) Burial of fossils by unnatural agencies
c) Transportation of fossils for a long distance by streams
d) None of the above

64. Era is subdivided into
a) Period, Epoch and Age
b) Series, Stage and Age
c) System, Series and Stage
d) Eon, Period and Epoch

65. Calceola is a
a) Coral
b) Bryozoan
c) Branchiopod
d) Trilobite

66. Which of the following shows hemimarginal development of shell in the Brachiopods ?

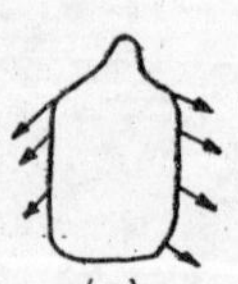
(a)

(b)

(c)

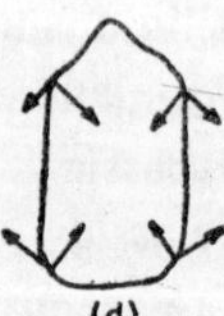
(d)

67. The Neotectonic movements occur between
a) Paleogene-Quaternary
b) Neogene-Quaternary
c) Paleogene-Neogene
d) Tertiary-Quaternary

68. Arrange the following in the decreasing order of susceptibility to weathering

1. Biotite
2. Hornblende
3. Pyroxene
4. Olivine

a) 4321
b) 4312
c) 1234
d) 1324

69. Insolation is a dominant process in the regions of
a) Low precipitation
b) Wide variation in temperature
c) Low precipitation and wide variation in temperature
d) Low precipitation and large wind velocity

70. Decollement is a type of
a Fold
b) Fault
c) Joint
d) Unconfirmity

71. Horizontal beds of Eocene fossils are overlain on the horizontal beds of Ordovician age and the contact is also conformable, then the unconformable contact is
a) Heterolithic unconfirmity
b) Disconfirmity
c) Local unconfirmity
d) Angular unconfirmity

72. If sedimentary rocks overlie plutonic igneous rocks, the unconfirmity is described as
a) Nonconfirmity
b) Angular unconfirmity
c) Disconfirmity
d) Heterolithic unconfirmity

73. Eperiogenesis is associated with
a) Cratons
b) Orogens
c) Geosynclines
d) Volcanoes

74. Which of the following is a primary joint ?

a) Exfoliation joint

b) Columnar joint

c) Sheeting

d) Extension joint

75. A reclined fold is one in which

a) The beds are sheared perpendicular to the strike of the beds

b) The beds are sheared parallel to the strike of the beds

c) The anticlines and synclines dip towards the top and bottom

d) None of the above

76. In the 'Pencil structure' associated with micaceous schists, the structure is parallel to

a) Axial planes of folds

b) Cleavage direction

c) Schistosity

d) None of the above

77. The dip separation is parallel to dip slip when the net slip is

a) Parallel to the dip of the faulted bed

b) Parallel to the of the fault

c) Perpendicular to the regional trend of the beds

d) Parallel to the strike of the bed

78. The best known section for the Permian-Triassic boundary in India is

a) Kashmir

b) Spiti

c) Salt range

d) Assam

79. Pick out the oldest formation among the following

a) Nimar sandstone

b) Daonella marls

c) Nodular limestone

d) Coral limestone

80. The flysch sequences typically consists of

a) Limestones and sandstones

b) Limestones and shales

c) Sandstones and shales

d) Sandstones and conglomerates

81. A bed dips 60° due ENE. Its apparent dip in the SSE direction is

a) 60°

b) 45°
c) 30°
d) 0°

82. The diagram below shows a superimposed fold. The sedimentary formation terminate at X due to

a) Facies change
b) Structural closure
c) Pinching
d) None of the above

83. The diagram below shows

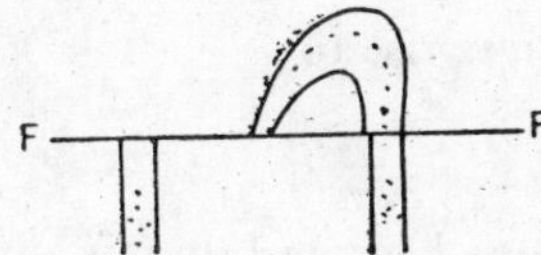


a) A plunging syncline effected by normal fault
b) A plunging syncline effected by a reverse fault
c) A syncline effected by a strike-slip fault
d) None of the above

84. The vesicles in the lava flows indicate
a) Age of the lava flows
b) Top and bottom of lava flows
c) Temperature of lava flows
d) None of the above

85. Which of the following can be used to determine the top and bottom of the beds ?
a) Current ripples
b) Graded bedding
c) Slump structures
d) Climbing ripples

86. The beds are overturned when
a) Bedding and cleavage dips in the same direction but the former dips more steeply
b) Bedding and cleavage dips in the same direction but the later dips more steeply

c) Bedding and cleavage dip in the opposite directions

d) None of the above

87. If a chlorite schist is subjected to plucking, the schistosity produced is known as

a) Slaty cleavage

b) Slip cleavage

c) Crenulation cleavage

d) Boudinage

88. Two parallel quartz veins have been subjected to shear, one is transformed into a microfold and the other to a boudin. This is due to

a) Horizontal stress in one direction and another horizontal stress perpendicular to the first

b) Horizontal stress in one direction and another horizontal stress inclined to the first

c) Horizontal stresses in both the directions

d) None of the above

89. The shear folding gives rise to

a) Thinning

b) Thickening

c) Thickening in some beds and thinning in others

d) Do not effect the beds

90. A large depression bounded by steep walls and having flat floors and formed at the karst regions is described as

a) Doline

b) Uvala

c) Polje

d) Hum

91. The narrow shifting boundary that lies between land and water is called

a) Shoreline

b) Coastline

c) Shelf

d) Beach

92. The Gutenberg discontinuity lies between

a) Outer core and lower mantle

b) Lower mantle and upper mantle

c) Outer and inner cores

d) Crust and the upper mantle

93. Match the following and choose the corret answer

A. Gravity faulting 1. Imbricate zone

B. Plumose markings — 2. Rift valleys
C. Schuppen structure — 3. Joints
D. Digitations — 4. Recumbent folds
5. Thrust faults

a) A—2 B—1 C—5 D—4
b) A—2 B—3 C—1 D—4
c) A—2 B—3 C—5 D—4
d) A—5 B—3 C—1 D—2

94. Which of the following diagrams shows a fold with vertical plunge at the axial surface ?

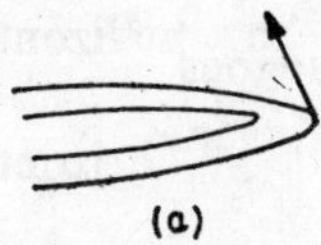
(a)
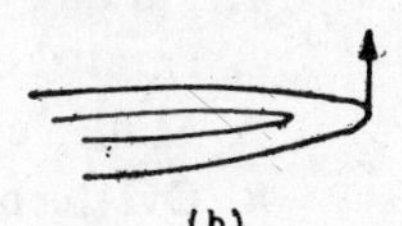
(b)

(c)

(d)

95. A stereographic diagram of a fold is given below. It shows that the fold is

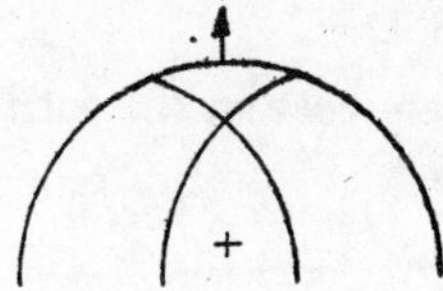

a) Vertical
b) Horizontal
c) Plunging
d) Non-plunging

96. The symmetry of some folds is given below. Match the following with their characteristic symmetry ?

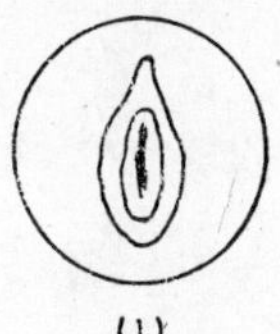
(1)

(2)

(3)

a) 1—Monoclinic, 2—Triclinic , 3—Triclinic
b) 1—Triclinic, 2—Monoclinic, 3—Monoclinic
c) 1—Orthorhombic, 2—Monoclinic, 3—Monoclinic
d) 1—Orthorhombic, 2—Triclinic, 3—Monoclinic

97. Arrange the following folds in the decreasing order of inter-limb angle

1. Open
2. Tight
3. Closed
4. Gentle
5. Isoclinal

a) 41325
b) 41235
c) 12354
d) 54132

98. Match the following and choose the correct answer

A. Conjugate folds — 1. Parallel layers
B. Similar folds — 2. Parallel dip isogons
C. Fan folds — 3. Multiple axial planes
D. Parallel folds — 4. Overturning

a) A—3 B—1 C—4 D—2
b) A—3 B—2 C—4 D—1
c) A—2 B—1 C—4 D—3
d) A—4 B—1 C—3 D—2

99. Choose the correct sequence in the formation of salt domes

1. Doming of strata
2. Intrusion of salts
3. Upliftment
4. Erosion of salted domes

a) 1234
b) 3124
c) 3214
d) 1324

100. Fossil excrements are described as

a) Casts
b) Coprolites
c) Moulds
d) Tracks

101. Deformation in which parallel lines remain parallel is described as

a) Homogenous
b) Heterogenous
c) Monocyclic
d) Polycyclic

102. The strength of a rock drecreases with

1. Increase in confining pressure
2. Increase in temperature
3. Increase in creep

a) 1 and 2 only
b) 1 and 3 only
c) 2 and 3 only
d) All of the above

103. Traingular facets give an indication of

a) Fault scarps
b) Thrust zones
c) Recumbent folds
d) Primary joints

104. Owen Fracture Zone lies in the

a) Pacific ocean
b) Atlantic ocean
c) Arabian sea
d) Bay of Bengal

105. Submergence of an oceanic crust is indicated by

a) Plate tectonics concept
b) Sea floor spreading
c) Island arcs
d) Seamounts and guyots

106. Helictite is a feature associated with

a) Karst regions
b) Glacial regions
c) Arid landforms
d) Fluvial landforms

107. Which of the following has the highest half life among the following ?

a) Rb—Sr
b) Sm—Nd
c) K—Ar
d) $U^{238}$

108. 'Hogbacks' are associated with

a) Stream erosion
b) Aeolian erosion
c) Glacial erosion
d) Solution action

109. Irregular and erratic topography formed by thrusting and crumpling is described as
   a) Hogback and cuesta
   b) Knob and kettle
   c) Basin and range
   d) Crag and tail

110. Formation is the term used to designate mappable units in the scale
   a) 1 : 10,000
   b) 1 : 100,000
   c) 1 : 50,000
   d) 1 : 500

111. The fundamental unit of lithostratigraphy is
   a) Supergroup
   b) Group
   c) Formation
   d) Member

112. Laterite and bauxite are formed in tropical climates with alternate wet and dry seasons. The intense chemical reaction leaches all the constituents downwards except
   a) K+Na
   b) K+Mg
   c) Na+Fe
   d) Fe+Al

113. The most important source for freshwater on the earth is
   a) Polar ice caps
   b) Seas
   c) River discharges
   d) Groundwater

114. Eparchaen unconfirmity is an unconfirmity between
   a) Archaen and Algonkian
   b) Algonkian and Paleozoic
   c) Paleozoic and Mesozoic
   d) Mesozoic and Cenozoic

Q. 115—Q. 118 are based on Reasoning and Assertion. For directions to answer these questions, refer Chapter-15 of Section—2.

115. Assertion : Diapiric material is mostly mafic/ultramafic

Reason : Gravity has a great influence in the formation of diapiric folds

116. Assertion : Cleavage that is found parallel to bedding is called bedding cleavage

Reason : Bedding is due to isoclinal folding

117. Assertion : Gravity faults are always associated with rift valleys

Reson : Rift valleys are formed by horizontal tension in the crust

118. Assertion : Recumbent folds cannot be described as either antiformal or synformal

Reason · Recumbent folds have horizontal axial surfaces

## ANSWERS

| | | | |
|---|---|---|---|
| 1. c | 2. c | 3. a | 4. a |
| 5. c | 6. a | 7. c | 8. a |
| 9. c | 10. a | 11. c | 12. b |
| 13. a | 14. a | 15. b | 16. b |
| 17. a | 18. c | 19. a | 20. c |
| 21. c | 22. a | 23. a | 24. b |
| 25. a | 26. a | 27. b | 28. c |
| 29. b | 30. a | 31. a | 32. d |
| 33. b | 34. c | 35. d | 36. b |
| 37. c | 38. d | 39. d | 40. b |
| 41. a | 42. a | 43. b | 44. a |
| 45. a | 46. b | 47 a | 48. b |
| 49. a | 50. c | 51. a | 52. c |
| 53. b | 54. a | 55. a | 56. a |
| 57. b | 58. c | 59. a | 60. b |
| 61. d | 62. a | 63. a | 64. a |
| 65. a | 66. b | 67. b | 68. b |
| 69. c | 70. a | 71. b | 72. a |
| 73. a | 74. b | 75. a | 76. c |
| 77. a | 78. a | 79. c | 80. c |
| 81. c | 82. b | 83. c | 84. c |
| 85. a | 86. a | 87. c | 88. a |
| 89. d | 90. c | 91. a | 92. a |
| 93. b | 94. b | 95. d | 96. a |
| 97. a | 98. a | 99. a | 100. b |
| 101. a | 102. c | 103. a | 104. d |
| 105. d | 106. a | 107. b | 108. a |
| 109. c | 110. c | 111. c | 112. d |
| 111 a | 114. b | 115. e | 116. b |
| 117. a | 118. a. | | |

## (B) PAPER—II

(Geochemistry, Crystallography, Mineralogy, Optical Mineralogy and Petrology)

1. The contact metamorphism of calcareous sediments gives rise to
   a) Argillites
   b) Hornfelses
   c) Skarn
   d) Tactites
2. In a Solid—→Solid+Liquid metamorphic reaction, the PT equilibrium curve between the reactant and product of an assemblage is a
   a) Straight line
   b) Curved line
   c) Hyperbola
   d) Parabola
3. A progressive increase in the metamorphic grade from spotted slate containing andalusite porphyroblast (outer zone), cordeirite and microcline, hornfels and cordeirite-hypersthene, potash felspar, hornblende (inner zone) is commonly seen in
   a) Basalt
   b) Gabbro
   c) Syenite
   d) Granite
4. The phase transformation between the three polymorphs of $Al_2SiO_5$ in the metamorphic terrane is due to
   a) Solid solution
   b) Dehydration
   c) Decarbonation
   d) Exsolution
5. The presence of which of the following minerals indicate the beginning of metamorphism
   a) Andalusite
   b) Chlorite
   c) Cordeirite
   d) Laumontite
6. In the Staurolite zone of regional metamorphism, staurolite is found as
   a) Staurolite mineral

b) Staurolite+Kyanite

c) Liquid melt

d) None of the above

7. Which of the following metamorphic combinations is odd ?

a) Kyanite—sillimanite—biotite—plagioclase—quartz

b) Sillimanite—albite—plagioclase—quartz

c) Sillimanite—garnet—aluminium—plagioclase—quartz

d) None of the above

8. The metamorphic rock with mineral assemblages Plagioclase+quartz+Andalusite+Cordeirite+Biotite+Hbl indicate

a) Contact metamorphism of low grade

b) Regional matamorphism of low grade

c) Contact metamorphism of high grade

d) Regional metamorphism of high grade

9. In the Paired Metamorphic belts when compared to the blueschist facies of LP/HT, the amphibolite facies occurs towards

a) Inland

b) Ocean

c) Continental rift

d) Oceanic rift

10. A porphyroclast is

a) A large clast in a sedimentary rock

b) Similar to porphyroblast

c) Large crystal in igneous rocks

d) None of the above

11. The structure which shows spotted appearance is termed as

a) Cataclastic

b) Maculose

c) Helicitic

d) Mylonite

12. According to the Plate Tectonics concept a paired metamorphic belt is composed of LP/HT blueschist facies and a HP/HT amphibolite facies. These are formed due to

(i) Subduction involving ocean-continent and arc collision

(ii) Subduction involving continental-arc or continent-continent or arc-arc collision

a) Both the above

b) (i) only

c) (ii) only

d) None of the above

13. The Indian rock which typically consists of Quartz+Garnet +Sillimanite+graphite is known as

a) Khondalite
b) Charnockite
c) Gondite
d) Kodurite

14. Match the following and choose the correct answer

| | | | |
|---|---|---|---|
| A. | Charnockite | 1. | Sillimanite+Qtz+Feldspar+garnet |
| B. | Khondalite | 2. | Quartz+Perthite+K-feldspar+Hyp |
| C. | Blueschist | 3. | Hornblende+Plagioclase |
| D. | Amphibolite | 4. | Glaucophane+Jadeite+soda amphibole |

a) A—2 B—1 C—4 D—3
b) A—1 B—2 C—3 D—4
c) A—2 B—4 C—1 D—3
d) A—1 B—4 C—2 D—3

15. Girnarite is associated with

a) Nepheline syenites
b) Alkali Olivine basalts
c) Gabbroic igneous complexes
d) Granites

16. Match the following and choose the correct answer

| | | | |
|---|---|---|---|
| A. | Lacustrine | 1. | Karewas |
| B. | Fluviatile | 2. | Siwaliks |
| C. | Marine | 3. | Spiti |

a) A—1 B—2 C—3
b) A—3 B—2 C—1
c) A—1 B—3 C—2
d) A—3 B—1 C—2

17. The Cherrapunji limestone of Meghalaya is of

a) Triassic age
b) Paleaozoic age
c) Proterozoic age
d) Tertiary age

18. The environment which is characteristically found below 7000 m is described as

a) Hydal
b) Abyssal
c) Neritic
d) Littoral

19. Which of the following minerals has the property of losing water on heating and again absorbing it when exposed to air ?
   a) Gypsum
   b) Calcium
   c) Quartz
   d) Nepheline

20. Aragonite is a polymorph of
   a) Calcite
   b) Gypsum
   c) Pyrite
   d) Quartz

21. The $C^{14}$ level of the atmosphere prior to 1944 has been changed and has since been on the increase. This is due to
   a) Neutron explosion in the atmosphere
   b) Death of many animals and plants
   c) Destruction of the ozone layer in the atmosphere
   d) None of the above

22. Primary geochemical dispersion is characterized by
   a) HP, HT and restricted circulation of fluids
   b) LP, HT and restricted circulation of fluids
   c) LP, HT and free circulation of fluids
   d) HP, HT and free circulation of fluids

23. Match the following geochemical association of minerals

| | |
|---|---|
| A. Rubidium | 1. Calcium |
| B. Strontium | 2. Potassium |
| C. Manganese | 3. Iron |
| D. Silicon | 4. Germanium |

   a) A—2 B—1 C—3 D—4
   b) A—1 B—2 D—4 D—3
   c) A—2 B—1 C—4 D—3
   d) A—1 B—2 C—3 D—4

24. Absorption of groundwater from soils by the root of a plant is an example of
   a) Primary biochemical dispersion
   b) Secondary biochemical dispersion
   c) Primary hydrological dispersion
   d) Secondary hydrological dispersion

25. A fan-shaped dispersal pattern is formed by

a) Wind
b) Rivers
c) Glaciers
d) Soil

26. The biogeochemical analysis involves
a) Visual observation of plants to locate ore deposits
b) Chemical analysis of plant parts
c) Chemical analysis of plant morphological features
d) Chemical analysis of dissolved plant products

27. The presence of zinc-rich soils is indicated by
a) Astragalus pattersoni
b) Aerocephalus roberti
c) Salsola nitraria
d) Viola calamensis

28. In the Diopside-Anorthite system, the eutectic composition reached between the two minerals is
a) $Di_{55}$ $An_{45}$
b) $Di_{60}$ $An_{40}$
c) $Di_{70}$ $An_{30}$
d) $Di_{80}$ $An_{20}$

29. If large crystals contain thin granules of pyroxene in the matrix, the texture is described as
a) Intergranular
b) Helictite
c) Sub-ophitic
d) Perthite

30. Which of the following is an ultra-alkalic rock ?
a) Adamellite
b) Toscanite
c) Transjhemite
d) Ijolite

31. Lamprophyres composed of barkivikite and augite are known as
a) Vogesite
b) Kersanite
c) Alnoiite
d) Camptonite

32. Which of the following lamprophyres is rich in orthoclase ?
a) Vogesite

b) Kersanite
c) Spessartite
d) Malchite

33. The serpentinite, keratophyre and ophiolites are
a) Pre-tectonic
b) Post-tectonic
c) Pre and post-tectonic
d) Syntectonic

34. Which of the following is an optical anomaly ?
a) Zoning in plagioclase
b) Undulose extinction in quartz
c) Schillar structure in hypersthene
d) Polysynthetic twinning in leucite

35. The exfoliation in vermiculite refers to
a) Its ability to break along cleavage surfaces
b) Expand on heating
c) Contract on heating
d) None of the above

36. Itacolomite is flexible due to
a) Presence of weak planes of muscovite
b) Planes of weaknesses containing Al/K
c) Planes of weaknesses containing biotite
d) None of the above

37. Match the following and choose the correct answer

| *Mineral* | *Extinction angle* |
|---|---|
| A. Augite | 1. 45° |
| B. Diopside | 2. 0° |
| C. Hedenbergite | 3. 38° |
| D. Enstatite | 4. 41° |

a) A—3 B—4 C—1 D—2
b) A—4 B—3 C—2 D—1
c) A—3 B—1 C—4 D—1
d) A—3 B—4 C—2 D—1

38. Which of the following denotes a biaxial negative mineral ?

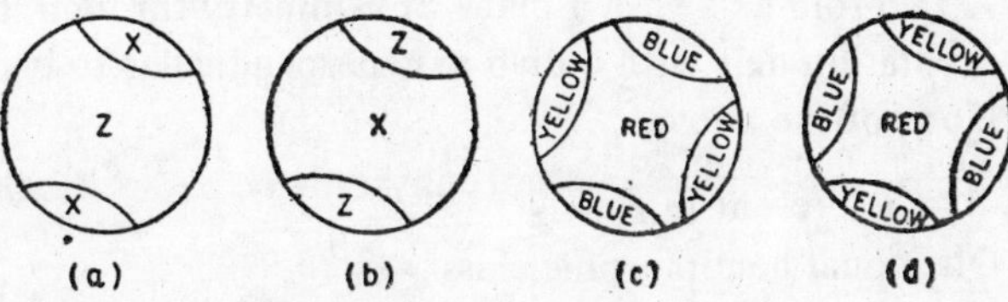

39. Identify the mineral on the basis of the diagram given below

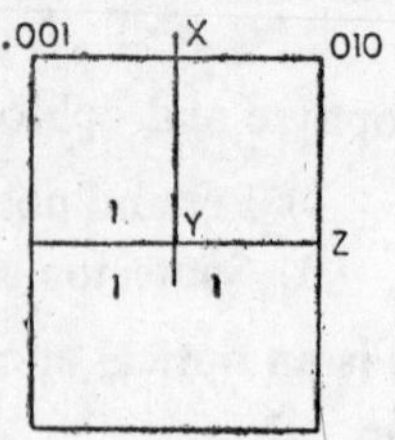


a) Pyroxene
b) Plagioclase
c) Garnet
d) Olivine

40. $\overline{6}$ can be represented as

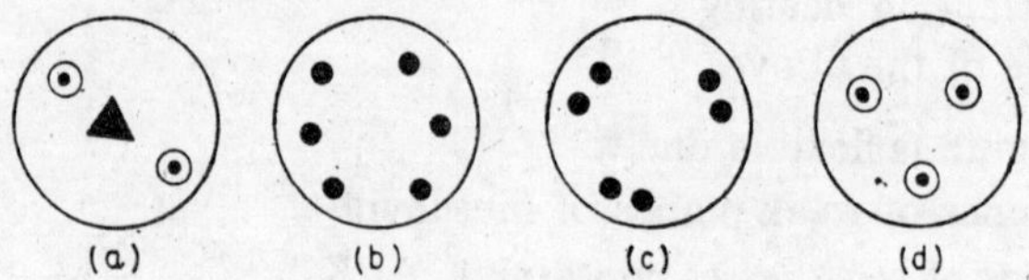


41. Which of the following is the symmetry exhibited by tourmaline ?

a) 3AP, $1^{111}$, No centre
b) 3AP, $3A_3$, Centre
c) 3AP, $3A_3$, No centre
d) 6AP, $1A_4$, Centre

42. A cubic crystal with i4 faccs is made up of

a) Cube+Octahedron
b) Octahedron+Dodecahedron
c) Dodecahedron+Cube
d) Cube+Rhombohedron

43. 2/m denotes

a) A two-fold axis normal to plane of symmetry
b) A two-fold axis with a plane of symmetry through it
c) A rotation axis with a diad axis perpendicular to it
d) None of the above

44. 3P, $1A_3$ is present in

a) Ditrigonal hemimorphic class

b) Hexagonal hemimorphic class
c) Trigonal Trapezohedral class
d) Hexagonal scalenohedral class

45. A mineral crystallizing in Orthorhombic system with sp. gr. 3.3 and hardness 8 is likely to be
a) Olivine
b) Sulphur
c) Topaz
d) Barytes

46. The proved number of Space Lattices is
a) 28
b) 14
c) 230
d) 7

47. PCIF space lattices are found in
a) Orthorhombic system
b) Monoclinic system
c) Triclinic system
d) Hexagonal system

48. In the Dauphine twin the twin plane is
a) $01\bar{1}2$
b) $02\bar{2}1$
c) $11\bar{2}0$
d) 100

49. Match the following twins and choose the correct answer

| | |
|---|---|
| A. Polysynthetic | 1. Twin lamellae cyclic |
| B. Contact | 2. Twins are knee-shaped |
| C. Geneculated | 3. Twins lie merely in contact |
| D. Cyclic | 4. Twin lamellae parallel |

a) A—4 B—3 C—2 D—1
b) A—1 B—3 C—2 D—4
c) A—4 B—2 C—3 D—1
d) A—1 B—2 C—3 D—4

50. The twin in which the twin plane and the composition plane is 001 and if the twin axis lies perpendicular to it, is called
a) Dovine twin
b) Carlsbad twin
c) Baveno twin

d) Manebach twin

51. Microcline shows a typical gridion structure due to two sets of polysynthetic twinning on
   a) Albite+Acline laws
   b) Albite+Pericline laws
   c) Pericline+Anorthite laws
   d) Albite+Manebach twin

52. The zone symbol is given by
   a) E=2U+L
   b) nd=2d sin $\phi$
   c) P+F=C+2
   d) Uh+Vk+Wl=0

53. In a sample $Rb^{87}/Sr^{87}$ are found in the ratio of 1 : 3. If the half-life of $Rb^{87}$ is 50 b.y., the age of the sample is
   a) 50 b.y.
   b) 100 b.y.
   c) 150 b.y.
   d) 200 b.y.

54. Sphene has the formula
   a) $CaTi_2O_5$
   b) $CaTiO_5$
   c) $Fe_2CaTi_2O_5$
   d) $FeCaTiO_5$

55. Aluminium is practically absent is
   a) Chlorite
   b) Pistacite
   c) Pinnacite
   d) Vredenbergite

56. Which of the following shows a solid solution series ?
   a) Diopside-Hedenbergite
   b) Anthophyllite-Cummingtonite
   c) Augite-Albite
   d) Albite-Anorthite

57. In the formula for Pyroxenes $(WXY)_2Z_2O_6$, if Aluminium is to be present it would be found in :
   a) X site
   b) Y site
   c) Z site
   d) Any of the above

58. Basaltic hornblende differs from ordinary hornblende in
   a) High $Fe^{2+}/Fe^{+3}$ and low hydroxyl content
   b) High $Fe^{+2}/Fe^{+3}$ and high hydroxyl content
   c) Low $Fe^{+2}/Fe^{+3}$ and low hydroxyl content
   d) None of the above

59. Incongruent melting is exhibited by
   a) Albite
   b) Nepheline
   c) Quartz
   d) Olivine

60. Which of the following amphiboles is characteristic of high temperatures ?
   a) Edenite
   b) Pargasite
   c) Tremolite
   d) Hornblende

61. Which of the following amphiboles is associated with Pyroxenites ?
   a) Paragonite
   b) Pargasite
   c) Gedrite
   d) Anthophyllite

62. Match the following and choose the correct answer

| | |
|---|---|
| A. Lithophiles | 1. As, Sb, Pb, Cu, Zn |
| B. Chalcophiles | 2. Co, Ni |
| C. Siderophiles | 3. Rb, Sr, Ca, Na |
| D. Atmophiles | 4. He, Ne, N |

   a) A—3 B—1 C—2 D—4
   b) A—3 B--2 C—! D—4
   c) A—2 B—1 C—3 D—4
   d) A—4 B—2 C—1 D—3

63. Geochemical classification of elements is based on its behaviour in
   a) Liquid-liquid equilibrium between melts
   b) Gas-gas equilibrium between melts
   c) Solid-solid equilibrium between melts
   d) Solid-liquid equilibrium between melts

64. The Dispersion of elements is a function of
   a) Density of minerals
   b) Atomic weights of minerals

c) Chemical composition

d) Geochemical affinities

65. Match of the following and choose the correct answer

| | |
|---|---|
| A. Nesosilicate | 1. Fayalite |
| B. Cyclosillicate | 2. Nepheline |
| C. Phyllosilicate | 3. Beryl |
| D. Tectosilicate | 4. Chlorite |

a) A—1 B—2 C—3 D—4

b) A—1 B—3 C—4 D—2

c) A—1 B—4 B—3 D—2

d) A—4 B—3 C—1 D—2

66. Which of the following is a high temperature polymorph of quartz ?

a) Coesite

b) Cristobalite

c) Tridymite

d) Opal

67. Potassium is not found in greater amount in sea water because

a) Na is more soluble than K

b) K is more soluble than Na

c) K is used in the formation of polyhalides

d) K is used in the formation of clay minerals

68. Sieve texture is exhibited by

a) Zeolites

b) Scapolites

c) Granites

d) None of the above

69. If a grain of a microcline crystal is focussed by a polarizing microscope and on raising the objective if the colours are found to move in then the grain must be bordered by

a) Nepheline

b) Plagioclase

c) Quartz

d) Garnet

70. If the granites show gradational contacts with the country rocks then the granites must have formed from

a) Metamorphic differentiation

b) Anatexis

c) Granitization

d) All of the above

## ANSWERS

| | | | |
|---|---|---|---|
| 1. d | 2. b | 3. b | 4. d |
| 5. c | 6. a | 7. a | 8. d |
| 9. b | 10. a | 11. b | 12. c |
| 13. a | 14. a | 15. c | 16. a |
| 17. d | 18. a | 19. a | 20. a |
| 21. b | 22. a | 23. c | 24. b |
| 25. b | 26. c | 27. d | 28. b |
| 29. c | 30. d | 31. d | 32. a |
| 33. b | 34. d | 35. b | 36. a |
| 37. d | 38. c | 39. b | 40. c |
| 41. d | 42. a | 43. a | 44. a |
| 45. c | 46. c | 47. a | 48. c |
| 49. a | 50. d | 51. b | 52. d |
| 53. b | 54. a | 55. d | 56. d |
| 57. b | 58. a | 59. a | 60. d |
| 61. a | 62. a | 63. a | 64. d |
| 65. b | 66. a | 67. d | 68. a |
| 69. d | 70. c | | |

## (C) PAPER—III

### (Economic Geology, Mineral Exploration, Mineral Economics and Mining Geology)

1. The pathfinder in gold deposits is
   a) Arsenic
   b) Molybdenum
   c) Mercury
   d) Silver
2. The Paleobotanical indicator for zinc deposits is
   a) Viola calamina
   b) Astragalus
   c) Clusia kosea
   d) None of the above
3. The only workable mines for diamond in India are in
   a) A.P.
   b) U.P.
   c) Bihar
   d) M.P.

4. Ceylon and Burma are important producers of
   a) Tin
   b) Uranium
   c) Silver
   d) Precious stones
5. The estimated world reserves of manganese are
   a) 1.0 BT
   b) 12.0 BT
   c) 3 BT
   d) 4 BT
6. Thorium was first discovered in syenites in
   a) USA
   b) USSR
   c) Brazil
   d) Norway
7. Match the following and choose the correct answer

| | |
|---|---|
| A. Manufacture of pottery | 1. Fire clays |
| B. Refractory works and foundry and steel work | 2. China clay |
| C. Absorbent | 3. Fuller's earth |
| D. Clay lying below the limonite in tropical countries | 4. Lithomarge clay |

   a) A—2 B—4 C—3 D—1
   b) A—2 B—1 C—3 D—4
   c) A—4 B—1 C—2 D—3
   d) A—2 B—1 C—4 D—3
8. The metallogenic epoch responsible for the deposition of gold in India is
   a) Mesozoic
   b) Paleozoic
   c) Tertiary
   d) Precambrians
9. Based on the following mineral zonation, the deposits are

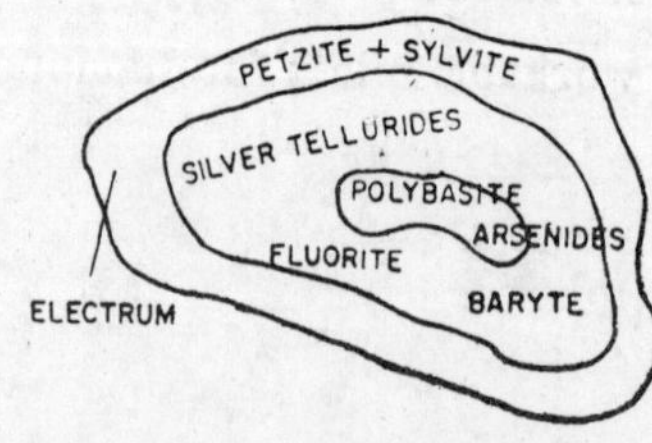

a) Epithermal
b) Mesothermal
c) Hypothermal
d) Xenothermal

10. Crustification is associated with
a) Vug fillings
b) Replacement
c) Contact metasomatism
d) Sedimentation

11. The genesis of the Kyanite deposits of Singhbhum is
a) Sedimentary
b) Plutonic igneous
c) Volcanic igneous
d) Metamorphic

12. The trap rocks in the petroleum deposits of Digboi oilfield are
a) Hansapung sandstone
b) Babapung sandstone
c) Tanda parbhat sandstone
d) Sylhet limestone

13. The oil and gas field of Bombay is associated with
a) Eocene sandstone
b) Miocene limestones
c) Pliocene shales
d) Miocene sandstones

14. Which of the following is true of Cambay and Ankleshwar oilfields ?
a) Cambay is rich in petroleum and Ankleshwar is rich in gas
b) Cambay is rich in gas and Ankleshwar is rich in oil
c) Both are rich in oil and gas
d) Both are rich in oil but poor in gas

15. The structure of Bombay High is a
a) Anticline
b) Syncline
c) Dome
d) Basin

16. The production of Kyanite was very low in the last decade due to
(1) Less international demand

(2) Less production from proved reserves

(3) Increase in the international competition

a) 1 only

b) 1 and 2 only

c) 2 and 3 only

d) All of the above

17. Gypsum deposits of commercial value are associated with

a) Nagaur series of Rajasthan

b) Tal/Krol series of Kashmir

c) Spiti Shales of Spiti

d) Cumbum Phyllites of A.P.

18. Phosphorites are associated with

a) Tal

b) Krol

c) Blaini

d) Jammu Limestone

19. In Tiruchinapalli Formation, phosphorus occurs in the form of

a) Nodules

b) Botryoidal masses

c) Lenses

d) Blankets

20. The manganese deposits of Srikakulum are formed due to the alteration of

a) Gondites

b) Kodurites

c) Khondalites

d) Charnockites

21. Silver and cobalt are found as co-products in

a) Zawar

b) Khetri

c) Singhbhum

d) Keonjhar

22. Which of the following has the richest bauxite deposits ?

a) East Coast

b) Dharwars

c) Archaens

d) Deccan Traps

23. The richest copper deposits are

a) Khetri
b) Majghawan
c) Mayurbhanj
d) Ghatsila

24. The Sausar group is associated with
a) Fe
b) Zn
c) Sn
d) Mn

25. Which of the following is the correct sequence of the dispersal of sulphide deposits ?
a) Cu-Pb-Zn
b) Pb-Cu-Zn
c) Zn-Cu-Pb
d) Zn-Pb-Cu

26. The sub-anthracite of Tertiary is found in
a) Neyveli
b) Jammu
c) Palana
d) Ranjit Valley

27. The Ranigunj coal is classified as high grade because of
1. High calorific value
2. High phosphorus
3. Low sulpur
a) 1 and 2 only
b) 1 and 3 only
c) 1, 2 and 3

28. The thickest coal seam in India is
a) Karo seam in Bokaro
b) Jingurdah in Singrauli
c) King seam in Singareni
d) None of the above

29. The agency which regulates the trade of minerals is
a) NMDC
b) MECON
c) GSI
d) MMTC

30. Which of the following are strategic to India ?
a) Asbestos—phosphorus—silver—non-ferrous metals

b) Asbestos—uranium—silver—non-ferrous metals
c) Asbestos—gold—silver—thorium
d) Thorium—uranium—Silver—lead

31. Which of the following minerals are classified as critical ?
a) Pb—Zn—Cu
b) Zn—S—Ag
c) U—Th—Pb
d) Fe—Mn—Ti

32. An Essential mineral is one which is
a) The essential ore in the mineral
b) Essential for the development of a country's economy
c) Essential for warfare
d) None of the above

33. Recycling of minerals is resorted due to
1. Environmental hazards
2. Conservation of minerals
3. Increased consumption in the country
4. Increased competition
a) 1, 2 and 3 only
b) 1, 2 and 4 only
c) 2 and 3 only
d) 2, 3 and 4 only

34. The export of manganese from India has been reduced due to
1. High shipping costs
2. Low grade of ores
3. Increased internal consumption
4. Increased competition with foreign countries
a) 1, 3 and 4 only
b) 2, 3 and 4 only
c) 3 and 4 only
d) All of the above

35. Which of the following minerals are imported by India ?
a) Ni—tin—copper
b) Copper—manganese—gypsum
c) Thorium—uranium—iron
d) Thorium—gold—mica

36. Which of the following is NOT matched correctly ?

| *Fluorite deposit* | *Host rocks* |
|---|---|
| 1. Amba Dungar—Gujarat | A. Nepheline syenites |

2. Mando Ki Pal—M.P. B. Dolomites
3. Chowkri—Chapoli—Rajasthan C. Archaen schists
4. Chandi—Dungri—M.P. D. Quartzites

a) 1—A 2—B 3—C 4—D
b) 1—B 2—A 3—D 4—C
c) 1—A 2—B 3—C 4—C
d) 1—C 2—A 3—D 4—B

37. Graphite is imported in large quantities from abroad because
a) Indigenous graphite is of poor quality
b) Graphite deposits are rare in India
c) Graphite deposits are not workable in India
d) The cost of production is high in India

38. Sphalerite resembles diamond in
a) Crystal structure
b) Crystal habit
c) Formation of similar polymorphs
d) Mode of occurrence

39. Wulfram deposits are found in
a) Wajrakarur, A.P.
b) Chingelput, Karnataka
c) Chenda Pathar, Benkura (W.B.)
d) None of the above

40. Which of the following is used as a moderator in nuclear reactors ?
a) Magnesium
b) Berylium
c) Strontium
d) Graphite

41. Conservation of mineral deposits is effected by
a) Putting a control on the production of minerals
b) Mining only richer portions of the deposits
c) Mining only lean portions of the deposits
d) Avoiding selective mining

42. At present the ilmenite deposits of India are being used
a) To extract titanium
b) To produce ferrotitanite
c) As atomic mineral
d) To produce high-grade steels

43. India is the largest producer of
   a) Aluminium
   b) Sheet mica
   c) Copper
   d) Iron
44. India has traditionally exported
   a) Copper
   b) Lead
   c) Iron
   d) Sheet mica
45. Which of the following states is the largest producer of sheet mica ?
   a) Rajasthan
   b) Bihar
   c) Andhra Pradesh
   d) Orissa
46. Which of the following minerals has substituted fluorspar in the steel industry for use as a flux ?
   a) Feldspar
   b) Bauxite
   c) Mica
   d) Pyrophyllite
47. Tonnage factor is calculated as
   a) $\frac{2000}{\text{Sp. gr.} \times 62.5}$
   b) $\frac{2000}{\text{Sp. gr.} \times 6.25}$
   c) $\frac{1000}{\text{Sp. gr.} \times 6.5}$
   d) $\frac{2600}{\text{Sp. gr.} \times 62.5}$
48. In the case of dispersed molybdenite deposits in the granite country rocks the best method of sampling would be
   a) Channel sampling
   b) Chute sampling
   c) Chip sampling
   d) Grab sampling
49. Which of the following is a high-pressure gangue mineral ?
   a) Rhombohedron calcite
   b) Zeolite

c) Zircon
d) Tourmaline

50. Which of the following minerals is found as a placer ?
a) Uraninite
b) Cinnabar
c) Smithsonite
d) Tetrahedrite

51. Saddle deposits are controlled by
a) Folds
b) Faults
c) Joints
d) Unconfirmities

52. Contact metasomatism is enhanced by
a) The intrusion of acid pluton in the calcareous rocks
b) The intrusion of a basic pluton in the calcareous rocks
c) Large-scale volcanic exhalations adjoining calcareous rocks
d) None of the above

53. Which of the following is a hypogene mineral ?
a) Pyrrhotite
b) Arsenopyrite
c) Sphalerite
d) Molybdenite

54. Match the following and choose the correct answer

| *Mechanism* | *Type of placers* |
|---|---|
| A. Concentration insitu during weathering | 1. Deluvial |
| B. Concentration during transport in liquid media | 2. Alluvial |
| C. Concentration during transport in solid media | 3. Aeolian |
| D. Concentration during transport in gaseous media | 4. Residual |
| | 5. Eluvial |

a) A—5 B—2 C—1 D—3
b) A—5 B—2 C—4 D—1
c) A—1 B—5 C—4 D—3
d) A—4 B—2 C—5 D—1

55. Pyrite is essential in the process of oxidation and sulphide enrichment because
a) It forms $SO_4^{-2}$ ions

b) It forms $S^{-2}$ ions

c) It facilitates the breakdown of copper oxides

d) It help in the concentration of sulphides

56. The gold deposits of Witwatersand are formed due to

a) Hydrothermal process

b) Magmatic process

c) Contact metasomatism

d) Mechanical concentration

57. Arrange the following states in the increasing order of their coal production

a) 1. Bihar

2. West Bengal

3. M.P.

4. Andhra Pradesh

a) 2314

b) 1234

c) 2134

d) 2143

58. A diamond bit is moved from 30.0 m to 30.5 m and the sample recovered is 60 cm. What is the % of recovery of the samples ?

a) 20%

b) 30%

c) 40%

d) 60%

59. The leading producer for manganese in the world is

a) USA

b) USSR

c) Brazil

d) Germany

60. The leading producer for chromite in the world is

a) South Africa

b) Urals

c) Turkey

d) Pakistan

61. Pick out the correct sequence of geochemical prospecting

1. Decomposition of sample

2. Estimation of reserves

3. Geochemical prospecting
4. Analysis of sample

a) 1342
b) 1432
c) 1423
d) 3412

62. Which of the following can be considered as a geochemical anomaly ?
a) Difference between peak value and regional threshold value
b) Difference between peak value and local threshold value
c) Difference between peak value and brink value
d) None of the above

63. The presence of such minerals as pervoskite in the Kimberlite pipes of Madhya Pradesh indicates that the Kimberlite pipes are
a) Derived from alkaline rocks
b) Rich in minerals such as Rb, Zr, Hf
c) Rich in minerals such as Nb, Tl
d) None of the above

64. Which of the following is known as 'Peacock ore' ?
a) Bornite
b) Chalcopyrite
c) Pyrrhotite
d) Pentlandite

65. In the crust of the earth the Fe : Mn ratio is
a) 20 : 1
b) 50 : 1
c) 30 : 1
d) 100 : 1

66. In the Aravallis the Pb—Zn deposits are associated with
a) Dolomitic limestones and grits
b) Shales and limestones
c) Conglomerates and dolomites
d) Quartzites

67. Cx in geochemical prospecting is used to describe
a) Complete dispersion
b) Cold dispersion
c) Percentage of dispersion
d) None of the above

68. "The general conditions of the ores is such that there is positively no risk, involved in the exploitation and in facilitating the mining department to give clearance". This statement refers to

a) Poor ores

b) Proved ores

c) Possible ores

d) Primary ores

69. Tippam formation consists of huge thickness of oil-bearing sandstone which lie with a disconformable contact on

a) Surmas

b) Barrails

c) Disang

d) Dihing

70. Which of the following mineral enriches the soil ?

a) Chlorite

b) Calcite

c) Gypsum

d) Corundum

71. Which of the following mineral is most resistant to weathering ?

a) Biotite

b) Hornblende

c) Muscovite

d) Quartz

72. Deuteric alteration refers to

a) Alteration of a mineral deposit from its own residual solutions

b) Alteration by means of solvents from hydrothermal solutions

c) Alteration at high temperature and pressures

d) None of the above

73. Ulvospinel has the formula

a) $FeAl_2O_4$

b) $Fe_2Al_2SiO_5$

c) $FeAlSiO_5$

d) None of the above

74. The solubility of calcite in water

a) Increases with an increase in $pCO_2$ and decrease in temperature

b) Increase with both $pCO_2$ and temperature
c) Decrease with both $pCO_2$ and temperature
d) Decrease with $pCO_2$ and increase in temperature

75. Transportation in desert regions is mostly effected by
a) Saltation+Creep
b) Suspension+Creep
c) Saltation+Suspension
d) Saltation+Suspension+Creep

## ANSWERS

| | | | |
|---|---|---|---|
| 1. a | 2. a | 3. d | 4. d |
| 5. b | 6. c | 7. b | 8. d |
| 9. a | 10. d | 11. b | 12. a |
| 13. c | 14. b | 15. a | 16. c |
| 17. a | 18. a | 19. a | 20. b |
| 21. a | 22. a | 23. a | 24. d |
| 25. a | 26. b | 27. b | 28. b |
| 29. d | 30. a | 31. a | 32. d |
| 33. c | 34. b | 35. a | 36. d |
| 37. a | 38. a | 39. b | 40. b |
| 41. d | 42. b | 43. b | 44. d |
| 45. b | 46. b | 47. d | 48. c |
| 49. d | 50. a | 51. a | 52. a |
| 53. a | 54. a | 55. a | 56. b |
| 57. a | 58. c | 59. b | 60. b |
| 61. a | 62. b | 63. b | 64. a |
| 65. b | 66. d | 67. c | 68. b |
| 69. c | 70. c | 71. d | 72. a |
| 73. b | 74. b | 75. d. | |

## (D) PAPER—IV (Hydrogeology)

1. The 'Karst' topography has been named from its type area in
a) Yugoslavia
b) Italy
c) USA
d) Norway

2. The Karst water is found in the terrains of
   a) Sandstones
   b) Granites
   c) Limestones
   d) Basalts

3. A 'lost river' is found in
   a) Granitic terranes
   b) Sandstone terranes
   c) Shale terranes
   d) Limestone terranes

4. Juvenile water is formed by
   a) Intrusive activity
   b) Sedimentation processes
   c) Evaporation in the seas
   d) Melting of ice

5. Water that is found in the sedimentary rocks at the time of their formation is called
   a) Connate water
   b) Meteoric water
   c) Vadose water
   d) Hygroscopic water

6. The water that is found in the zone of aeration is known as
   a) Ground water
   b) Vadose water
   c) Capillary water
   d) Primary water

7. The water that is trapped in a sediment or bed against gravity is called
   a) Capillary water
   b) Gravitational water
   c) Hygroscopic water
   d) None of the above

8. The water that is available to the soil is known as
   a) Pellicular water
   b) Capillary water
   c) Hygroscopic water
   d) Gravitational water

9. Which of the following is NOT a mode of precipitation ?

a) Rainfall
b) Hail
c) Snow melt
d) Snow fall

10. Artesian conditions are found in
a) Cuddalore sandstone
b) Dwaraka beds
c) Tirupati sandstones
d) None of the above

11. Which of the following places has shown a great utilization of hot thermal springs ?
a) Puga valley
b) Spiti valley
c) Narmada valley
d) Tapi valley

12. A confined acquifer becomes an unconfined acquifer when the piezometric surface
a) Goes above the confining beds
b) Goes below the confining beds
c) Coincides with the confining beds
d) None of the above

13. According to Darcy's law the Infiltration Index is given as
a) $V=KI$
b) $V=KIS^{-1}$
c) $V=K^{-1}I$
d) $V=KI/S^2$

14. Which of the following formations show primary permeability :
a) Sandstone
b) Gravel
c) Limestone
d) Shale

15. The permeability in the case of clays can be determined by
a) Constant-head permeameter
b) Falling-head permeameter
c) Non-discharging permeameter
d) None of the above

16. The property of the rocks which is given by the total volume of open spaces is termed
a) Porosity

b) Permeability
c) Surface tension
d) Viscosity

17. Periodic and short term fluctuation in Groundwater are caused due to
a) Change in the atmospheric pressure
b) Recharge and discharge by streams
c) Tides effecting piezometric level
d) None of the above

18. If groundwater is withdrawn from a well adjacent to a river stream
a) It does not effect the recharge
b) It has no effect on the water table
c) It leads to induced recharge
d) None of the above

19. Groundwater in hard rock terranes can be tapped by
a) Digging tube wells
b) Building tanks
c) Digging open wells of large diameters
d) None of the above

20. In granitic rocks, the groundwater is mostly located at
a) Crests of the folds
b) Planes of schistocity
c) Fracture zones
d) None of the above

21. The presence of saline water is indicated by
a) Phreatophytes
b) Halophytes
c) Hydrophytes
d) Xerophytes

22. The resistivity of sandstones in Ohm-m ranges from
a) $10-10^{-3}$
b) $10-100$
c) $10^{2}-1$
d) $1-1/10$

23. The resistivity value of gravels differ from that of clays by about
a) 10,000
b) 100,000

c) 1000,000

d) 500,000

24. The seismic velocities in groundwater depends on its

a) Salinity of groundwater

b) Quantity of groundwater

c) Elastic constants and density of groundwater

d) None of the above

25. The major constituents of groundwater are

a) Ca, Na, Mg, K

b) Ca, Na, Mg, Al

c) Ca, K, Na, Al

d) Na, Mn, Fe, Al

26. The bulk of the freshwater on the surface of the earth is found in

a) Glaciers

b) Rivers and streams

c) Seas

d) Groundwater

27. Water discharge is given as

a) $m^3/s$

b) $ms^{-1}$

c) $m^2/s$

d) $m/s^2$

28. The concept of water budget does not involve

a) Precipitation

b) Surface run-off

c) Infiltration

d) Percolation

29. Isohytes are the points joining the areas of equal

a) Pressure

b) Temperature

c) Rainfall

d) Salinity

30. Groundwater containing iron has

a) Low $_pH$ and high $E_h$

b) High $_pH$ and high $E_h$

c) High $_pH$ and low $E_h$

d) Low $_pH$ and low $E_h$

31. Which of the following constituents is enriched in the groundwater by biochemical pollution ?
   a) Nitrates
   b) Potash
   c) Sulphur
   d) Chlorides

32. Bacteria which can exist only in the presence of dissolved oxygen are known as
   a) Nitrifying bacteria
   b) Denitrifying bacteria
   c) Anaerobic bacteria
   d) Aerobic bacteria

33. Which of the following places suffers with huge water logging due to intense irrigation ?
   a) Haryana
   b) Uttar Pradesh
   c) Madhya Pradesh
   d) Orissa

34. The main reason for low groundwater resources in India is
   a) High utilisation
   b) Low surface run-off
   c) Low precipitation
   d) Tropical climate

35. Which of the following constituents is hazardous even when present in very small quantities ?
   a) Zinc
   b) Boron
   c) Sulphur
   d) Lead

36. According to the Darcy's law infiltration index is
   a) Directly proportional to head gradient
   b) Inversely proportional to head gradient
   c) Unrelated to the head gradient
   d) None of the above

37. The geomorphic feature that helps in groundwater recharge is
   a) Cuesta
   b) Pediment
   c) Pedeplain
   d) River terrace

38. Tracers are used to
   a) Determine the trace elements in groundwater
   b) Determine the salinity of groundwater
   c) Determine the age of the groundwater
   d) None of the above
39. High Na-rich waters are classified as 'unfit' for irrigation because
   a) Na decoagulates into a thick mass
   b) Na flocculates into a thick mass
   c) Na develops a thin clay layer over the soil
   d) None of the above

## ANSWERS

| | | | |
|---|---|---|---|
| 1. a | 2. c | 3. d | 4. c |
| 5. a | 6. b | 7. c | 8. a |
| 9. c | 10. a | 11. a | 12. b |
| 13. a | 14. b | 15. c | 16. a |
| 17. b | 18. c | 19. c | 20. c |
| 21. b | 22. b | 23. c | 24. c |
| 25. a | 26. c | 27. a | 28. d |
| 29. c | 30. a | 31. a | 32. a |
| 33. a | 34. c | 35. d | 36. a |
| 37. b | 38. c | 39. b | |

## 6. Geologists' Examination. 1987

### GEOLOGY PAPER—I

(General Geology, Geomorphology, Structural Geology, Stratigraphy and Palaeontology)

N.B. : **Mark tne correct answer (a), (b), (c) or (d)**

1. Chondrites are dated at
   a) 500 Ma
   b) 1050 Ma
   c) 2500 Ma
   d) 4500 Ma

2. The base of the Upper Mantle is drawn at the seismic discontinuity at
   a) 1000 km
   b) 650 km
   c) 400 km
   d) 250 km

3. The Secondary (transverse) waves do not travel through the outer Core because its
   a) Density is high
   b) Bulk modules is low
   c) Density is low
   d) Shearing (rigidity) modulus is low

4. Current views are that the Earth
   a) Grew out of an aggregation of fragments derived from the disruption of two stars
   b) Grew out of accretion of particles in space
   c) Was produced out of a gaseous filament drawn out of a primitive sun by an advancing star
   d) Developed out of the breaking up of a disc rotating round a primitive sun

5. Continental Shelf is designated as a miogeocline and described as an analogue of miogeosyncline and continental slope and rise is designated as a eugeocline and described as an analogue of eugeosyncline by
   a) Aubouin
   b) Dietz
   c) Kay
   d) Stille

6. Continential rise is absent along the active continental margins because of the presence of

a) Ridge
b) Transform fault
c) Continental slope
d) Trench

7. The Polar Wandering theory
   a) Supports the continental drift theory
   b) Shows polar migration paths of different continents
   c) Traces the coherent movement paths of the major continents
   d) Explains the contraction theory

8. The rate of ocean floor spreading is highest in
   a) East Pacific Rise
   b) Mid-Atlantic Ridge
   c) Indian Ocean Ridge
   d) Juan de Fuca Ridge

9. Strong proof for the ocean floor spreading hypothesis is provided by the observation that
   a) The Eastern coastline of South America fits with the Western Coastline of Africa
   b) The sediments near the continental margins are older than those near the oceanic ridges
   c) There is a regular magnetic pattern of ocean floor rocks which increase in age away from the oceanic ridges
   d) The sediments near the continental margins are younger than those near the oceanic ridges

10. The present day ocean floor does not contain sediments older than the
    a) Cretaceous
    b) Devonian
    c) Eocene
    d) Triassic

11. The Mid-Atlantic Ridge system is an area of
    a) Shear plate boundary
    b) Consuming plate boundary
    c) Accreting plate boundary
    d) Stable plate boundary

12. Asthenosphere which underlies the lithosphere differs from the latter in its
    a) Higher rigidity
    b) Lower rigidity and higher thermal state
    c) Higher thermal state and higher rigidity

d) Lower rigidity, higher thermal state and a lowering in the velocity of seismic waves

13. Subduction zones are usually marked by

a) Deep oceanic trenches

b) Deep oceanic trenches and volcanic island arcs

c) Volcanic island arcs

d) Deep oceanic trenches, volcanic island arcs and deep-focus earthquakes

14. According to the Plate Tectonic Theory

a) India and Antartica may drift towards each other

b) African rift will give birth to a new ocean

c) The Mediterranean will not widen up

d) The Indian ocean will disappear

15. The difference between the observed and predicted gravity values is called

a) Bouger anamoly

b) Isostatic gravity anomaly

c) Topographic correction

d) Bouger correction

16. Blocks of various materials, having different lengths and densities, but equal in cross-section and in total weight, are immersed in Hg (sp. gr. 13.6). It is observed that these blocks sink to equal depth.

Which one of the following explain the observed isostatic equilibrium ?

a) Airy's hypothesis

b) Heiskanen's hypothesis

c) Pratt's hypothesis

d) Phase-change theory

17. Magnetic anamoly bands like those if the mid-oceanic ridges are

a) Found in the trenches

b) Absent in the trenches

c) Found flanking the trough zone

d) Randomly oriented

18. The Palaeomagnetic dip (I) is related to the latitude ($\lambda$) of the locality as

a) $\sin I = 2 \sin \lambda$

b) $\cos I = 2 \cos \lambda$

c) $\tan I = 2 \tan \lambda$

d) $2 \tan I = \tan \lambda$

19. Which one of the following is most important in radioactive dating of archaeological remains ?

a) U-238

b) Rb-87

c) K-40

d) C-14

20. Chronostratigraphic position of a Tertiary strata is best determined by studying its

a) Fossil assemblage

b) $C^{14}/C^{12}$ Ratio

c) Remnant magnetism

d) Minerals

21. The velocity of the P waves in the inner core is

a) 20.45 km/sec

b) 18 km/sec

c) 15.20 km/sec

d) 11.23 km/sec

22. Which one of the following indicates that the Earth's core is in the liquid state, in contrast to the surrounding mantle, which is in a solid state ?

a) Simple S waves are not received on the side of the globe opposite the earthquake focus

b) The surface waves travel along the ground surface

c) Rigid continental plates ride over the plastic asthenosphere

d) P-waves are first to be recorded by the Seismograph

23. The continental crust is the thickest under the

a) Shield areas

b) Platform areas

c) Tertiary folded belt area

d) Proterozoic folded belt areas

24. According to the Kay's classification of tectonic elements, geosyncline located between the cratons are of two types, one is actively subsiding with associated volcanics and the other is less active with no volcanies. They are, respectively

a) Exogeosyncline and Autogeosyncline

b) Eugeosyncline and Miogeosyncline

c) Polygeosyncline and Mesogeosyncline

d) Parageosyncline and Embayment

25. Hawaiin volcanism belongs to which one of the following types ?
   a) Mid Oceanic Ridge volcanism
   b) Island arc volcanism
   c) Geosynclinal volcanism
   d) Intraplate volcanism
26. Which of the following belongs to an island arc ?
   a) Ireland
   b) Japan
   c) Madagascar
   d) Sri Lanka
27. Seismicity associated with mid oceanic ridges is
   a) Shallow focus
   b) Intermediate focus
   c) Deep focus
   d) Randomly focused
28. Active volcanism at the end of aseismic ridges are some of the most prolific volcanoes in the world whose lavas are mostly
   a) Trachytic
   b) Rhyolitic
   c) Basaltic
   d) Andesitic
29. The deep oceanic trenches are marked by
   a) Negative gravity anamoly
   b) Extinct volcanoes
   c) Folded sediments
   d) Faulted sediments
30. The increase in seismic velocity from 4.5 km/sec to 6.7 km/sec as one proceeds away from the ridge crest is attributed by Hess to
   a) Consolidated volcanic rocks
   b) Cooling and healing up of the fracture by slight recrystallization
   c) Sedimentary layers
   d) The presence of magma
31. The 'Andesite Line'
   a) Runs along the Mediterranean
   b) Marks the Himalayan axes
   c) Generally separates the continental crust from the oceanic crurt
   d) Runs in the Atlantic Ocean

32. Reef building corals flourish best in the zone lying between
   a) 40°N and 40°S latitudes
   b) 60°N and 60°S latitudes
   c) 80°N and 80°S latitudes
   d) 30°N and 30°S latitudes

33. The Hercynian Orogeny has taken place during the Late
   a) Carboniferous
   b) Cambrian
   c) Devonian
   d) Precambrian

34. Moraines deposited during the retreating stages of a glacier are known as
   a) Recessional moraines
   b) Terminal moraines
   c) Lateral moraines
   d) Ground moraines

35. Which of the following pairs is NOT correctly matched ?

| | |
|---|---|
| a) Massive reef building along the shore | Fringing Reef |
| b) Waves drag off fragments of the reef and pile them up above sea level on the lagoon ward side, in time, land is thus formed, vegetation gets started and a ring like structure is formed | Guyots |
| c) Coral reefs which are separated from the shore by a channel or a lagoon | Barrier reef |
| d) Coral reefs which are enclosing a lagoon and are more or less circular in shape | Atoll |

36. The portion of the sea enclosed within an atoll
   a) Bay
   b) Coral lake
   c) Fringing reef
   d) Lagoon

37. Clastic sediments on the continental shelves usually exhibit which one of the following patterns of Distribution in the slope direction ?
   a) Sandy mud-Sand-Mud
   b) Sand-Mud-Sandy Mud
   c) Mud-Sandy Mud-Sand
   d) Mud-Sand-Sandy Mud

38. Lapies are

a) Glacial forms
b) Volcanic forms
c) Solution etched and fluted landforms in Karst areas
d) Structural landforms

39 Travertine is a
a) Lava flow
b) Glacial moraine
c) Deposits formed by waves and currents
d) Deposit of $CaCO_3$ from hot springs

40. Sapropelite is a sedimentary rock which is a kind of
a) Alluvial deposit
b) Colluvial deposit
c) Glacial deposit
d) Lacustrine deposit

41. The residual hills standing out over the end surfaces in an arid landscape are called
a) Monadnocks
b) Magotes
c) Inselbergs
d) Hums

42. When a stream is flowing over a very low gradient and it is carrying a coarse bed load, it begins to
a) Meander
b) Develop Braided channels
c) Flow straight
d) Cut gorge in the bed rock

43. Peneplain is a type of plain formed by
a) Slope retreat in mountainous region mainly by wave erosion
b) Weathering of table land and removal of the wave residium by the denudation
c) Lateral erosion by streams with attendent flood plain construction
d) Denudation of rocks of varying resistance by continental ice sheets

44. Match the following and choose the correct code

| *(Process)* | *(Landforms)* |
|---|---|
| A. Volcanic action | 1. Oxbow lake |
| B. Oceanic deposition | 2. Kame Terraces |
| C. Glacial action | 3. Calderas |

D. Stream action — 4. Atolls

5. Batholiths

a) A—4 B—2 C—3 D—1
b) 3421
c) 5342
d) 2134

45. Which one of the following environments is confined between the low tidal level and the high tidal level
a) Sub-littoral
b) Continental slope
c) Littoral
d) Abyssal

46. Fresh or salt water intertrapped in sediments during their deposition is called
a) Connate water
b) Juvenile water
c) Meteoric water
d) Spring water

47. Matching the following and then choose the correct answer

| *(Process)* | *(Landforms)* |
|---|---|
| A. River action | 1. Exfoliation domes |
| B. Glacial action | 2. Karst areas |
| C, Weathering | 3. Natural bridge |
| D. Ground water | 4. Eskers |

a) A—1 B—2 C—3 D—4
b) 2413
c) 3412
d) 4321

48. An antecedent drainage is
a) Younger than the structure
b) Older than the structure
c) Contemporaneous with the structures
d) Not related to the structure

49. Which one of the following deposits can be considered to be the best example of wind action ?
a) Levees
b) Loess and dunes
c) Moraines
d) Drumlins

50. Podsols are
    a) Lateritic soils of the tropics
    b) Grey soil of the temperate
    c) Red earths of the deserts
    d) Black cotton soils

51. A rock body is said to be under tension when
    a) The external forces tend to pull it apart
    b) Equal forces operate in opposite directions in the same plane
    c) Forces act equally in all directions
    d) Forces act in a single direction

52. Stress in a rock body is
    a) Yielding to external pressure
    b) Accumulation of pressure
    c) The mutual action and reaction of pressure along a plane
    d) Forces with which the constituent particles are held in a rock body

53. The property of a substance that determines whether it will behave as a solid or a fluid is its
    a) Elasticity
    b) Plasticity
    c) Rheidity
    d) Rigidity

54. The rock whose material properties and stress-strain responses are NOT the same in all directions are said to be
    a) Isotropic
    b) Anisotropic
    c) Homogenous
    d) Inhomogenous

55. Under conditions of extreme rock deformation a substance tested at a confining pressure of 1000 kg/cm² could not support a compressive stress greater than 3900 kg/cm². When the substance was tested at a confining pressure of 4000 kg/cm² it would support a compressive stress of
    a) 1000—2000 kg/cm²
    b) 2000—3000 kg/cm²
    c) 3000—3900 kg/cm²
    d) Over 3900 kg/cm²

56. On a plane containing any two principal axes of the stress Ellipsoid the shear stress is
    a) Zero

b) Maximum
c) Greater than zero
d) Lesser than zero

57. If the orientation of the primary axes of a strain ellipsoid changed during deformation, process would be
a) Pure shear
b) Simple shear
c) Homogenous deformation
d) Heterogenous deformation

58. The ratio of the transverse strain to axial strain is called
a) Modulus of rigidity
b) Bulk modulus
c) Rigidity modulus
d) Poisson's ratio

59. A rock specimen under stress is said to be ductile when
a) Rupture takes place before elastic strain
b) Rupture takes place after a short interval between elastic limit and rupture
c) Rupture takes place after short or long interal between elastic limit and rupture
d) Rupture may or may not occur beyond elastic limit

60. Which one of the following represents a triaxial stress
a) $\sigma_1$ $\sigma_2$ $\sigma_3$ have non zero values
b) $\sigma_1$ or $\sigma_3$ is non-zero, the other two being zero
c) Two are non-zero and the other is zero
d) $\sigma_1 = -\sigma_3$ is non zero $\sigma_2 = 0$

61. The ratio of the shearing stress to shearing strain is
a) Poisson's ratio
b) Incompressibility
c) Young's modulus
d) Rigidity modulus

62. Match the following and choose the correct answer

| | |
|---|---|
| A. Buckling | 1. Shear zone |
| B. Shear folding | 2. Parallel folding |
| C. Mylonite | 3. Similar folding |
| D. Puckering | 4. Crenulation |

a) A—3 B—1 C—4 D—2
b) A—2 B—3 C—1 D—4
c) A—4 B—2 C—3 D—1
d) A—3 B—4 C—2 D—1

63. Which one of the following relates to pure shear ?

a) $\lambda_1 > \lambda_2 = \lambda_3$

b) $\lambda_1 = \lambda_2 > \lambda_3$

c) $\lambda_1 > \lambda_2 = 1 > \lambda_3$

d) $\lambda_1 > \lambda_2 > \lambda_3$

64. Rocks which have been thrust a long distance are called

a) Allocthonous

b) Windows

c) Autochthonous

d) Klippen

65. Which one of the following faults have comparatively straight trace at the surface of the ground ?

a) Gravity fault

b) Thrust fault

c) Strike-slip fault

d) Dip-slip fault

66. Joints developed at right angles to a fold axes of folded strata are known as

a) (hol) joints

b) a-c joints

c) (okl) joints

d) Conjugate joints

67. Which one of the following statements gives the most significant characteristic of enechelon fractures ?

a) No displacement along joint planes

b) Creation of a number of closely-spaced joints

c) Viccinity of fault

d) Pressence of tension gashes

68. Which of the following is a magnetozone ?

a) Gothlanetan

b) Keuper

c) Olduvai

d) Sakmarian

69. Fossil large foraminifers are most abundant in

a) Carbonaceous deposits

b) Calcareous deposits

c) Arenaceous deposits

d) Argillaceous deposits

70. The Globotruncana is

a) Planktonic

b) Benthonic
c) Pelagic
d) Lacustrine

71. Nummulites is a characteristic fossil of
a) Blaini
b) Krol
c) Shali
d) Subathu

72. Ptillophyllum is a characteristic fossil of
a) L. Gondwanas
b) M. Gondwanas
c) U. Gondwanas

73. Moeritheres is of the------age
a) Eocene
b) Miocene
c) Oligocene
d) Pliocene

74. Parahippus appeared in
a) Eocene
b) Miocene
c) Cretaceous
d) Pliocene

75. Apart from that rich mammalian fossils, the Siwalik rocks also contain abundant fossils of
a) Echinoderms
b) Brachiopods
c) Foraminifers
d) Reptiles

76. Black pyritic shales have yielded abundant fossils of
a) Trilobites
b) Graptolites
c) Echinoids
d) Corals

77. In highly organized echinoids water is introduced through a special plate called
a) Madreporite
b) Ampullae
c) Stone canals
d) Ambulacra

78. Fusulinas are characteristic foraminifers of
a) Permian

b) Jurassic

c) Eocene

d) Pliocene

79. When the ocular plate in a occulogenital ring is in contact with periproct it is termed as ?

a) Insert

b) Exsert

c) Endarch

d) Exarch

80. Aristotle's Lantern in Echinoids is used as an organ of

a) Reproduction

b) Mastication

c) Locomotion

d) Respiration

81. The Lower and Middle Triassic deposits were dominated by ammonoids of

a) Goniatite suture

b) Creatite suture

c) Nautiloid suture

d) Ammonitic suture

82. In trilobites when the pygidium is not equal in size to the cephalon it is described as

a) Isopygous

b) Micropygous

c) Apygous

d) Mesopygous

83. The most distinctive feature of Gasteropods when compared with other classes of the Mollusca is

a) Shell consists of one valve

b) Shell with one opening

c) Shell made up of $CaCO_3$

d) Unchambered shell

84. Cameral deposit in Cephalopods was used for

a) Sinking in the sea

b) Floating

c) Swimming

d) Horizontal mode of life

85. Which of the following features is not present in Gasteropods ?

a) Nuclear whorls

b) Septa

c) Radula
d) Opercula

86. The post-embryonic growth in Trilobites is accompained by
a) Accretion of existing parts
b) Addition of new parts
c) Molting
d) Modification

87. The Ordovician is known as the age of
a) Brachiopods
b) Corals
c) Crinoids
d) Graptolites

88. The earliest known echinoids are from
a) U. Cambrian
b) Ordovician
c) L. Silurian
d) U. Silurian

89. Plant fossils are less than animal fossils because
a) Plants are younger than animals
b) Plants do not have too many hard parts
c) Plants are washed away by streams

90. Which one of the following is not of marine habitat ?
a) Pelecypods
b) Cephalopods
c) Crinoids
d) Brachiopods

91. Impressions indicating the movement of organisms are called
a) Index fossils
b) Tracks
c) Coprolites

92. The formation which has yielded Makrana marbles
a) Gwalior series
b) Railo series
c) Aravalli series
d) Ajabgarh series

93. The Vindhyan group of rocks is famous for the occurrence of
a) Silver deposits
b) Manganese deposits

c) Diamond deposits

d) Copper deposits

94. Which of the following has been most successfully used in geochemistry ?

a) Trilobites

b) Gasteropods

c) Echinoids

d) Corals

95. Which of the following is a type section of Permian rocks ?

a) USA

b) USSR

c) UK

d) UAR

96. The marine Upper Paleozoic rocks in the Peninsular India are exposed in

a) Tamil Nadu

b) Karnataka

c) Madhya Pradesh

d) Maharashtra

97. Top and bottom determination by mudcracks is made possible because

a) Mudcracks are found only at the top

b) Limestone prism structures are useful in the recognition of the top of the beds

c) The polygonal surfaces point downwards

98. Calderas are

a) Isolated plutonic bodies projecting from the main body of igneous intrusion

b) Openings of the volcanoes

c) Glacial deposits

d) Volcanic mounts

99. Which of the following is commonly referred as slaty cleavage

a) Axial Plane foliation

b) Fracture cleavage

c) Bedding cleavage

d) Slip cleavage

100. Assertion : (A) The lineation on a folded schistose rock sequence is curved

Reason : (R) Folding deforms both schistosity and lineation

a) Both A and R are correct and R is the corret interpretation of A

b) Both A and R are correct are R is the correct explanation
c) A is false, R is ture
d) A is true, R is false

## ANSWERS

| | | | |
|---|---|---|---|
| 1. d | 2. c | 3. d | 4. b |
| 5. c | 6. d | 7. b | 8. c |
| 9. c | 10. a | 11. c | 12. d |
| 13. d | 14. b | 15. a | 16. a |
| 17. c | 18. c | 19. d | 20. a |
| 21. d | 22. a | 23. a | 24. b |
| 25. b | 26. b | 27. a | 28. d |
| 29. a | 30. a | 31. c | 32. b |
| 33. a | 34. b | 35. b | 36. d |
| 37. a | 38. c | 39. d | 40. d |
| 41. c | 42. a | 43. c | 44. b |
| 45. c | 46. a | 47. c | 48. b |
| 49. b | 50. b | 51. a | 52. c |
| 53. c | 54. b | 55. d | 56. a |
| 57. d | 58. d | 59. c | 60. a |
| 61. d | 62. b | 63. a | 64. d |
| 65. a | 66. d | 67. b | 68. c |
| 69. b | 70. b | 71. d | 72. c |
| 73. b | 74. a | 75. d | 76. b |
| 77. b | 78. a | 79. a | 80. b |
| 81. b | 82. b | 83. a | 84. b |
| 85. c | 86. b | 87. d | 88. b |
| 89. b | 90. c | 91. b | 92. b |
| 93. c | 94. d | 95. b | 96. a |
| 97. c | 98. b | 99. a | 100. a |

# 7. ONGC Examination for Recruitment of Executive Trainees, 1990

1. The average heat flow in the ocean or continental crust is equal to
   a) $5.5 \times 10^{-2}$ joules $m^{-2}$ $s^{-1}$
   b) $5.5 \times 10^{-3}$ joules $m^{-2}$ $s^{-1}$
   c) $5.5 \times 10^{-4}$ joules $m^{-2}$ $s^{-1}$
   d) None of the above
2. The Ocean-Continent collision results in
   a) The subduction of oceanic plate beneath the continental plate
   b) The subduction of continental plate beneath the oceanic plate
   c) The subduction of both the plates into the mantle
   d) None of the above
3. Red sea is example of
   a) Accreting plate margin
   b) Consuming plate margin
   c) Conservative plate margin
   d) None of the above
4. The 90°E Ridge is located in
   a) Pacific Ocean
   b) Indian Ocean
   c) Atlantic Ocean
   d) Arctic Ocean
5. The Earth is composed of————major plates
   a) 21
   b) 18
   c) 8
   d) 6
6. Collision of India with Eurasia took place about———ago
   a) 170 MY
   b) 110 MY
   c) 60 MY
   d) 50 MY
7. In Micas, the Si : O ratio is
   a) 1 : 3
   b) 1 : 2

c) 2 : 5

d) 2 : 3

8. In the Interstitial solid solution

a) Ions are incorporated in the intertices of crystal lattice

b) Extensive ionic replacement takes place

c) Crystal lattices serves as an avenue for ionic replacement

d) None of the above

9. Which of the following unit cell dimension of amphiboles is roughly twice that of corresponding Pyroxenes ?

a) a-cell

b) b-cell

c) c-cell

d) None of the above

10. Piezoelectricity is a

a) Scalar property

b) Vector property

c) Both scalar and vector property

d) None of the above

11. Which of the following shows the highest bond strength ?

a) Ionic bonds

b) Covalent bonds

c) Metallic bonds

d) Vander Waal's bonds

12. A biaxial mineral is positive when

a) X is acute bisectrix

b) X is obtuse bisectrix

c) Y is acute bisectrix

d) Z is acute bisectrix

13. In the Fresnel's ellipsoid the length represents the

a) R.I. of light

b) Wavelength of light

c) Velocity of light

d) None of the above

14. Bijaigarh shales belongs to

a) Lower Kaimurs

b) Upper Kaimurs

c) Semri group

d) Bijawar group

15. Mount Everest Limestone belongs to
    a) Ordovician
    b) Silurian
    c) Devonian
    d) Carboniferous

16. The Singhbhum granite was emplaced about——ago
    a) 3200 MY
    b) 2700 MY
    c) 2000 MY
    d) 950 MY

17. Gangamopteris beds of Kashmir belongs to
    a) Mississipean
    b) Pennsylvanian
    c) Artinskian
    d) Tartarian

18. In the Karauli area, the rocks equivalent to the Rohtas stage are described as
    a) Tirohan limestones
    b) Nimbahera limestones
    c) Suket shales
    d) Khardeola grits

19. The Deccan Traps range in age from
    a) L. Cretaceous—U. Eocene
    b) M. Cretaceous to U. Eocene
    c) U. Cretaceous—U. Eocene
    d) U. Cretaceous to L. Eocene

20. The Old Red Sandstone consists of
    a) Shallow marine facies
    b) Deep marine facies
    c) Fluvio-lacustrine facies
    d) Lagoonal-continental facies

21. The Indian Mesozoic Gondwanas have been classified biostratigraphically into——assemblage zones
    a) Three
    b) Four
    c) Six
    d) Two

22. Which of the following statements is incorrect ?
    a) Paleozoic is the oldest and the longest era in the earth's history

b) Beds and veins of gypsum are found in the Bhander series

c) Subathu group of rocks belongs to lower eocene

d) None of the above

23. Variegated series of northern Kumaon belongs to

a) Cambrian

b) Ordovician

c) Silurian

d) Devonian

24. The biostratigraphic classification of the Gondwana rocks of India was proposed by

a) T.H. Holland

b) E. Vredenburg

c) Verma and Gupta

d) Shah and Shastri

25. The age of the Gunflint formation of the Lake Superior region of North America is

a) 2800—3000 MY

b) 2500—3000 MY

c) 2100—2400 MY

d) 1600—2000 MY

26. Which of the following group of animals is exclusively marine ?

a) Molluscs

b) Coelentrates

c) Trilobites

d) Echinoderms

27. Remains of Dinosaurs have been found from which of the formation in the Godavari valley

a) Maleri

b) Yerapalli

c) Gangapur

d) Kota

28. Organic evolution would not have been taken place on the earth if

a) Individuals had not shown genetic variation

b) Individuals had not inherited somatic variation

c) Somatic variations were not transferred to genetic variations

d) None of the above

29. Peat and lignite may contain pollen grains. These are obtained by dissolving in

a) Wultz solution

b) Schulze solution

c) Clerici's solution

d) $CCl_4$

30. The common forms of Foraminifer tests ranges in size from

a) 0.1 to 5 mm

b) 0 01 to 0.5 mm

c) 0.1 to 1 mm

d) 1.0 to 5.0 mm

31. Dadoxylon indicum is characteristic of

a) Barakar series

b) Ranigunj series

c) Rajmahal series

d) Mahadeva series

32. The statement "Stress is proportional to strain within the elastic limit" refers to

a) Bulk modulus

b) Hooke's law

c) Rigidity modulus

d) Coloumb's law

33. The main characteristic of fault-line scarps is

a) Presence of triangular facets

b) Presence of normal faults in a parallel alignment

c) Presence of fault scarps

d) Close correlation between structure and topography

34. The movement in which there is no rotation of blocks relative to one another is described as

a) Rotational movement

b) Translational movement

c) Spiral movement

d) None of the above

35. An eroded dome shows

a) An increase in radius with an increase in age

b) A decrease in radius with an increase in sediment age

c) An increase in radius with the decreasing age of sediments

d) None of the aboue

36. Joints which are parallel to the axial planes of folds are described as

a) Extension joints
b) Release joints
c) Transverse joints
d) Longitudinal joints

37. Joints which are perpendicular to the fold axes are described as
a) Release joints
b) Extension joints
c) Cross joints
d) Columnar joints

38. In the case of tear faults
a) Dip slip is zero and the net slip is 90°
b) Dip slip is zero and the strike slip equals the net slip
c) Dip slip equals the strike slip
d) Dip slip and the net slip have zero values

39. A reverse fault dipping less than 45° is known as
a) Overthrust
b) Underthrust
c) Upthrust
d) Thrust

40. In an overthrust, the relation F/W denotes
a) Displacement
b) Coefficient of friction
c) Compression coefficient
d) Amount of thrust

41. Which of the following is an unstratified sedimentary deposit ?
a) Glacial tillite
b) Biogenic limestone
c) Sandstone
d) Shale

42. A group of cross-bedded units showing a uniform current-flow direction is given the term
a) Foreset
b) Coset
c) Laminae
d) None of the above

43. Sediments precipitated from solutions within the basin of deposition and have later been moved within the basin are called

a) Orthochemical
b) Allochemical
c) Epiclastic
d) Periclastic

44. The difference between Oolites and Pellets lies in
a) Size
b) Composition
c) Internal structure
d) Mode of occurrence

45. The Molasse association belongs to
I. Continental environments
II. Marine environments
III. Lagoonal environments
a) I only
b) I and II only
c) I and III only
d) II and III

46. Which of the following facies show close association with tectonic activity ?
a) Arkosic
b) Molassic
c) Flysch
d) All of the above

47. The Talchir tillites were formed under
a) Glacial environment
b) Fluvio-glacial environment
c) Periglacial environment
d) Glacial and periglacial environments

48. Euxinic facies is a synonym for
a) Red shale facies
b) Blueschist facies
c) Black shale facies
d) Evaporite facies

49. The Sand-Shale ratio is given as

a) $\frac{\text{Shale} + \text{Conglomerate}}{\text{Sandstone} + \text{Siltstone}}$

b) $\frac{\text{Sandstone+Shale}}{\text{Conglomerate+Siltstone}}$

c) $\frac{\text{Sandstone+Siltstone}}{\text{Shale+Conglomerate}}$

d) $\frac{\text{Sandstone+Conglomerate}}{\text{Shale+Siltstone}}$

50. The detritus which has been formed elsewhere and is brought into a sedimentary basin from outside is described as
a) Authigenic
b) Allogenic
c) Epigenetic
d) Phyllogenetic

51. In the Shand's classification, the Per-aluminous rocks are represented as
a) $Al_2O_3 < Na_2O + K_2O$
b) $Al_2O_3 = Na_2O + K_2O$
c) $Al_2O_3 > Na_2O + K_2O$
d) $Al_2O_3 > Na_2O + K_2O + CaO$

52. The silica saturated end member of tholeiitic basalt is
a) Phonolite
b) Mugearite
c) Rhyolite
d) Andesite

53. Cone sheet dykes dip towards a common centre at the angle of
a) 20°—30°
b) 30°—40°
c) 40°—50°
d) 50°—60°

54. Which of the following is synonymous with kimberlites ?
a) Mica peridotite
b) Quartz monzonite
c) Hypersthene
d) Amphibolite

55. Syenite, nepheline syenite and limestone association is suggestive of
a) Magmatic assimilation
b) Magmatic differentiation
c) Assimilation
d) Deuteric alteration

56. The difference between Picrite and Essexite is that
   a) Picrite has more olivine and lesser mafics than Essexite
   b) Picrite has more mafics but lesser olivine than Essexite
   c) Picrite is totally devoid of olivine and mafics
   d) None of the above

57. Geothermal gradient in the regions of Precambrian terrain is
   a) 5°/km
   b) 8°/km
   c) 10°/km
   d) 15°/km

58. Pahoehoe is a type of
   a) Stony meteorite
   b) Volcanic crator
   c) Ropy lava
   d) Blocky lava

59. More active solvents and mineralizers in the metamorphism, except carbondioxide are of
   a) Direct magmatic origin
   b) Igneous origin
   c) Metamorphic origin
   d) Sedimentary origin

60. Pyrometamorphism involves
   a) HP changes
   b) HT changes
   c) LP/LT changes
   d) HP/HT changes

61. If a Gneiss is formed from igneous rocks due to metamorphism, it is described as
   a) Metagneiss
   b) Orthogneiss
   c) Paragneiss
   d) None of the above

62. Greenschist facies suggests
   a) Low-grade metamorphism
   b) Medium-grade metamorphism
   c) High-grade metamorphism
   d) Ultrametamorphism

63. **Which of the following is the mineral of the highest zone of metamorphism ?**
a) Almandine
b) Kyanite
c) Sillimanite
d) Biotite

64. **The mineralogical phase rule is given by the formula**
a) P+F=C+2
b) P+F=C—2
c) P—F=C+2
d) F=C+P—2

65. **Which of the following shows the correct transition in the composition of garnets with the increasing pressure and temperature ?**
a) Mn—→Mg—→Fe
b) Mg—→Mn—→Fe
c) Mn—→Fe—→Mg
d) Mg—→Fe—→Mn

66. With the increase in the grade of metamorphism
a) Na+K/Mg of muscovite increases
b) Fe/Mg of muscovite increases
c) Fe+Mg/K of muscovite decreases
d) Fe+Mg of muscovite decreases while Al increases

67. The characteristic minerals in the zone of Oxidation are
a) Sulphides
b) Oxides
c) Native metals
d) Sulphates

68. The mercury content of wall rocks and soil acts as a common path finder in the geochemical prospecting for
a) Vein type gold ores
b) Fissure type gold ores
c) Pb—Zn—Ag ores
d) Cu—Pb—Zn ores

69. China clay deposit of Singhbhum is an example of
a) Residual concentration
b) Mechanical concentration
c) Placer deposit
d) Hydrothermal deposit

**70. Kolar gold deposit is an example of**

**a) Fissure vein deposit**

**b) Shear zone deposit**

**c) Ladder vein deposit**

**d) Stockwork deposit**

## ANSWERS

| | | | |
|---|---|---|---|
| 1. a | 2. a | 3. a | 4. b |
| 5. d | 6. d | 7. c | 8. a |
| 9. c | 10. a | 11. b | 12. d |
| 13. c | 14. a | 15. b | 16. c |
| 17. d | 18. a | 19. d | 20. c |
| 21. a | 22. a | 23. c | 24. d |
| 25. d | 26. d | 27. d | 28. a |
| 29. b | 30. a | 31. a | 32. b |
| 33. d | 34. b | 35. c | 36. b |
| 37. b | 38. b | 39. d | 40. b |
| 41. a | 42. b | 43. b | 44. c |
| 45. b | 46. d | 47. d | 48. c |
| 49. d | 50. b | 51. d | 52. d |
| 53. b | 54. a | 55. a | 56. a |
| 57. c | 58. c | 59. a | 60. b |
| 61. b | 62. a | 63. c | 64. a |
| 65. c | 66. d | 67. c | 68. c |
| 69. a | 70. a | | |

# 8. ONGC Examination for Recruitment of Executive Trainees, 1989

1. The average depth of MOHO below the oceans is
   a) 8 km
   b) 9 km
   c) 11 km
   d) 16 km
2. Choose the correct statement from the following
   a) Gondwana land has a greater area than the Laurasia
   b) Laurasia had a greater area than the Gondwana land
   c) Both Gondwana land and Laurasia had equal areas
   d) None of the above
3. The most recent theory explaining Plate Tectonics is
   a) Magnetic theory
   b) Triple point theory
   c) Slab-pull theory
   d) The expanding Earth hypothesis
4. Plates are made up of
   a) Crust only
   b) Crust and whole of the mantle
   c) Crust and part of the mantle
   d) Crust, mantle and part of the core
5. When two continental plates collide they form ?
   a) Oceanic volcanoes
   b) Island arcs
   c) Great mountain chains
   d) Both oceanic volcanoes and Islands Arcs
6. Main Boundary fault is a
   a) Normal fault of length 150 km
   b) Normal fault of length 250 km
   c) A series of normal faults aligned along the entire length of the Himalayas
   d) A series of reverse faults aligned along the entire length of the Himalayas
7. The Triple Junction nearest to India is located in the
   a) Bay of Bengal
   b) Indian Ocean
   c) Arabian Sea
   d) Rann of Kutch

8. The coastline of India is about
   a) 3100 km
   b) 4100 km
   c) 5100 km
   d) 6100 km
9. The mineralogy of Pyrolite varies as a function of
   a) Temperature
   b) Pressure
   c) Both temperature and pressure
   d) None of the above
10. Layers or massifs or rocks with a temperature above 0°C surrounded by frozen layer is termed as
    a) Hommocks
    b) Taliks
    c) Permafrost ridge
    d) Naleds
11. A common term for the depositional feature in a limestone cavern which develop vertically, horizontally, obliquely, upward and downward directions is
    a) Stalacitite
    b) Stalagmite
    c) Helictite
    d) Flow stone
12. Which of the following minerals does not show alteration ?
    a) Biotite
    b) Olivine
    c) Feldspar
    d) Quartz
13. Exfoliation occurs due to
    a) Increase of lateral pressure
    b) Chemical reactions
    c) Excessive heating
    d) Biogenic activity
14. The topography of oceanic bottoms is depicted by means of
    a) Hypsographic curves
    b) Contour maps
    c) Hydrostatic curves
    d) Choropleth maps
15. A ramifying or branching tube consisting of fused silica or silicates formed due to the action of lightning striking a sandy soil is

a) Fusain
b) Protore
c) Fulgurite
d) Riedals

16. The gas found dissolved in the sea in an amount greater than its amount in the atmosphere is
a) Nitrogen
b) Oxygen
c) Carbondioxide
d) Argon

17. Aquenoxes rocks are formed by the
a) Action of water
b) Cooling of hot masses
c) Deposition of sediments on ocean beds
d) Intrusive action of igneous plutons

18. Enantiomorphic crystal forms possess
a) Only a plane of symmetry
b) Only a centre of symmetry
c) Both a plane and a centre of symmetry
d) Neither a plane nor a centre of symmetry

19. Which of the following is NOT a stoichiometric crystal defect ?
a) Bertholide compound
b) Schottley defect
c) Frenkel defect
d) Zerkel defect

20. Which of the following possesses maximum density ?
a) Spessartite
b) Hydrogrossularite
c) Almandine
d) Pyrope

21. Monticellite is a
a) Ca—Fe olivine
b) Mg—Fe olivine
c) Ca—Mg olivine
d) Ca—Al—Fe olivine

22. The structure of Olivine consists of layers which are parallel to
a) (100)

b) (001)
c) (010)
d) (111)

23. **The cleavage pattern in diamond is described as**
a) Dodecahedral
b) Octahedral
c) Rhombohedral
d) Rhomdodecahedral

24. **Which of the following is composed of calcium carbonate ?**
a) Satinspar
b) Icelandspar
c) Topaz
d) Dog-tooth spar

25. **Which of the following is a strongly pleochroic mica consisting of abundant ferric ions ?**
a) Lepidolite
b) Muscovite
c) Lepidomelane
d) Paragonite

26. **The mineral epidote is formed due to the process of**
a) Serpentinisation
b) Saussuritisation
c) Sericitisation
d) Kaolinisation

27. **An acute bisectrix section is one which has been cut**
a) Parallel to acute bisectrix
b) Normal to acute bisectrix
c) Normal to obtuse bisectrix
d) Inclined to acute bisectrix

28. **A "Triaxial ellipsoid" is also known as**
a) Uniaxial indicatrix
b) Triaxial indicatrtx
c) Biaxial indicatrix
d) None of the above

29. **According to the Oddo-Harkins rule**
a) Elements of odd atomic numbers are more abundant than those of even atomic numbers
b) Elements of even atomic numbers are more abundant than those of odd atomic numbers

c) Both are equally abundant

d) None of the above

30. The lines joining the areas of equal atmospheric pressure are called

a) Isobars

b) Isohytes

c) Isotherms

d) Isothels

31. Comouflage type atomic substitution occurs in elements having

a) Similar charge and similar ionic radius

b) Similar charge and higher ionic radius

c) Higher charge and higher ionic radius

d) Higher charge and similar ionic radius

32. The phenomenon in which trivalent ions of rare earth element decreased in radius with increasing atomic number due to building of an inner electron shell instead of the addition of a new shell, is known as

a) Nuclear fission

b) Nuclear fusion

c) Lanthanide contraction

d) Wultz reaction

33. The second and third periods in the Periodic Table

a) Starts with an alkali metal and ends with an inert gas

b) Begins with an inert gas and ends with an alkali metal

c) There is an increase in the atomic numbers from left to right

d) There is a decrease in the atomic numbers from left to right

34. The average $_{P}$H of sea water lies between

a) 3—5

b) 5—8

c) 8—8.2

d) 12—14

35. Colloidal solutions are formed due to

a) Increase in the surface area of dispersed phase

b) Increase in the volume of dispersion medium

c) Decrease in the surface tension of dispersed surface

d) Increase in surface area and surface tension of dispersed phase

36. Lachi series of Mount Everest region belongs to
   a) Silurian
   b) Cambrian
   c) Carboniferous
   d) Permian

37. Which of the following Palaeozoic successions of Ladakh is correct ?
   a) Muthquartzite, Lignite series, Fenestella shale
   b) Muthquartzite, Fenestella shale, Lignite series
   c) Muthquartzite, Fenestella shales, Sarchu Lignite series
   d) None of the above

38. The fundamental unit in lithostragraphic classification is
   a) Group
   b) Formation
   c) Supergroup
   d) Member

39. In the Nepal-Sikkim area, the gneissose group of Archaen rocks is described as
   a) Jutogh series
   b) Vaikrita system
   c) Darjeeling series
   d) Dahling series

40. Which of the following stratigraphic successions of the Siwaliks is correct ?
   a) Dagshai—Kasauli—Subathu
   b) Kasauli—Dagshai—Subathu
   c) Kasuali—Dagshai—Subathu
   d) Subathu—Dagshai—Kasauli

41. Palaeontological evidence in favour of organic evolution is beautifully illustrated by the Fauna of
   a) Merychippus
   b) Darwins finches
   c) Rhinoceras
   d) Archeopterix

42. Which of the following theories is NOT associated with the formation of coral reefs ?
   a) Darwin-Dana theory
   b) Wallac's theory
   c) Semper Murray theory
   d) Daly Glacial theory

43. The extent of Siwaliks was
   a) L. Miocene to U. Pliestocene
   b) M. Miocene to L. Pliestocene
   c) Miocene to Pliocene
   d) Eocene to L. Pliestocene

44. A limited area of older rocks surrounded by younger rocks are called
   a) Outlier
   b) Inlier
   c) Overlap
   d) Offlap

45. Two sets of joints nearly at right angles to one another are known as
   a) Diagonal joints
   b) Dilation joints
   c) Cross joints
   d) None of the above

46. Orocline is
   a) A type of fold
   b) Formed when an orogenic belt takes a sharp bent
   c) Formed from the bending of the whole folded belt
   d) All of the above are correct

47. Petrofabric analysis
   a) Ignores the non-directional texture
   b) Gives complete study of rock fabrics
   c) Gives the megascopic structure only
   d) Does not ignore the scalar fabric of rock

48. Reverse rotary drill is used for
   a) Drilling blast hole for seismic prospecting
   b) Groundwater drilling
   c) Oilwell drilling
   d) All of the above

49. For drilling groundwater wells in extremely hard formation the following is used
   a) Reverse circulation
   b) Down the hole hammer
   c) Diamond drill
   d) Cable tool drilling

50. Arrange the following in the order of their increasing specific yield

a) Sand—Gravel—Clay
b) Silt—Gravel—Clay
c) Clay—Sand—Gravel
d) Gravel—Clay—Sand

51. The term "Stratum" is used for the layers of thickness
a) >1 cm
b) <1 cm
c) Much greater than 1.7 cm
d) =1 cm

52. The "Azimuth" is
a) The vertical angle between the meridian and the horizontal projection of the dip line of the foreset
b) The horizontal angle between the meridian and the horizontal projection of the dip line of the topset
c) The horizontal angle between the meridian and the norizontal projection of the dip line of the bottomset
d) The horizontal angle between the meridian and the horizontal projection of the dipline of the foreset

53. COFROLITES are
a) Smaller objects with an origin different from fecal pellets
b) Larger objects with an origin similar to that of fecal pellets
c) Spherical objects with radial structures
d) Remains of plants and animals

54. Pelagic deposits found at the oceanic bottoms are composed of
a) Remains of animals and plants
b) Volcanic ash
c) Shelly gravels
d) Alluvial silts and clays

55. When a tool impacts on a sediment surface at regular intervals a set of markings are produced, these are known as
a) Skip and Roll Marks
b) Bush Marks
c) Bounce Marks
d) Prod Marks

56. Peat has its greatest development in
a) Tropical regions
b) Subtropical regions
c) Temperate and Cold regions
d) The regions of great rainfall

57. In an isochemical metamorphic rock there is

a) No addition and subtraction of material from outside
b) An addition and subtraction of material from outside
c) A great change in the bulk chemical composition of rock
d) No change in the bulk chemical composition of rock

58. The characteristic vermicular quartz in the myrmekite structure is found in the granitic rocks. They are formed by
a) Replacement of K-felspar
b) Replacement of plagioclase
c) Both the above are correct
d) Magmatic differentiation

59. Hornfelsic rocks are a result of
a) Regional metamorphism
b) Contact metamorphism
c) Cataclastic metamorphism
d) Plutonic metamorphism

60. In the zoned ring complexes, the core is composed of
a) Hbl—Pyroxenite
b) Wohrlite
c) Dunite
d) Olivine—Pyroxenite

61. The gravel-sized particles are described as
a) Argiliaceous deposits
b) Rudaceous deposits
c) Arenaceous deposits
d) Arenaceous and Rudaceous deposits

62. Cavity filling deposit in which mass of rock body is intercalated by a network of small closely spaced ore-bearing veinlets is known as
a) Stockwork
b) Ladder vein
c) Saddle reef
d) Box work

63. Which of the following method is used for detailed exploration of oil and gas ?
a) Gravity method
b) Seismic refraction
c) Seismic reflection
d) Magnetic method

64. Sampling which is done in order to obtain information about

the structure of the mineral, interfering elements and chemical and physical factors which are liable to help, or hinder recovery is

a) Grab sampling

b) Borehole sampling

c) Channels sampling

d) Chip sampling

65. Which of the following types of drilling is used for oil prospecting ?

a) Diamond drilling

b) Churn drilling

c) Reverse rotary drilling

d) Turbo drilling

66. For sampling large homognneous, flat lying deposits of even textural ?

a) Percussion drilling

b) Churn drilling

c) Calyx drilling

d) Diamond drilling

67. In certain group of minerals of local importance or minerals which can be developed on small scale for the benefit of local people have been specifically declared as

a) Minor minerals

b) Major minerals

c) Mineral concessions

d) Exominerals

68. The rotary drill is most suited for

a Drilling blast hole in seismic prospecting

b) Ground water exploration

c) Oil well drilling in soft areas

d) All of the above

69. Leaching in soils denotes

a) Removal of soluble salts from the soil

b) Enrichment of the soil with soluble salts

c) Breakdown of soil particles

d) None of the above

70. In a Flocculation structure the soil particles

a) Attract each other

b) Repel each other

c) Attract or repel each other

d) None of the above

71. Which of the following denotes a method of converting a spherical surface into a plane surface ?

a) Photogrammetry

b) Projections

c) Cartography

d) Geodesy

72. The method of determining the altitude of various points by using a thermobarometer is called

a) Pyrometry

b) Hypsometry

c) Hysometry

d) Altometry

73. In a geological map the doubly plunging folds are represented as

a) Hyperobolas

b) Parabolas

c) Concentric ellipses

d) Concentric circles

74. The power of a telescope to produce sharp image depends upon the accuracy with which the lenses are ground and polished. This known as

a) Illumination

b) Aplanation

c) Definition

d) Magnification

75. The method of measuring the horizontal angle with high precision, in which several angles are measured successively and finally the horizon is closed is known as

a) Reiteration method

b) Repetition method

c) Direct angle method

d) Deflection angle method

76. Absence of spherical aberration in a telescope is described as

a) Achromatism

b) Aplanchian

c) Reitteration

d) Precision

77. The excess quantity of rock that is broken and removed from a proposed tunnel is known as

a) Caving
b) Overbreak
c) Excavation
d) Caving and excavation both

78. The point vertically below the camera lenses on the ground is called
a) Ground point
b) Ground nadir point
c Perspective center
d) Isocenter

79. In the vertical photograph the principle point, nadir point and the isocentre
a) Always coincides
b) Rarely coincides
c) Never coincides
d) None of the above

80. A number of parallel ridges in an aerial photograph indicates
a) Tightly folded strata
b) Plunging anticline
c) Eroded dome
d) Steeply dipping beds

81 Oval-shaped ridge in an aerial photograph indicates a
a) Dome
b) Doubly plunging fold
c) Fault
d) Shear zone

82. For taking the photographs from an aircraft the cameras generally used are having a shutter speed in the range of
a) 1.01 second to 1.25 second
b) 0.01 second to 0.001 second
c) 0.1 second to 0.005 second
d) None of the above

83. The vertical distance from the ground to the perspective centre is the
a) Flying height
b) Distance between ground and photo nadir point
c) Total height of the axis
d) None of the above

84. Which of the following statements is not correct ?
a) An aerial photograph is a central perspective

b) Low oblique photographs are sometimes used to compile reconnaissance maps of inaccessible areas

c) In true vertical photograph the principle point, nadir point and isocenter do not coincide

d) The vertical distance from the ground to the perspective centre is the flying height

85. The Lamellibranchs reached their acme in

a) Carboniferous

b) Permian

c) Mesozoic

d) Cenozoic

86. Which of the following regions suffers most from soil erosion ?

a) Maharashtra

b) Madhya Pradesh

c) NE region

d) Gujarat

87. Silicon is identified by

a) Burning it to high Temperature

b) Treating it with HCl+HF

c) Treating it with conc. KOH

d) Treating it to conc. $H_2SO_4$+HCl

88. For taking aerial photographs a plane generally flies in the

a) Troposphere

b) Stratosphere

c) Ionosphere

d) Exosphere

89. Which of the following is NOT a mixture ?

a) Natural gas

b) Coal

c) Marsh gas

d) Gasoline

90. Which of the following is referred as 'heavyspar' ?

a) Barytes  c) Calcite

b) Galena  d) Bornite

91. The tensile strength of rocks is least in sections along a/an

a) Dome  c) Fault plane

b) River valley  d) Unconfirmity

92. If glycol-based liquifiers are used in drilling

a) The pressure of the gas is increased

b) The pressure of the gas is decreased

c) The temperature of the gas is increased

d) None of the obove

93. 20,000 bb/day of petroleum is represented as

a) 1 MMT PD
c) 1.5 MMT PD
b) 2 MMT PD
d) 0.2 MMT PD

94. In rhombic crystal packing the porosity is

a) 20%
b) 30%
c) Atleast 50%
d) >50%

95. When a bed suffers several phases of deformation, the axial planes of folds are

a) Concurrent
c) Parallel
b) Mutually perpendicular
d) None of the above

## ANSWERS

| | | | |
|---|---|---|---|
| 1. c | 2. a | 3. c | 4. c |
| 5. c | 6. c | 7. b | 8. d |
| 9. c | 10. b | 11. c | 12. d |
| 13. c | 14. a | 15. c | 16. c |
| 17. a | 18. d | 19. a | 20. c |
| 21. c | 22. a | 23. d | 24. b |
| 25. c | 26. d | 27. b | 28. c |
| 29. b | 30. a | 31. a | 32. c |
| 33. a | 34. c | 35. a | 36. d |
| 37. a | 38. b | 39. c | 40. d |
| 41. d | 42. b | 43. b | 44. c |
| 45. c | 46. d | 47. b | 48. b |
| 49. b | 50. c | 51. a | 52. d |
| 53. b | 54. a | 55. a | 56. c |
| 57. d | 58. a | 59. b | 60. c |
| 61. b | 62. a | 63. c | 64. d |
| 65 b | 66. d | 67. a | 68. d |
| 69. a | 70. a | 71. b | 72. c |
| 73. c | 74. c | 75. a | 76. b |
| 77. b | 78. b | 79. a | 80. a |
| 81. b | 82. b | 83. a | 84. c |
| 85. c | 86. c | 87. b | 88. b |
| 89. c | 90. a | 91. d | 92. d |
| 93. b | 94. c | 95. a | |

## 9. O.N.G.C. Examination, July 1987

### GEOLOGY

1. Rhodocrosite is
   a) Oxide of Manganese
   b) Silicate of Manganese
   c) Carbonate of Manganese
   d) Sulphide of Manganese

2. The composition of chromite is
   a) $FeCrO_4$
   b) $Fe_2Cr_2O_4$
   c) $FeCr_2O_4$
   d) $Cr_2O_3$

3. Pentlandite is an ore mineral of
   a) Copper
   b) Iron
   c) Nickel
   d) Sulphur

4. Titanium deposits are associated with
   a) Igneous rocks
   b) Metamorphic rocks
   c) Sedimentary rocks

5. Calcite is distinguished from quartz in their sections by
   a) Its absenee of cleavage and twinning
   b) Its presence of cleavage and twinning
   c) Its colour and shape

6. Labradorite is distinguished from quartz by
   a) Cleavage
   b) Specific gravity
   c) Lustre
   d) Hardness

7. In this sections Tourmaline is characterized by
   a) Isotropy
   b) Anisotropy and Second Order Interference colours
   c) Anisotropy and Inclined Extinction
   d) Anisotropy and absence of cleavage

8. The 'Gondite' rocks of India are associated with
   a) Manganese deposits
   b) Iron deposits
   c) Copper deposits
   d) Zinc deposits

9. In India oil and natural gas are found together in
   a) The Tertiary rocks of Assam—Punjab
   b) The Tertiary rocks of Himalayas
   c) The Sind-Baluchistan Belt
   d) None of the above

10. The Lignite deposits of Neyveli are of
    a) Eocene age
    b) Miocene-Pliocene age
    c) Oligocene age
    d) None of the above

11. The richest deposits of coal in the Gondwanas is found in
    a) Panchet series
    b) Damudar series
    c) Jabalpur series
    d) Talchir series

12. The Black Cotton soil of the Central India is formed by
    a) Vesicular acidic rocks
    b) Vesicular basic rocks
    c) Trap rocks
    d) Sedimentary rocks

13. Garnet crystallizes in
    a) Cubic system
    b) Tetragonal system
    c) Hexagonal system
    d) Orthogonal system

14. The Manganese nodules in the deep seas are formed due to
    a) Inorganic reaction of the deep sea sediments
    b) Organic reactions of the deep sea sediments
    c) Transportation of the rivers and stream

15. 9 Planes 13 Axes and a Centre is present in
    a) Normal Class of Cubic System
    b) Normal Class of Tetragonal System
    c) Normal Class of Hexagonal System

d) Normal Class of Orthagonal System

16. The difference between Isometric and Hexagonal System lies in

a) Length of the axes

b) Axial disposition

c) Shape of the minerals

d) Crystal Structure of the mineral

17. The Mineral Axinite has

a) 6 Planes of Symmetry

b) 1 Plane of Symmetry

c) 3 Planes of Symmetry

d) No Plane of Symmetry

18. Which of the following formations is found at the base of the Cretaceous succession of Trichinapolly ?

a) Trichinapolly

b) Ariyalur

c) Ninniyur

d) Uttatur

19. The Pericambrian rocks of India have given rise to two distinct Groups of rocks which are

a) The Cuddapah System and the Vindhyans System

b) The Kurnool System and the Vindhyans System

c) The Kurnool System and the Cuddapah System

d) The Cuddapah System and the Dharwar System

20. Which of the following is the youngest rock series of the Assam ?

a) Tippam

b) Surma

c) Dihing

d) Jaintia

21. In an outcrop if a bed repeats itself and the beds on the two sides of the bed dip in opposite directions then it indicates

a) Folding only

b) Faulting only

e) Folding and faulting

d) Jointing

22. An anticline can be differentiated from the syncline by one of the following characteristics

a) Anticline is concave upwards

b) Syncline is convex upwards
c) In a syncline the dips on the two sides of the limbs are opposite
d) In anticlines the beds dip towards each other

23. Geosynclines are the sites of
a) Future seismic belts
b) Future oceans and seas
c) Future complex mountains
d) Future stable areas

24. Rossiferol scale is used to determine
a) The velocity of earthquakes
b) The salinity of sea water
c) The intensity of earthquakes
d) None of the above

25. The part of the crust beneath the continents that is composed of silica and magnesium is known as
a) Sial
b) Sima
c) Lithosphere
d) Barysphere

26. The seismicity in the earth is concentrated in
a) Outer Core
b) Inner Core
c) Mantle
d) The Crust and the U. Mantle

27. The Dust-Cloud hypothesis explains
a) The origin of the earth
b) The origin of the solar system
c) The origin of the planets
d) None of the above

28. A plastic asthenosphere exists under a rigid lithosphere and above a rigid mantle because
a) The rigidity of the lithosphere is constant
b) The rigidity of the mantle is not constant
c) The temperature above the asthenosphere is more and the pressure below is less
d) The minerals at the asthenosphere are at their melting points

29. The continents are able to move at the top of the asthenosphere due to

a) Convection currents generated within the asthenosphere itself

b) The push associated with the oceanic ridges

c) Various mechanisms causing plate motions

d) None of the above

30. The alteration of pyroxenes and amphiboles form

a) Micas

b) Kaolinite

c) Limonite

d) Iron oxides

31. A coarse-grained, rudaceous, siliceous rock is known as

a) Conglomerate

b) Breccia

c) Arkose

d) Grit

32. Dolerite is

a) Volcanic basic rock

b) Plutonic basic rock

c) Hypabyssal basic rock

d) Plutonic acidic rock

33. In porphyritic texture

a) Smaller grains form the groundmass and the bigger grains form the phenocrysts

b) The smaller grains are enclosed in the bigger grains

c) The bigger grains of plagioclase are enclosed in pyroxenes

d) None of the above

34. Which of the following is a volcanic lava of acidic composition ?

a) Andesite

b) Trachyte

c) Rhyolite

d) Basalt

35. Marbles are formed from limestones due to

a) Thermal metamorphism

b) Contact and Dynamothermal metamorphism

c) Cataclastic metamorphism

d) Plutonic metamorphism

36. In the Kolar Gold Field, the Champion lode belongs to

a) The Precambrian Dharwar rocks

b) The Vindhyan sandstones
c) The Cuddapah group of rocks
d) The Gondwana group of rocks

37. A 'Pinacoid' form cuts
a) One axis and remains parallel to the other two
b) Two axes and remains parallel to the other
c) All the three axes
d) No axes and remains parallel to all

38. Confined water is found
a) Above the water table
b) Below the water table
c) Between an aquiclude and aquifer
d) Between an aquifuge and aquifer

39. A rock that is very porous but practically impervious is known as
a) Aquifer
b) Aquiclude
c) Aquifuge
d) Aquitard

40. Silicification and mineralization is associated with
a) Folding
b) Faulting
c) Jointing
d) Unconfirmities

41. If the two groups of rocks differ structurally i.e. in the mode of origin, it indicates a/an
a) Angular UC
b) Disconfirmity
c) Non-confirmity
d) Local UC

42. A gently sloping feature found at the base of the continental slope is known as
a) Continental rise
b) Continental shelf
c) Submarine canyon
d) Continental margin

43. The Precambrian era covering about 3.9 b.y. occupies what percentage of the geologic time ?

a) About 90%
b) About 60%
c) About 50%
d) About 40%

44. 'Cheyair Series' belongs to
a) Cuddapah group
b) Vindhyan group
c) Dharwar group
d) Gondwana group

45. Dripstones are
a) Stalactites in the limestone caves
b) Stalagmites in the limestone caves
c) Huge solid columns found in limestone caves
d) None of the above

46. The parent element for the stable $Sr^{87}$ is
a) $U^{238}$
b) $U^{235}$
c) $Rb^{119}$
d) $Rb^{87}$

47. The deepest parts on an oceanic surface is
a) Oceanic ridge
b) Oceanic trench
c) Continental slope
d) Continental rise

48. The point of origin of the earthquakes is known as
a) Epicentre
b) Focus
c) Centre
d) Seismic vertical

49. The chromite deposits of Mysore are associated with
a) Vindhyan sandstones
b) Dharwar rocks
c) Tertiary rocks
d) Cretaceous rocks

50. The vertical component of the apparent displacement of a bed, measured along the direction of the dip of a fault is known as
a) Strike slip
b) Dip slip
c) Heave
d) Throw

## ANSWERS

| | | | |
|---|---|---|---|
| 1. c | 2. c | 3. c | 4. c |
| 5. b | 6. a | 7. b | 8. a |
| 9. a | 10. b | 11. b | 12. c |
| 13. a | 14. a | 15. a | 16. b |
| 17. d | 18. d | 19. c | 20. c |
| 21. a | 22. d | 23. c | 24. c |
| 25. b | 26. d | 27. b | 28. d |
| 29. c | 30. a | 31. a | 32. c |
| 33. b | 34. c | 35. a | 36. a |
| 37. a | 38. d | 39. b | 40. b |
| 41. c | 42. a | 43. a | 44. a |
| 45. c | 46. d | 47. b | 48. a |
| 49. b | 50. b | | |

# 10. Joint UGC–CSIR Junior Research Fellowship Examination, January 1989

## EARTH SCIENCES

N.B. Each question carries 4 marks. All questions are compulsoıy.

1. Which of the following countries have tropical forests, tropical deserts and tropical monsoon forests in its natural vegetation ?
   a) Phillipines
   b) Indonesia
   c) Malaysia
   d) Thailand
2. The salinity of sea waters can be determined by the ratios of
   a) Na/K
   b) Ca/Na
   c) C/C
   d) Rb/Sr
3. The Depth of water can be determined by
   a) He method
   b) $C^{14}$ method
   c) Salinity-Temperature Diagram
   d) None of the above
4. If the half-life of an isotope is 1 hour, then after 4 hours what part of the original isotope would remain ?
   a) 1/4
   b) 1/8
   c) 1/16
   d) 1/32
5. Which of the following features is NOT associated with the Karst topography ?
   a) Polje
   b) Blind valley
   c) Sink holes
   d) Kettle
6. Iceland is an exposed part of
   a) Mid-oceanic ridges
   b) Island arcs
   c) Shield areas
   d) Plateaus
7. Earth possesses an atmosphere whereas the Moon does not since

a) Moon has got one core

b) Moon has no water

c) Moon has got less gravitational force

d) Moon has got less oxygen

8. MOHO is the discontinuity between

a) Lithosphere and Asthenosphere

b) Asthenosphere and Mesosphere

c) Crust and Mantle

d) Homosphere and Asthenosphere

9. The least important method for determining the age of pre-cambrian rocks is

a) Sr—Rb

b) Pb—Pb

c) $C^{14}$

d) U—Pb

10. Conservation of natural resources means

a) Complete utilization of natural resources

b) Less utilization of natural resources

c) Less and effective utilization of natural resources

d) Rational utilization of natural resources

11. List out the mineral which is abundantly present in the epi-continents when compared to the earth as a whole

a) K

b) Na

c) Si

d) Mn

12. Monsoons are present in which of the following regions

a) North America

b) South America

c) Europe

d) Australia

13. Felspars change into clayey sediments in the following climate

a) Tropical

b) Arid

c) Humid

d) Dry

14. At the present times, earthquakes are normally measured by the modified Mercalli Scale. This scale determines the

a) Subjective intensity of the earthquakes with the epicentre

b) Objective intensity of the earthquakes with the epicentre

c) Subjective intensity of the earthquakes

d) Objective intensity of the earthquakes

15. In the Bhakra Nagal Dam the underthrust of water is controlled by
    a) Gravity of concrete and cement
    b) Elasticity of concrete and cement
    c) Rigidity of concrete and cement
    d) Thickness of concrete and cement

16. Which of the following lavas showed continued eruption throughout the year ?
    a) Rhyolite
    b) Basalt
    c) Trachyte
    d) Andesite

17. A Geostropic wind has two components which are
    a) Coriolis force and Pressure gradient
    b) Coriolis force and Centrifugal force
    c) Centrifugal force and Centripetal force
    d) Centrifugal force and Pressure gradient

18. We observe through the radiation to the distant galaxy (Doppler Effect) in order to gain knowledge of their
    a) Temperature
    b) Chemical composition
    c) Density
    d) Position and size of the object in the galaxy

19. Bombay South Wind starts from
    a) South
    b) North
    c) East
    d) West

20. The rate of soil moistures is determined by the difference between
    a) Percolation and infiltration
    b) Percolation and run off
    c) Percolation and potential evaporation
    d) Percolation and transpiration

21. Which of the following rocks is least resistant for use as a building material ?
    a) Marble
    b) Slate
    c) Quartzite

d) Basalt

22. At the poles the sun shines throughout the day for six months due to

a) Tilt of the earth's axis of rotation
b) Thick caps of ice
c) Bulge at the equator
d) Ellipse of the sun's axis

23. Rounded, pitted and well-sorted grains are formed due to the action of

a) River
b) Glaciers
c) Oceans
d) Wind

24. The angular velocity of the earth is

a) $3.65 \times 10^{-8}$
b) $4.19 \times 10^{-7}$
c) $2.6 \times 10^{-4}$
d) None of the above

25. Smoke from a chimney is going in the direction of N 20°E and the inclination of the chimney itself is 17° due east. The final direction of the smoke is

a) N 3°E
b) 3°E
c) 3°W
d) N 17°W

26. In the Flotation Process different desired minerals are separated from each other by taking into consideration the differences in their

a) Paramagnetism
b) Piezoelectricity
c) Density
d) Salinity

27. Which of the following minerals are abundantly found in the ferromagnesian polymetallic nodules ?

a) Fe, Mn, Co, Ni
b) Fe, Mn, Li, Ni
c) Li, Si, Ni, Mn
d) Fe, Mn, Mg, Pb

28. The snowline at a place changes with a change in

a) Vegetation

b) Latitude

c) Altitude

d) Sunshine

29. The upwelling of the seawater is due to

a) Convergence of two streams of water

b) Water parallel to the coast when the coast is towards right in the southern hemisphere

c) Water parallel to the coast when the coast is towards left in the Northern Hemisphere

d) None of the above

30. The distance between two latitudes over the longitude of $\theta$ angle is

a) Tan $\theta$

b) Cos $\theta$

c) Sin $\theta$

d) $\theta$

## ANSWERS

| | | | |
|---|---|---|---|
| 1. c | 2. a | 3. c | 4. c |
| 5. d | 6. a | 7. c | 8. c |
| 9. c | 10. d | 11. a | 12. b |
| 13. a | 14. b | 15. d | 16. b |
| 17. b | 18. d | 19. d | 20. c |
| 21. d | 22. d | 23. a | 24. d |
| 25. a | 26. c | 27. a | 28. c |
| 29. d | 30. b | | |

# 11. Joint UGC-CSIR Junior Research Fellowship Examination, March 1988

## EARTH SCIENCES

N.B. : All questions are compulsory

1. Although the abundance of K and Na in river waters is almost same, K is highly depleted in sea waters because

   a) K is insoluble in sea water

   b K is inert to erosion

   c) K is used up in the formation of clay mineral

   d) K is never carried to the seas by the rivers

2. Which of the following is not suitable for dating Tertiary rocks ?

   a) U—Th

   b) K—Ar

   c) Rb—Sr

   d) C—C

3. The maximum age of the oceanic sediments would be

   a) Paleozoic

   b) Cenozoic

   c) Mesozoic

   d) Proterozoic

4. Atmosphere is transparent to sun's radiation of the wavelength

   a) 18—120 nm

   b) 180—1200 nm

   c) 1.8—12 nm

   d) 0.18—1.2 nm

5. Thrust faults indicates

   a) Footwall has gone up in relation to the hanging wall

   b) Hanging wall has gone up in relation to the footwall

   c) Shortening of the crust

   d) Extension of the crust

6. 'Leap second' is a comparatively new concept on the lines of 'Leap year'. Choose the correct statement about these two

   a) Both indicate revolution of the earth

   b) Both indicate rotation of the earth

   c) Leap second indicate rotation while leap year indicate revolution

d) Leap second indicates revolution while leap year indicates rotation

7. Mountain system is the most fragile of the ecosystems because
   a) They are subjected to frequent glacial erosion
   b) They are characterized by many unstable slopes
   c) They are characterized by vegetation having different root lengths
   d) None of the above

8. Rivers have run-off directing effect. Given below are three rivers X, Y and Z. Arrange them in the increasing order of run off deflecting effect

   Vegetation at water divides—X

   Vegetation at valleys—Y

   Swamps—Z

   a) XYZ
   b) YXZ
   c) XZY
   d) YZX

9. Considering that the precipitation is uniform over all the areas, which of the following will have maximum underground water potential ?
   a) Valley fill
   b) River terrace
   c) Moraine
   d) Badlands

10. Consider that about 4000 M.Y. ago the earth had the same atmospheric conditions, then the b.p. of water would have been
    a) More than 100°C
    b) Less than 100°C
    c) 100°C only
    d) 0°C

11. The intensity of chemical weathering is maximum in
    a) Ladakh region
    b) Shillong plateau
    c) Rajasthan
    d) Siwaliks

12. The P waves cover the earth's diameter in 20 minutes. Its velocity in the U. Crust is 5 km/sec. Which of the following conclusions can be deduced from the above statements ?

a) P waves velocity progressively increases in all the layers
b) P wave velocity progressively decreases in all the layers
c) Atleast in some of the layers of the earth, the velocity is greater than that in the U. Crust
d) None of the above

13. Assume that in the tropical regions the sun's radiation is uniform. This condition is achieved in

a) Wetlands
b) Drylands
c) Open oceans
d) Snow surface

14. As the very shallow surface waves of the ocean approach the coast their velocity

a) Increases towards the coast
b) Decreases towards the coast
c) Remains the same
d) None of the above

## ANSWERS

| | | | |
|---|---|---|---|
| 1. c | 2. c | 3. c | 4. b |
| 5. c | 6. c | 7. b | 8. a |
| 9. a | 10. c | 11. a | 12. c |
| 13. c | 14. b | | |